MRI Atlas of Human White Matter

MRI Atlas of Human White Matter

Second Edition

Kenichi Oishi

Department of Radiology and Radiological Sciences, Division of MRI Research, Neurosection Johns Hopkins University School of Medicine, Baltimore, MD, USA

Andreia Faria

Department of Radiology and Radiological Sciences, Division of MRI Research, Neurosection Johns Hopkins University School of Medicine, Baltimore, MD, USA and Department of Radiology, University of Campinas, Brazil

Peter CM van Zijl

Department of Radiology and Radiological Sciences, Division of MRI Research, Neurosection Johns Hopkins University School of Medicine and F.M. Kirby Research Center for Functional Brain Imaging, Hugo Moser Research Institute at Kennedy Krieger, Baltimore, MD, USA

Susumu Mori

Department of Radiology and Radiological Sciences, Division of MRI Research, Neurosection Johns Hopkins University School of Medicine and F.M. Kirby Research Center for Functional Brain Imaging, Hugo Moser Research Institute at Kennedy Krieger, Baltimore, MD, USA

Collaborator:
Dr. Barbara J. Crain
Johns Hopkins University School of Medicine
Department of Pathology

2011

ELSEVIER

Amsterdam • Boston • Heidelberg • London • New York • Oxford
Paris • San Diego • San Francisco • Singapore • Sydney • Tokyo

Academic Press is an imprint of Elsevier
32 Jamestown Road, London NW1 7BY, UK
30 Corporate Drive, Suite 400, Burlington, MA 01803, USA
525 B Street, Suite 1800, San Diego, CA 92101-4495, USA

First edition 2005
Second edition 2011

British Library Cataloguing-in-Publication Data
A catalogue record for this book is available from the British Library

Library of Congress Cataloging-in-Publication Data
A catalog record for this book is available from the Library of Congress

ISBN : 978-0-12-382081-5

For information on all Academic Press publications
visit our website at www.elsevierdirect.com

Typeset by MPS Ltd, a Macmillan Company, Chennai, India
www.macmillansolutions.com

Printed and bound in China

10 11 12 13 10 9 8 7 6 5 4 3 2 1

Contents

A companion website to support this book can be found at:
http://booksite.academicpress.com/oishi/

Preface

Five years have past since the first edition of this atlas book was introduced. During this period, the field of MRI has witnessed a tremendous expansion of studies based on diffusion tensor imaging, which, in turn, triggered renewed interest in white matter anatomy and its pathology. Another important development during this period was the adoption of quantitative image analysis referenced to an anatomical template (atlas), as well as the use of image warping technologies to allow co-registration to these atlases for a wide range of neuroimaging approaches. The availability of a stereotaxic atlas with structure segmentation can play an important role in brain research. First, such an atlas can be used to automatically segment patient brains by warping it to the shape of the individual patient's brain. Second, voxel-based analysis results can be registered to the atlas and used to identify anatomical structures of interest. For example, voxel-based analysis can define brain regions of interest in the template coordinates (x, y, z), but it is not always straightforward to translate the coordinates to precise anatomical knowledge, such as the identification of the "corona radiata." A stereotaxic atlas with anatomical annotation would be a useful resource to link between the coordinates and anatomical entities. Third, if the same atlas is used for multiple studies with multiple imaging modalities, it can serve as a resource to connect different types of anatomical and functional information within the same coordinate system.

In order to better realize some of the premises of the atlas, we have introduced two major modifications in this Second Edition. First, the atlas is now in stereotaxic coordinate systems. We adopted one of the most widely used Montreal Neurological Institute (MNI) coordinates as a primary coordinate system. The atlas was normalized to the ICBM-152 brain in MNI coordinates. For the secondary coordinate system, we have also provided the classical Talairach coordinate system, which ensures the connection between previous works and more updated anatomical knowledge. Second, this new edition now provides full segmentation of 220 white matter and gray matter structures with boundaries based on MRI/DTI contrasts. In addition to these two major modifications, there are several important additions to the atlas. For example, the atlas contains a modern cytoarchitectonic map based on recent work by Drs Zilles, Amunt, Eickhoff, and their colleagues. The new edition defines not only the core white matter structures, as in the First Edition, but also the superficial white matter, the deep gray matter, and the cortex. The nomenclature is consistent with the *Atlas of the human brain*, Third Edition (Mai et al., Academic Press, 2008) as much as possible.

We hope that our readers will enjoy the new additional information we have provided in this Second Edition.

Acknowledgments

The authors of the second edition would like to sincerely acknowledge the work of the first edition authors, Dr Setsu Wakana and Dr Lidia M. Nagae-Poetsher. We owe enormous gratitude to Dr George Paxinos, Dr Karl Zilles, Dr Katrin Amunts, and Dr Simon B. Eickhoff for valuable comments. Dr Zilles, Dr Amunts, and Dr Eickhoff kindly provided us an electric form of the cytoarchitectonic map. This publication was made possible by grant number P41 RR015241 from the National Center for Research Resources (NCRR), a component of the National Institutes of Health (NIH). Its contents are solely the responsibility of the authors and do not necessarily represent the official view of NCRR or NIH. This research was also supported by NIH grants R21AG033774, P50AG005146, PO1EB001955, F05NS059230, and RO1AG20012. We had immense help from Ms Terri Brawner, Mr Joseph Gillen, Ms Kathleen Kahl, and Ms Ivana Kusevic for MRI studies. We gratefully acknowledge the contributions of our colleagues Ms Yajing Zhang and Mr Kegang Hua for creating probabilistic tractography (Chapter 4), Ms Xin Li, Dr Hangyi Jiang, and Dr Jiangyang Zhang for the software development, Ms Kazi Akhter for image pre-processing, Mr Johnny Hsu, Mr Zhipeng Hou, and Ms Xin Xu for the data management, and Ms Mary McAllister for help with manuscript editing. The highly sophisticated image transformation studies were made possible by help from our colleagues Dr Michael I. Miller, Dr Can Ceritoglu, Mr Anthony Kolasny, Mr Timothy Brown, and Mr Mike Bowers. Finally, a special thanks to Dr Johannes Menzel, Ms Mica Haley, Ms Caroline Jones, and Ms Melissa Turner at Elsevier for their help during the preparation and production of this book.

CHAPTER 1

Introduction

We present a human brain atlas based on co-registered T_1-weighted imaging and diffusion tensor imaging (DTI). The conventional T_1-weighted MRI provides a high-resolution view of gray matter anatomy, while the DTI provides unique contrast to decipher the complicated architecture of the white matter. In order to appreciate the images in this atlas, it is necessary to understand how these contrasts are generated. In this section, we provide some general MRI background, followed by the basics of DTI, and an overview of the principles and limitations of fiber tracking using this methodology.

1. Magnetic resonance imaging (MRI)

Magnetic resonance (MR) images consist of individual picture elements (pixels) with different intensities (brightness). When evaluating MR images, two important parameters need to be considered: spatial resolution and contrast. In modern MRI, the spatial resolution (pixel size) is on the order of 1–3 mm or even smaller, revealing a large amount of detail about brain anatomy. Contrast is created by differences in pixel intensities among different areas of the brain. Conventionally, MR contrast has been mostly based on differences in tissue water relaxation times, such as T_1 and T_2, which can be used to distinguish various brain regions, such as the cortex, the deep gray matter, and the white matter. However, conventional MR has not been successful in providing good contrast within the white matter, which usually looks rather homogeneous, no matter how high the spatial resolution. From an MRI point of view, the white matter generally appears as a fluid-like homogeneous structure, which, of course, is not the case.

The white matter consists of axons that connect different areas of the brain. These axons tend to form "bundles," together with other axons, and, depending on the destinations, the diameters of these bundles (called white matter tracts) can be as large as a few centimeters. Some of them, such as the corpus callosum and the anterior commissure at the mid-sagittal level, are clearly visible on conventional MRI. However, most of these bundles cannot be individually identified by MRI or even with postmortem brain slices. This is because the majority of these bundles have similar chemical compositions and MR signatures (T_1 and T_2 relaxation times). It is very difficult to appreciate where the specific white matter tracts of interest are and how they are spatially related to one another. Actually, conventional MRI is also unable to provide a high level of contrast for intra-gray matter structures. Differences in cytoarchitecture among different cortical layers or between cortical areas cannot be distinguished. Nevertheless, gyral and sulcal patterns can be used as visual clues for cortical anatomy discrimination. Unfortunately, equivalent visual clues are often not available for the identification of white matter tracts.

2. Diffusion tensor imaging (DTI)

2.1. The diffusion tensor and the diffusion ellipsoid

MRI is an imaging technique that detects proton signals from water molecules. Images thus reflect water density and properties as a function of position in space. In addition, MRI can be used to measure the local chemical and physical properties of water. Two such properties are molecular diffusion and flow. MRI can assess both, but the measurement methods are different for each, and they should not be confused.

The diffusion process is a reflection of thermal Brownian motion. This process can be understood in a simple manner by comparing it to the evolution in the shape of a stain after ink is dropped on a piece of paper. Usually, the drop turns into a circle (Gaussian distribution) that grows over time. The faster the diffusion is, the larger the diameter of the circle, and the extent of the diffusion can be estimated from this. Because the extent of the stain is equivalent in all directions, the diffusion is called isotropic. However, if the paper consists of a special fabric that is woven with dense vertical fibers and sparse horizontal fibers, the stain will have an oval shape elongated along the vertical axis. This is called anisotropic diffusion. A similar process happens in the brain where water tends to diffuse preferentially along axonal fibers. If ink were injected inside the brain white matter, the shape of its

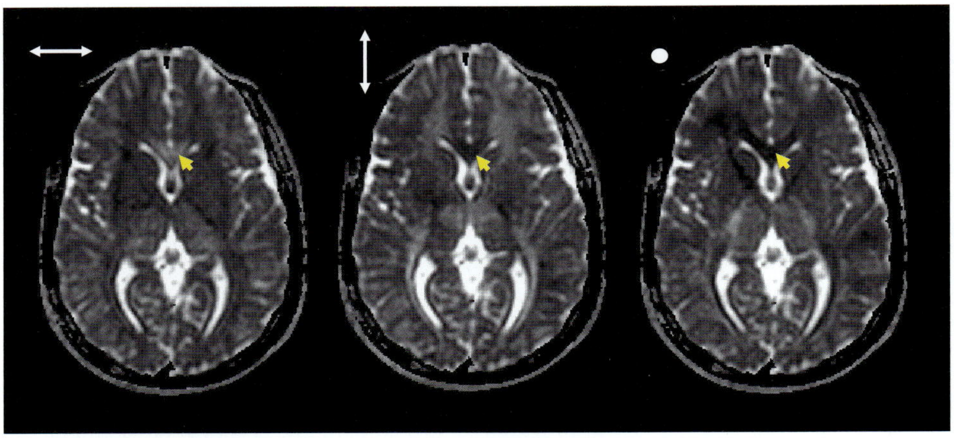

Fig. 1. Diffusion constant maps of the same slice measured along three different orientations (indicated by arrows). For the third image, the orientation is perpendicular to the plane. In these images, brightness represents the extent of diffusion (magnitude of the diffusion constant). Diffusion in white matter is more anisotropic than that in gray matter. For example, the corpus callosum (yellow arrows) has a high diffusion constant when water diffusion is measured along the horizontal axis (first image), but a low diffusion constant when applying diffusion weighting along the vertical axis (second image) or perpendicular to the plane (third image).

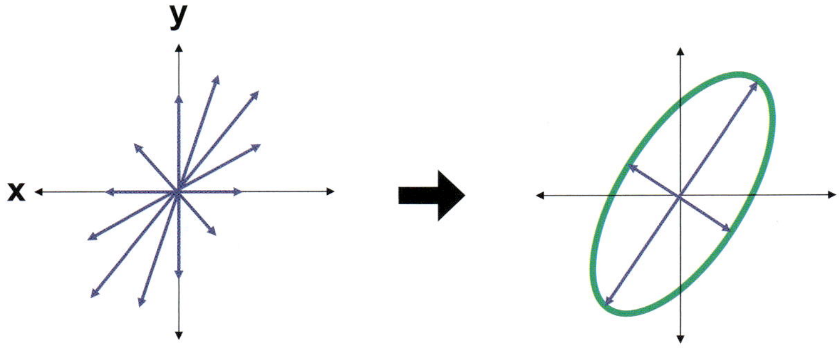

Fig. 2. Using diffusion measurements along multiple axes (left, the length of blue arrows represent the magnitude of diffusion constants), the anisotropic diffusion process can be delineated in detail. In diffusion tensor imaging, the measurement results are fitted to a simple symmetric 3D ellipsoid (or oval in this 2D simplified representation).

distribution would be elongated along the axonal tracts. Inside the gray matter, the diffusion process is more random because of the lack of aligned fiber structures, and the shape of the ink would be more spherical.

One of the unique aspects of diffusion measurements by MRI is that the extent of water diffusion can be measured along one predetermined axis and this axis can be set arbitrarily. Diffusion measurements can be combined with imaging, from which the effective diffusion constant at each pixel can be measured (called an apparent diffusion constant map). Examples of such diffusion constant maps are shown in Fig. 1. From these images, it can be clearly seen that water diffusion inside the brain is anisotropic: the extent of the measured diffusion constants depends on the measurement orientation, especially in the white matter.

The extent of anisotropy of the water diffusion can be delineated in more detail by measuring the diffusion constant along multiple axes. Due to the complexity of the underlying cytoarchitecture (in addition to noise), the measurements may yield a very complex pattern of anisotropy. However, if there is a homogeneous cytoarchitecture within a pixel (typically $2 \times 2 \times 2\,mm^3$ or larger, the magnitude of diffusion (or the shape of ink in our analogy) can be described by a simple symmetric 3D ellipsoid (Fig. 2).

Diffusion tensor imaging is a technique in which diffusion is measured in a series of different spatial directions, from which the shape and orientation of the diffusion ellipsoid is determined at each image pixel by fitting the measurement results obtained in the reference frame of the scanner to the ellipsoid that describes the average local cytoarchitecture. During this measurement and fitting process, we use a 3×3 tensor, from which the name, diffusion tensor imaging, is derived. The fitting process for determining the ellipsoid is a mathematical process called diagonalization of the tensor. The 3×3 diffusion tensor has nine elements, but there are only six independent numbers because the diffusion tensor is a symmetric tensor, which makes sense, given the fact that the tensor uniquely determines the diffusion ellipsoid. Six parameters are needed to completely describe the magnitude and orientation of the 3D ellipsoid (Fig. 3), and the determination of these six parameters is the target of DTI. Thus, determination of the tensor elements requires measurement of diffusion constants along at least six spatial directions.

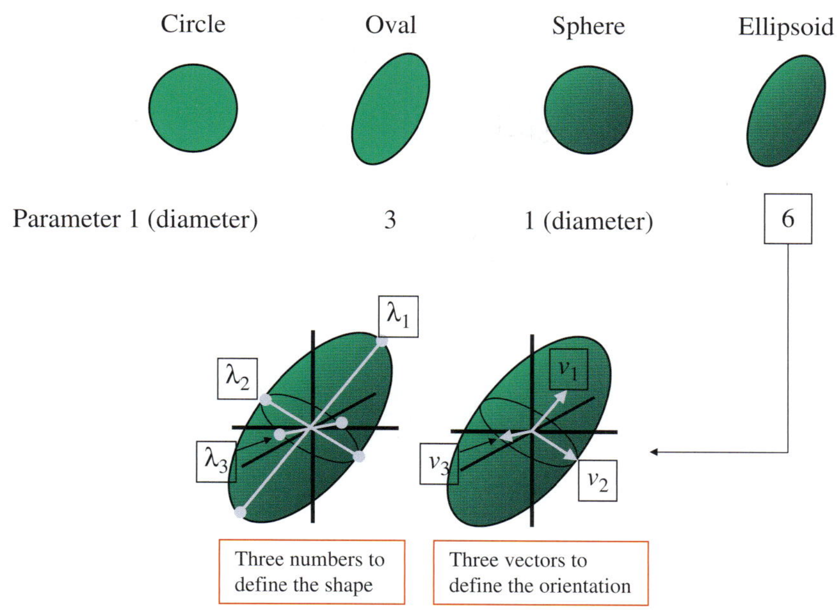

Fig. 3. Parameters required to mathematically describe a circle, oval, sphere, and ellipsoid. The six parameters required for an ellipsoid are three eigenvalues (λ1, λ2, and λ3) that define the shape of the ellipsoid and three eigenvectors ($v1$, $v2$, and $v3$) that define the orientation of the ellipsoid.

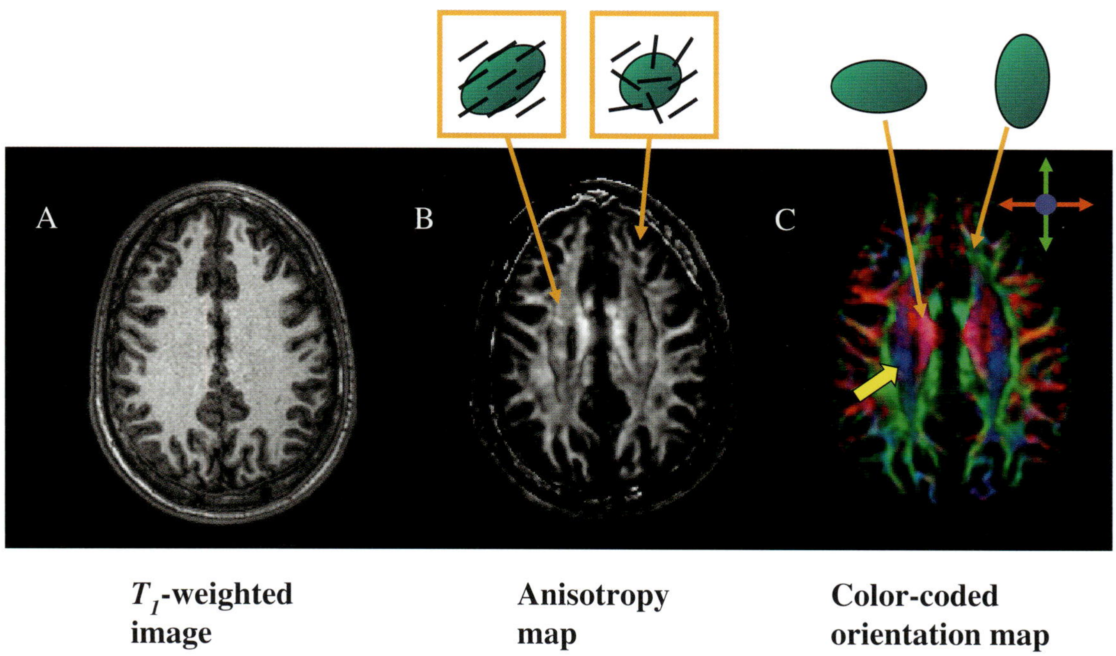

Fig. 4. Examples of (A) conventional MRI (T_1-weighted image), (B) anisotropy map, and (C) color-coded orientation map. For the color-coded images, the intensity (brightness) is proportional to the anisotropy information (same as in anisotropy map, B) and the different colors, red, green, and blue, represent fibers running along the right–left, anterior–posterior, and inferior–superior axes, respectively.

2.2. *Two-dimensional visualization of DTI results*

Once the six tensor parameters are determined at each pixel, the information is usually reduced to produce different types of images that can be appreciated visually. It is almost impossible to visualize all six parameters in an intuitive way in a single image, and there are two types of presentations that are generally used. One is the anisotropy map and the other is the orientation map or color map. The anisotropy map provides information about the extent of elongation (anisotropy) of the diffusion ellipsoid (Fig. 4B). In this kind of map, the white matter appears brighter (more anisotropic) than the gray matter, showing brain areas with coherent fiber structures. It is believed that such factors as axonal density, myelination, and homogeneity in the axonal orientation affect the degree of diffusion anisotropy. In the color map, the angle of the longest axis of the ellipsoid ($v1$ in Fig. 3) is visualized using three

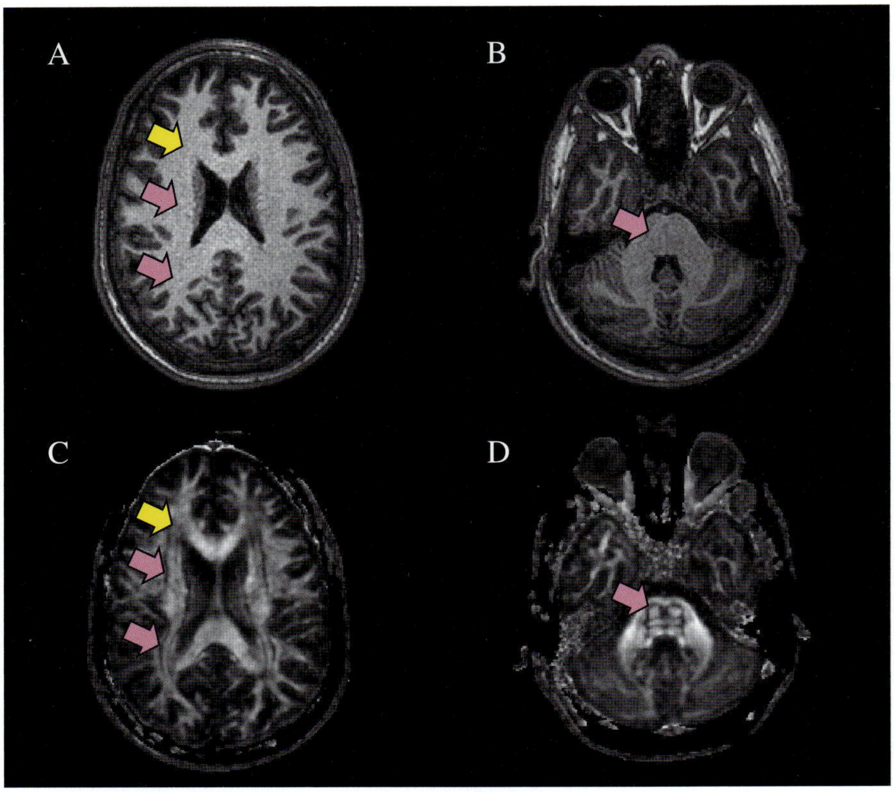

Fig. 5. Typical white matter regions where low anisotropy is observed due to a mixture of axonal tracts bearing different orientations in the same pixel. Images in (A) and (B) are T_1-weighted images and (C) and (D) are anisotropy maps of the corresponding slice locations.

principal colors (red (R), green (G), and blue (B)), under the assumption that this is indicative of the dominant fiber orientation (Fig. 4C). The signal intensity is weighted by the diffusion anisotropy. In this kind of map, any arbitrary angle can be represented by a mixture of RGB colors. For example, in Fig. 4, fibers running 45° between the right–left (red) and anterior–posterior (green) axes would be labeled by a yellow (red + green) color.

By comparing these three images, it is clear that the color-coded orientation map carries, by far, the greatest amount of information regarding white matter anatomy. In the conventional T_1-weighted MR image on the left, the white matter presents as a homogeneous field, whereas many internal structures are visible in the color map. For example, the blue-colored fiber indicated by a yellow arrow in Fig. 4C is the corona radiata, which could not be identified in the T_1-weighted image. Thus, DTI can be used to parcellate the white matter into substructures.

There is one important limitation of DTI that is essential to understand when reading color maps. This is due to the DTI assumption that there is a homogeneous fiber structure within a pixel. Because of the relatively large size of the pixels in DTI data (2–3 mm), a pixel often contains axonal tracts with multiple orientations. The inclusion of tracts with different orientations causes the pixel to lose anisotropy. In fact, gray matter displays low anisotropy, not because there are no axonal fibers there, but, because the fiber architecture is convoluted with respect to the pixel size. A similar situation can occur in the white matter as well. One of the most notable regions for this is at the junction between the corpus callosum and the corona radiata (yellow arrow in Fig. 5C), causing a dark region to occur in the anisotropy map. These low anisotropy strings in the white matter occur in many other boundaries between prominent fibers of different orientations, as indicated by the pink arrows in Fig. 5C and 5D. These areas do not have any representation in T_1-weighted images (Fig. 5A and 5B) in which the gray and white matter contrast is provided predominantly by myelin water concentration and relaxation properties. These low anisotropy areas should not be interpreted as areas with fewer axonal tracts.

2.3. *Three-dimensional reconstruction of white matter tracts*

In chemical tracer techniques, the tracer may travel the entire length of an axonal projection. Using DTI, such a long-range tracing of a single axon is not possible, as one can obtain only pixel-by-pixel information about water diffusion. Many axons may enter a pixel and exit from it. Consequently, the information about a specific trajectory degenerates within this pixel. However, if many axonal tracts form a large bundle (i.e., white matter tract) and the bundle is sufficiently large to contain many pixels across its diameter, a 3D rendering of such a bundle can be reconstructed based on the DTI data simply by connecting the fiber orientation information

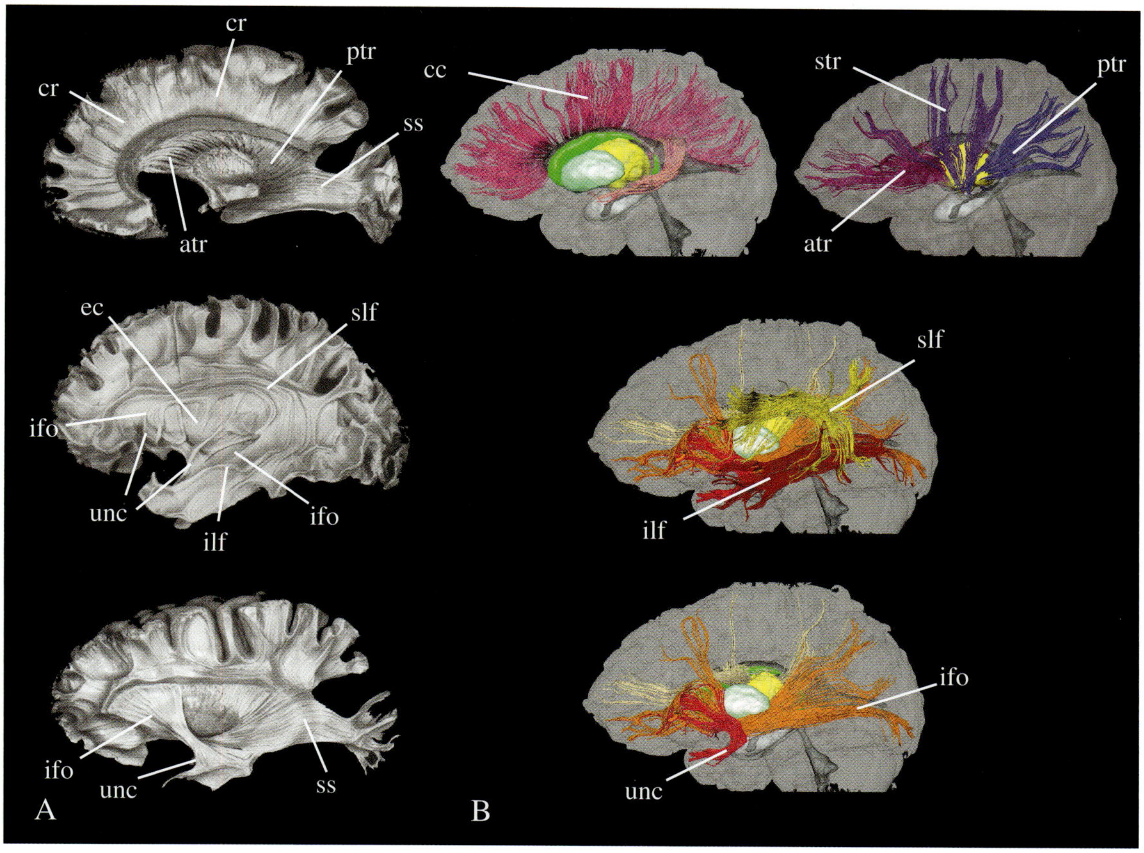

Fig. 6. Comparison between postmortem preparation (A) and DTI-based reconstruction results (B). Abbreviations are: atr, anterior thalamic radiation; cc, corpus callosum; cr, corona radiata; ec, external capsule; ifo, inferior fronto-occipital tract; ilf, inferior longitudinal fasciculus; ptr, posterior thalamic radiation; ss, sagittal stratum; slf, superior longitudinal fasciculus; str, superior thalamic radiation; unc, uncinate fasciculus.

pixel-by-pixel (for more information see Chapter 2). An example of a 3D tract reconstruction and comparison with a postmortem specimen is shown in Fig. 6. The strong similarity of the two images suggests that the DTI-based fiber reconstruction technique can reveal the macroscopic anatomy of white matter tracts. On the other hand, it is also true that neither DTI images nor postmortem dissections can trace the trajectory of specific axons. For example, one may postulate that the superior longitudinal fasciculus (slf), which appears as a long tract connecting the frontal and temporal lobes, is actually an assembly of short-range tracts merging in and exiting out at many points along the tract. DTI cannot distinguish whether this is true. Therefore, DTI-based 3D tract reconstruction results should be treated as a macroscopic delineation of tract configurations or a white matter parcellation tool.

CHAPTER 2

Three-Dimensional Reconstruction of White Matter Tracts

1. Data acquisition

A 1.5 T Philips GyroscanNT was used. DTI data were acquired using a single-shot echo planar imaging (EPI) sequence with parallel imaging (SENSitivity Encoding factor, R = 2.5). The imaging matrix was 112 × 112, with a field of view of 246 × 246 mm (native resolution of 2.2 mm), which was zero-filled to a 256 × 256 matrix. The image orientation was axial, with a 2.2 mm slice thickness, which was aligned parallel to the anterior–posterior commissure line. A total of 55 slices covered the entire cerebral hemispheres and the brainstem. The diffusion weighting was encoded along 30 independent orientations, with a maximum b = 700 mm^2/s. Five additional images with minimal diffusion weighting were also acquired. The scanning time of one dataset was about 6 min. In order to enhance the signal-to-noise ratio, the scan was repeated six times. A co-registered T_1-weighted image (Magnetization-Prepared Rapid Gradient Echo, MPRAGE) with 1.1 mm nominal resolution was also recorded for anatomical guidance.

2. Data processing

The DTI datasets were transferred to a PC workstation, re-aligned for bulk motion, and processed using DtiStudio, a component of MRI studio (H. Jiang & S. Mori, Johns Hopkins University, www.mristudio.org, http://cmrm.med.jhmi.edu or http://godzilla.kennedykrieger.org). All diffusion-weighted images were visually inspected for apparent artifacts due to subject motion and instrumental malfunction. Then, the six elements of the diffusion tensor were calculated for each pixel using multivariant linear fitting. After diagonalization, three eigenvalues (λ_1, λ_2, λ_3) and three eigenvectors (v_1, v_2, v_3) were obtained. For the anisotropy map, the so-called fractional anisotropy (FA) parameter was calculated, which is scaled from 0 (isotropic) to 1 (anisotropic):

$$FA = \sqrt{\frac{1}{2}} \frac{\sqrt{((\lambda_1 - \lambda_2)^2 + (\lambda_2 - \lambda_3)^2 + (\lambda_3 - \lambda_1)^2)}}{\sqrt{\lambda_1^2 + \lambda_2^2 + \lambda_3^2}}$$

The eigenvector associated with the largest eigenvalue (v_1) was utilized as an indicator for the fiber orientation.

3. Reconstruction of tracts

3.1. Algorithm

The 3D tract reconstruction was performed using the FACT (Fiber Assignment by Continuous Tracking) method (Fig. 1), which performs a straightforward linear line propagation based on the $v1$ vector angle. An anisotropy threshold of FA >0.2 was used, while the angle of progression was restricted by using an inner product of more than 0.75 between consecutive eigenvectors.

Fig. 2 shows a schematic diagram of the fiber reconstruction process. Suppose there is a tract with the structure shown in Fig. 2A. DTI measurement provides the fiber orientation information shown in Fig. 2B. For fiber reconstruction, a tract of interest first must be identified and marked by a region of interest (ROI) (bold box). Two possible reconstruction approaches are demonstrated. In the first, the so-called "from-ROI" approach (Fig. 2C), tracking originates from the ROI. In this simple example, the ROI contains only one pixel and, as a result, only one line is produced in the "from-ROI" approach, which restricts the labeling to merely a segment of the tract. In the second, so-called "brute-force" approach, all pixels are visited and examined and all tracts originating from such pixels and penetrating the ROI are tracked. This approach is much more time-consuming, but leads to a more comprehensive delineation of the tract(s). Fiber reconstruction in this atlas is therefore based on the brute-force approach.

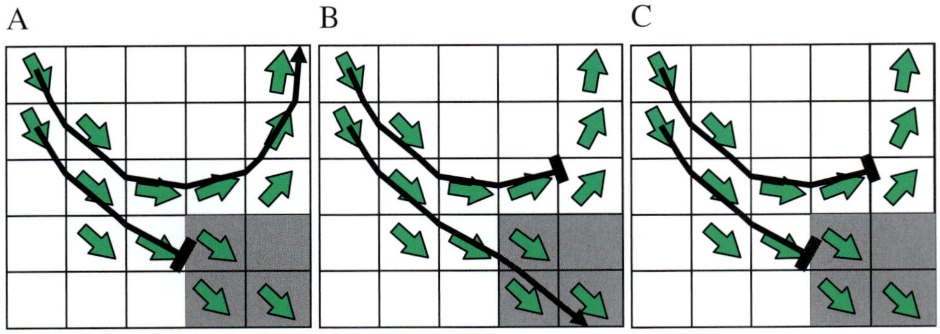

Fig. 1. Schematic diagram of FACT fiber tract reconstruction based on DTI data. Once the fiber orientation ($v1$) is estimated at each pixel, putative projections are traced by propagating a line along the estimated fiber orientations. The propagation terminates either when it enters an area with anisotropy lower than a threshold (A: dark boxes), or when the trajectory has a turn judged as too sharp by an inner product between two connected pixels (B). In FACT, both criteria are applied (C).

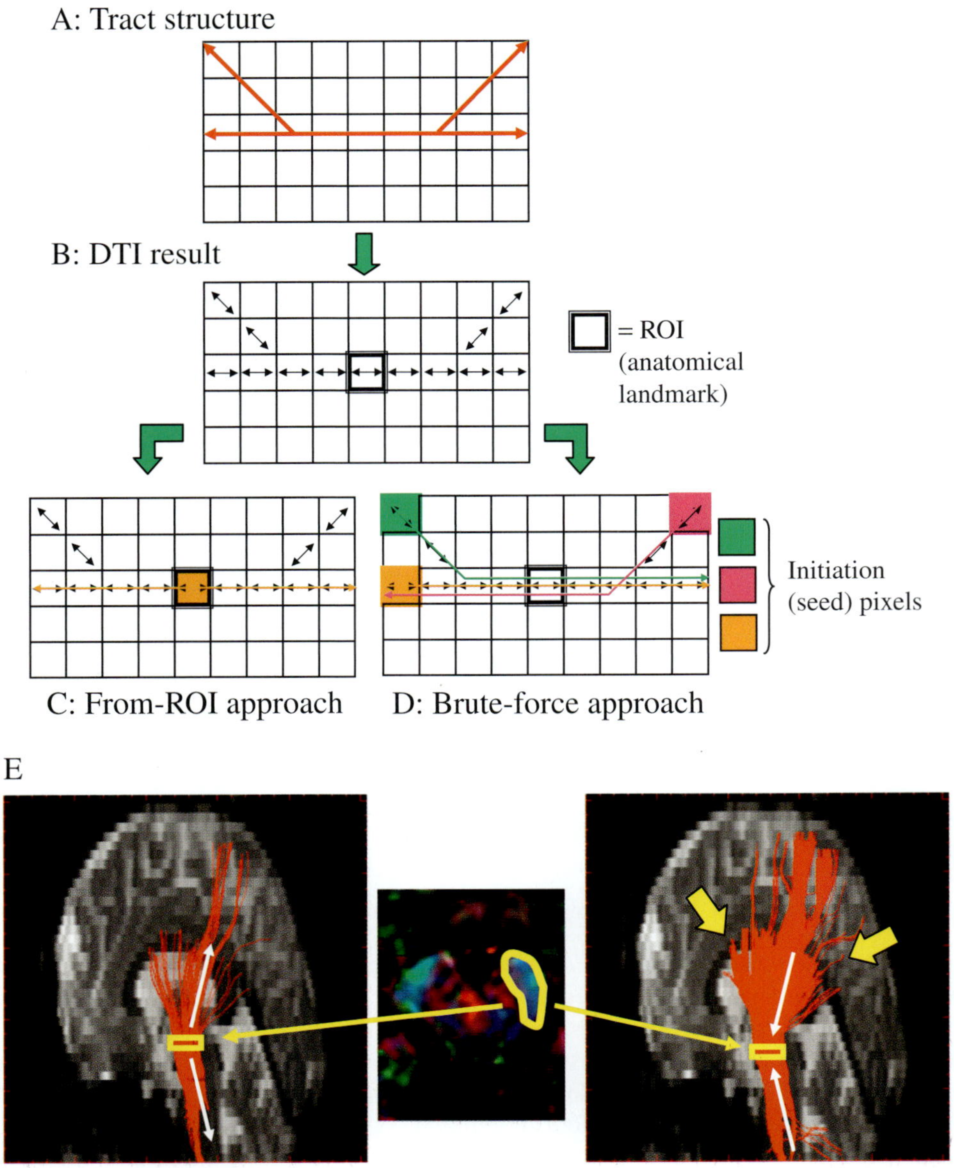

Fig. 2. Principles of tract reconstruction using the "from ROI" and the "brute-force" approaches. (A) Example of a tract structure with two branching points. (B) Results of DTI measurement, with a vector field that shows the fiber orientation at each pixel. A bold box shows the anatomical landmark where an ROI is defined. (C) Results of the tract reconstruction using the "from ROI" approach, in which tracking is started from the ROI. This generally leads to an incomplete delineation of the tract. (D) Results of the "brute force" approach, in which tracking is started from all pixels. Several initiation (seed) pixels from which the tracking can lead to the same ROI are demonstrated. (E) Results of actual tracking using the cerebral peduncle as a reference ROI. The left and right panels show results for the "from ROI" and "brute force" approaches, respectively.

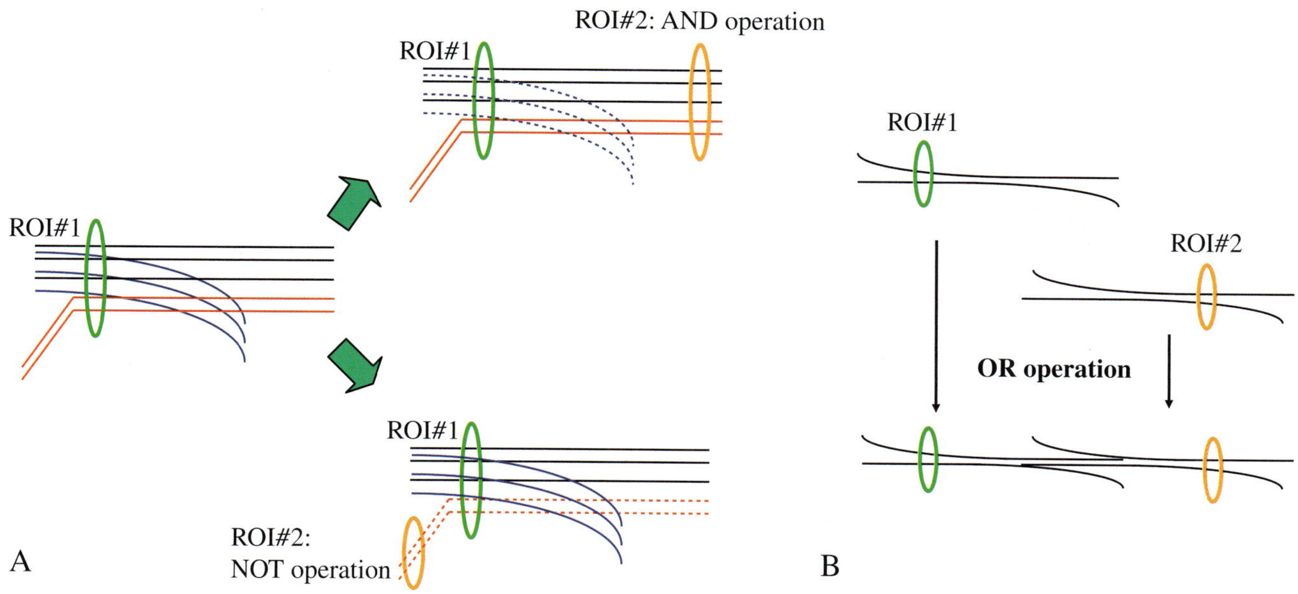

Fig. 3. Schematic diagram of three ROI operations. When the first ROI is drawn, all tracts that penetrate the ROI are retrieved (black, red, and blue fibers) (A). If the second ROI is applied as an "AND" operation, only the fibers that penetrate both ROIs are retained (black and red fibers). If a "NOT" operation is applied to the second ROI, a subset of the fibers penetrating the first ROI (ROI #1), but not the second ROI (ROI #2), is selected (black and blue fibers). When the "OR" operation is used, multiple tracking results are combined (B).

3.2. Editing of reconstruction results using multiple ROIs

In the present atlas, we used several fiber-editing techniques that employ multiple ROIs to delineate tracts of interest. There were three types of multi-ROI operations; AND; OR; and NOT (Fig. 3). These operations were employed diversely depending on the characteristic trajectory of each tract. The most commonly used function was the "AND" operation, applied to identify specific fibers that connect more than one anatomical landmark depicted by ROIs. The "OR" operation was used for fibers in the limbic system, which tend to have a narrow tubular shape for which a single-tract tracking result sometimes failed to delineate the entire length. Use of the "NOT" operation was sometimes necessary to remove a subset of projections from a reconstruction result. As mentioned in the previous section, tracts in this atlas were reconstructed as faithfully as possible based on classical *a priori* anatomical knowledge from postmortem studies. DTI-based fiber tracking often leads to trajectories that have not been well-characterized previously. These unexpected tracts might correspond to reality, as well as to an artifact, which, presently, cannot be validated. In this atlas, such undocumented fibers were removed by the "NOT" operation. The "NOT" operation was also used to remove the "relayed" projections often observed in the thalamus. For example, when the anterior thalamic radiation is reconstructed using two ROIs, one at the thalamus and the other at the frontal lobe, a small number of projections that connect the frontal lobe to the thalamus penetrate the thalamus and proceed to various brain regions such as the brainstem and cerebellum. Some of these trajectories are observed reproducibly and could possibly correspond to real connections, but they were chosen to be removed using the "NOT" operation as they apparently do not belong to the classical description of the anterior thalamic radiation.

3.3. ROI drawing strategy

Once all the data have been preprocessed, the tracts must be reconstructed by the proper choice of reference ROIs and subsequently assigned. In Fig. 4, an example of actual fiber reconstruction of the inferior fronto-occipital fasciculus using multiple ROIs is shown. These ROIs are strategically located, based on known fiber trajectories. From this example, it can be seen that multiple ROIs pose a strong constraint, and large ROIs that surround the target areas result in clean definition of the tracts of interest. The actual locations and sizes of ROIs used in this book follow our previous publication for fibers in the brainstem[1] as well as in the cerebral hemispheres.[2, 3]

3.4. Limitations of DTI-based reconstruction

Compared to invasive chemical-tracer-based techniques, DTI-based tract reconstruction has several important limitations. First, it does not distinguish afferent from efferent projections. Second, because of a limitation in the spatial resolution, which is typically 2–3 mm for each pixel dimension, specific connectivity information about individual axons degenerates. This point is illustrated in

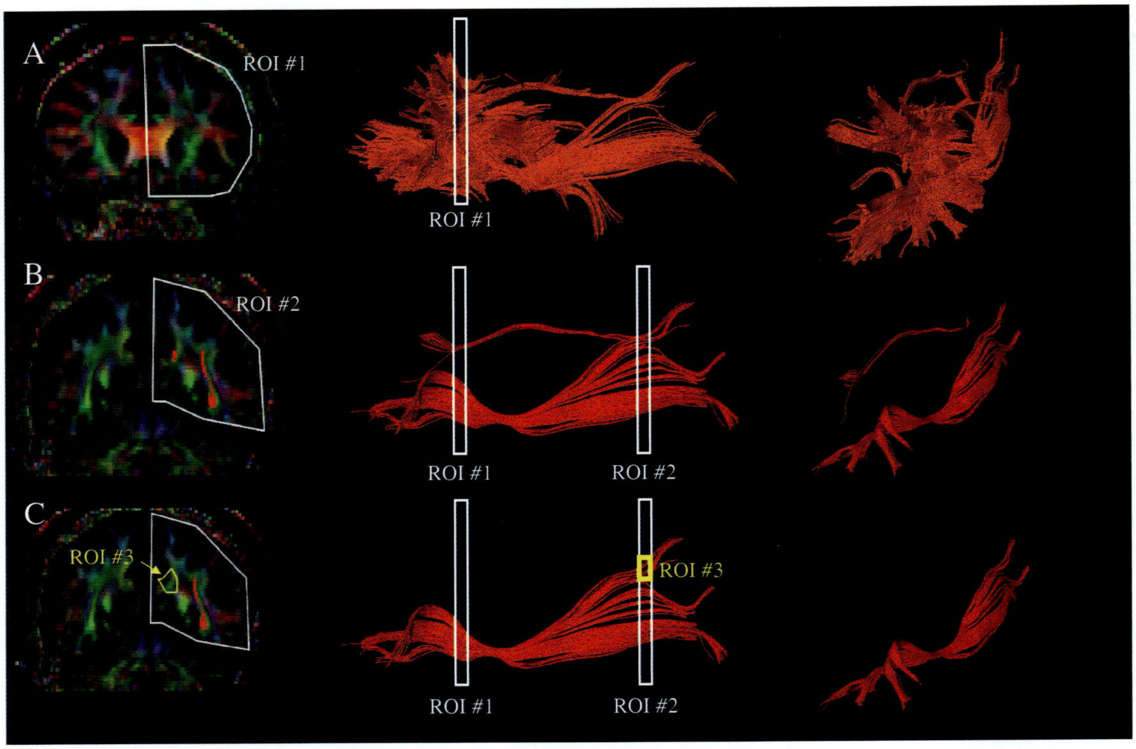

Fig. 4. An example of fiber editing using the "AND" and "NOT" operations to reconstruct the inferior fronto-occipital fasciculus (IFO). First, a large ROI that delineates the entire frontal lobe is drawn at a coronal slice level. This leads to a large number of trajectories that penetrate the ROI (A), including callosal connections to the contralateral hemisphere, the anterior thalamic radiation, and various association fibers. The second ROI is drawn at the occipital lobe (B) and an "AND" operation is used to search only for trajectories that penetrate both ROIs. This leads to delineation of the IFO. However, it appears that part of the cingulum also penetrates the same ROIs. This cingulum component is removed by a "NOT" operation (C, yellow ROI), which results in an IFO representation that is similar to that of classical drawings by neuroanatomists.

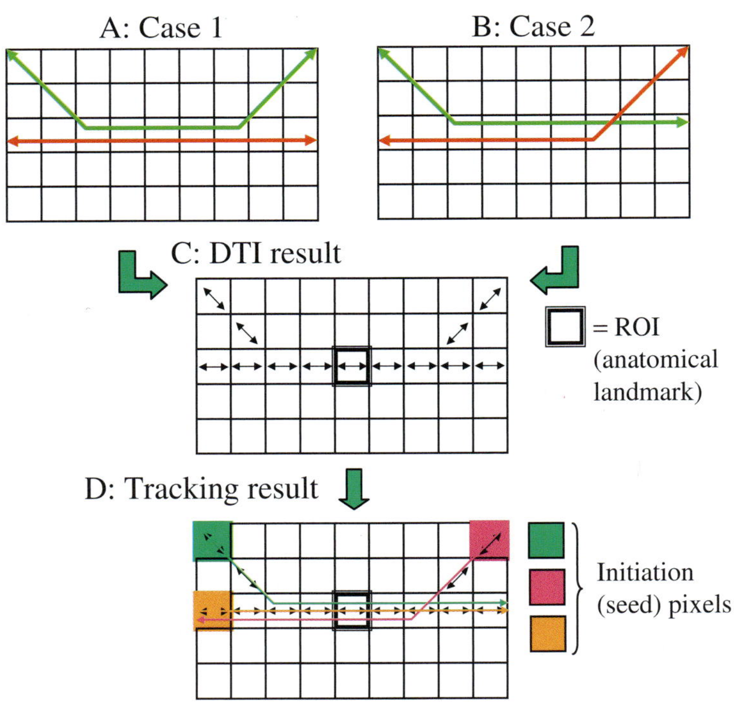

Fig. 5. Schematic diagram of the degeneration of connectivity information. (A and B) Two hypothetical cases of tract structures that consist of two axonal tracts. (C) The vector map obtained from DTI measurement. (D) Results of "brute force" tract reconstruction.

Fig. 5, where two hypothetical structures of tract systems, leading to similar DTI results, are shown. The "brute force" approach can reveal the overall configuration of the tract by collectively displaying all tracking results, but each tracking result (e.g., the green, pink, or orange line in Fig. 5D) does not necessarily represent a real axonal connectivity.

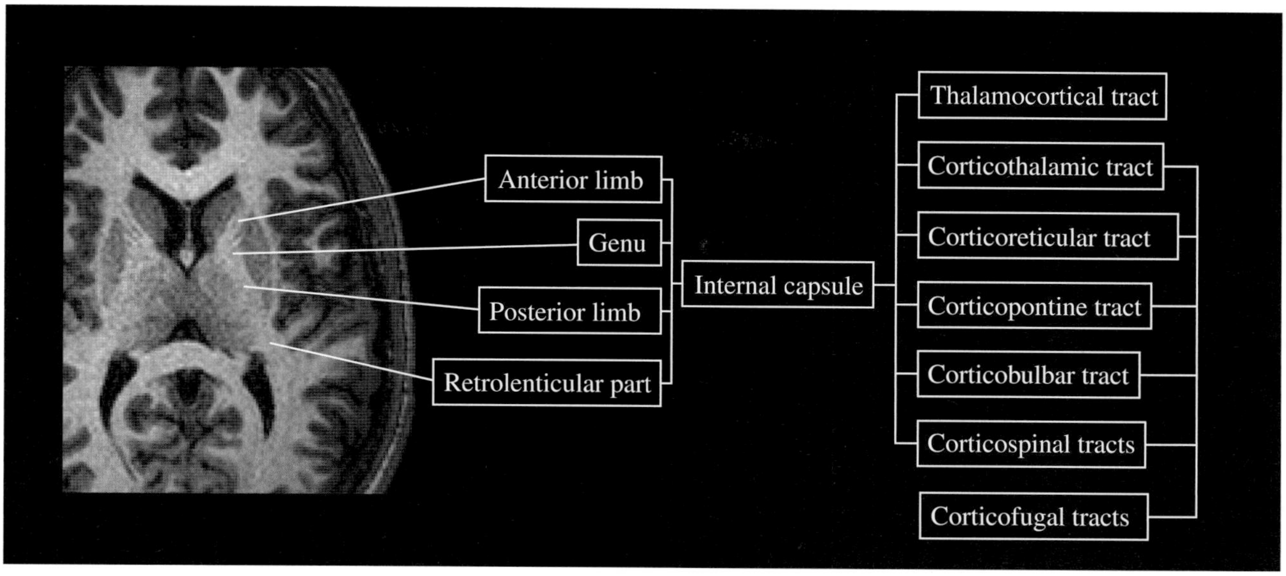

Fig. 6. Relationship between various nomenclatures of white matter tracts in the internal capsule. Names in the left column are related to anatomical locations while those in the right column are connectivity-based.

The third important limitation occurs when there are more than two tract families with different orientations within a pixel. For example, if there is a dominant population of one tract mixed with a small number of tracts bearing different fiber orientations, the DTI results can provide only the pathway for the dominant fibers. If there are two equal populations of tracts with different orientations within a pixel, the pixel tends to have low anisotropy, as previously discussed (Fig. 5, Chapter 1). This leads to a termination of the tracking before delineating the entire length of the tract. For example, low anisotropy areas due to merging of projection fibers and the corpus callosum are shown in Fig. 5, Chapter 1. Because of these areas of low anisotropy, tracking results of the projection tracts in Fig. 2E terminate prematurely before they reach the cortical areas (indicated by yellow arrows). This type of possible "lack of labeling" is an important limitation.

4. 3D visualization

In the 3D atlas (Chapter 3), the trajectories of the white matter tracts are visualized together with the ventricles and several other gray matter structures, including the caudate, the lentiform nucleus, the thalamus, and the hippocampus (including the amygdala), which were manually defined using the co-registered MPRAGE images.

5. Nomenclature and annotation

Nomenclature for white matter structures is sometimes confusing. Probably most straightforward is the assignment of fibers that include the two connected anatomical regions in their names (e.g., corticospinal tract, corticopontine tract, inferior fronto-occipital fasciculus), but not all tracts follow this nomenclature (e.g., fornix, superior longitudinal fasciculus, uncinate fasciculus). There is also a group of names that is not necessarily related to specific tract connectivity. One example of this is the internal capsule (Fig. 6), which has subdivisions (anterior limb, genu, posterior limb, and retrolenticular part) based on anatomical locations. These areas collectively contain corticoafferent pathways, such as thalamocortical tracts, as well as corticoefferent (corticofugal) pathways, such as corticothalamic, corticoreticular, corticopontine, corticobulbar, and corticospinal tracts.

In this atlas, many white matter tracts were reconstructed as faithfully as possible using multiple regions of interest that were chosen based on existing anatomical knowledge. Names of white matter structures that were not 3D-reconstructed or not related to single specific tracts are listed below. These structures are shown in Chapter 4.

Corona radiata

This structure contains the corticopetal (corticoafferent: thalamocortical fibers), long corticofugal (corticoefferent: corticothalamic, corticostriatal, corticopontine, corticobulbar, corticospinal, and corticoreticular), and commissural fibers. In the 2D atlas (Chapter 4), the corona radiata is divided into anterior (acr), superior (scr), and posterior (pcr) parts, although the boundaries are arbitrarily defined.

Internal capsule

The corona radiata converges into the internal capsule, which contains the thalamocortical and long corticofugal tracts. It can be divided into the anterior limb (aic), posterior limb (pic) and retrolenticular part (rlic), as shown in Fig. 6.

Cerebral peduncle

At the midbrain level, the internal capsule converges into the cerebral peduncle, which contains the long corticofugal tracts.

Sagittal stratum

This structure seems contiguous to the posterior region of the corona radiata (Fig. 6, Chapter 1), its main constituent being the optic radiation (a part of the thalamocortical tract). The sagittal stratum also contains commissural fibers from the splenium of the corpus callosum and corticofugal fibers, as well as association fibers, most notably the inferior fronto-occipital fasciculus and the inferior longitudinal fasciculus.

External capsule

Seen as a thin layer of white matter lateral to the lentiform nucleus, the external capsule contains association fibers, such as the superior longitudinal fasciculus, inferior fronto-occipital fasciculus, and uncinate fasciculus. With current imaging resolution for DTI, the extreme and external capsules cannot be discriminated.

References

[1] B. Stieltjes, W.E. Kaufmann, P.C.M. van Zijl, K. Fredericksen, G.D. Pearlson, S. Mori, Diffusion tensor imaging and axonal tracking in the human brainstem, Neuroimage 14 (2001) 723–735.

[2] S. Mori, W.E. Kaufmann, C. Davatzikos, B. Stieltjes, L. Amodei, K. Fredericksen, et al., Imaging cortical association tracts in human brain using diffusion-tensor-based axonal tracking, Magn. Reson. Med. 47 (2002) 215–223.

[3] S. Wakana, A. Caprihan, M.M. Panzenboeck, J.H. Fallon, M. Perry, R.L. Gollub, et al., Reproducibility of quantitative tractography methods applied to cerebral white matter, Neuroimage 36 (3) (2007) 630–644.

CHAPTER 3

Three-Dimensional Atlas of Brain White Matter Tracts

White matter tracts were classified into four groups: tracts in the brainstem and projection, association, and commissural tracts in the cerebral hemispheres. Projection fibers connect cortical and subcortical gray matter, association tracts connect two cortical areas, and commissural tracts connect right and left hemispheres. In the following sections, 3D architectures of these tracts are reconstructed.

1. Tracts in the brainstem

Five major white matter tracts were reconstructed in the brainstem. These are: the superior, middle, and inferior cerebellar peduncles, the corticospinal tract and the medial lemniscus. A schematic diagram and 3D trajectories of these tracts are shown in Fig. 1 and Fig. 2. The global trajectories of these tracts have been described in many neuroanatomy textbooks and the DTI-based fiber tracking could faithfully reconstruct them. Unlike histology-based techniques, the MRI-based reconstruction allows viewing of the structures from multiple orientations and adding or removing structures of interest as desired (Fig. 3).

The superior cerebellar peduncle (scp) is the main efferent pathway from the dentate nucleus of the cerebellum toward the thalamus (Fig. 4). The tract reconstruction result shows that the one side of the superior cerebellar peduncle originates at the deep cerebellar nuclei and the other at the thalamus. In 2D slices, it can be discretely identified in an axial slice at the level of the dentate nuclei (slice #2 in Fig. 4). At a superior level (slices #1 and #4 in Fig. 4), its decussation can be found as a circular red (right–left orientation) structure. Although the majority of the tract crosses the midline at the decussation, the reconstructed trajectory remained in the same hemisphere. This is due to limitation of the fiber tracking technique, which tends to provide > < (kissing) solution when two fibers have X (crossing) trajectories.

The inferior cerebellar peduncle (icp) contains afferent and efferent connections to the cerebellum (Fig. 5). It originates in the caudal medulla, traverses the pons, and branches into the cerebellar cortex. In the color map, it can be readily identified at the dorsal area of the medulla (slice #5 in Fig. 5) and the pons (slices #3 and #4 in Fig. 5).

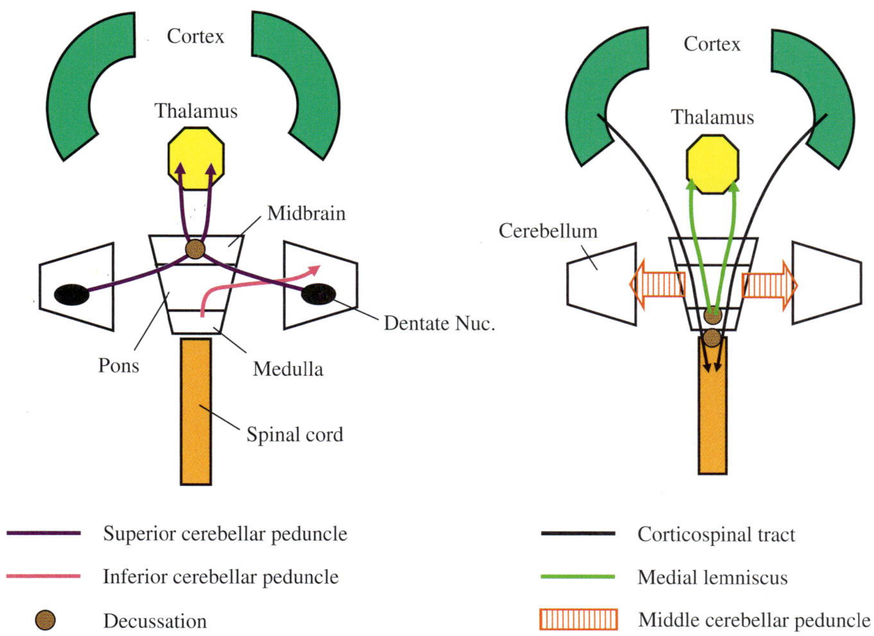

Fig. 1. Schematic diagram of the tracts in the brainstem. Decussation indicates that of the superior cerebellar peduncle.

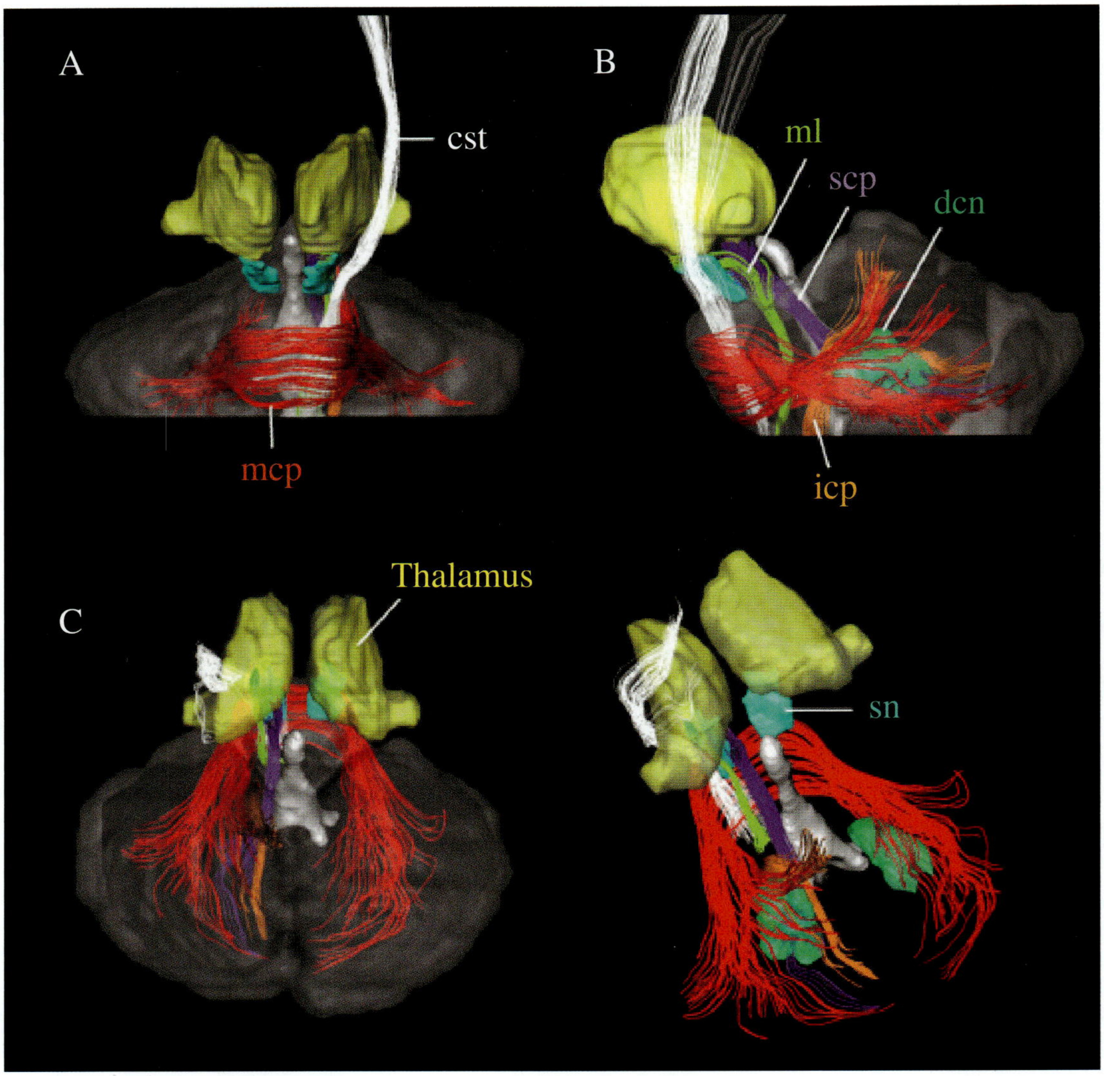

Fig. 2. 3D reconstruction results of five major tracts in the brainstem. Tracts are viewed from the anterior (A), left (B), superior (C), and oblique (left-superior–posterior) (D) orientations. The cerebellum is represented by a semi-transparent gray structure in A, B, and C. The cerebral aqueduct and the IV ventricle are represented by a white 3D reconstructed structure. Abbreviations are: cst, corticospinal tract; dcn, deep cerebellar nuclei; icp, inferior cerebellar peduncle; mcp, medial cerebellar peduncle; ml, medial lemniscus; scp. superior cerebellar peduncle; sn, substantia nigra.

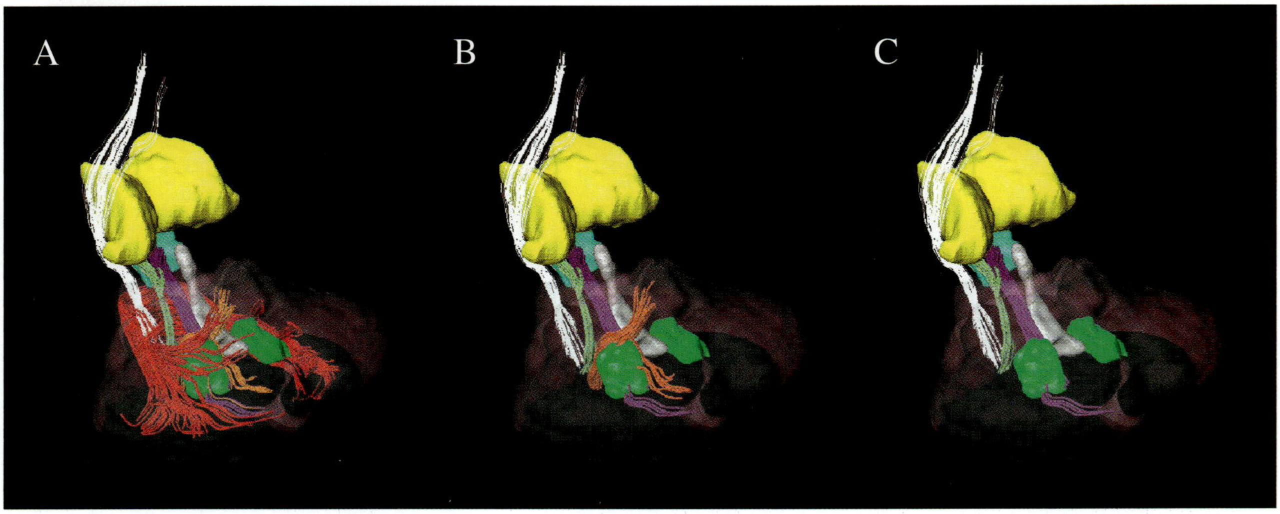

Fig. 3. Examples of structure stripping to reveal inner structures. In B, the mcp was excluded, allowing better visualization of the cst, ml, icp, and scp. In C, the icp was removed, allowing better assessment of the relationship between the scp, dnc, and IV ventricle.

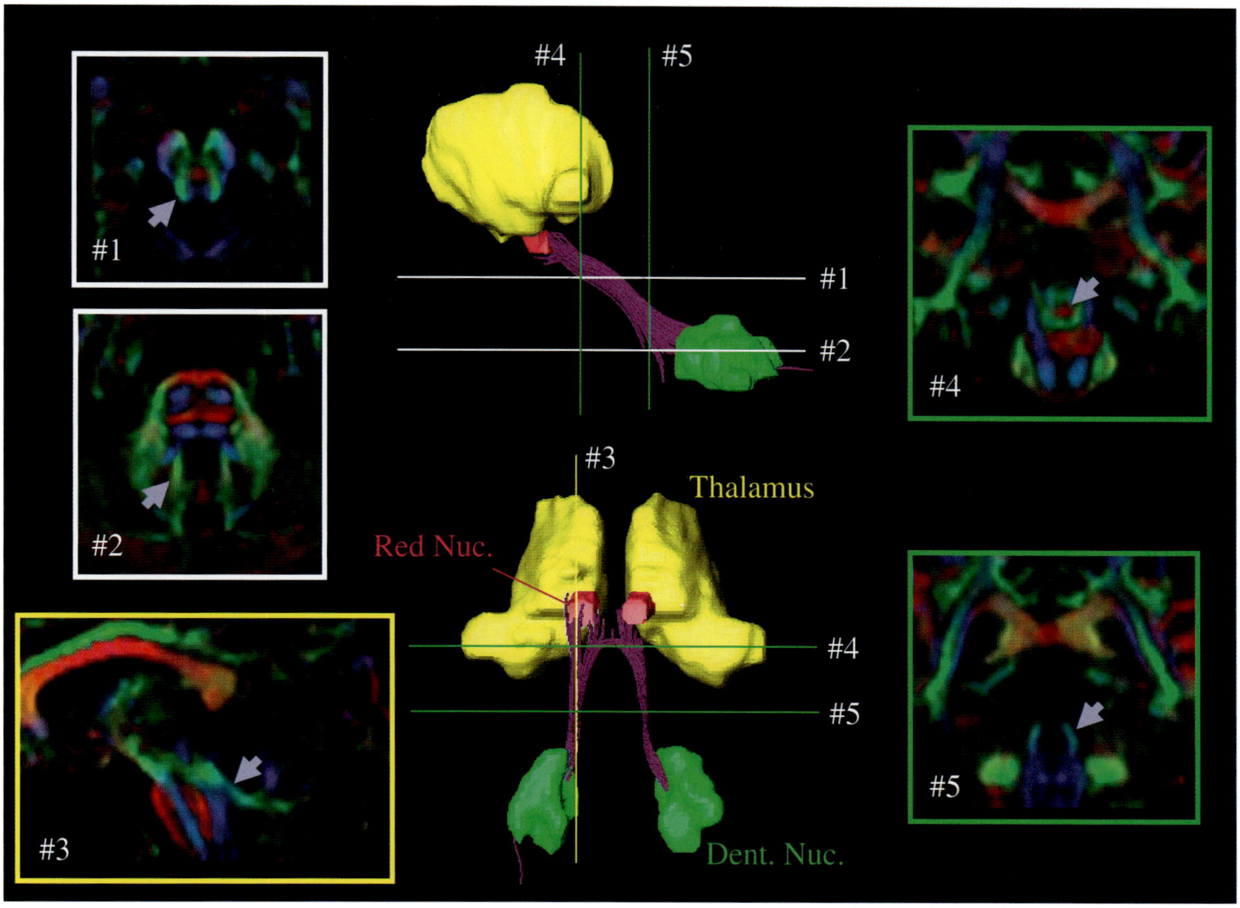

Fig. 4. The trajectory of the scp and its identification in color maps at various slice locations and orientations. Locations of 2D slices are indicated in the 3D reconstruction viewed from the left (upper panel) and inferior (bottom panel) orientations.

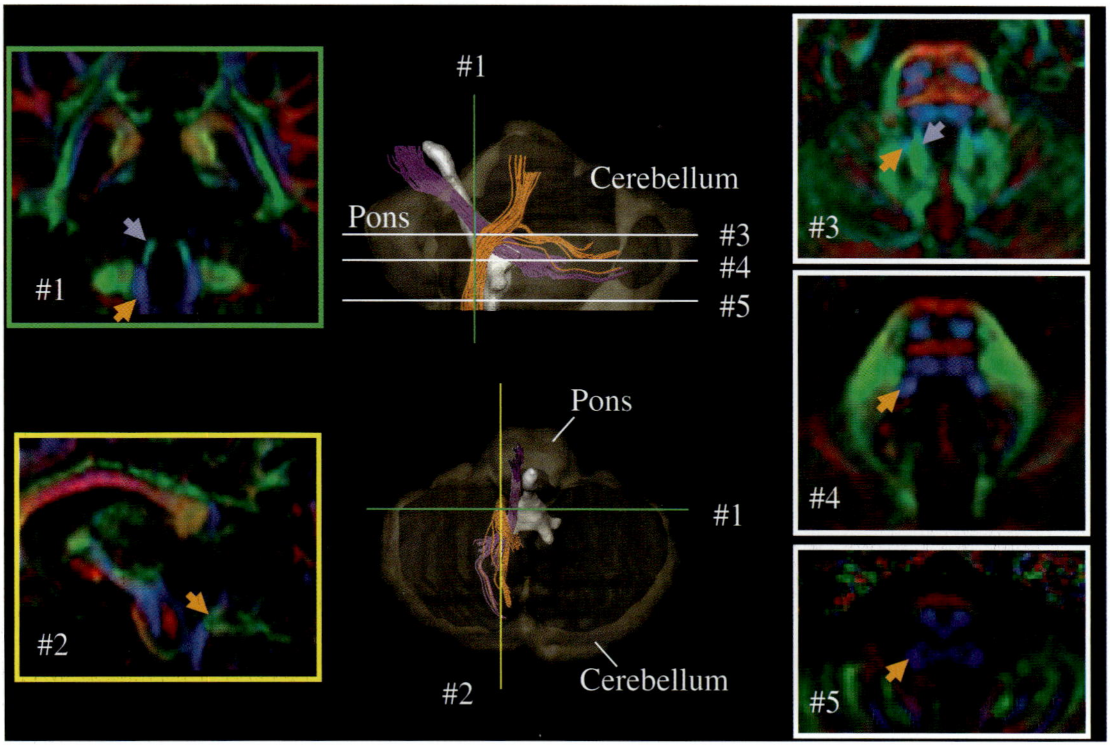

Fig. 5. The trajectory of the icp (orange) and its identification in color maps at various slice levels and orientations. The icp is indicated by orange arrows and the scp by purple arrows. The semi-transparent gray object is the cerebellum and the white structure includes the cerebral aqueduct and the IV ventricle. Locations of 2D slices are indicated in the 3D reconstruction viewed from the left (upper panel) and superior (bottom panel) orientations.

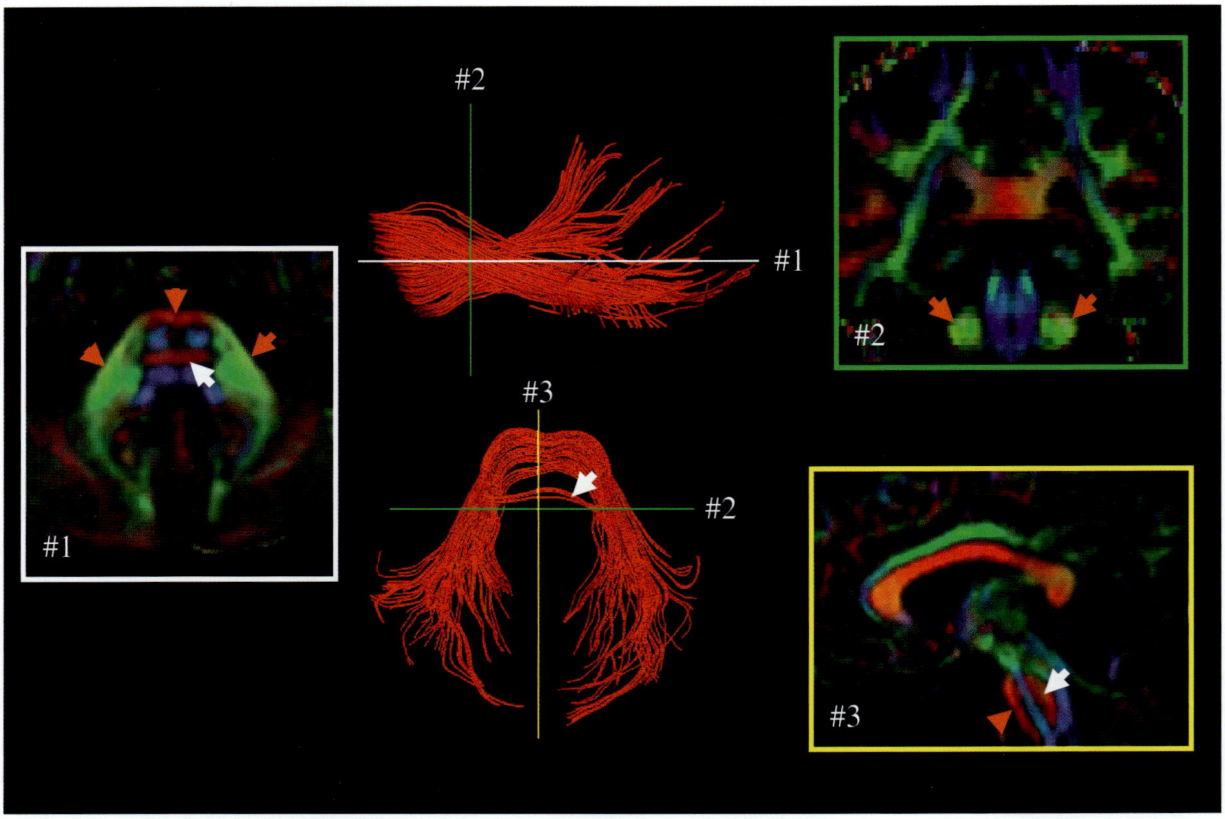

Fig. 6. The trajectory of the MCP and its identification in color maps (red arrows) at three slice locations and orientations. Locations of the 2D slices are indicated in the 3D reconstruction view from the left (upper panel) and superior (bottom panel) orientations. The transverse pontine fibers (white arrows) are part of the mcp.

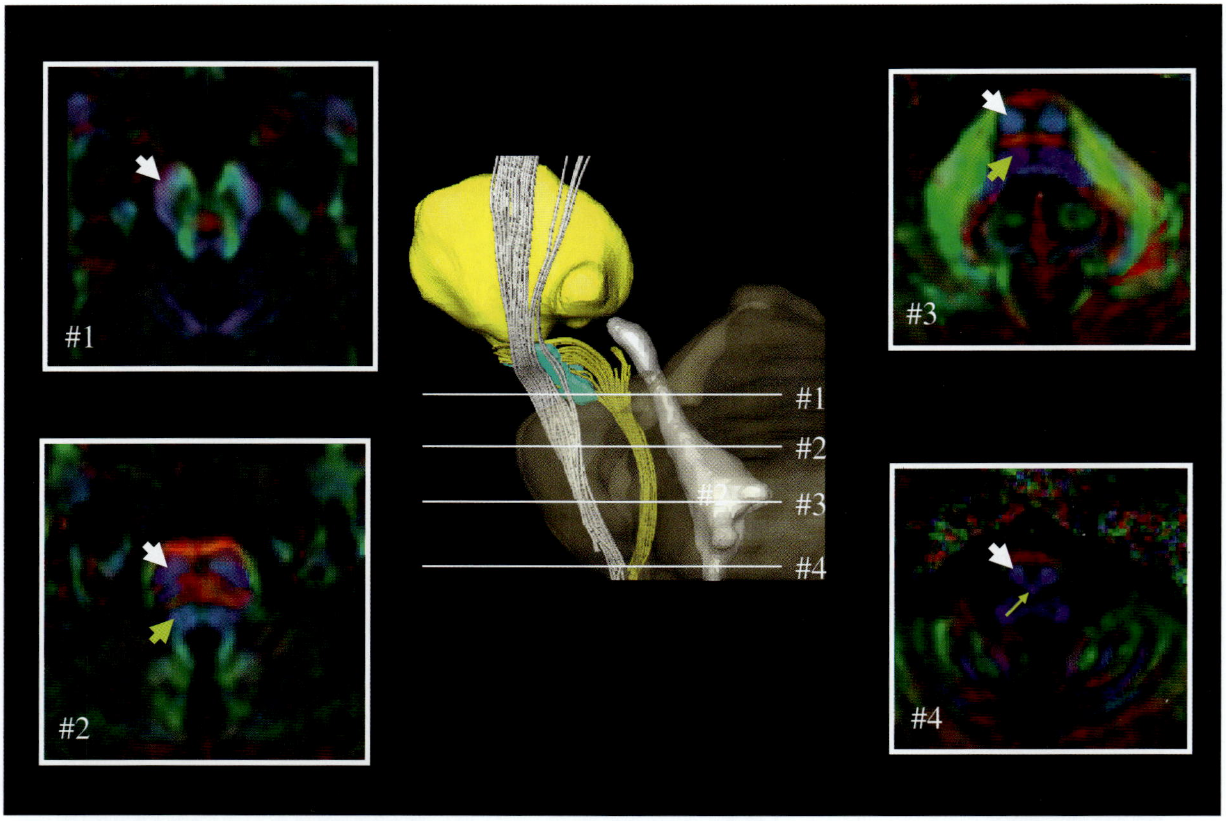

Fig. 7. The trajectory of the corticospinal tract (cst, white) and medial lemniscus (ml, light green) and their identification in color maps at various axial slice levels. White and light green arrows in the color maps indicate the locations of the cst and ml, respectively. The levels of the 2D slices are indicated in the 3D reconstruction viewed from the left side. The thalamus (yellow), substantia nigra (blue), ventricle (white), and cerebellum (gray) are also shown in the 3D reconstruction.

The middle cerebellar peduncle (mcp) also contains afferents from the pons to the cerebellum (pontocerebellar tracts), forming a massive sheet-like structure that wraps around the pons (Fig. 6). The cross-section of this tract can be readily identified by coronal planes (slice #2 in Fig. 6) through the pons. The transverse pontine fibers indicated by white arrows in slices #1 and #3 in Fig. 6 are also part of the pontocerebellar tracts.

The medial lemniscus (ml) is a major pathway for ascending sensory fibers (Fig. 7). It decussates at the level of the ventral medulla (slice #4 in Fig. 7). As it ascends the pons, it travels along the dorsal aspect of the pons and midbrain. The right and left tracts can be identified at slice levels #2 and #3 and it decussate at the slice #4 in Fig. 7. At the midbrain level (slice #1), it becomes too dispersed for discrete identification.

The corticospinal tract (cst) is a descending pathway from the cortex; it penetrates the cerebral peduncle (indicated by the white arrow in slice #1) and is readily identifiable in the pons and medulla (slices #2 – #4).

2. Projection fibers

Two classes of projection fibers were reconstructed for this atlas: the corticothalamic/thalamocortical fibers (collectively called thalamic radiations) and the long corticofugal (corticoefferent) fibers. The corticofugal fibers include such fibers as the corticopontine, corticoreticular, corticobulbar, and corticospinal tracts. A schematic diagram of their trajectories is shown in Fig. 8.

In Fig. 9, the 3D trajectories of these reconstructed projection fibers are shown. They all penetrate the internal capsule either between the thalamus and the putamen or between the caudate and the putamen. When approaching to the cortex, they fan out to form the corona radiata (see also Fig. 6 in Chapter 1).

In Fig. 10, the locations of these projection fibers are indicated in several 2D slices. The thalamus is known to have reciprocal connections to a wide area of the cortex (see Fig. 10C). In this atlas, the different parts of the thalamic radiation that penetrate the anterior limb, posterior limb, and retrolenticular part of the internal capsule are labeled as anterior, superior (central), and posterior thalamic radiations, respectively. The posterior thalamic radiation should include the optic radiation, which connects the lateral geniculate nucleus and the occipital lobe. Although these three components of the thalamic radiation are differently color-coded in this reconstruction, there is no precise distinction between the anterior and superior or superior and posterior thalamic radiations and, thus, their boundaries should not be considered as definite for assignment of these fibers.

The internal capsule also contains descending pathways from the cortex (long corticofugal pathways) toward the brainstem. The vast majority of these pathways terminate in some pontine nuclei and are called corticopontine tracts (light blue in Fig. 10). Depending upon their area of origin in the cortex, the corticopontine tracts are subdivided into fronto-, parieto-, occipito-, and temporo-pontine tracts. The pontine nuclei then send projections to the contralateral cerebellum through the middle cerebral peduncle,

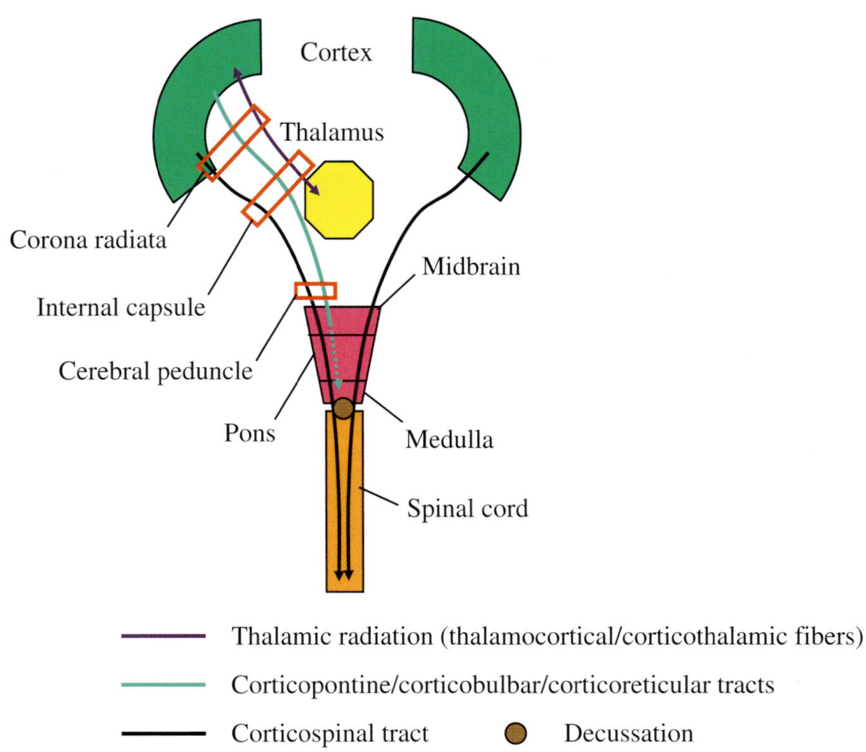

Fig. 8. A schematic diagram of trajectories of projection fibers reconstructed in this atlas. The decussation is that of the corticospinal tract.

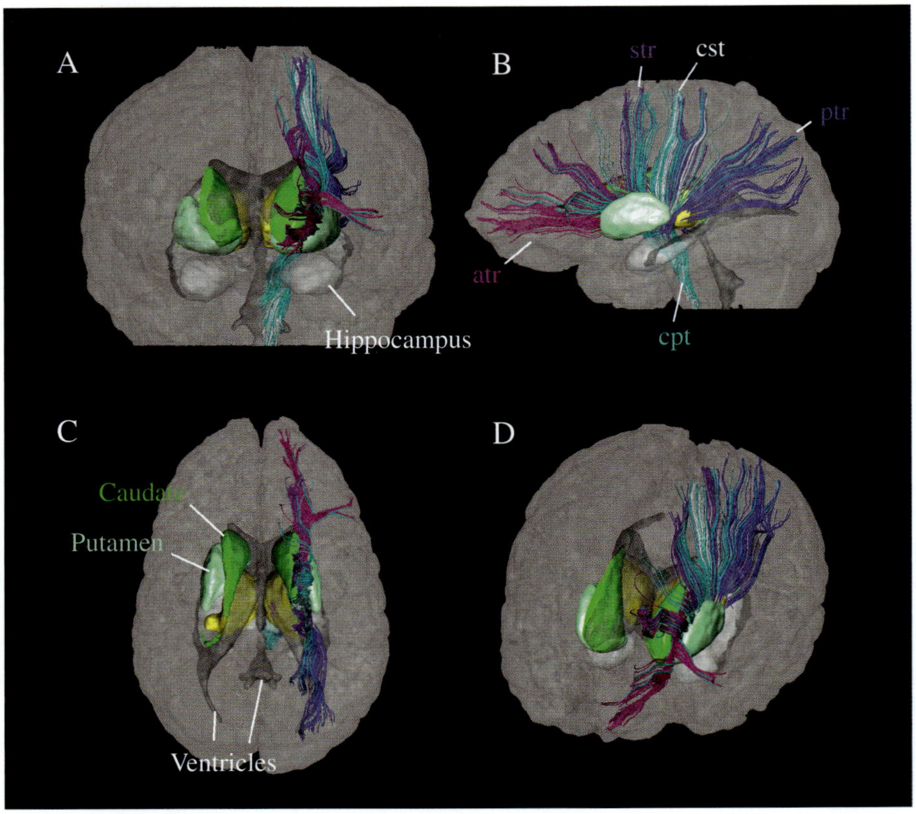

Fig. 9. 3D reconstruction results of projection fibers. Tracts are viewed from the anterior (A), left (B), superior (C), and oblique (left-superior–anterior) (D) orientations. The hemispheres are delineated in semi-transparent gray. Abbreviations are: atr, anterior thalamic radiation; cpt, corticopontine tract; cst, corticospinal tract; ptr, posterior thalamic radiation; str, superior thalamic radiation.

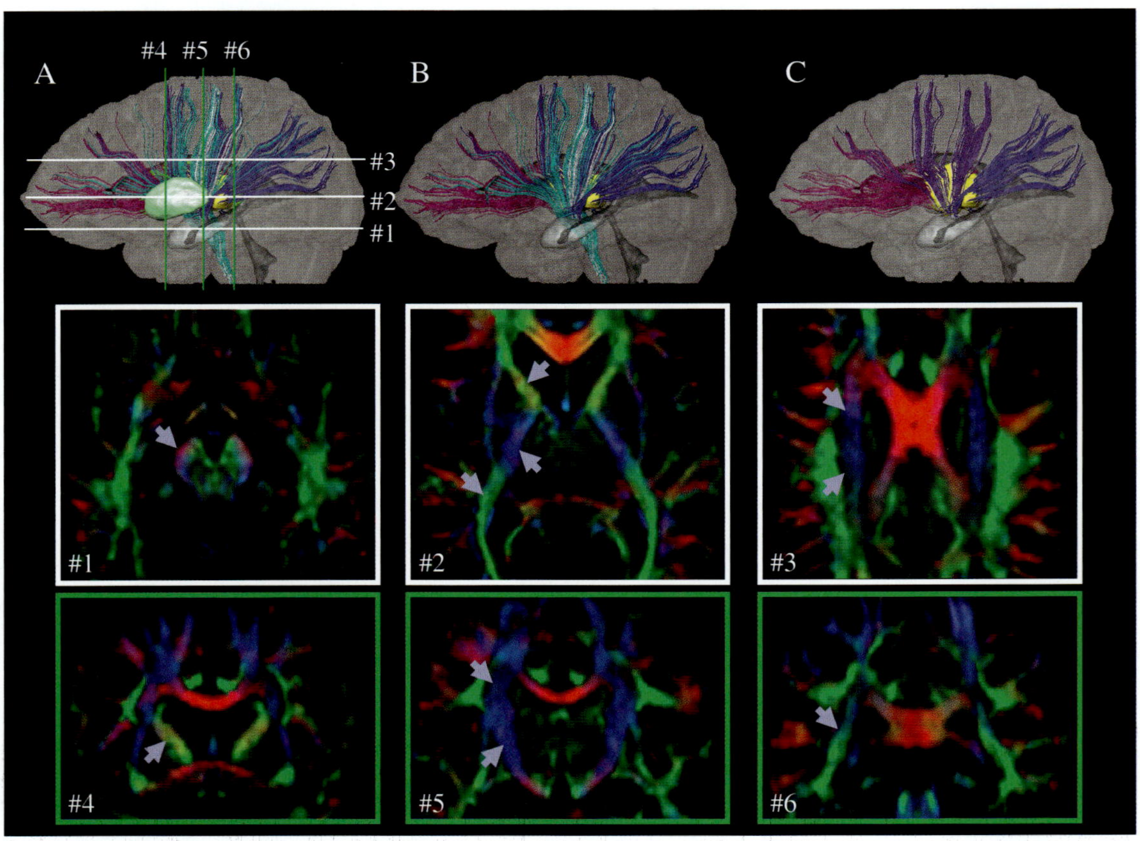

Fig. 10. 3D reconstruction results of the projection fibers (A–C). In (B), the putamen was removed to better visualize the tracts, and in (C), the corticopontine (light blue) and corticospinal (white) tracts were removed for better assessment of the thalamic radiations. Locations of the projection tracts are indicated by purple arrows in the six color maps.

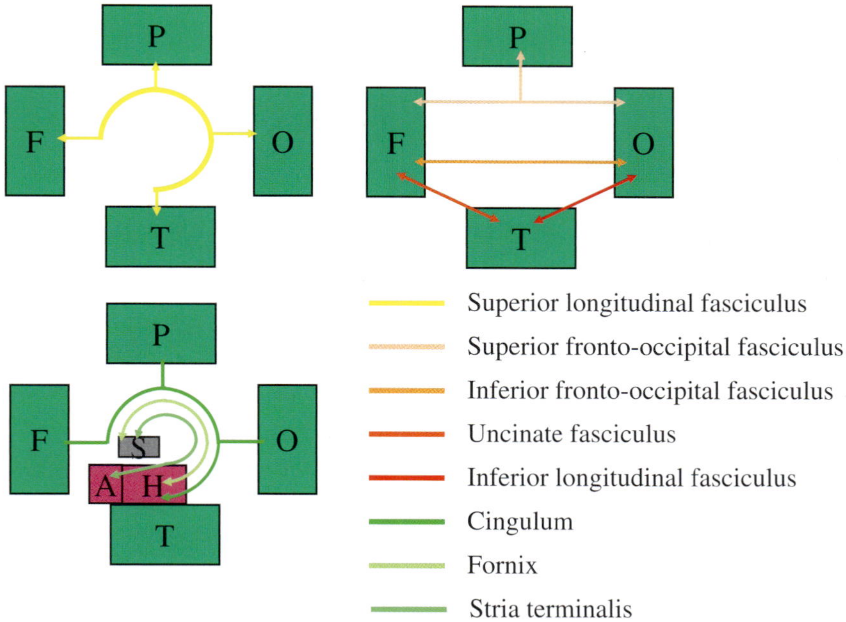

Fig. 11. A schematic diagram of some cortico-cortical connections by association fibers. Abbreviations are F, frontal, P, parietal, O, occipital, and T, temporal cortices, A, amigdala, H, hippocampus, S, septal area.

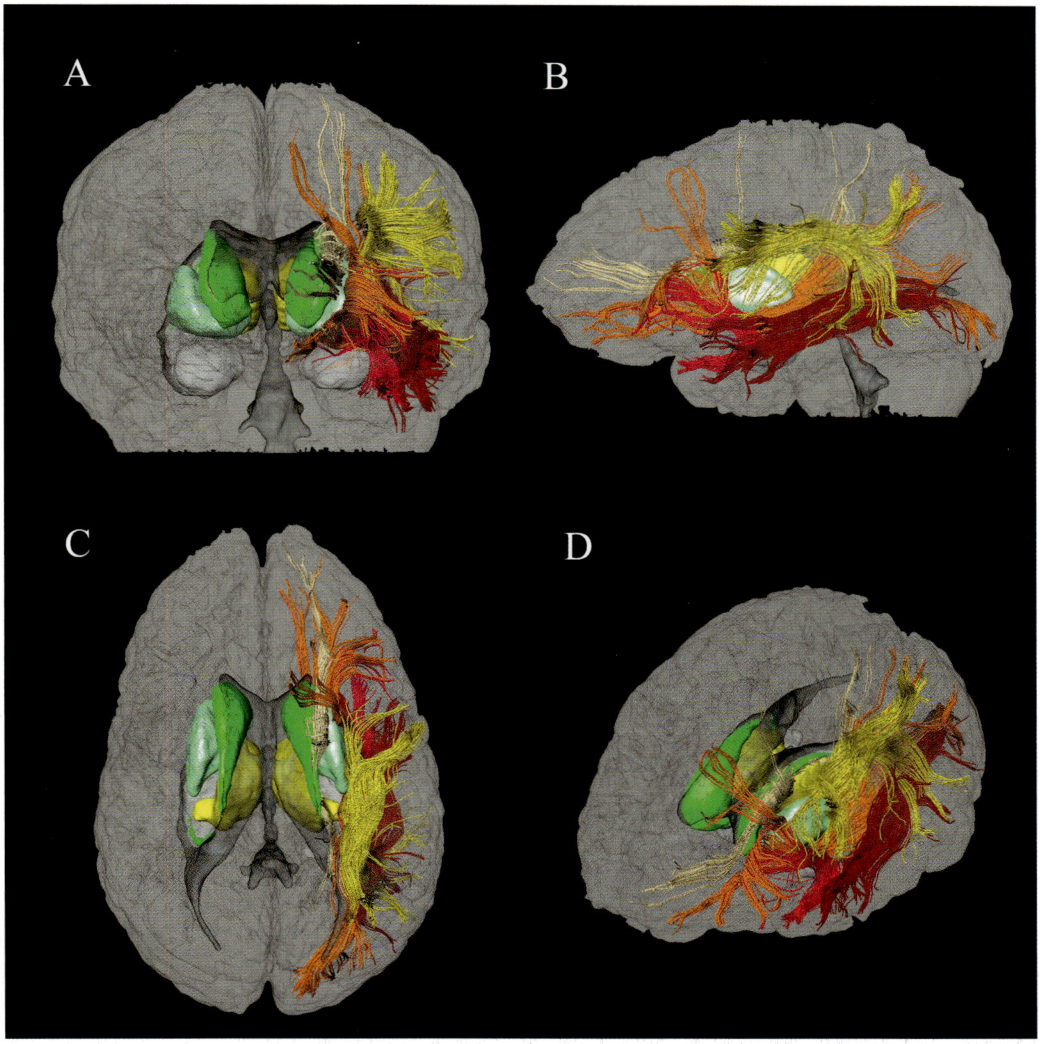

Fig. 12. 3D reconstruction results of some association fibers. Tracts are viewed from the anterior (A), left (B), superior (C), and oblique (left-antero-superior) (D) orientations. Color coding: slf is yellow, ifo is orange, unc is red, and ilf is brown. Cerebral hemispheres are delineated by semi-transparent gray. Thalami are yellow, ventricles are gray, caudate nuclei are green and lentiform nuclei are light green.

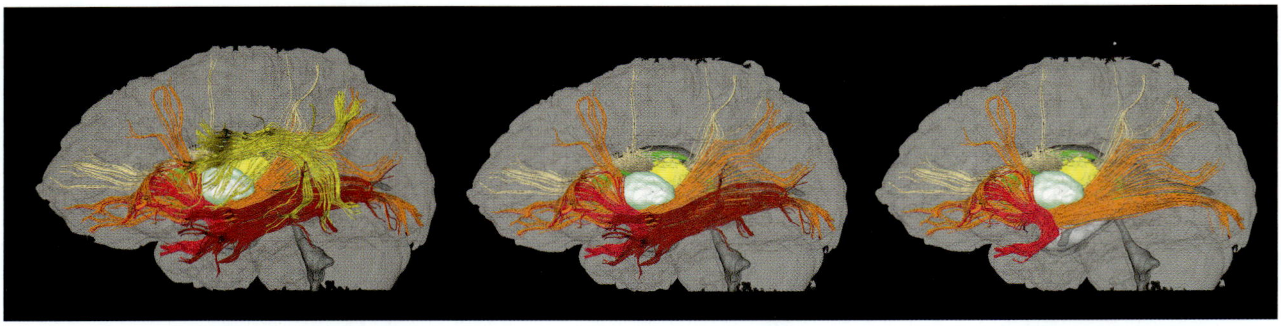

Fig. 13. Structural stripping to reveal inner structures. In the middle and right panels, the superior longitudinal fasciculus (yellow) and inferior longitudinal fasciculus (brown) are removed sequentially, revealing the uncinate (red) and inferior fronto-occipital (orange) fasciculus.

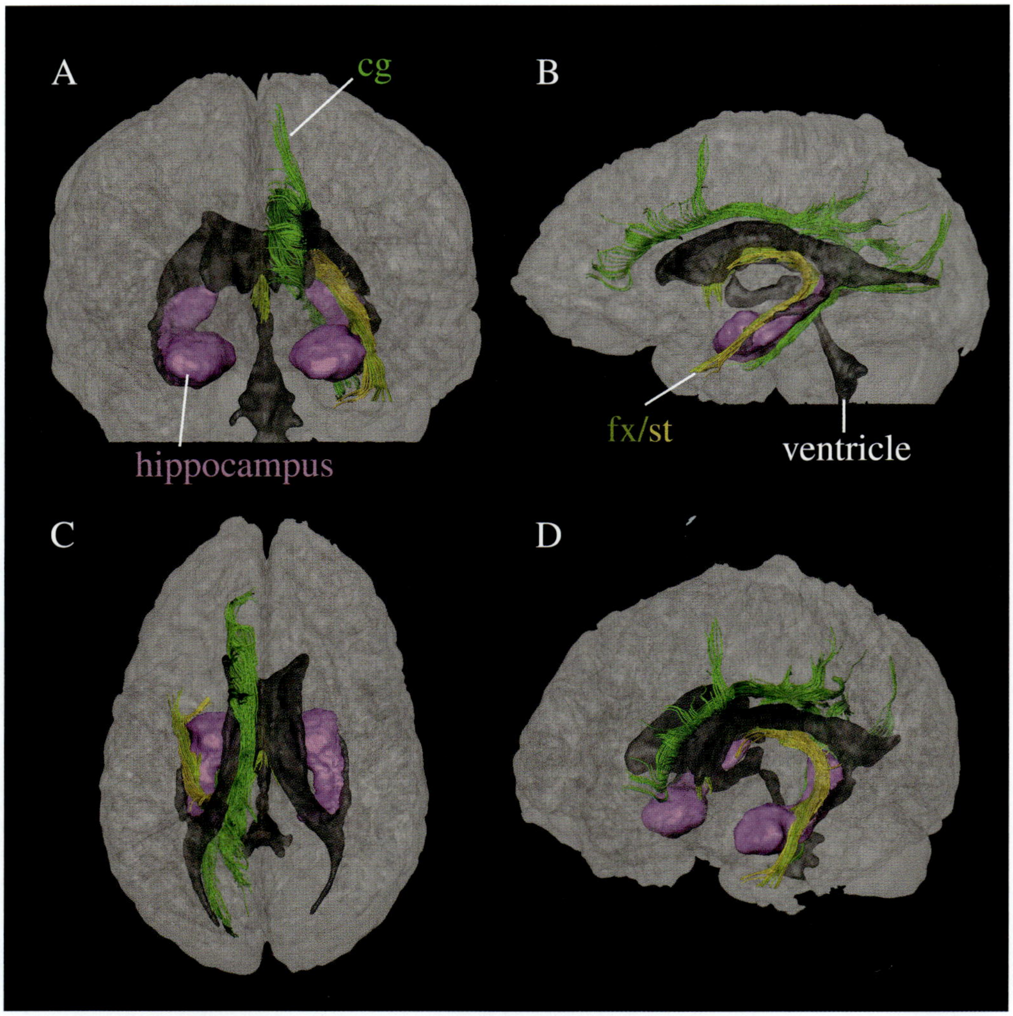

Fig. 14. 3D reconstruction results of association fibers in the limbic system. Tracts are viewed from the anterior (A), left (B), superior (C), and oblique (left-anterior) (D) orientations. Abbreviations are: cg, cingulum; fx, fornix, and st, stria terminalis. The cerebral hemispheres are delineated in semi-transparent gray. The ventricular system is depicted in gray and the hippocampi in purple.

forming the massive cortico-ponto-cerebellar pathway. The corticopontine tracts pass through both the internal capsule (slice #2 in Fig. 10) and the cerebral peduncle (slice #1 in Fig. 10). Strictly speaking, this reconstruction also contains other cortico-brainstem pathways such as the corticoreticular and corticobulbar tracts. Because DTI-based tract reconstruction cannot differentiate these pathways from the corticopontine tracts, they were included within the corticopontine tracts.

The middle portion of the cerebral peduncle is believed to contain tracts that pass through the brainstem and reach the spinal cord, it is therefore called the corticospinal tract (white in Fig. 10). In this tract, descending projections from the precentral (motor) area are the dominant constituent, although some branches from the somatosensory and parietal cortices are also included.

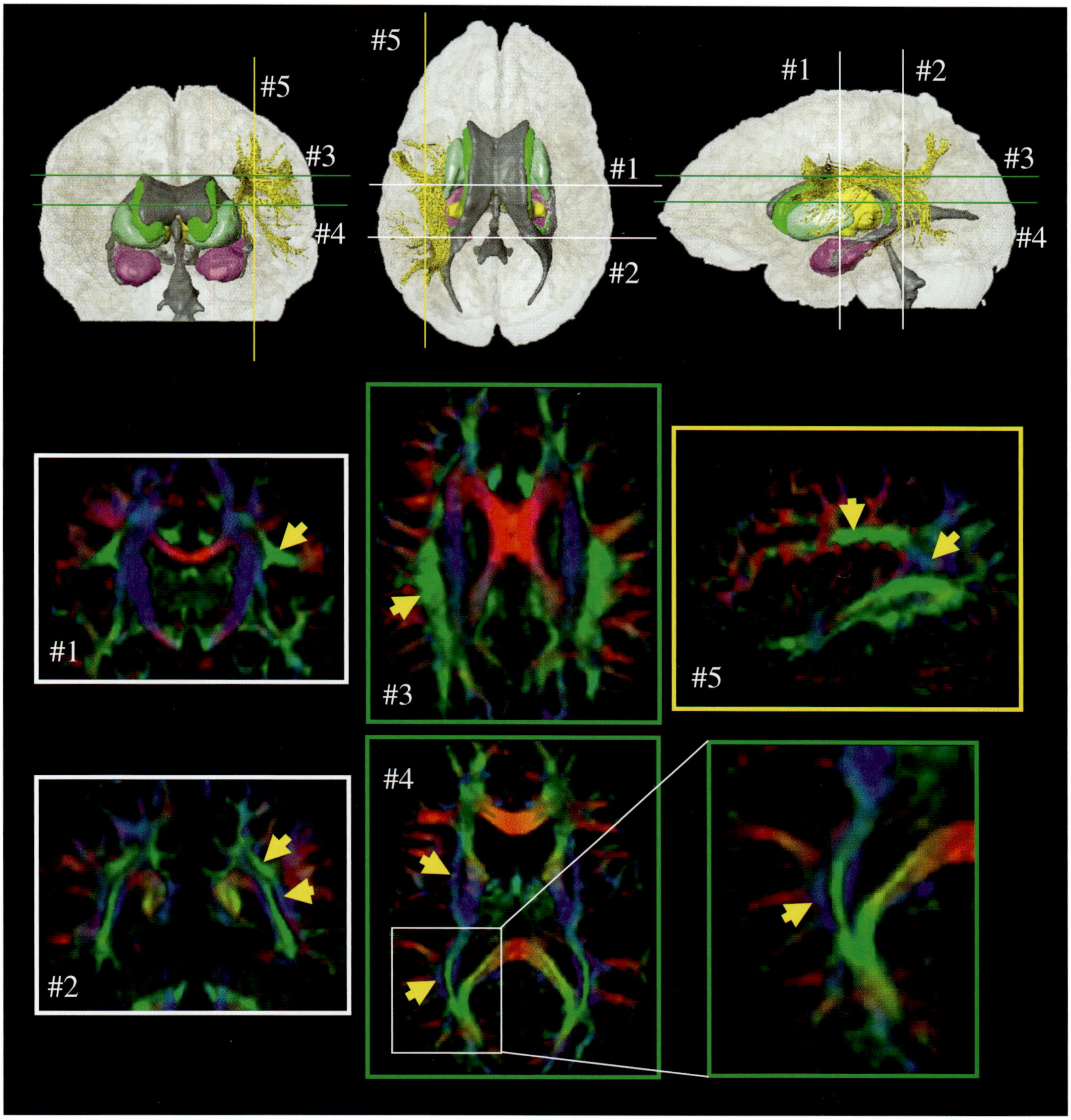

Fig. 15. The trajectory of the superior longitudinal fasciculus and its identification in the color maps at various slice levels and orientations. Locations of 2D slices are indicated in the 3D panels. The thalamus (yellow), ventricular system (gray), caudate (green), and putamen (light green) are also shown in the 3D reconstruction.

3. Association fibers

Association fibers connect different areas of the cortex and are classified into short and long association fibers. The former connect areas within the same cerebral hemisphere and include the fibers connecting adjacent gyri, called the U-fibers. The long association fibers connect different lobes, forming prominent fiber bundles. Fig. 11 shows a schematic diagram of the trajectories of some association tracts. Their approximate locations and trajectories have been well documented in the literature. In the present atlas, these major long association fibers were reconstructed based on previous anatomical descriptions. These include the superior longitudinal fasciculus (slf), inferior longitudinal fasciculus (ilf), superior fronto-occipital fasciculus (sfo), inferior fronto-occipital fasciculus (ifo), and uncinate fasciculus (unc) (Figs. 12 and 13). Three major association fibers connecting the limbic system (hippocampus), namely the cingulum (cg), fornix (fx), and stria terminalis (st), are also presented in Fig. 14.

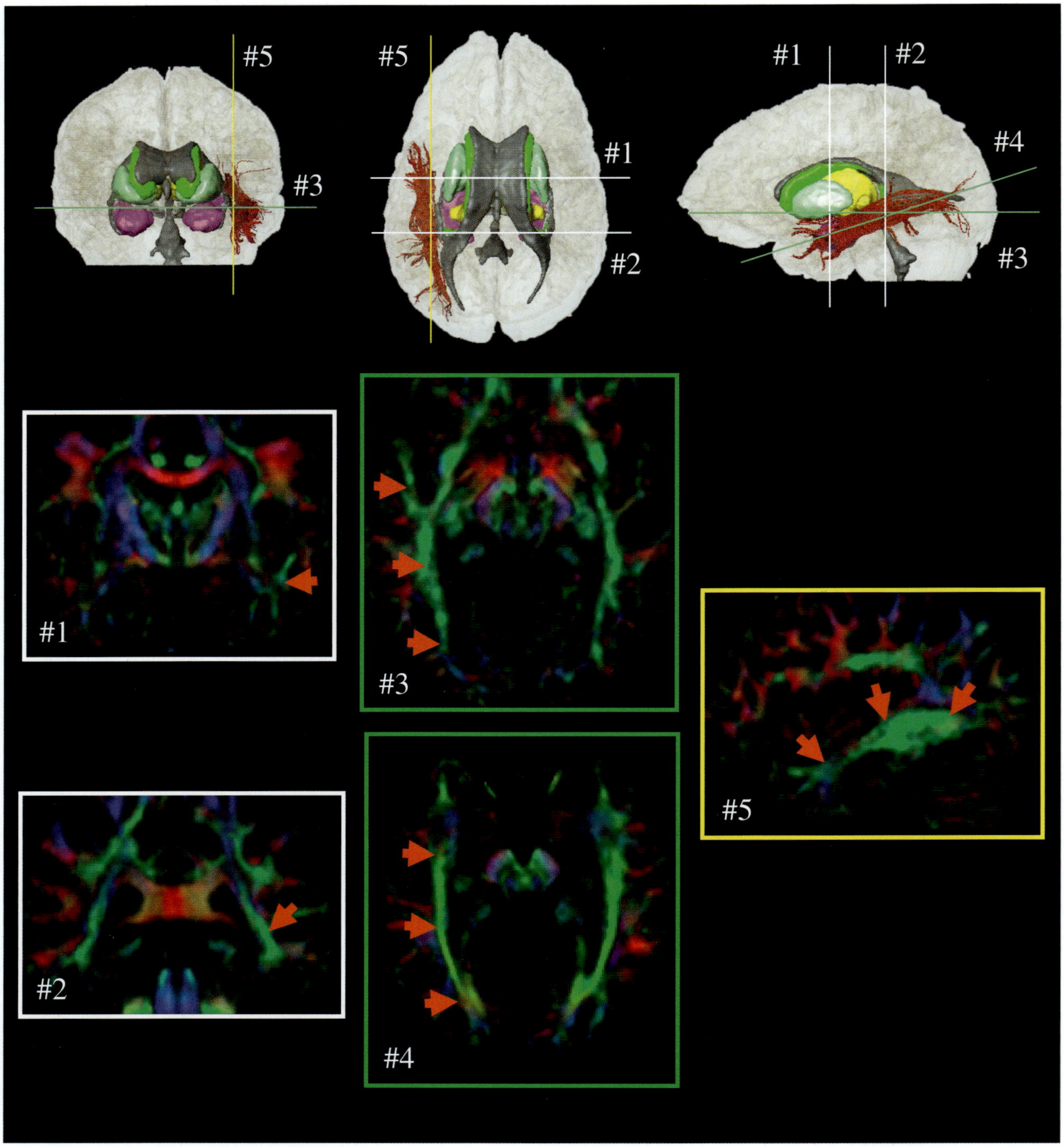

Fig. 16. The trajectory of the inferior longitudinal fasciculus and its identification in color maps at various slice levels and orientations. Locations of the 2D slices are indicated in the 3D panels. The thalamus (yellow), ventricular system (gray), caudate (green), and putamen (light green) are also shown in the 3D reconstruction.

Superior longitudinal fasciculus

This tract is located at the supero-lateral side of the putamen and forms a large arc (also called the arcuate fasciculus), sending branches into the frontal, parietal, occipital, and temporal lobes. The tract can be easily located in many coronal slices as a large and intense (high anisotropy) fiber bundle running along the superior edge of the insula (Fig. 15).

Inferior longitudinal fasciculus, inferior fronto-occipital fasciculus, and uncinate fasciculus

If a coronal slice is chosen at the retrolenticular part of the internal capsule, the sagittal stratum (green fiber bundle indicated by red arrow in Fig. 16, slice #2) can be identified in the color map, which contains two large association fibers. These are the inferior longitudinal fasciculus (Fig. 16), connecting the occipital and temporal lobes, and the inferior fronto-occipital fasciculus (Fig. 17), connecting the frontal and occipital lobes. The inferior fronto-occipital fasciculus has a characteristic shape and its most compact

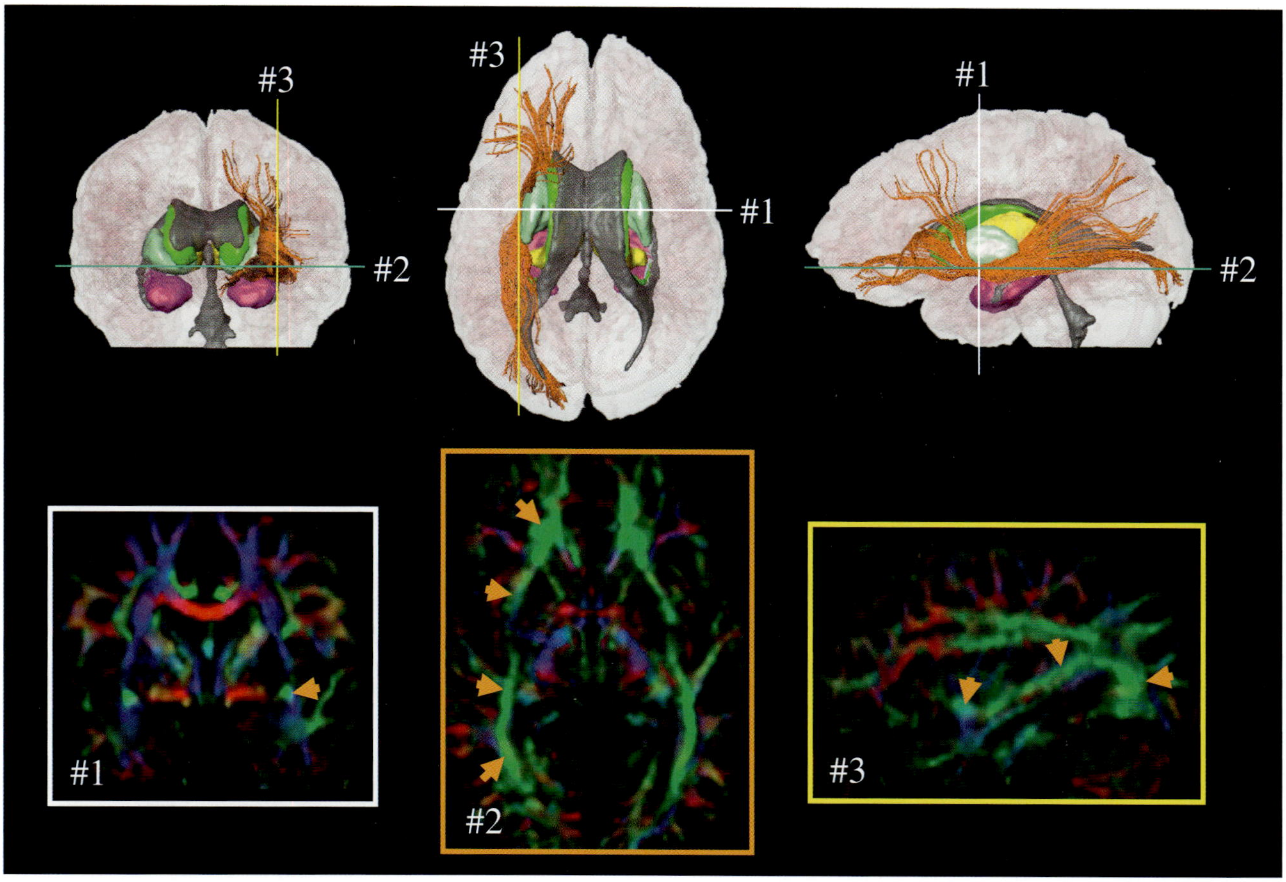

Fig. 17. The trajectory of the inferior fronto-occipital fasciculus and its identification in color maps at various slice levels and orientations. Locations of the 2D slices are indicated in the 3D panels. The thalamus (yellow), ventricular system (gray), caudate (green), and putamen (light green) are also shown in the 3D reconstruction.

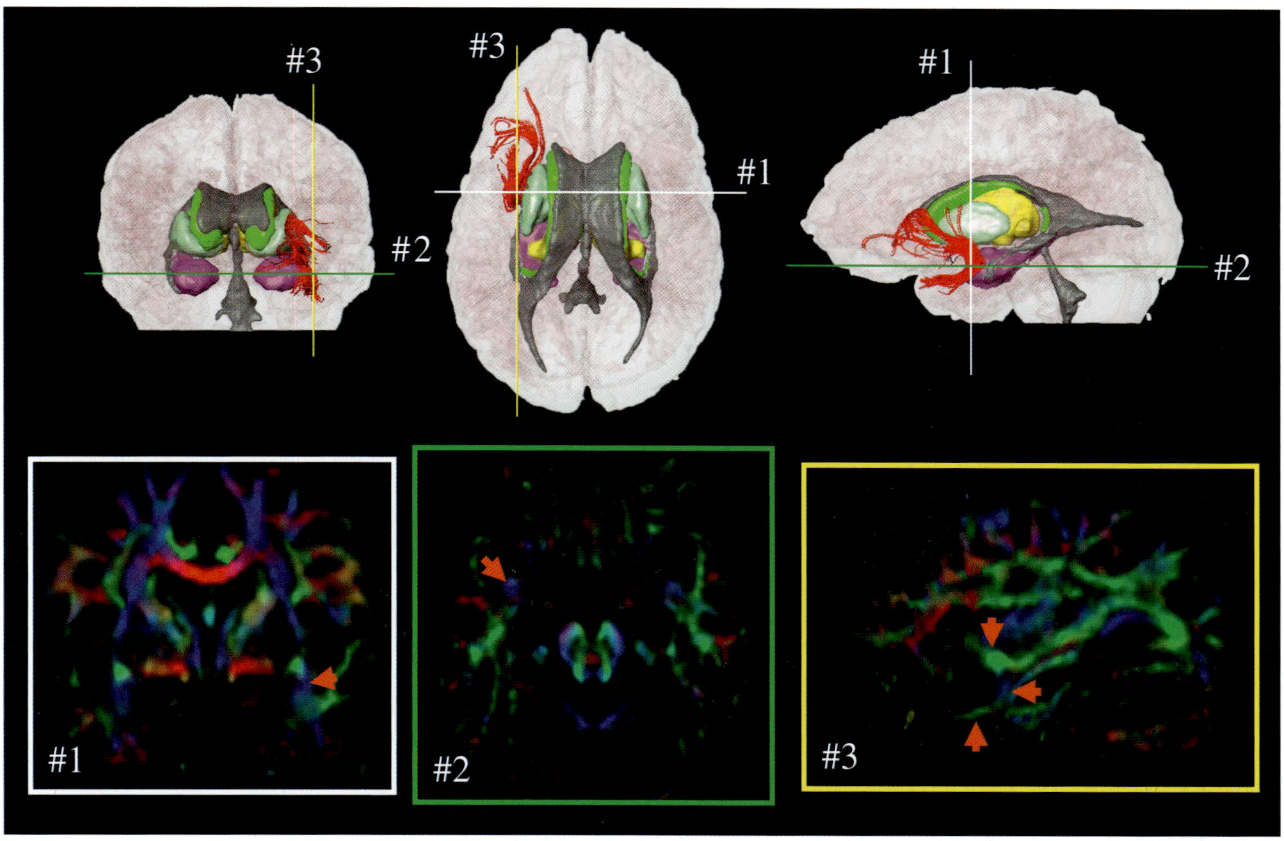

Fig. 18. The trajectory of the uncinate fasciculus and its identification in color maps at various slice levels and orientations. Locations of the 2D slices are indicated in the 3D panels. The thalamus (yellow), ventricular system (gray), caudate (green), and putamen (light green) are also shown in the 3D reconstruction.

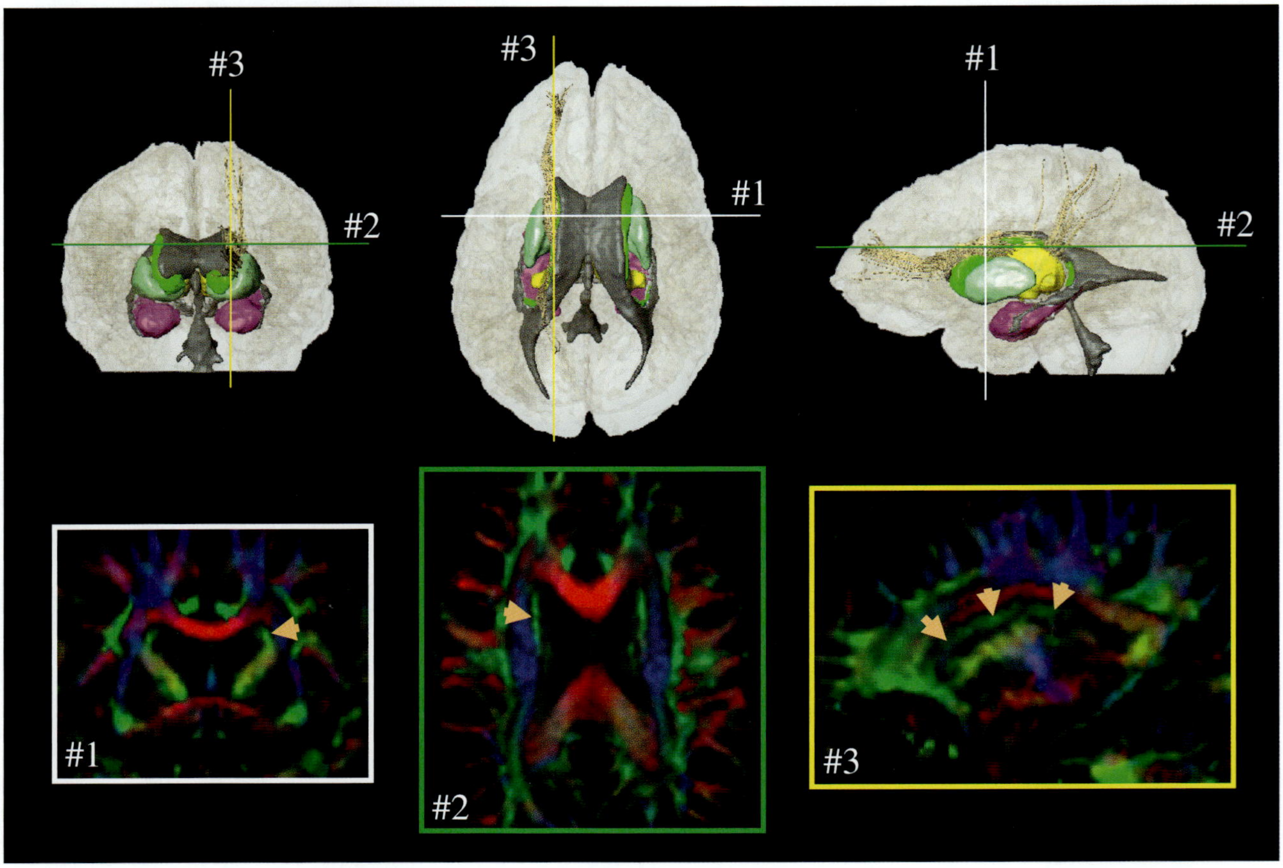

Fig. 19. The trajectory of the superior fronto-occipital fasciculus and its identification in color maps at various slice levels and orientations. Locations of the 2D slices are indicated in the 3D panels. The thalamus (yellow), ventricular system (gray), caudate (green), and putamen (light green) are also shown in the 3D reconstruction.

area can be identified in the color map at the horizontal caudal portion of the external capsule on the coronal plane, displaying high anisotropy (Fig. 17, slice #1). This region is also shared with the uncinate fasciculus, which connects the anterior temporal lobe (including the hippocampal formation) to the orbital cortex (Fig. 18).

Superior fronto-occipital fasciculus

The superior fronto-occipital fasciculus has been previously described, however its location is subject to debate among neuroanatomy textbooks. Association fibers that most closely resemble the superior fronto-occipital fasciculus can be found at the superolateral area of the caudate (Fig. 19). Anteriorly, this fiber bundle occupies the superior edge of the anterior limb of the internal capsule and projects into the frontal lobe. Its distinction from the anterior thalamic radiation is unclear. Posteriorly, it travels along the caudate and seems to merge into the corona radiata. Tracking results lead mostly to the parietal lobe, contrary to its name.

Cinglulum, fornix and stria terminalis

The trajectories and locations of these limbic fibers are shown in Fig. 20. The cingulum (cg) runs beneath the cingulate gyrus along its entire length. It collects axons from the cingulate gyrus that travels immediately dorsal to the corpus callosum and along the ventral face of the hippocampus, forming a large C-shape trajectory (Fig. 11). It carries afferent connections from the cingulate gyrus to the entorhinal cortex. Because of its narrow tubular shape, it is often difficult to reveal its entire length by a single reconstruction result. In this atlas, the cingulum was reconstructed for the dorsal and ventral parts separately and combined by an "OR" operation.

The fornix (fx) includes both afferent and efferent pathways between the hippocampus and the septal area and the hypothalamus. Similar to the cingulum, the fornix has a C-shaped trajectory (smaller C-shape than the cingulum), which consists of the column, the body, and the crus. Its anterior part is formed by the column of the fornix, which branches into two pathways near the anterior commissure (Fig. 20, slice #6). The precommissural part of the fornix projects to/from the septal area and postcommissural part to/from the hypothalamus and mammillary body. The middle part of the C-shape is the body of the fornix (Fig. 20, slice #5). At the body of the fornix, the right and left fornix are attached to each other, being distinguishable by high-resolution MR anatomical scans but not by DTI with the current resolution (2.2 mm). At the crus of the fornix, the right and left fornix are separated and project into the dorsal regions of the hippocampi. In this atlas, two regions of the fornix (column + body *vs* crus parts) were separately reconstructed and combined by an "OR" operation.

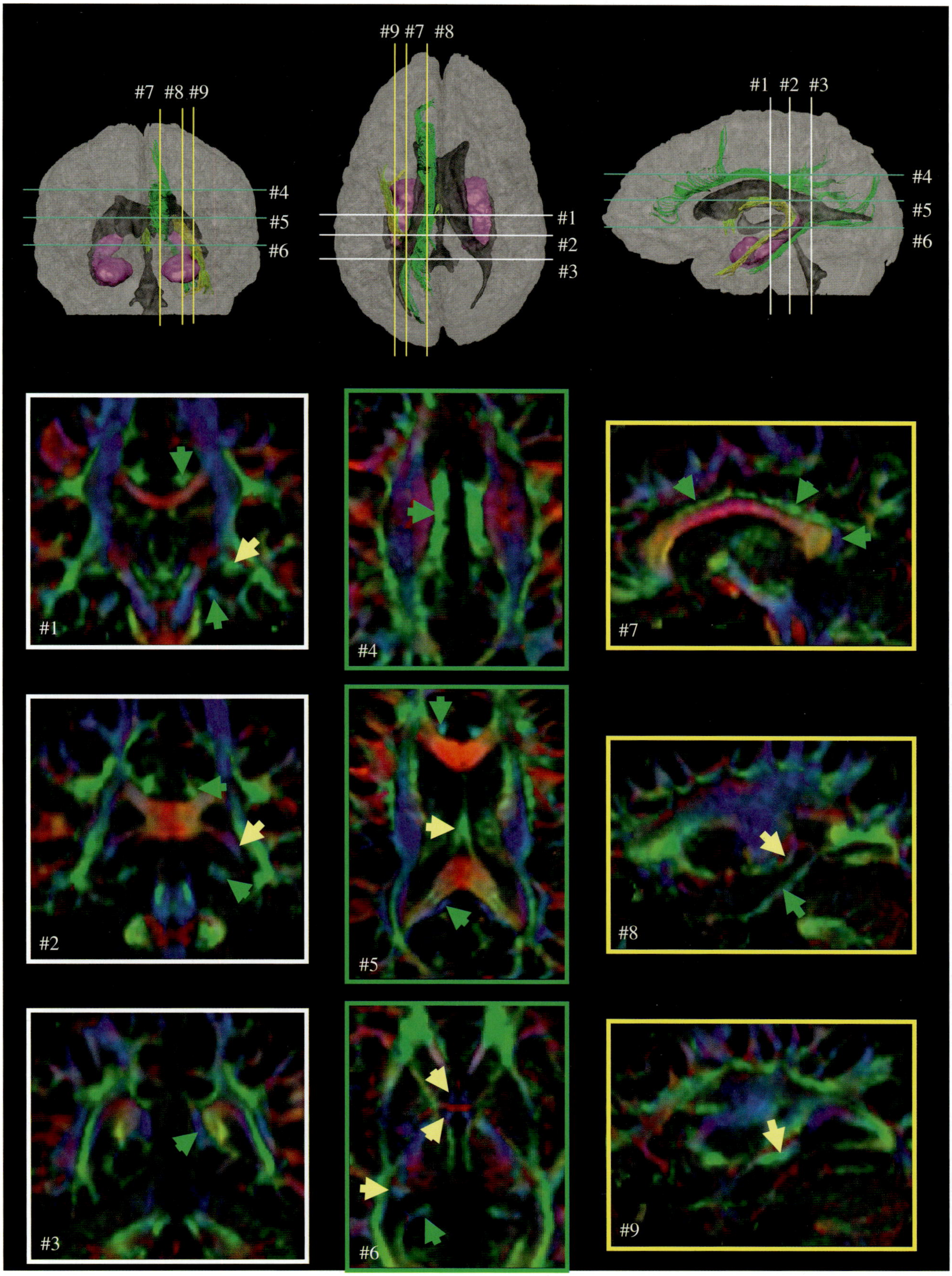

Fig. 20. Trajectories of the cingulum (green) and fornix/stria terminalis (light yellow) and their identification in color maps at various slice levels and orientations. Locations of the 2D slices are indicated in the 3D panels. Ventricular system (gray) and hippocampus (purple) are also shown in the 3D reconstruction.

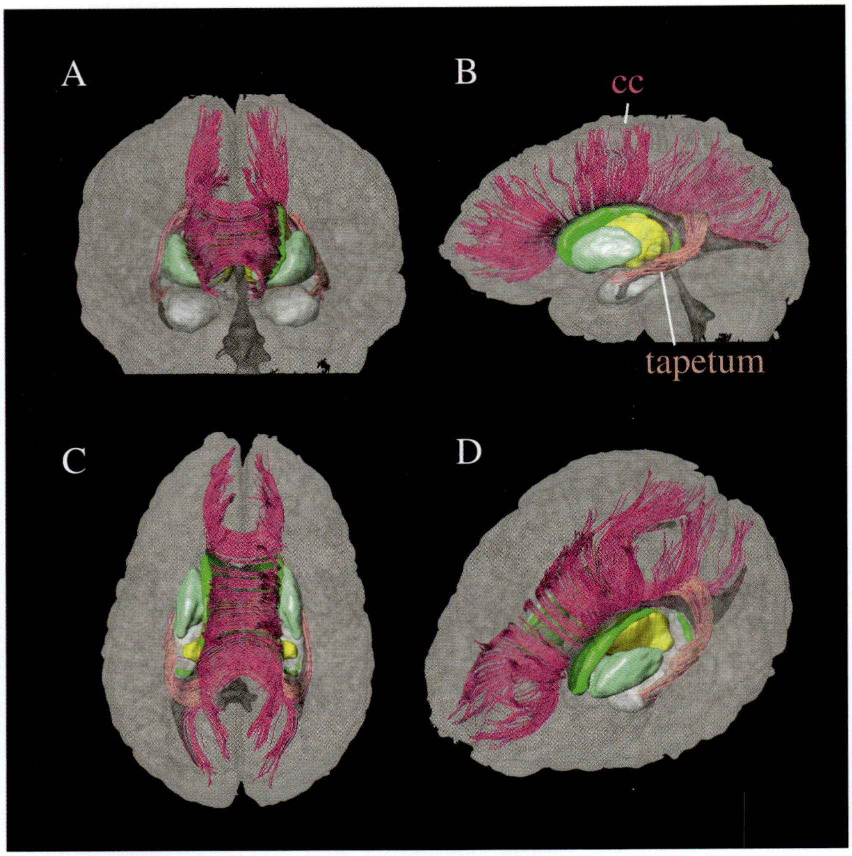

Fig. 21. 3D reconstruction results of commissural fibers. Tracts are viewed from the anterior (A), left (B), superior (C), and oblique (left-antero-superior) (D) orientations. Color coding is magenta representing the corpus callosum (cc), and peach representing the tapetum.

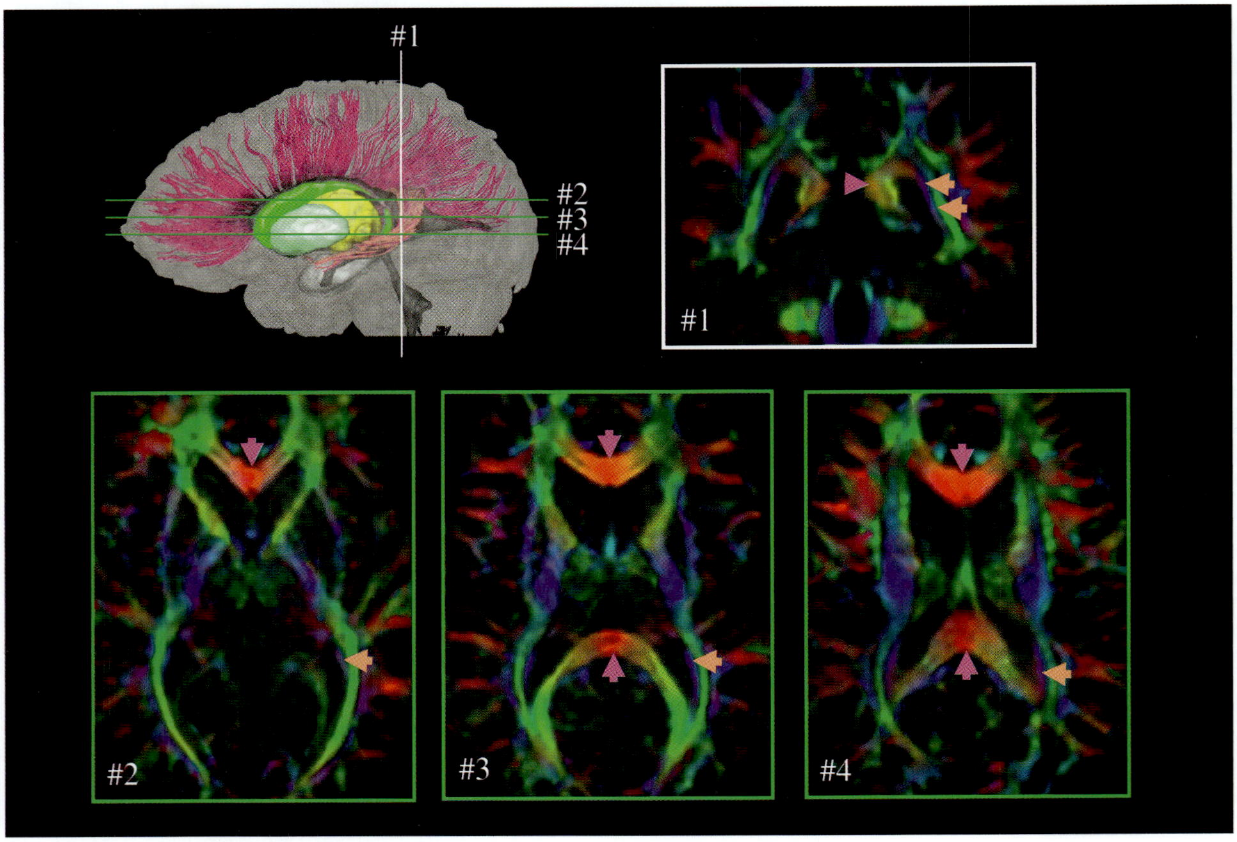

Fig. 22. Trajectories of the corpus callosum (magenta) and tapetum (peach) and their identification in color maps at various slice levels and orientations. Locations of the 2D slices are indicated in the 3D panel. The thalamus (yellow), ventricular system (gray), caudate (green), and putamen (light green) are also shown in the 3D reconstruction.

The stria terminalis is another limbic tract with a C-shaped trajectory (the innermost C-shaped trajectory of the three). This tract provides afferent and efferent pathways between the amygdala and the septal area and the hypothalamus. In the hippocampal area, distinction from the fornix is difficult with the current resolution.

4. Commissural fibers

The corpus callosum, which interconnects the two cerebral hemispheres, contains more than 300 million axons and is by far the largest fiber bundle in the human brain. Most of these fibers interconnect homologous cortical areas in roughly mirror-image sites, but a substantial number have heterotopic connections, ending in asymmetrical areas. Fig. 21 and Fig. 22 show 3D and 2D registration of the corpus callosum. Because of the distinctive commissural trajectories to the temporal lobes through the tapetum, these were assigned a different color for prompt identification. From degeneration studies in monkeys, it has been demonstrated that the entire cortex is connected by the commissural fibers. However, DTI-based tractography often fails to reveal commissural connections to the lateral areas of the hemispheres. This is probably because of massive projection fibers and long association fibers located lateral to the corpus callosum, which prevent the reconstruction tracking from penetrating these areas. This is one example of a limitation of the DTI-based tract reconstruction.

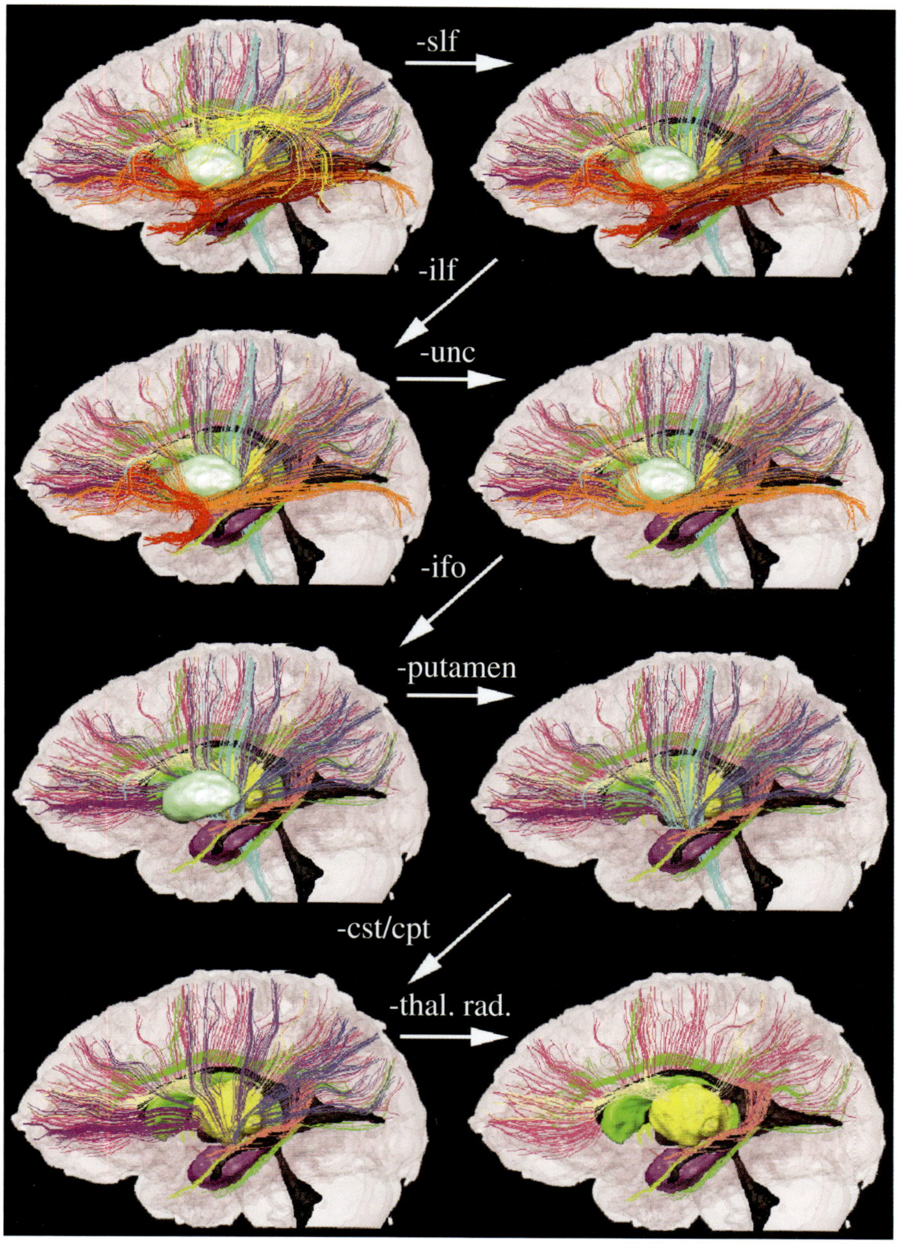

Fig. 23. Tract stripping for all the reconstructed fibers in this atlas. Tracts were removed one by one from lateral to medial regions. For visualization purpose, the density and thickness of tracts were reduced compared to other 3D figures in this atlas.

5. Overall view of reconstructed fibers

In Fig. 23, all reconstructed tracts are shown simultaneously (the image at upper left) and the tracts were removed one by one from the lateral to medial regions. This should help to understand spatial relationship among different tracts as well as between tracts and other brain structures such as gray matter nuclei and ventricles.

References

Brain anatomy

Anatomical assignments and annotations of white matter tracts in this atlas followed descriptions in these references:

[1] E. Ludwig, J. Klingler, Atlas Cerebri Humani: The Internal Structure of the Brain, S. Karger, Basel, 1956.
[2] E. Crosby, T. Humphrey, E. Lauer, Correlative Anatomy of the Nervous System, Macmillan, New York, 1962.
[3] M. Carpenter, Human Neuroanatomy, Williams & Wilkins, Baltimore, MD, 1976.
[4] R. Nieuwenhuys, J. Voogd, C. van Huijzen, The Human Central Nervous System, Springer, Berlin, 1983.
[5] J. Nolte, The Human Brain: An Introduction to its Functional Anatomy, Mosby–Year Book, St Louis, 1988.
[6] T.H. Williams, N. Gluhbegovic, J.Y. Jew, The Human Brain: Dissections of the Real Brain, Virtual Hospital, University of Iowa, 1997 < www.vh.org/Providers/Textbooks/BrainAnatomy/> and <www.brain-university.com/>.

DTI-based anatomical studies

DTI for brain anatomy

[7] M.E. Moseley, Y. Cohen, J. Kucharczyk, J. Mintorovitch, H.S. Asgari, M.F. Wendland, J. Tsuruda, D. Norman, Diffusion-weighted MR imaging of anisotropic water diffusion in cat central nervous system, Radiology 176 (1990) 439–445.
[8] R. Henkelman, G. Stanisz, J. Kim, M. Bronskill, Anisotropy of NMR properties of tissues, Magn. Reson. Med. 32 (1994) 592–601.
[9] P.J. Basser, J. Mattiello, D. Le Bihan, MR diffusion tensor spectroscopy and imaging, Biophys. J. 66 (1994) 259–267
Diffusion tensor theory used in this atlas is based on this publication.
[10] C. Pierpaoli, P.J. Basser, Toward a quantitative assessment of diffusion anisotropy, Magn. Reson. Med. 36 (1996) 893–906
Diffusion anisotropy contrasts used in this atlas is based on this publication.
[11] N. Makris, A.J. Worth, A.G. Sorensen, G.M. Papadimitriou, T.G. Reese, V.J. Wedeen, T.L. Davis, J.W. Stakes, V.S. Caviness, E. Kaplan, B.R. Rosen, D.N. Pandya, D.N. Kennedy, Morphometry of in vivo human white matter association pathways with diffusion weighted magnetic resonance imaging, Ann. Neurol. 42 (1997) 951–962
The color-coded orientation map (colormap) used in this atlas is based on this publication and one by Pajevic and Pierpaoli [13].
[12] C. Beaulieu, P.S. Allen, Determinants of anisotropic water diffusion in nerves, Magn. Reson. Med. 31 (1994) 394–400.
[13] S. Pajevic, C. Pierpaoli, Color schemes to represent the orientation of anisotropic tissues from diffusion tensor data: application to white matter fiber tract mapping in the human brain, Magn. Reson. Med. 42 (1999) 526–540.
[14] A. Ulug, P.C.M. van Zijl, Orientation-independent diffusion imaging without tensor diagonalization: anisotropy definitions based on physical attributes of the diffusion ellipsoid, J. Magn. Reson. Imaging 9 (1999) 804–813.
[15] C. Beaulieu, The basis of anisotropic water diffusion in the nervous system – a technical review, NMR Biomed. 15 (7–8) (2002) 435–455.

3D fiber reconstruction

[16] S. Wakana, H. Jiang, L. Nagae-Poetscher, P.C.M. van Zijl, S. Mori, A fiber tract based atlas of human white matter anatomy, Radiology 2004 230: 77–87.
[17] S. Mori, B.J. Crain, V.P. Chacko, P.C.M. van Zijl, Three dimensional tracking of axonal projections in the brain by magnetic resonance imaging, Ann. Neurol. 45 (1999) 265–269
The 3D reconstruction algorithm used in this atlas is based on this publication.
[18] R. Xue, P.C.M. van Zijl, B.J. Crain, M. Solaiyappan, S. Mori, In vivo three-dimensional reconstruction of rat brain axonal projections by diffusion tensor imaging, Magn. Reson. Med. 42 (1999) 1123–1127.
[19] T.E. Conturo, N.F. Lori, T.S. Cull, E. Akbudak, A.Z. Snyder, J.S. Shimony, R.C. McKinstry, H. Burton, M.E. Raichle, Tracking neuronal fiber pathways in the living human brain, Proc. Natl. Acad. Sci. USA 96 (1999) 10422–10427
The brute-force and multi-ROI approach for fiber reconstruction used in this atlas were introduced by this paper.

[20] B. Stieltjes, W.E. Kaufmann, P.C.M. van Zijl, K. Fredericksen, G.D. Pearlson, S. Mori, Diffusion tensor imaging and axonal tracking in the human brainstem, NeuroImage 14 (2001) 723–735
Detailed description about ROI drawing strategy, locations, and sizes are provided in this and the following three publications [21–23].

[21] S. Mori, W.E. Kaufmann, C. Davatzikos, B. Stieltjes, L. Amodei, K. Fredericksen, G.D. Pearlson, E.R. Melhem, M. Solaiyappan, G.V. Raymond, H.W. Moser, P.C.M. van Zijl, Imaging cortical association tracts in human brain using diffusion-tensor-based axonal tracking, Magn. Reson. Med. 47 (2002) 215–223.

[22] M. Catani, R.J. Howard, S. Pajevic, D.K. Jones, Virtual in vivo interactive dissection of white matter fasciculi in the human brain, NeuroImage 17 (2002) 77–94.

[23] S. Wakana, A. Caprihan, M.M. Panzenboeck, J.H. Fallon, M. Perry, R.L. Gollub, K. Hua, J. Zhang, H. Jiang, P. Dubey, A. Blitz, P. van Zijl, S. Mori, Reproducibility of quantitative tractography methods applied to cerebral white matter, Neuroimage 36 (3) (2007) 630–644.

[24] D.K. Jones, A. Simmons, S.C. Williams, M.A. Horsfield, Non-invasive assessment of axonal fiber connectivity in the human brain via diffusion tensor MRI, Magn. Reson. Med. 42 (1999) 37–41.

[25] P.J. Basser, S. Pajevic, C. Pierpaoli, J. Duda, A. Aldroubi, In vitro fiber tractography using DT-MRI data, Magn. Reson. Med. 44 (2000) 625–632.

[26] C. Poupon, C.A. Clark, V. Frouin, J. Regis, L. Bloch, D. Le Bihan, J.F. Mangin, Regularization of diffusion-based direction maps for the tracking of brain white matter fascicules, NeuroImage 12 (2000) 184–195.

[27] G.J. Parker, K.E. Stephan, G.J. Barker, J.B. Rowe, D.G. MacManus, C.A. Wheeler-Kingshott, R.E. Ciccarelli O' Passingham, R.L. Spinks, R.N. Lemon, R. Turner, Initial demonstration of in vivo tracing of axonal projections in the macaque brain and comparison with the human brain using diffusion tensor imaging and fast marching tractography, Neuroimage 15 (4) (2002) 797–809.

[28] S. Mori, P.C. Van Zijl, Fiber tracking: principles and strategies – a technical review, NMR Biomed. 15 (7–8) (2002) 468–480 (Review).

[29] N.F. Lori, J.S. Akbudak, T.S. Shimony, R.K. Snyder, T.E. Conturo, Diffusion tensor fiber tracking of brain connectivity: Reliability analysis and biological results, NMR Biomed. 15 (2002) 494–515.

CHAPTER 4

MRI/DTI Atlas of the Human Brain in the ICBM-152 Space

In this atlas, two types of 2D presentations were used to delineate the white matter anatomy: DTI-based, color-coded orientation maps (color maps); and maps of probabilistic distributions of the major white matter tracts (probabilistic tractography maps). In the latter, the 3D reconstructed probabilistic trajectories of the white matter tracts are superimposed on co-registered T_1-weighted images.

The color map is created directly from the diffusion tensor vector data and various tracts can be identified based on the pixel-by-pixel orientation information; red is assigned to the right–left, green to the anterior–posterior, blue to the superior–inferior orientation, and mixtures of the three principal colors for the oblique orientations. Because the color map contains a large amount of white matter anatomical information, it has great potential for clinical use. However, interpretation of color maps is far from trivial. Two different tract systems may have the same color if their in-plane fiber orientations are the same, while the same tract system can change color because the orientation changes.

To present the probabilistic tractography, each white matter tract is assigned a specific color regardless of its orientation, simplifying the interpretation. This information is superimposed on the right side of the T_1-weighted image.

For anatomical guidance, a brain parcellation map is overlaid on the left side of the MR images, which defines the boundaries of each brain structure. The parcellation was based on DTI and T_1-weighted contrast, as described in section 2.4.

This atlas is created in the ICBM-152 space. Two standard reference systems are widely used in the neuroimaging community to map areas of the human brain. One is a reference system proposed by Talairach et al.[1] (Talairach system). Using this system, Talairach and Tournoux created a brain atlas,[2] based on postmortem brain sections from a 60-year-old woman. This atlas space is often called a Talairach space or Talairach coordinates. The other system is the Montreal Neurological Institute (MNI) coordinates. This is a reference system used by a multi-institutional study group, The International Consortium for Brain Mapping (ICBM), which is dedicated to creating population-averaged atlases based on *in vivo* MRI of the human brain (ICBM atlases, www.loni.ucla.edu/ICBM/Downloads/Downloads_Atlases.shtml). ICBM atlases have been adopted as standard brain templates in many software packages for brain image analyses. One of the ICBM atlases, ICBM-152, is based on high-resolution T_1-weighted images of 152 normal subjects, which were linearly normalized and averaged.

The purpose of our DTI-based atlas is to provide coordinate information for the major white matter structures in these widely used coordinate systems. The size of the atlas is linearly normalized to that of the ICBM-152 atlas. The atlas also contains the brain coordinate system adopted by the Talairach atlas to ensure maximum compatibility with the existing atlases.

1. Reference systems

1.1. The Talairach system and cytoarchitectonic map

The Talairach system is based on a proportional grid system defined by three reference lines: the anterior–posterior commissure line (AC–PC line); the vertical traversing line of the anterior commissure (VAC line); and the midline. In addition, six points are identified: the highest point of the parietal cortex; the most posterior point of the occipital cortex; the lowest point of the temporal cortex; the most anterior point of the frontal cortex; and the most lateral points (left and right) of the parietotemporal cortex. The axial plane is set parallel to the AC–PC line and perpendicular to the VAC line. The coronal plane is set parallel to the VAC line and perpendicular to the AC–PC line. The sagittal plane is set parallel to the midline. Using these references, the total volume of the brain is divided as follows. By horizontal lines: above the AC–PC line in eighths and below the AC–PC line in quarters. By vertical lines: anterior to the VAC line in quarters, and posterior to the VPC line (line parallel to the VAC line that passes the PC) in quarters. The space between the VAC and VPC lines is subdivided into three zones. By right–left axis: lateral to the midline in quarters. These result in 1056 compartments ($8 \times 11 \times 12$) in the space.

There are several issues that should be stressed when the Talairach coordinates are used. The compartmentalization of the brain into the 1056 areas is performed to define corresponding brain areas across different human brains. When the same Talairach coordinates are compared between the original Talairach brain and our atlas, one may notice that the brain anatomy is not matched exactly. For example, Talairach coordinate [d, C, 8] contains part of the inferior frontal gyrus in the original Talairach atlas, while this structure is located in [d-c, C, 8] in our atlas, slightly medially. We attribute these differences to individual variability, but we also suspect severe fixation distortion in the histology sample used for the Talairach atlas. Another potential discrepancy is the cytoarchitectonic map assignment of the cortex. In the Talairach atlas, this was performed subjectively by transferring the original Brodmann's map to the atlas. In our atlas, we adopted the modern cytoarchitectorial map being created by Drs Eickhoff, Zilles, and Amunt.[3–6] Also, there are areas that have not been assigned yet, as described in section 2.6 of this chapter in detail.

1.2. MNI coordinates

The MNI coordinate system is based on the physical distances from the anterior commissure, which is defined as the origin of the coordinates [0, 0, 0]. Similar to the Talairach atlas, the brain is oriented along the AC–PC line. However, the AC of the ICBM-152 is not exactly at [0, 0, 0] of the MNI coordinates, but, rather, approximately 4 mm below and 2 mm anterior. Because our atlas is based on a single-subject brain MRI linearly normalized to the ICBM-152, the AC of this atlas is also shifted to [0, 2, −2.5]. ICBM-152 is a population-averaged atlas and each structure is not sharply defined due to structural blurring. The degree of the blurring reflects the amount of individual variability within the population. As opposed to the ICBM-152, our atlas is based on a single subject, in which each structure is clearly defined, but their locations and shapes contain the subject-specific anatomy. Linear normalization of the size of the brain would reduce the degree of such anatomical variability, but superimposition of our single-subject atlas on the ICBM-152 would still reveal a certain degree of mismatch, and, similar to other atlases, it is not perfect.

2. Creation of the atlas

2.1. MRI acquisition

Data were obtained from a 32-year-old healthy female participant without known brain or neurological disorders. A 1.5 T MR unit (Gyroscan NT, Philips Medical Systems) was used. DTI data were acquired as described in Chapter 2, except this time, a total of 60 slices were acquired to cover the entire brain. Co-registered double-echo (40 and 100 ms), Fast Spin-Echo (FSE) images, Fluid-attenuated Inversion Recovery (FLAIR) images, and a T_1-weighted image (MPRAGE) were also recorded for anatomical guidance.

2.2. Data processing

The raw diffusion-weighted images were corrected for motion and eddy current distortion using a 12-parameter affine transformation of Automated Image Registration (AIR).[7] B0-susceptibility distortion was corrected by non-linear warping using the least diffusion-weighted image of the FSE images (echo time: 100 ms), and then re-sliced to 1 mm isotropic resolution (181 × 217 × 181 matrix). The six elements of the diffusion tensor were calculated for each pixel, as detailed in Chapter 2. The eigenvector ($v1$) associated with the largest eigenvalue was used as an indicator of fiber orientation. In the color map, red (R), green (G), and blue (B) colors were assigned to right–left, anterior–posterior, and superior–inferior orientations, respectively. For the color presentation, a 24-bit color scheme was used, in which each of the RGB colors had an eight-bit (0–255) intensity level. The vector $v1$ ($= [v1_x, v1_y, v1_z]$) is a unit vector that always fulfills a condition, $v1_x^2 + v1_y^2 + v1_z^2 = 1$. Intensity values of $v1_x \times 255$, $v1_y \times 255$, $v1_z \times 255$ were assigned to the R channel, G channel, and B channel, respectively. In order to suppress orientation information in isotropic brain regions (there should not be a preferential orientation in isotropic areas and the calculated orientations in such areas are dominated by noise), the 24-bit color value was multiplied by the FA value.

2.3. Normalization

The MRIs (DTI, FSE, FLAIR, and MPRAGE) were transformed to the ICBM-152 template to create a single-participant atlas based on ICBM coordinates. A 12-parameter affine transformation of AIR was used since the template (ICBM-152) was also created in the same manner. First, the MPRAGE was normalized to the ICBM-152. Then, the transformation matrix was applied to the co-registered FSE, distortion-corrected DTI and FLAIR using the method described by Parker and Alexander[8] and Xu et al.[9] The entire normalization process was performed by DiffeoMap, which is a component of MriStudio (www.mristudio.org).

2.4. Parcellation of the gray and white matter

The boundary of the cortex and WM of our atlas was defined by MPRAGE. Each gyrus and subcortical gray matter was assigned manually, based on MPRAGE. For this procedure, we inspected all three slice orientations using the RoiEditor module of MriStudio. For the parcellation of the deep WM (DWM), we adopted our previous WM parcellation map (www.loni.ucla.edu/Atlases, lbam.med.jhmi.edu, and mri.kennedykrieger.org), in which 28 deep WM regions were manually defined.[10] Parcellation of the superficially located WM (SWM) was accomplished according to our previous publication.[11] Namely, we defined the SWM as the area between the cortex and the DWM, and manually parcellated the SWM into regions corresponding to 42 cerebral cortical areas, bilaterally.

This atlas also contains parcellation of the basal ganglia, as well as sub-parcellation of the thalamus. The basal ganglia were defined by the combination of MPRAGE and DTI contrasts, while the thalamus parcellation was based solely on the tissue orientation information in DTI. As described in previous papers,[12, 13] there are several nuclei that can be identified by the characteristic axonal orientations within the nuclei. However, care must be taken because the boundaries defined by the tissue orientation may not be identical to those defined by classical cytoarchitecture. In this respect, the nuclei names assigned in this atlas should be treated as approximations.

2.5. Probabilistic tractography (population averaged white matter tracts)

Probabilistic maps of the ten tract trajectories were created from 20 normal subjects, previously incorporated into our atlas for automated tract-specific quantification. The reconstructed tracts were: the anterior thalamic radiation; the cingulum cingular part; the cingulum hippocampal part; the corticospinal tract; the forceps major; the forceps minor; the inferior fronto-occipital fasciculus; the inferior longitudinal fasciculus; the superior longitudinal fasciculus; and the uncinate fasciculus. DTI data were acquired on Siemens 1.5 T MR units at the University of California Los Angeles under the ICBM collaboration. Single-shot, echo-planar imaging sequences with SENSE and a parallel imaging factor of 2.0[14] were used. The imaging matrix was 96×96, with a field of view of $240\,\text{mm} \times 240\,\text{mm}$ (nominal resolution: 2.5 mm). Transverse sections of 2.5 mm thickness were acquired parallel to the AC–PC line. A total of 60 sections covered the entire hemisphere and brainstem without gaps. Diffusion weighting was encoded along 30 independent orientations,[15] and the b-value was $1000\,\text{s/mm}^2$. Five additional images with minimal diffusion weighting were also acquired. To enhance the signal-to-noise ratio, imaging was repeated twice. Image processing was performed in a manner identical to that described in section 4.2.2. Three-dimensional tract reconstruction was performed as described in Chapter 2. After the fiber tracts were constructed in the subject data space, the tract coordinates were normalized to the atlas, using the affine transformation and subsequent large deformation diffeomorphic metric mapping (LDDMM). The normalized fiber streamlines of each tract were converted to binary images, which were averaged across 20 subjects to generate the probabilistic maps. We used a probability threshold of 50% to visualize the tracts on the atlas.

2.6. Brodmann's map assignment

An electronic form of the probabilistic cytoarchitectonic map was provided courtesy of Dr Eickhoff, Dr Zilles, and Dr Amunt (personal communication, November 2009). This was created from postmortem adult human brains normalized to the ICBM Single-Subject MRI Anatomical Template (Colin 27 brain) (www.loni.ucla.edu/ICBM/Downloads/Downloads_ICBMtemplate.shtml). To adjust the slight differences of the brain shape between the Colin 27 T_1 brain and our T_1 atlas, we transformed the Colin 27 T_1 brain to our T_1 atlas using LDDMM. We overlaid the transformed cytoarchitectonic map to our atlas. The map includes five areas in the frontal lobe [4a, 4p, 6, 44, and 45], 18 areas in the parietal lobe [1, 2, 3a, 3b, PF, PFcm, PFm, PFop, PFt, PGa, PGp, 5Ci, 5L, 5M, 7A, 7M, 7P, and 7PC], four areas in the operculum [OP1, OP2, OP3, and OP4], four areas in the temporal lobe [TE 3 (22 posterior), Te 1.0 (41), Te 1.1 (41), and Te 1.2 (41)], and five areas in the occipital lobe [17 (v1), 18 (v2), hOC3v (V3v), jOC4v (V4), and hOC5 (V5)].

2.7. Electronic versions of the atlas and software packages for image analyses

The atlas contains a T_1- and T_2-weighted atlas, a FLAIR atlas, and a DTI atlas.[16] All atlases as well as the software packages used to create them (DtiStudio, DiffeoMap, and RoiEditor) are downloadable from the MRI studio website (www.mristudio.org/). In this book, we present the details of the T_1-weighted atlas and the DTI atlas.

3. Nomenclature

This atlas adopts the nomenclature used in *Atlas of the human brain*, Third Edition (Mai atlas)[17] for most of the brain structures. For white matter structures not depicted in the Mai atlas, we used nomenclature based on our past publications.[10, 11] Note that there are several areas in which we combined different gyri to form a single parcel. Those areas are: (1) the straight gyrus and the inferior rostral gyrus; (2) the intermediate orbital gyrus, the lateral orbital gyrus, and the posterior orbital gyrus; and (3) the superior temporal gyrus, the planum polare, the transverse temporal gyrus, the planum temporale, and the semilunar gyrus; (4) the fushiform gyrus and O4; and (5) the supramarginal gyrus and the parietal operculum.

References

[1] J. Talairach, G. Szikla, P. Tournoux, A. Prossalentis, M. Bordas-Ferrer, L. Covello, Atlas D'anatomie Stereotaxique Du Telencephale, Masson, Paris, 1967.

[2] J. Talairach, P. Tournoux, Co-Planar Stereotaxic Atlas of the Human Brain: 3-Dimensional Proportional System: An Approach to Cerebral Imaging, Thieme, Stuttgart, 1988.

[3] K. Amunts, P.H. Weiss, H. Mohlberg, P. Pieperhoff, S. Eickhoff, J.M. Gurd, *et al.*, Analysis of neural mechanisms underlying verbal fluency in cytoarchitectonically defined stereotaxic space – the roles of Brodmann areas 44 and 45, Neuroimage 22 (1) (2004) 42–56.

[4] S.B. Eickhoff, S. Heim, K. Zilles, K. Amunts, Testing anatomically specified hypotheses in functional imaging using cytoarchitectonic maps, Neuroimage 32 (2) (2006) 570–582.

[5] S.B. Eickhoff, T. Paus, S. Caspers, M.H. Grosbras, A.C. Evans, K. Zilles, *et al.*, Assignment of functional activations to probabilistic cytoarchitectonic areas revisited, Neuroimage 36 (3) (2007) 511–521.

[6] S.B. Eickhoff, K.E. Stephan, H. Mohlberg, C. Grefkes, G.R. Fink, K. Amunts, *et al.*, A new SPM toolbox for combining probabilistic cytoarchitectonic maps and functional imaging data, Neuroimage 25 (4) (2005) 1325–1335.

[7] R.P. Woods, S.T. Grafton, C.J. Holmes, S.R. Cherry, J.C. Mazziotta, Automated image registration: I. General methods and intrasubject, intramodality validation, J. Comput. Assist. Tomogr. 22 (1) (1998) 139–152.

[8] G.J. Parker, D.C. Alexander, Probabilistic Monte Carlo based mapping of cerebral connections utilising whole-brain crossing fibre information, Inf. Process Med. Imaging 18 (2003) 684–695.

[9] D. Xu, S. Mori, D. Shen, C. Davatzikos, Statistically-based reorientation of diffusion tensor field. IEEE International Symposium on Biomedical Imaging, Washington, DC, IEEE; 2002.

[10] S. Mori, K. Oishi, H. Jiang, L. Jiang, X. Li, K. Akhter, *et al.*, Stereotaxic white matter atlas based on diffusion tensor imaging in an ICBM template, Neuroimage 40 (2) (2008) 570–582.

[11] K. Oishi, K. Zilles, K. Amunts, A.V. Faria, H. Jiang, X. Li, *et al.*, Human brain white matter atlas: identification and assignment of common anatomical structures in superficial white matter, Neuroimage 43 (3) (2008) 447–457.

[12] S. Wakana, L.M. Nagae-Poetscher, H. Jiang, P. van Zijl, X. Golay, S. Mori, Macroscopic orientation component analysis of brain white matter and thalamus based on diffusion tensor imaging, Magn. Reson. Med. 53 (3) (2005) 649–657.

[13] M.R. Wiegell, D.S. Tuch, H.B. Larsson, V.J. Wedeen, Automatic segmentation of thalamic nuclei from diffusion tensor magnetic resonance imaging, Neuroimage 19 (2 Pt 1) (2003) 391–401.

[14] K.P. Pruessmann, M. Weiger, M.B. Scheidegger, P. Boesiger, SENSE: sensitivity encoding for fast MRI, Magn. Reson. Med. 42 (5) (1999) 952–962.

[15] D.K. Jones, M.A. Horsfield, A. Simmons, Optimal strategies for measuring diffusion in anisotropic systems by magnetic resonance imaging, Magn. Reson. Med. 42 (1999) 515–525.

[16] K. Oishi, A. Faria, H. Jiang, X. Li, K. Akhter, J. Zhang, *et al.*, Atlas-based whole brain white matter analysis using large deformation diffeomorphic metric mapping: application to normal elderly and Alzheimer's disease participantstlas, Neuroimage 46 (2) (2009) 486–499.

[17] J.K. Mai, G. Paxinos, T. Voss, Atlas of the Human Brain, second ed., Academic Press, San Diego, 2008.

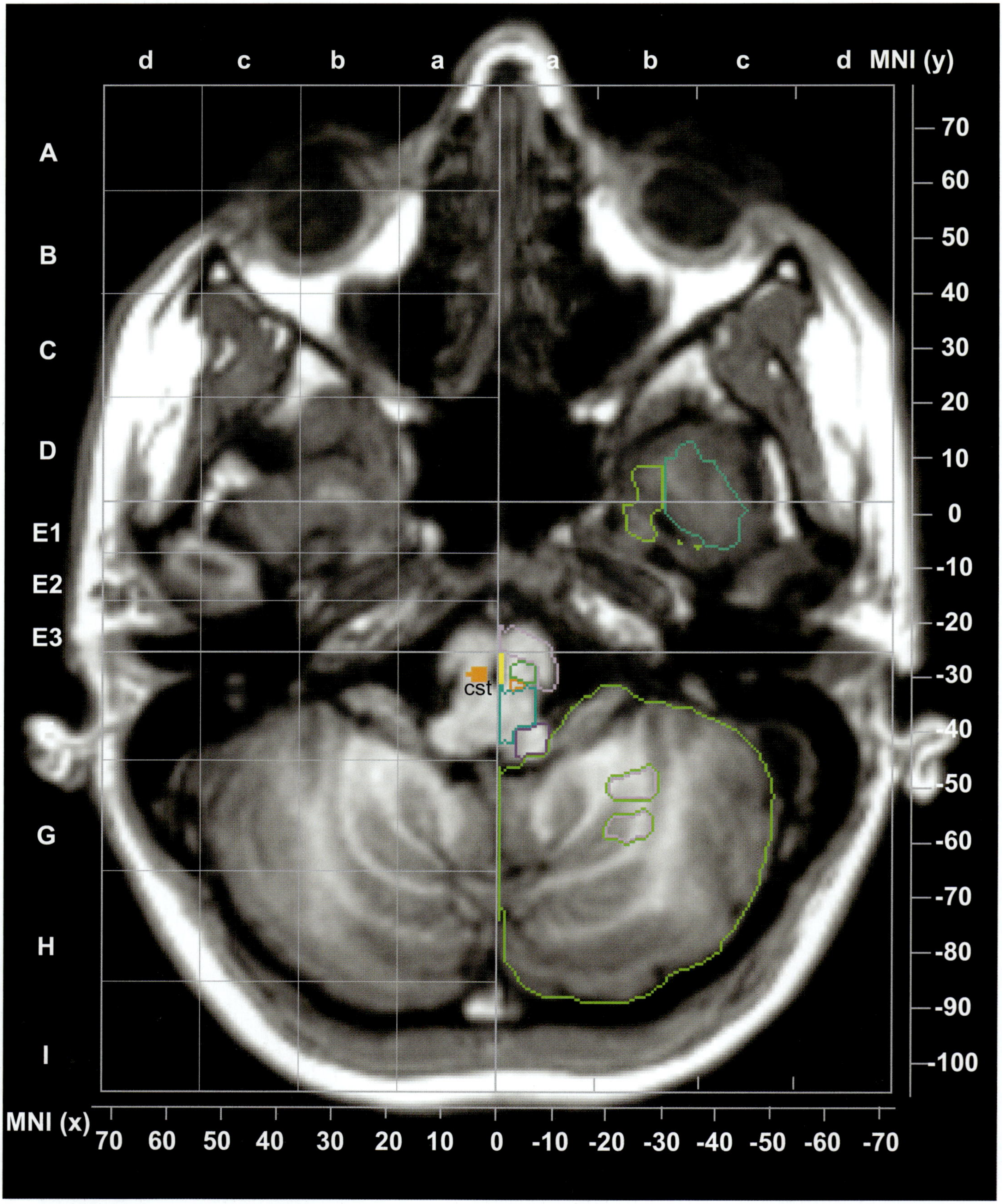

cst - corticospinal tract

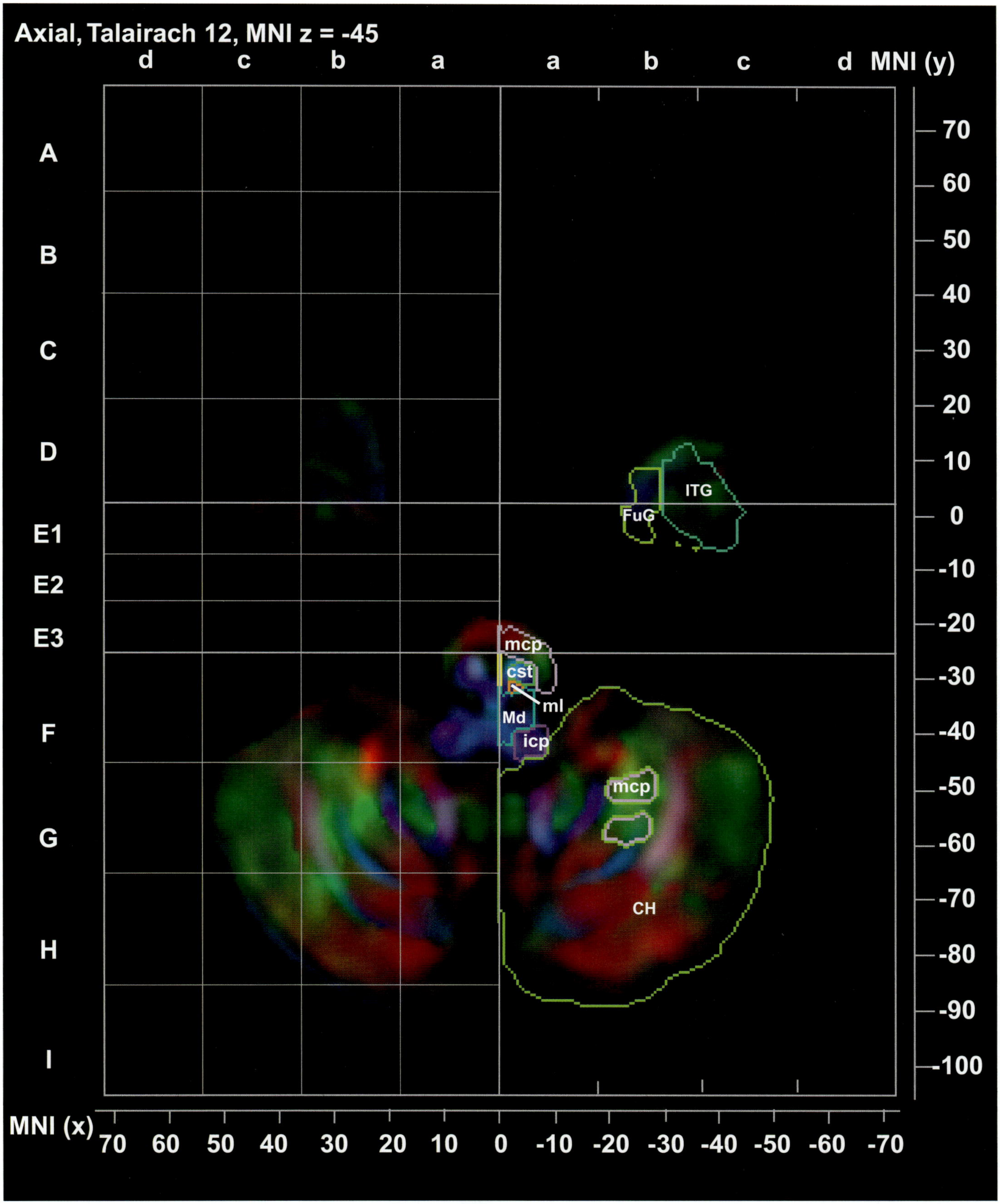

CH - cerebellar hemisphere
ml - medial lemniscus
Md - medulla oblongata
mcp - middle cerebellar peduncle
ITG - inferior temporal gyrus
icp - inferior cerebellar peduncle
FuG - fusiform gyrus
cst - corticospinal tract

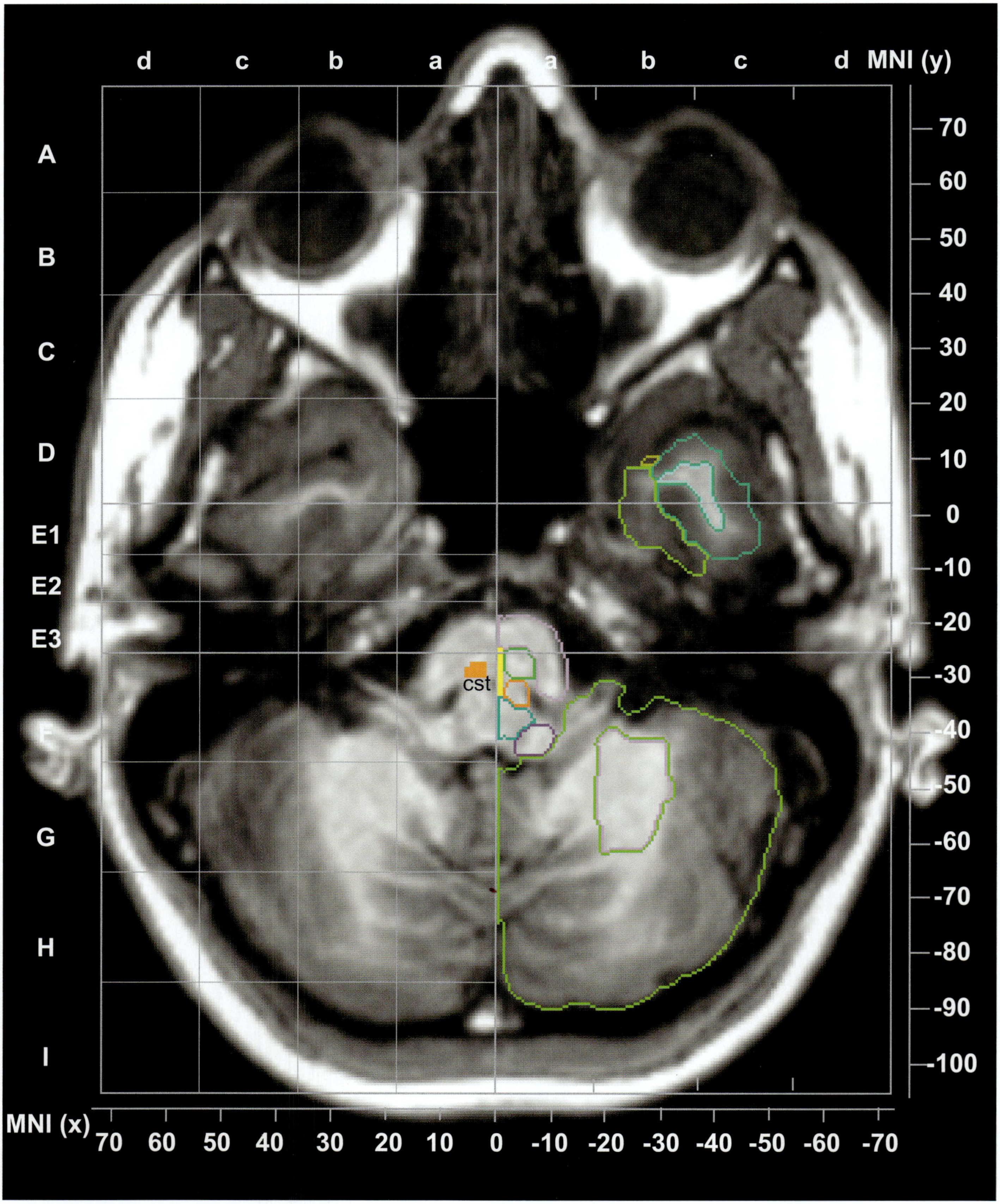

cst - corticospinal tract

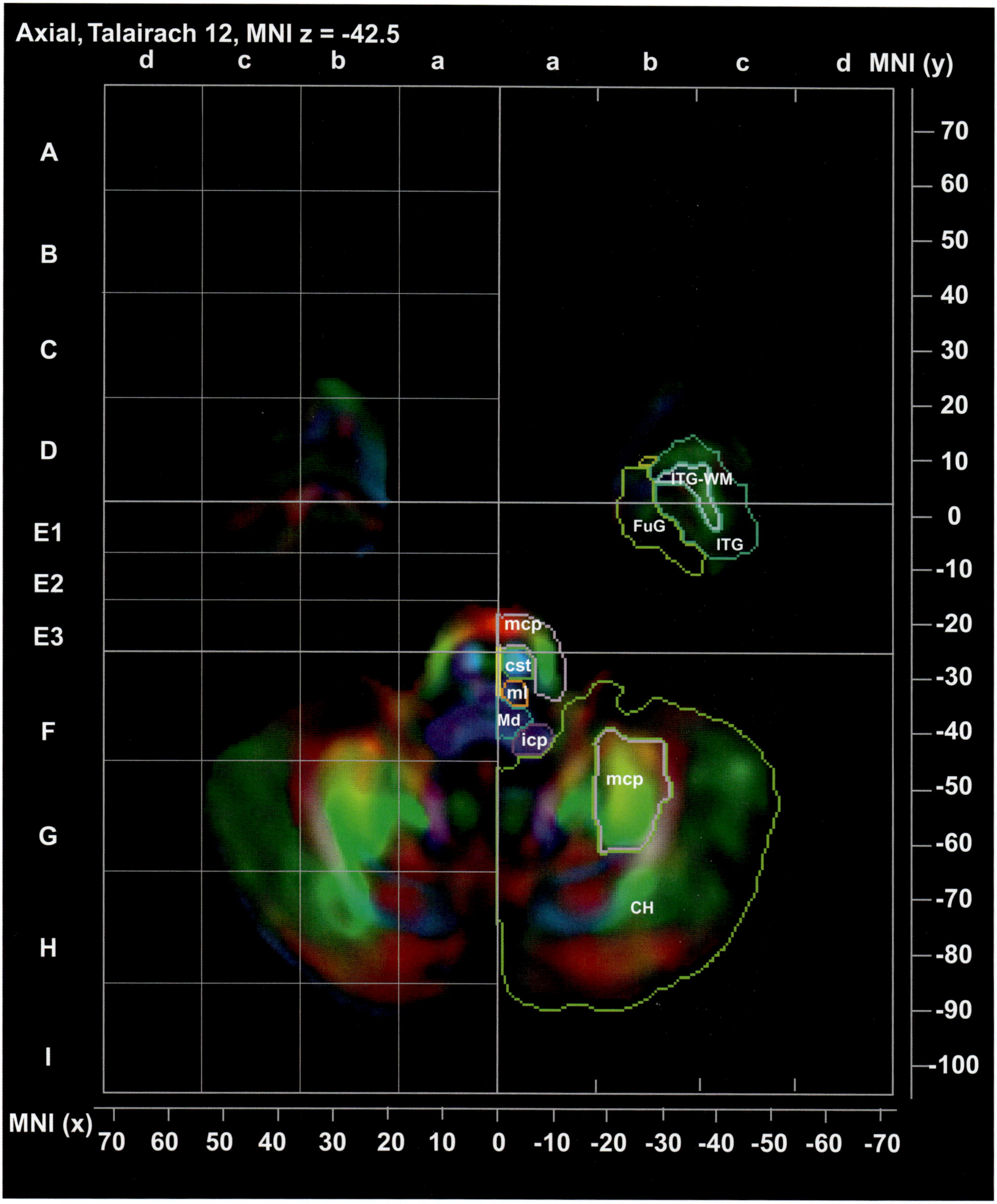

CH - cerebellar hemisphere
cst - corticospinal tract
FuG - fusiform gyrus
icp - inferior cerebellar peduncle
ITG - inferior temporal gyrus
ITG-WM - inferior temporal gyrus white matter
mcp - middle cerebellar peduncle
Md - medulla oblongata
ml - medial lemniscus

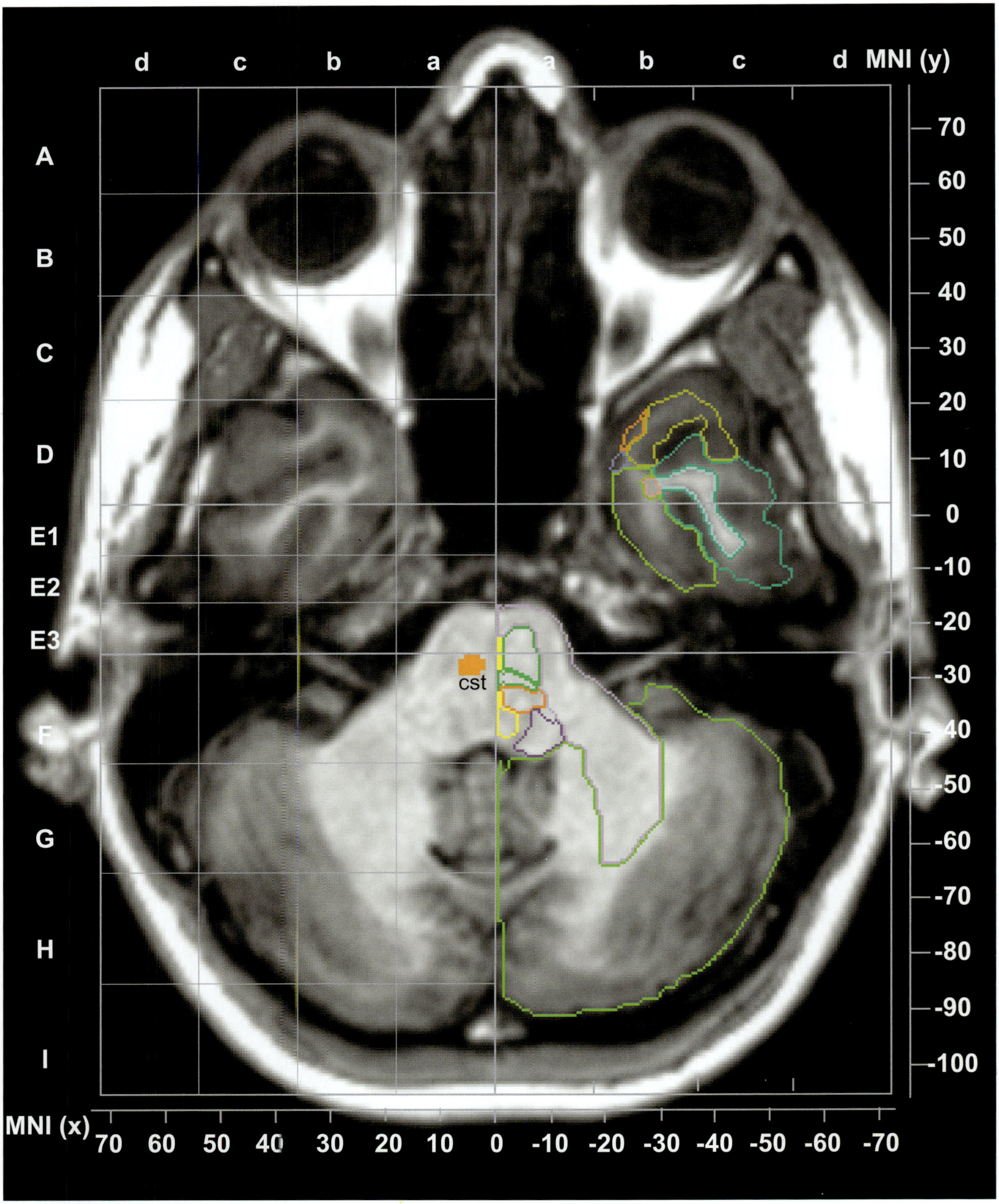

cst - corticospinal tract

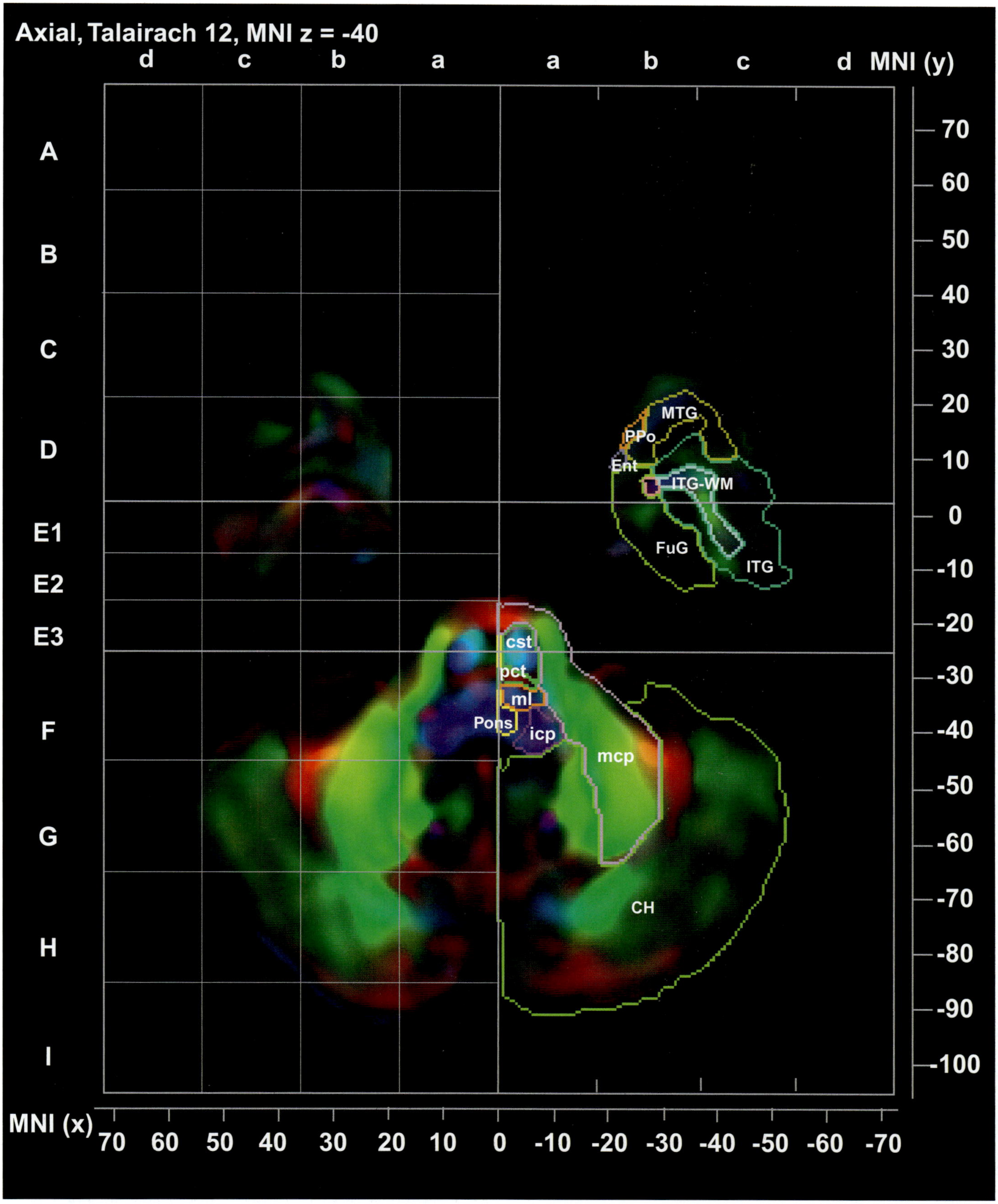

CH	- cerebellar hemisphere	ITG	- inferior temporal gyrus	MTG	- middle temporal gyrus
cst	- corticospinal tract	ITG-WM	- inferior temporal gyrus white matter	pct	- pontine crossing tract (a part of mcp)
Ent	- entorhinal cortex	mcp	- middle cerebellar peduncle	Pons	- pons
FuG	- fusiform gyrus	ml	- medial lemniscus	PPo	- planum polare
icp	- inferior cerebellar peduncle				

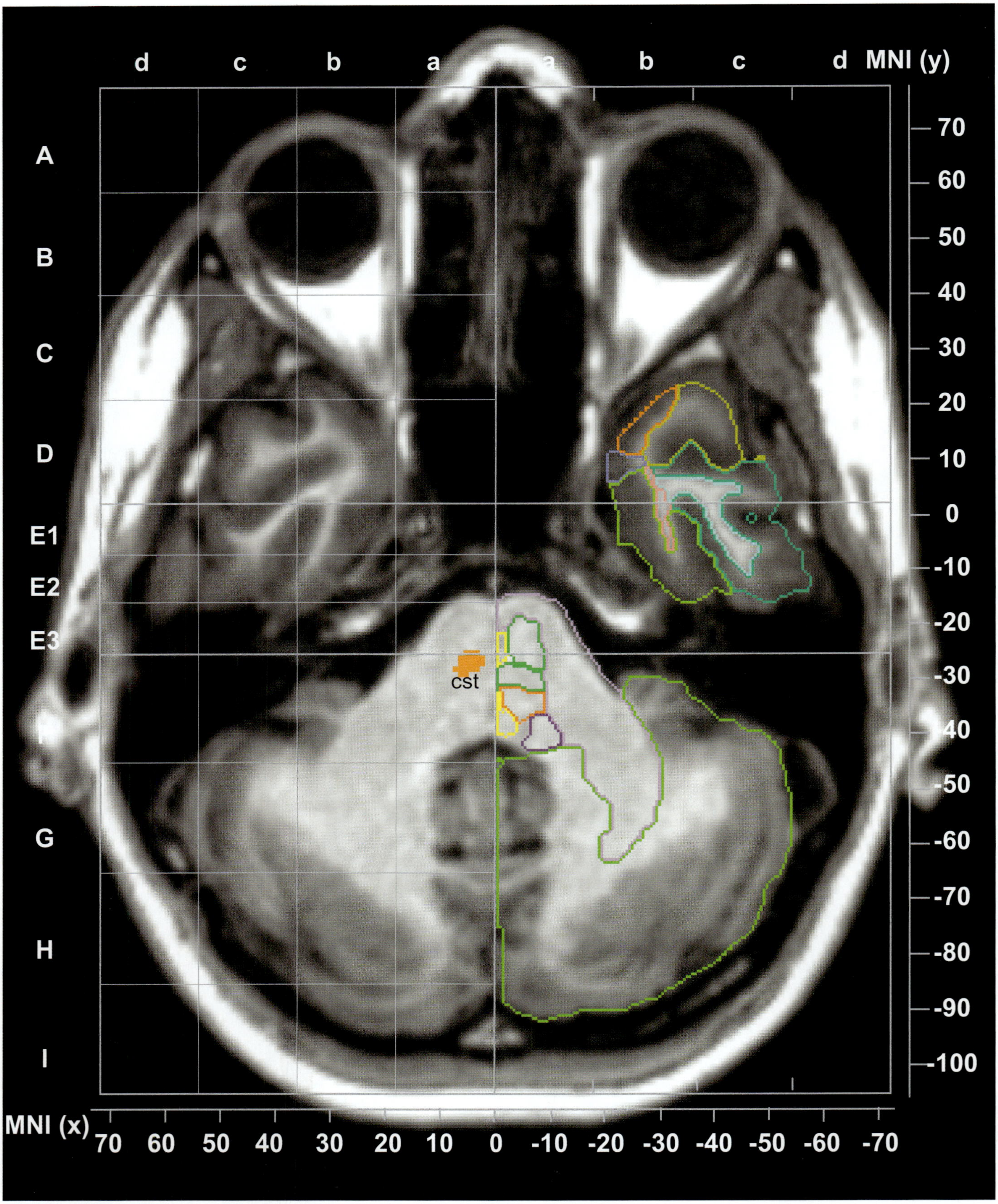

cst - corticospinal tract

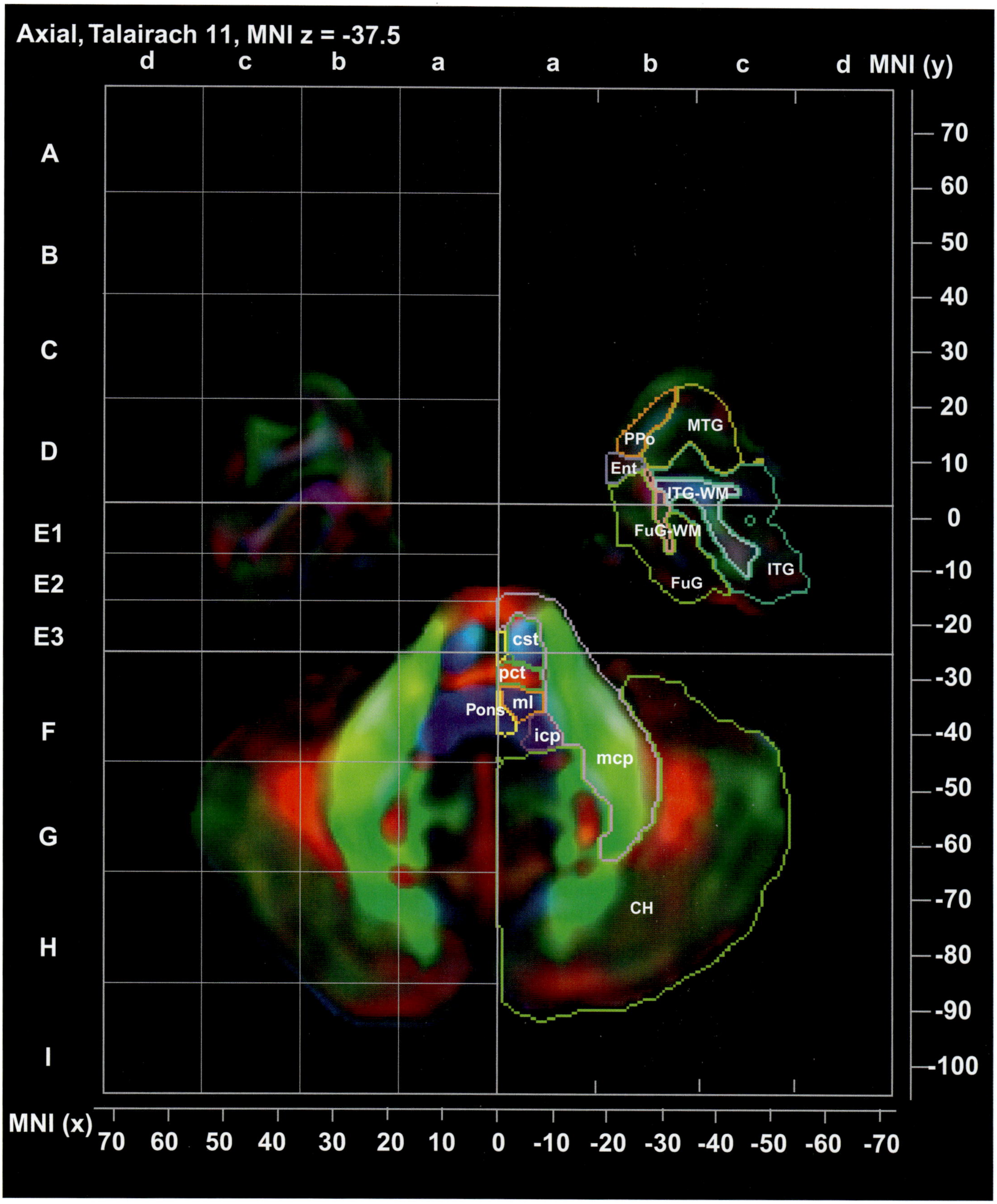

CH	-	cerebellar hemisphere
cst	-	corticospinal tract
Ent	-	entorhinal cortex
FuG	-	fusiform gyrus
FuG-WM	-	fusiform gyrus white matter
icp	-	inferior cerebellar peduncle
ITG	-	inferior temporal gyrus
ITG-WM	-	inferior temporal gyrus white matter
mcp	-	middle cerebellar peduncle
ml	-	medial lemniscus
MTG	-	middle temporal gyrus
pct	-	pontine crossing tract (a part of mcp)
Pons	-	pons
PPo	-	planum polare

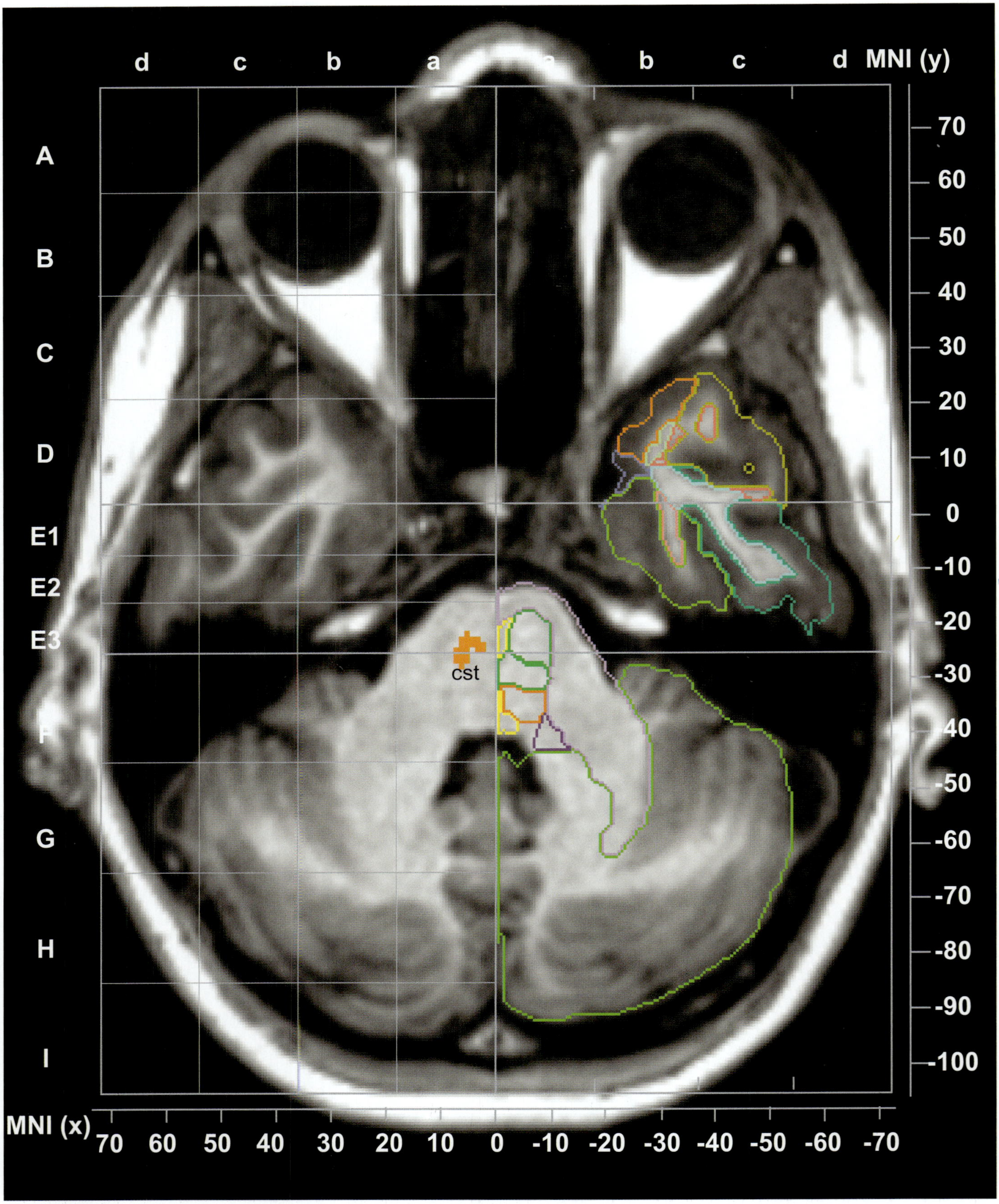

cst - corticospinal tract

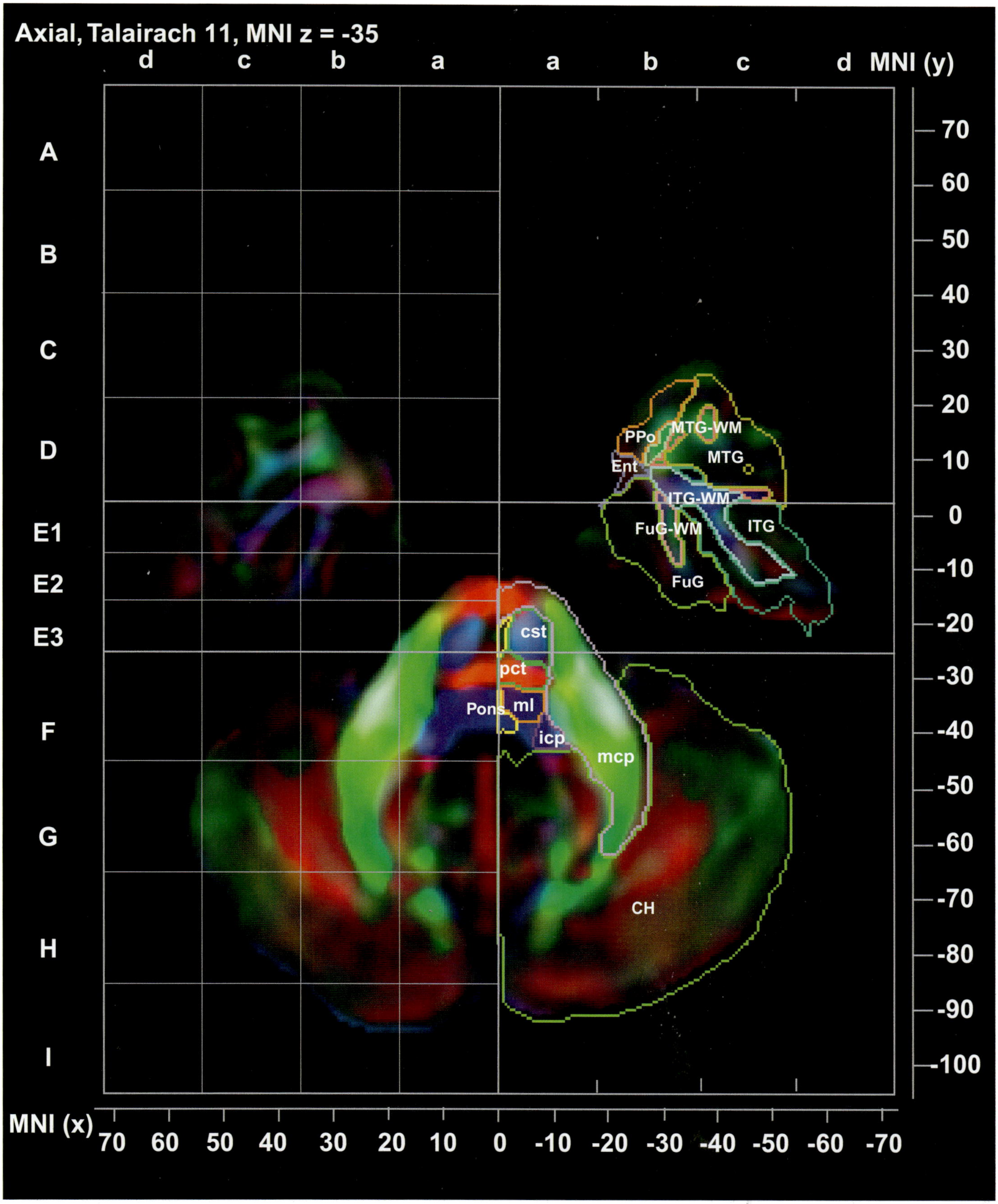

CH - cerebellar hemisphere
cst - corticospinal tract
Ent - entorhinal cortex
FuG - fusiform gyrus
FuG-WM - fusiform gyrus white matter
icp - inferior cerebellar peduncle
ITG - inferior temporal gyrus
ITG-WM - inferior temporal gyrus white matter
mcp - middle cerebellar peduncle
ml - medial lemniscus
MTG - middle temporal gyrus
MTG-WM - middle temporal gyrus white matter
pct - pontine crossing tract (a part of mcp)
Pons - pons
PPo - planum polare

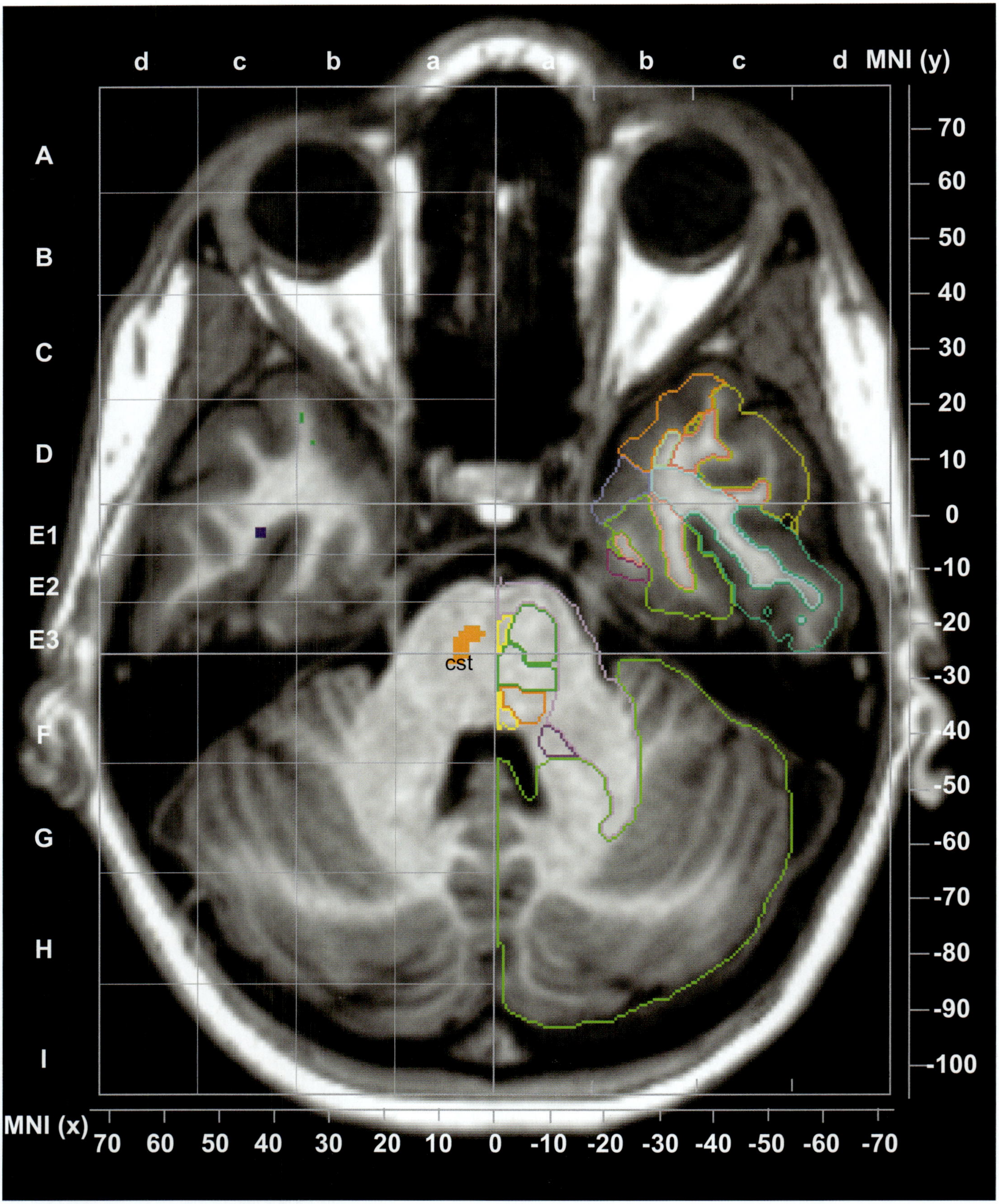

cst - corticospinal tract

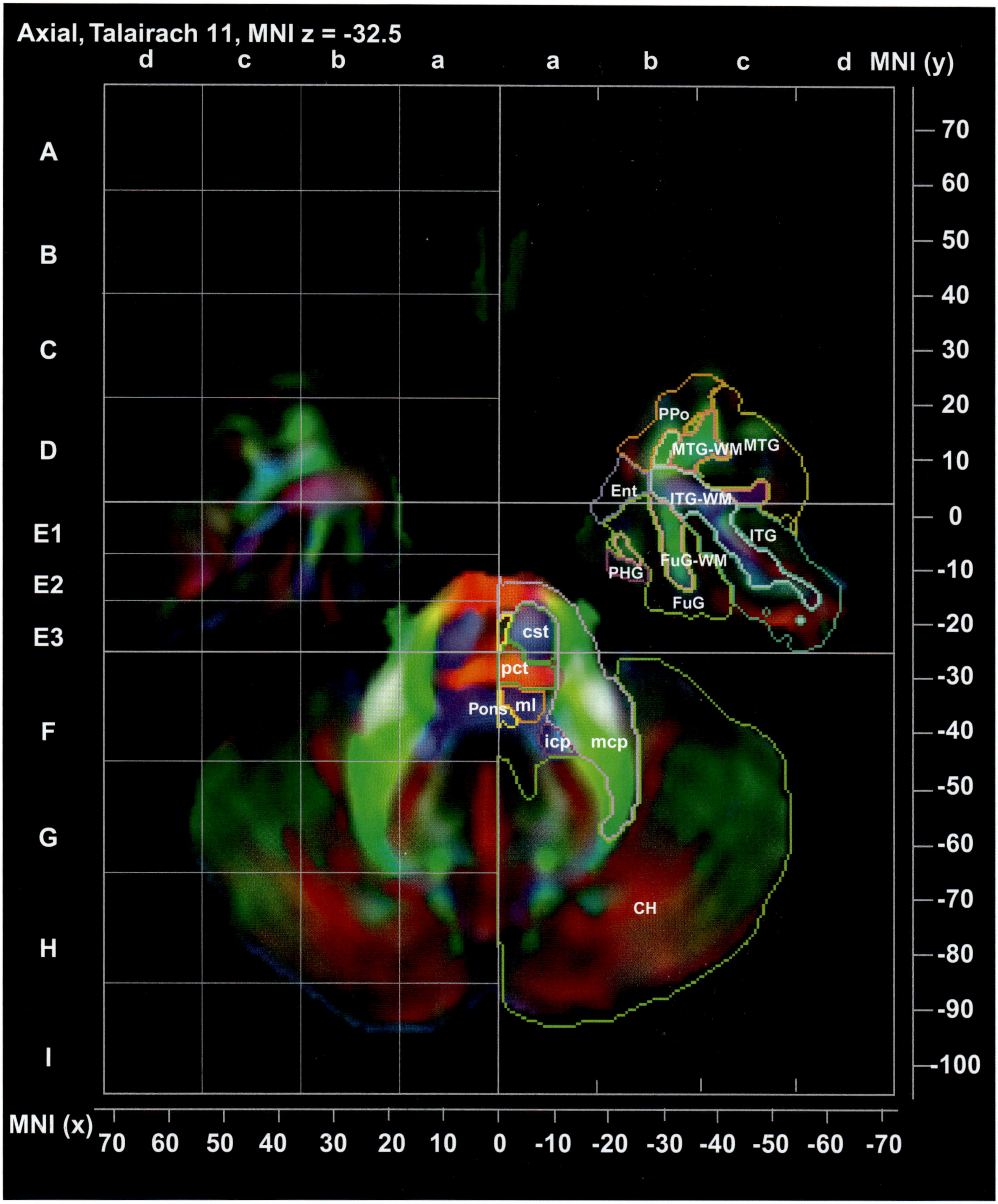

CH	-	cerebellar hemisphere
cst	-	corticospinal tract
Ent	-	entorhinal cortex
FuG	-	fusiform gyrus
FuG-WM	-	fusiform gyrus white matter
icp	-	inferior cerebellar peduncle
ITG	-	inferior temporal gyrus
ITG-WM	-	inferior temporal gyrus white matter
mcp	-	middle cerebellar peduncle
ml	-	medial lemniscus
MTG	-	middle temporal gyrus
MTG-WM	-	middle temporal gyrus white matter
pct	-	pontine crossing tract (a part of mcp)
PHG	-	parahippocampal gyrus
Pons	-	pons
PPo	-	planum polare

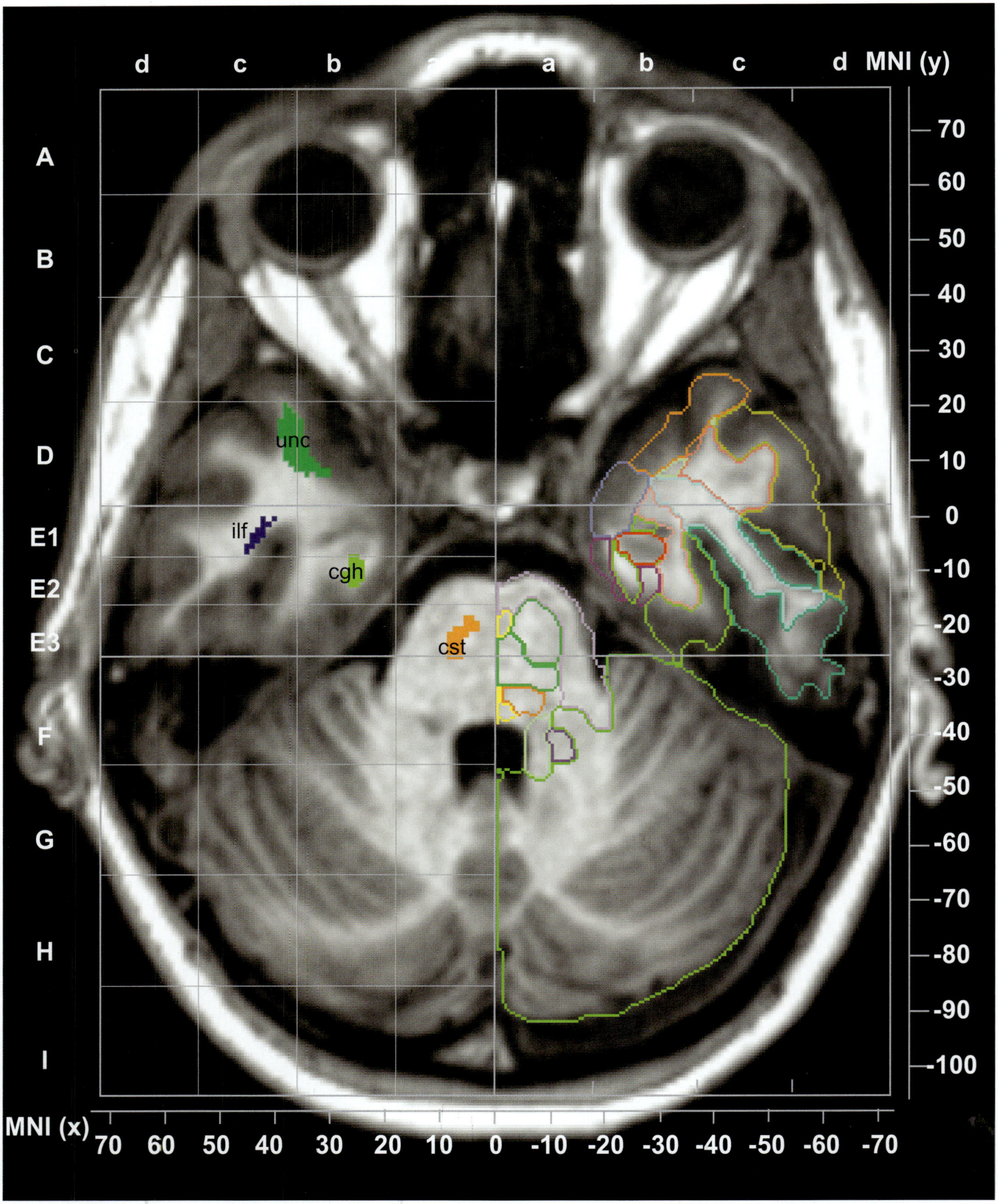

cgh - cingulum (hippocampal part)
CH - cerebellar hemisphere
cst - corticospinal tract
Ent - entorhinal cortex
FuG - fusiform gyrus
FuG-WM - fusiform gyrus white matter
Hi - hippocampus
icp - inferior cerebellar peduncle
ilf - inferior longitudinal fasciculus
ITG - inferior temporal gyrus
ITG-WM - inferior temporal gyrus white matter
mcp - middle cerebellar peduncle
ml - medial lemniscus
MTG - middle temporal gyrus
MTG-WM - middle temporal gyrus white matter

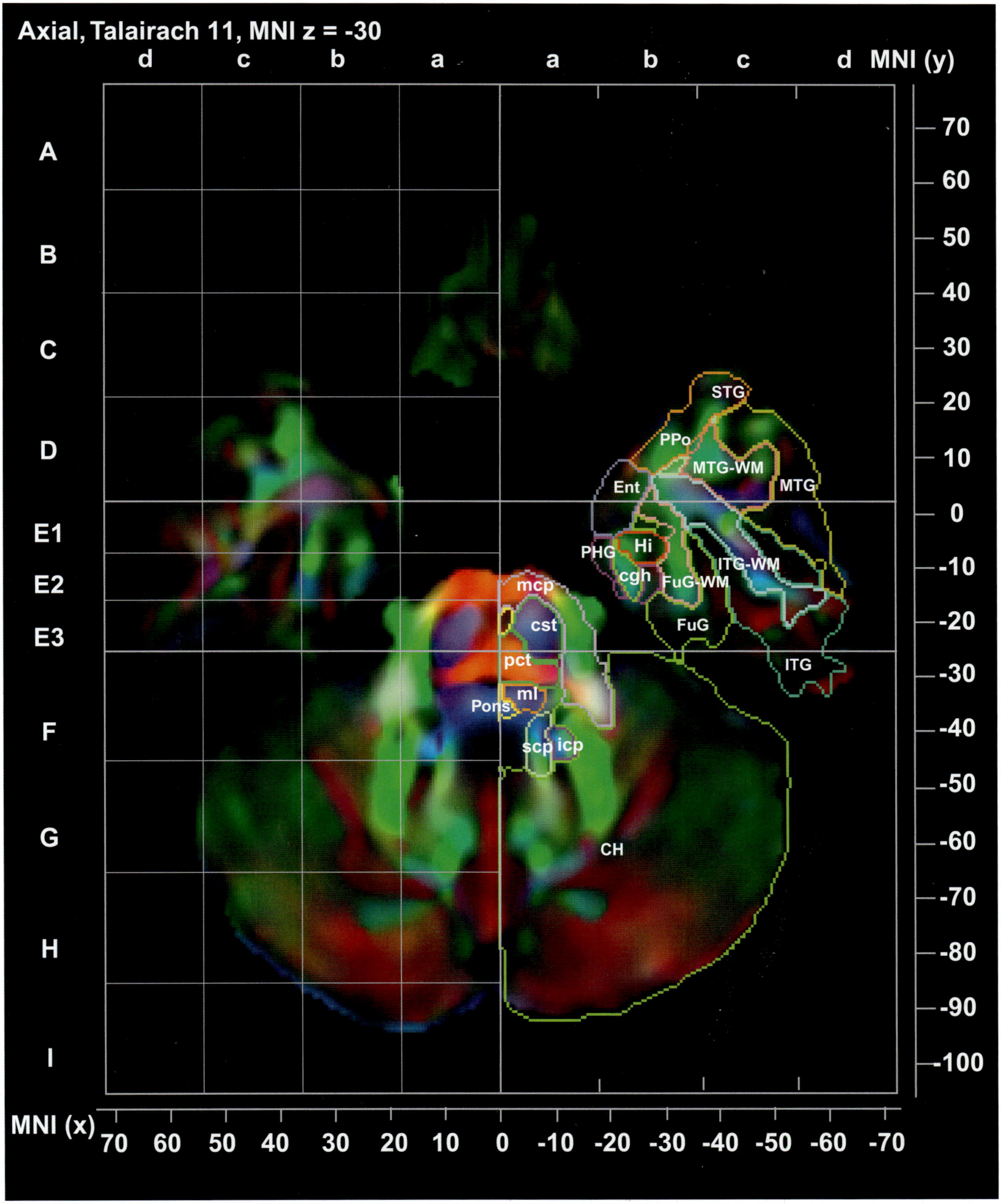

pct - pontine crossing tract (a part of mcp)
PHG - parahippocampal gyrus
Pons - pons
PPo - planum polare
scp - superior cerebellar peduncle
STG - superior temporal gyrus
unc - uncinate fasciculus

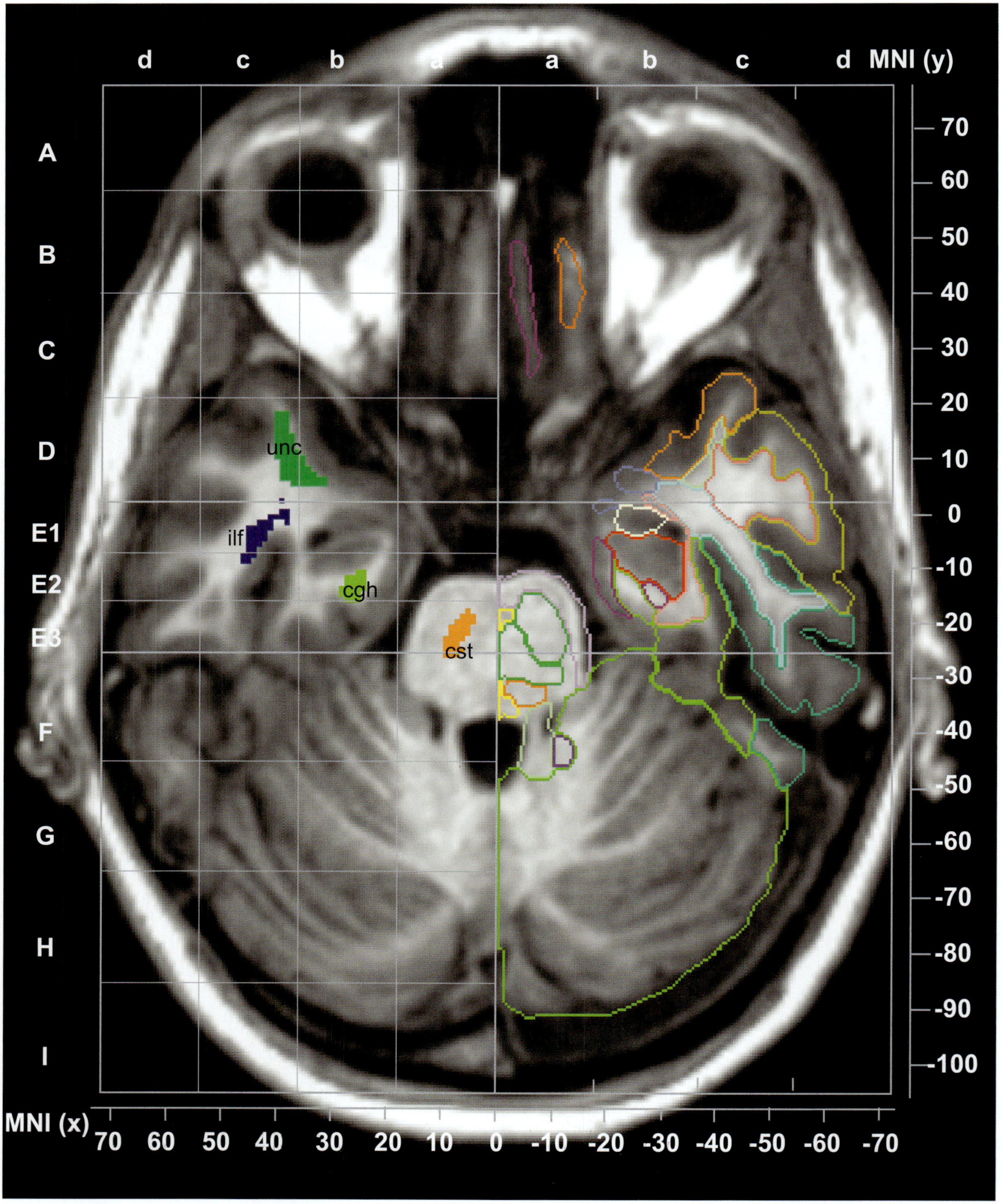

Amg - amygdala
cgh - cingulum (hippocampal part)
CH - cerebellar hemisphere
cst - corticospinal tract
Ent - entorhinal cortex
FuG - fusiform gyrus
FuG-WM - fusiform gyrus white matter
Hi - hippocampus
icp - inferior cerebellar peduncle
ilf - inferior longitudinal fasciculus
ITG - inferior temporal gyrus
ITG-WM - inferior temporal gyrus white matter
mcp - middle cerebellar peduncle
ml - medial lemniscus
MOrG - medial orbital gyrus

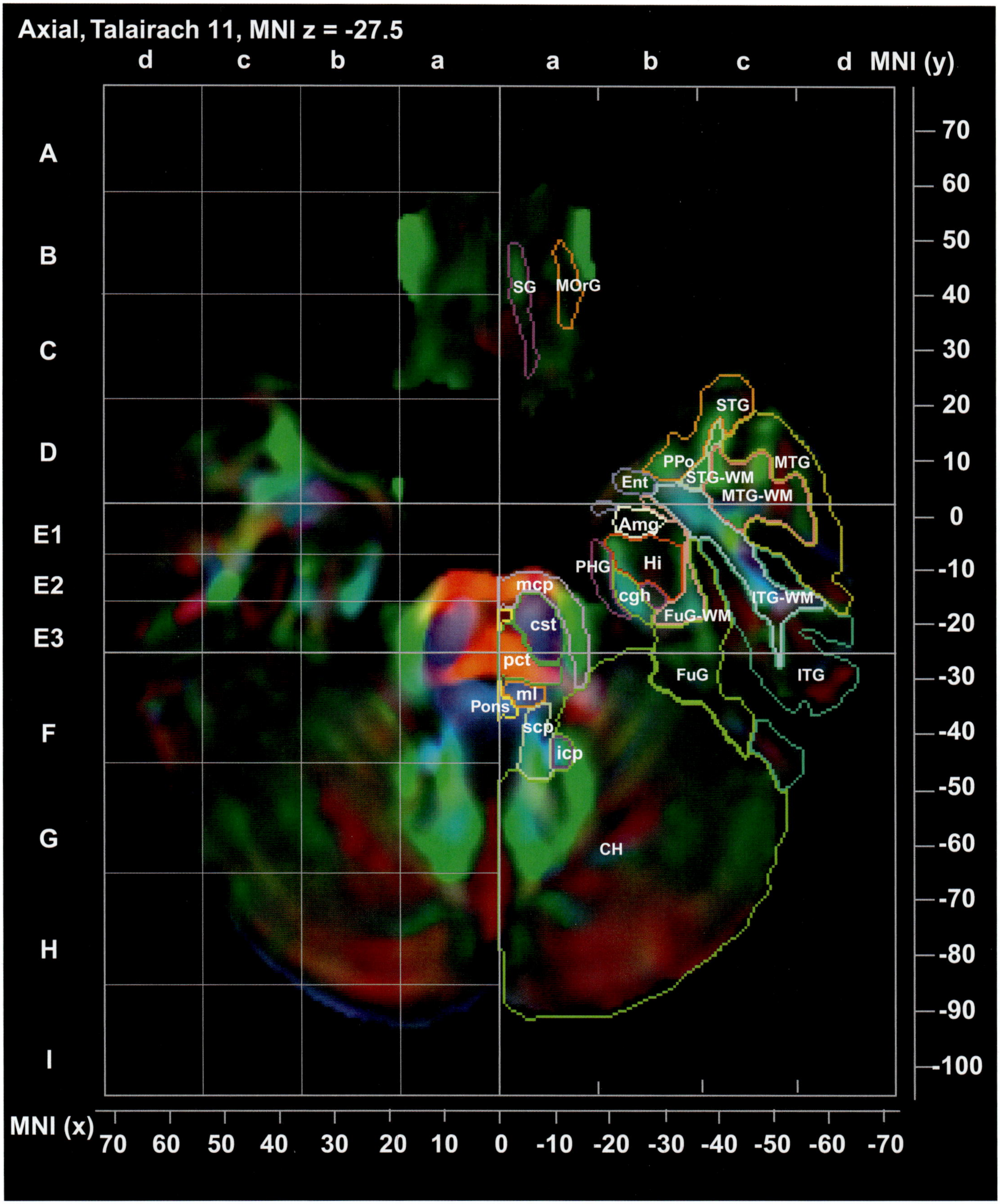

MTG - middle temporal gyrus
MTG-WM - middle temporal gyrus white matter
pct - pontine crossing tract (a part of mcp)
PHG - parahippocampal gyrus
Pons - pons
PPo - planum polare
scp - superior cerebellar peduncle
SG - straight gyrus
STG - superior temporal gyrus
STG-WM - superior temporal gyrus white matter
unc - uncinate fasciculus

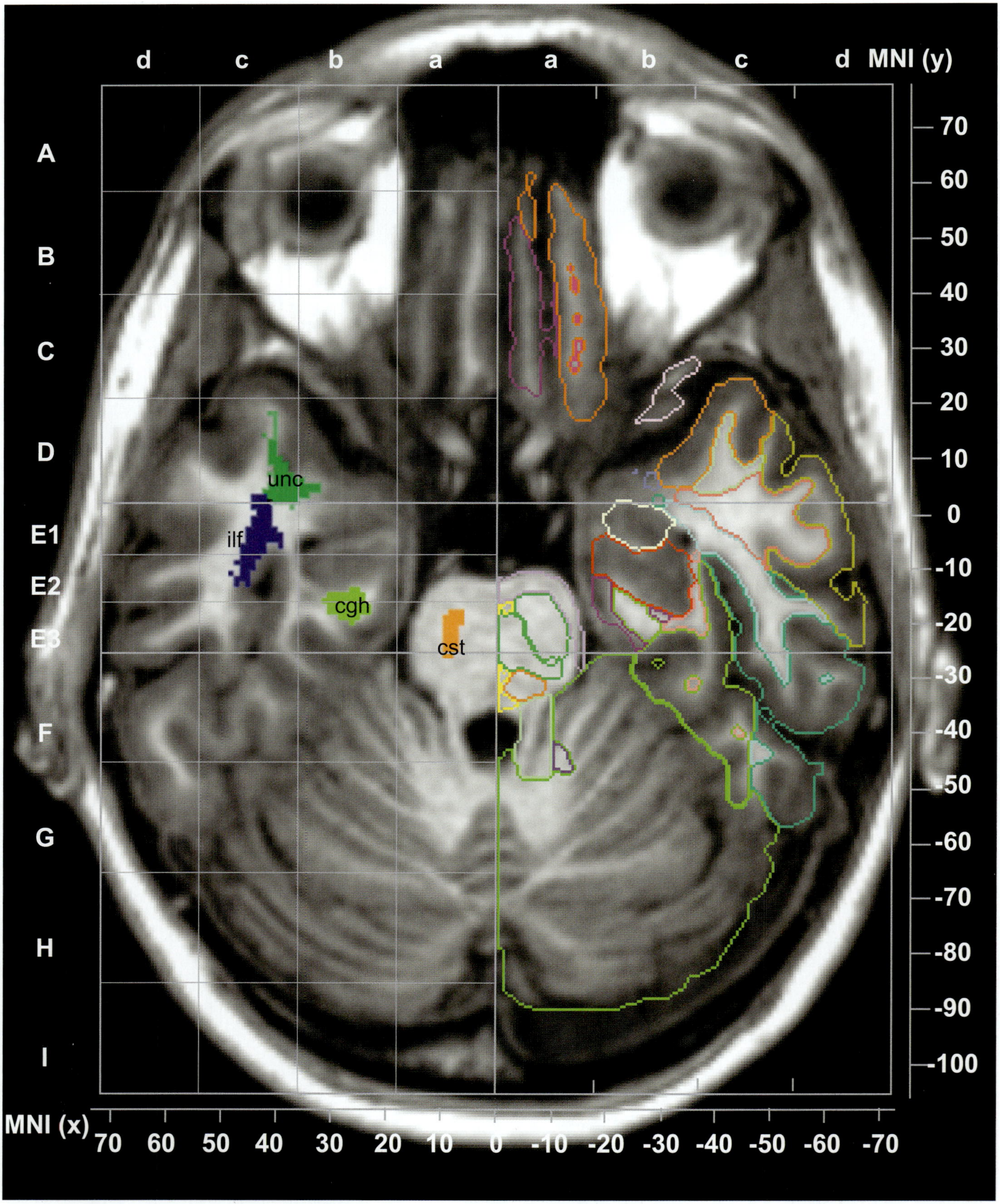

Amg - amygdala	Ent - entorhinal cortex	ilf - inferior longitudinal fasciculus
cgh - cingulum (hippocampal part)	FuG - fusiform gyrus	ITG - inferior temporal gyrus
	FuG-WM - fusiform gyrus white matter	ITG-WM - inferior temporal gyrus white matter
CH - cerebellar hemisphere	Hi - hippocampus	mcp - middle cerebellar peduncle
cst - corticospinal tract	icp - inferior cerebellar peduncle	ml - medial lemniscus

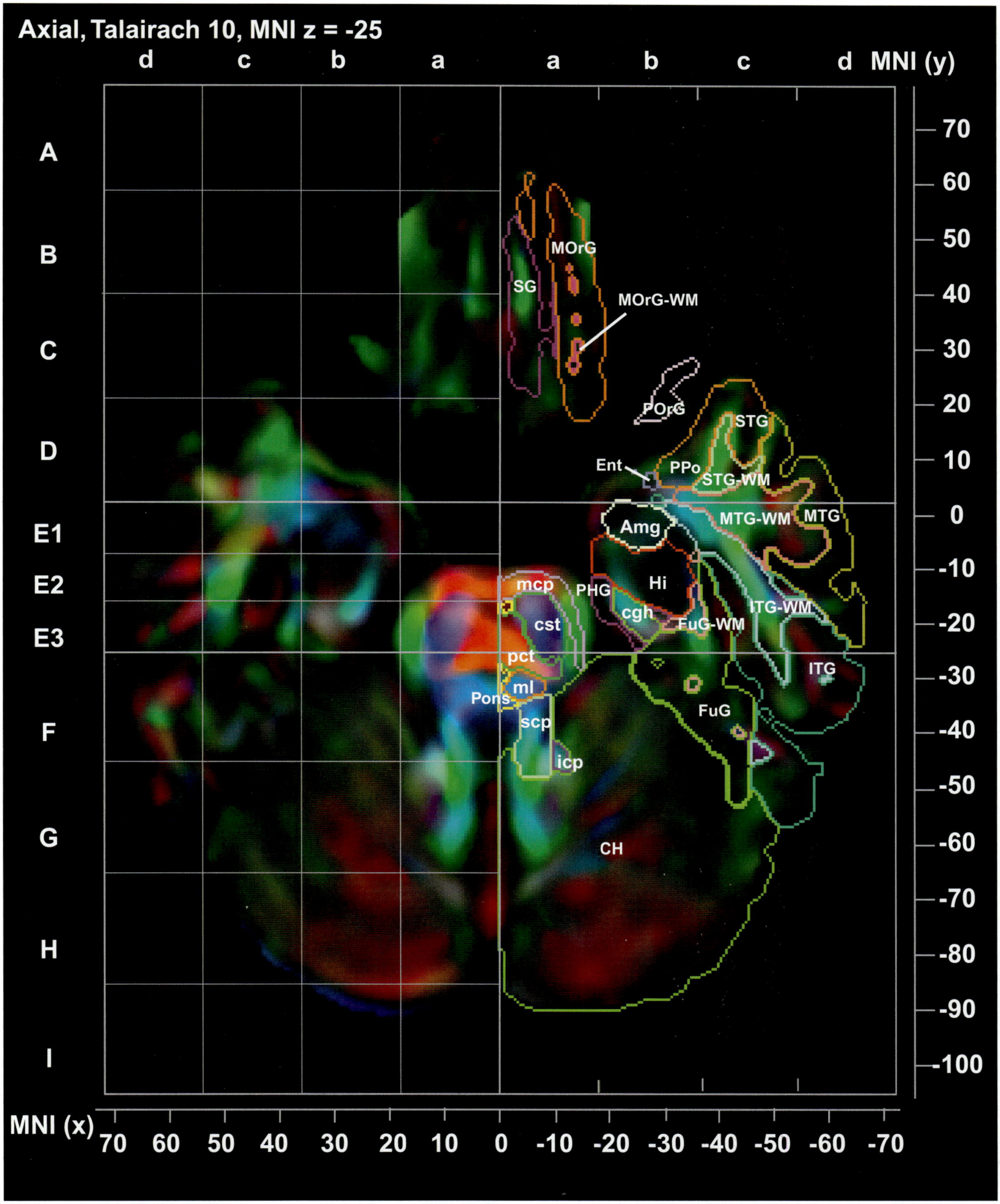

MOrG - medial orbital gyrus
MOrG-WM - medial orbital gyrus white matter
MTG - middle temporal gyrus
MTG-WM - middle temporal gyrus white matter
pct - pontine crossing tract (a part of mcp)
PHG - parahippocampal gyrus
Pons - pons
POrG - posterior orbital gyrus
PPo - planum polare
scp - superior cerebellar peduncle
SG - straight gyrus
STG - superior temporal gyrus
STG-WM - superior temporal gyrus white matter
unc - uncinate fasciculus

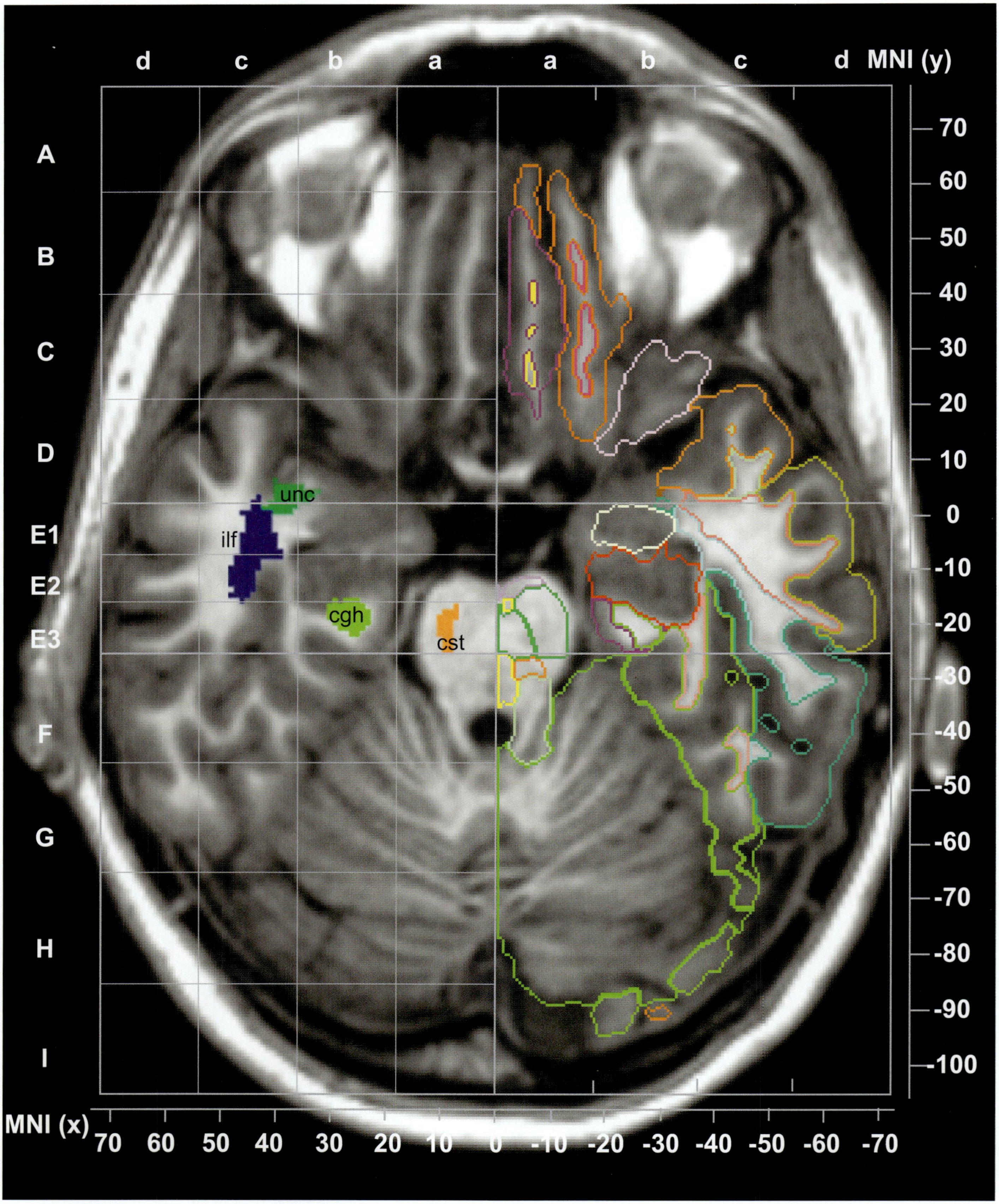

Amg	- amygdala	FuG-WM	- fusiform gyrus white matter	mcp	- middle cerebellar peduncle
cgh	- cingulum (hippocampal part)	Hi	- hippocampus	ml	- medial lemniscus
CH	- cerebellar hemisphere	ilf	- inferior longitudinal fasciculus	MOrG	- medial orbital gyrus
cst	- corticospinal tract	ITG	- inferior temporal gyrus	MOrG-WM	- medial orbital gyrus white matter
FuG	- fusiform gyrus	ITG-WM	- inferior temporal gyrus white matter	MTG	- middle temporal gyrus

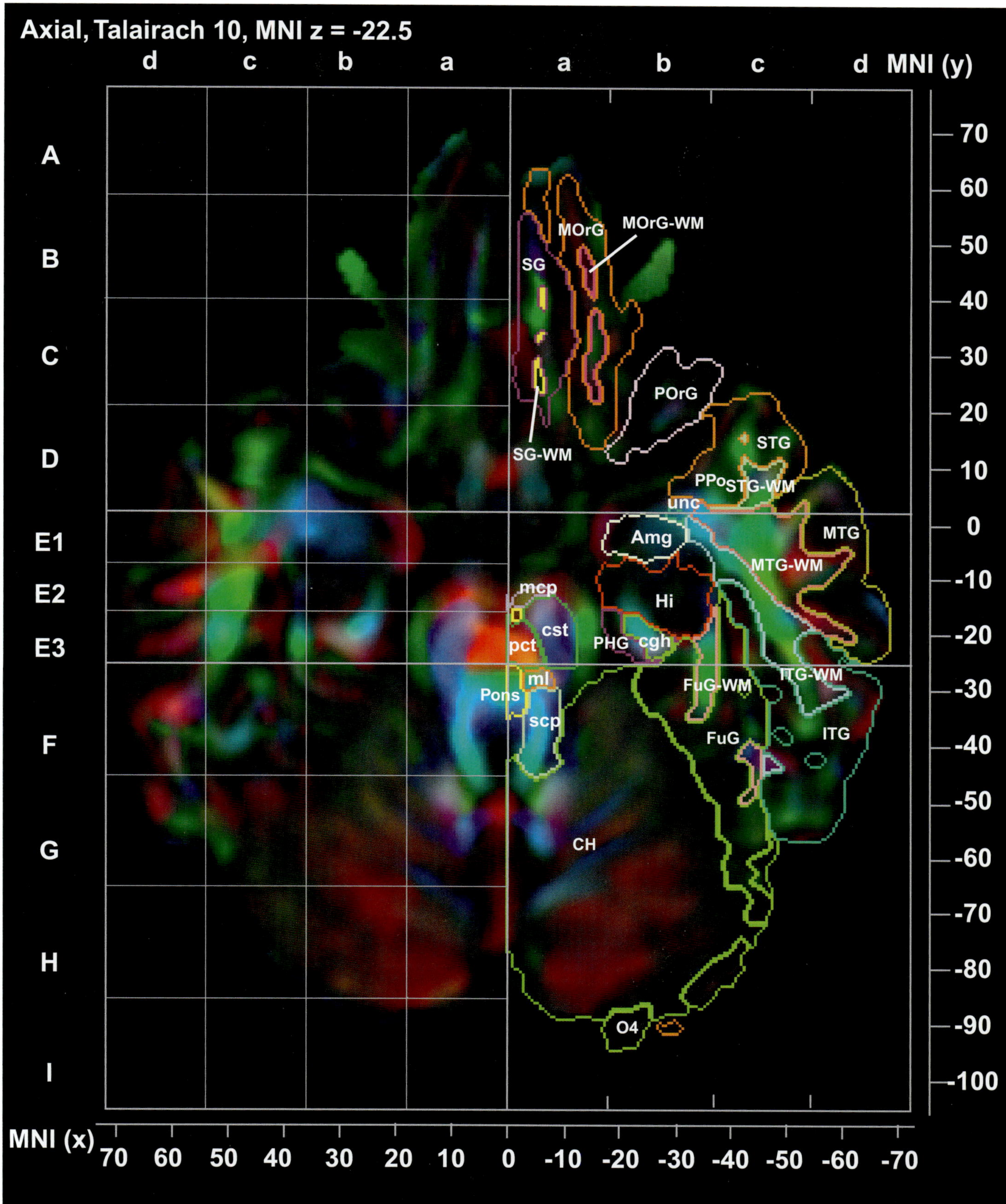

MTG-WM - middle temporal gyrus white matter
O4 - fourth occipital gyrus (part of the fusiform gyrus)
pct - pontine crossing tract (a part of mcp)
PHG - parahippocampal gyrus
Pons - pons
POrG - posterior orbital gyrus
PPo - planum polare
scp - superior cerebellar peduncle
SG - straight gyrus
SG-WM - straight gyrus white matter
STG - superior temporal gyrus
STG-WM - superior temporal gyrus white matter
unc - uncinate fasciculus

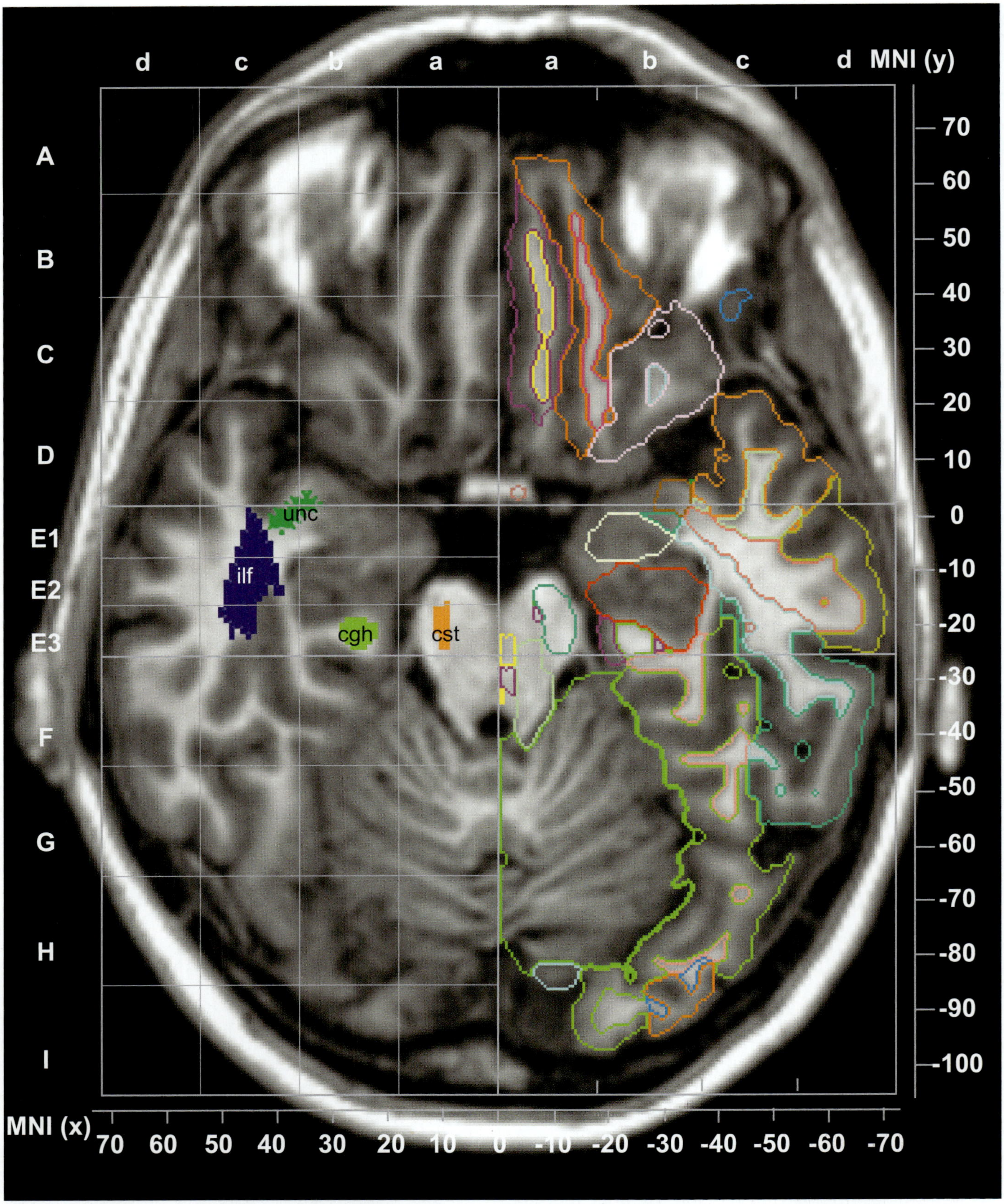

Amg - amygdala	Hi - hippocampus	MOrG-WM - medial orbital gyrus white matter
cgh - cingulum (hippocampal part)	IFG - inferior frontal gyrus	MTG - middle temporal gyrus
CH - cerebellar hemisphere	ilf - inferior longitudinal fasciculus	MTG-WM - middle temporal gyrus white matter
cp - cerebral peduncle	ITG - inferior temporal gyrus	O3 - inferior occipital gyrus
cst - corticospinal tract	ITG-WM - inferior temporal gyrus white matter	O3-WM - inferior occipital gyrus white matter
FuG - fusiform gyrus	M - midbrain	O4 - fourth occipital gyrus (part of the fusiform gyrus)
FuG-WM - fusiform gyrus white matter	MOrG - medial orbital gyrus	O5 - lyngual gyrus

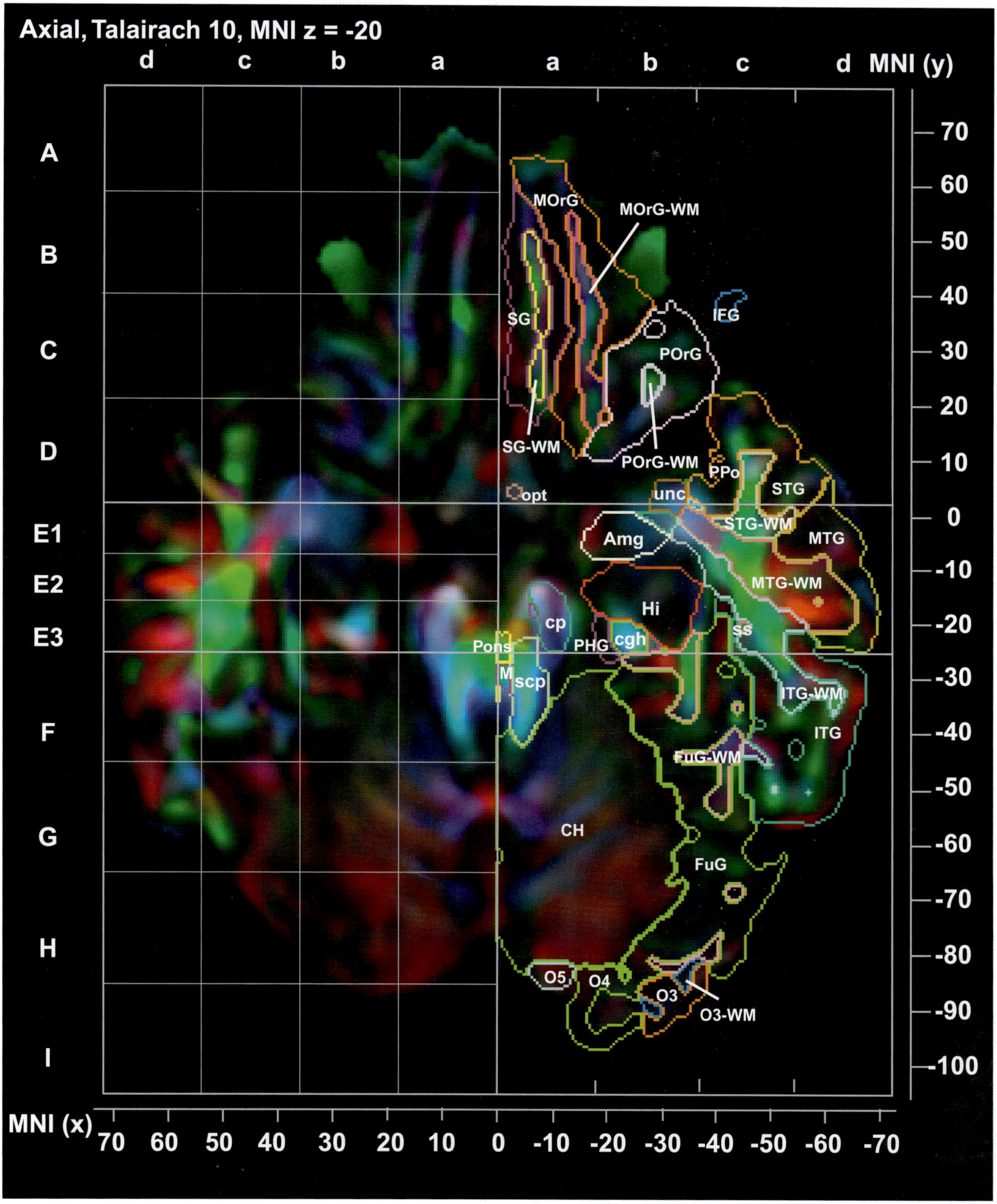

opt - optic tract
PHG - parahippocampal gyrus
Pons - pons
POrG - posterior orbital gyrus
POrG-WM - posterior orbital gyrus white matter
PPo - planum polare
scp - superior cerebellar peduncle
SG - straight gyrus
SG-WM - straight gyrus white matter
ss - sagittal stratum (include inferior longitidinal fasciculus and inferior fronto-occipital fasciculus)
STG - superior temporal gyrus
STG-WM - superior temporal gyrus white matter
unc - uncinate fasciculus

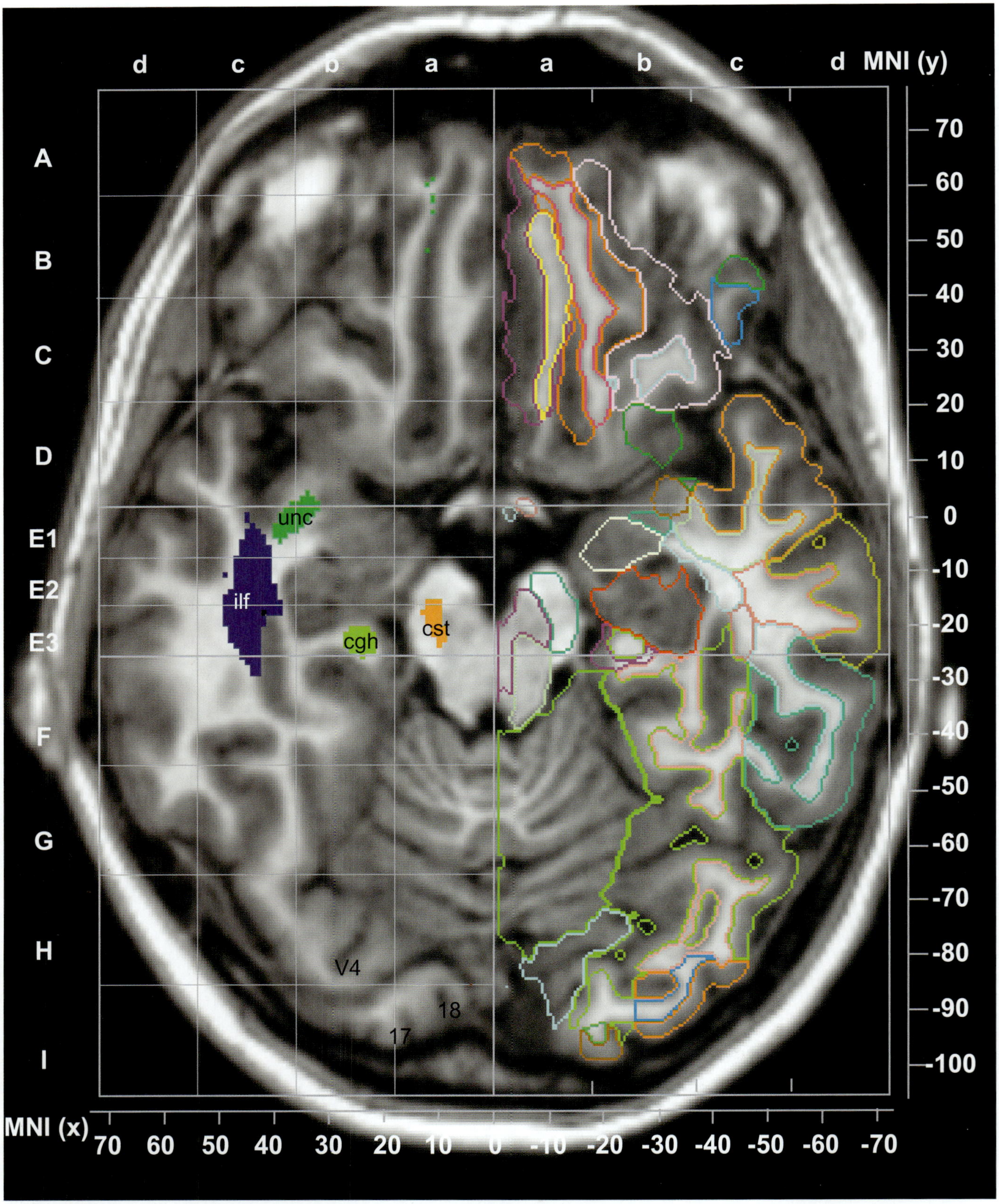

Amg - amygdala
cgh - cingulum (hippocampal part)
CH - cerebellar hemisphere
cp - cerebral peduncle
cst - corticospinal tract
FuG - fusiform gyrus
FuG-WM - fusiform gyrus white matter
Hi - hippocampus
Hy - hypothalamus
IFG - inferior frontal gyrus
IG - insular gyrus
ilf - inferior longitudinal fasciculus
IOrG - intermediate orbital gyrus
ITG - inferior temporal gyrus
ITG-WM - inferior temporal gyrus white matter
M - midbrain
MFG - middle frontal gyrus
MOrG - medial orbital gyrus
MOrG-WM - medial orbital gyrus white matter
MTG - middle temporal gyrus
MTG-WM - middle temporal gyrus white matter
O2 - middle or lateral occipital gyrus
O3 - inferior occipital gyrus
O3-WM - inferior occipital gyrus white matter
O4 - fourth occipital gyrus (part of the fusiform gyrus)
O5 - lyngual gyrus

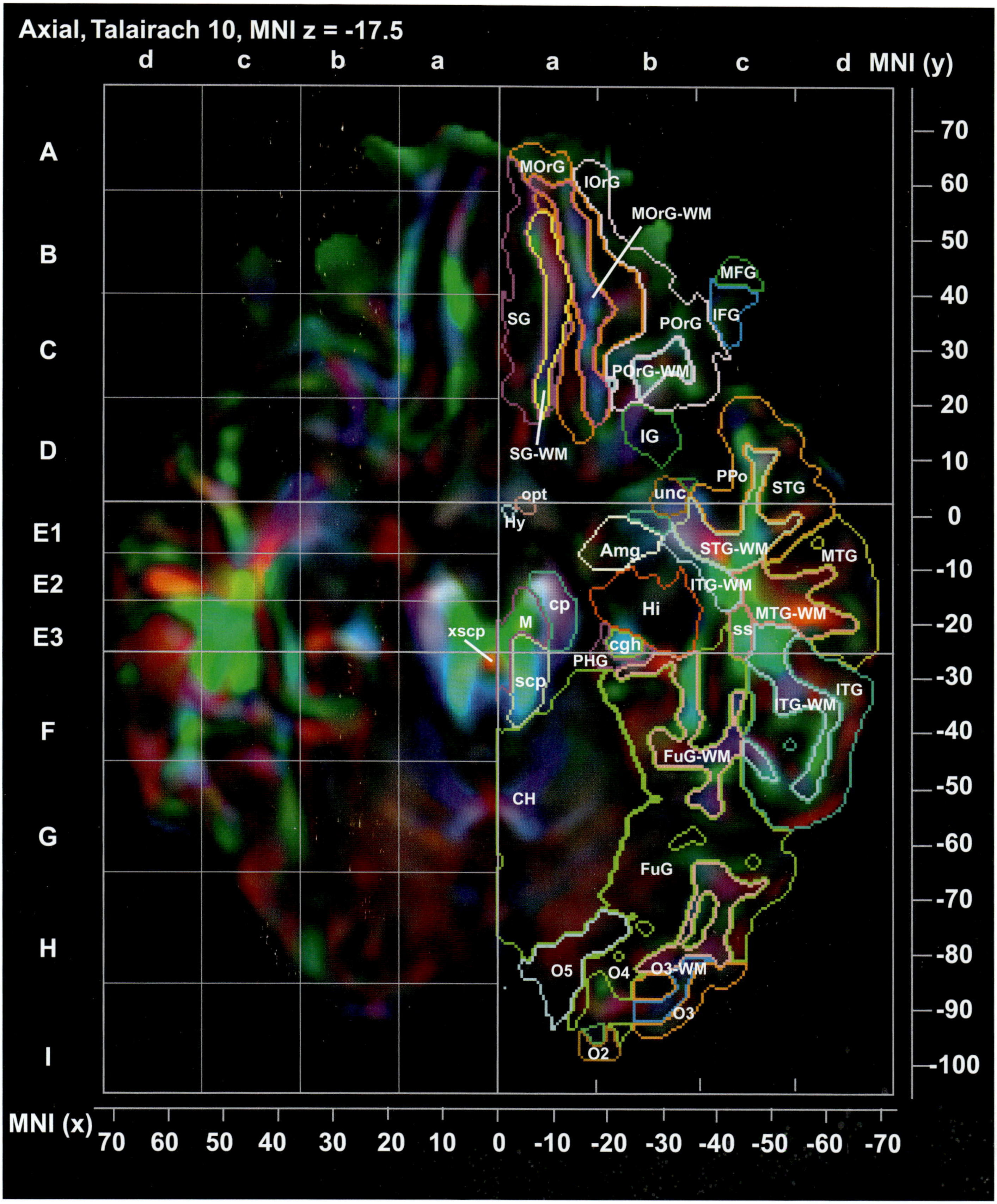

opt - optic tract
PHG - parahippocampal gyrus
POrG - posterior orbital gyrus
POrG-WM - posterior orbital gyrus white matter
PPo - planum polare
scp - superior cerebellar peduncle

SG - straight gyrus
SG-WM - straight gyrus white matter
ss - sagittal stratum (include inferior longitidinal fasciculus and inferior fronto-occipital fasciculus)

STG - superior temporal gyrus
STG-WM - superior temporal gyrus white matter
unc - uncinate fasciculus
xscp - decussation of the superior cerebellar peduncle

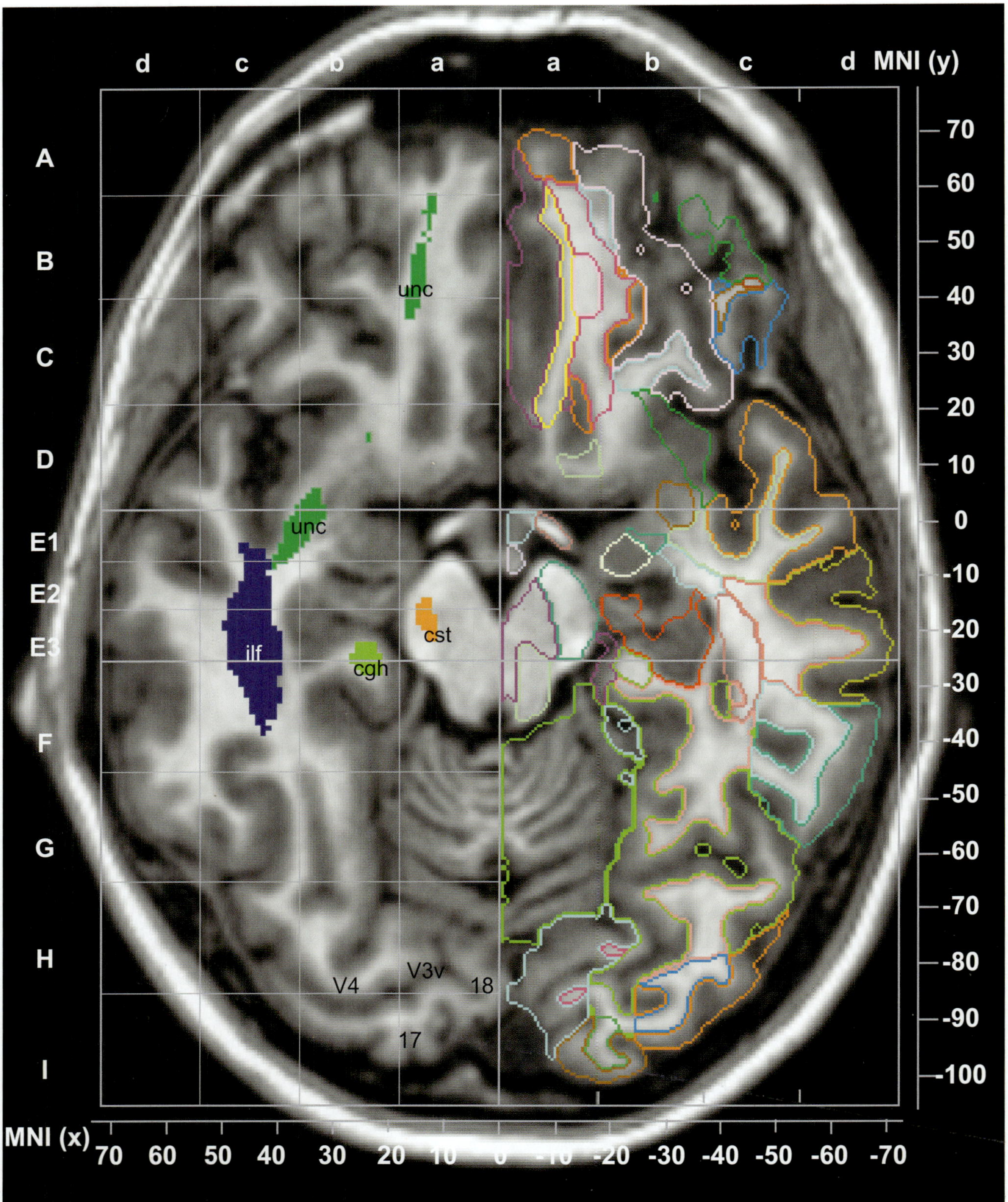

acr	-	anterior corona radiata
Amg	-	amygdala
cgh	-	cingulum (hippocampal part)
CH	-	cerebellar hemisphere
cp	-	cerebral peduncle
cst	-	corticospinal tract
FuG	-	fusiform gyrus
FuG-WM	-	fusiform gyrus white matter
Hi	-	hippocampus
Hy	-	hypothalamus
IFG	-	inferior frontal gyrus
IFG-WM	-	inferior frontal gyrus white matter
IG	-	insular gyrus
ilf	-	inferior longitudinal fasciculus
IOrG	-	intermediate orbital gyrus
IOrG-WM	-	intermediate orbital gyrus white matter
ITG	-	inferior temporal gyrus
ITG-WM	-	inferior temporal gyrus white matter
LOrG-WM	-	lateral orbital gyrus white matter
M	-	midbrain
MB	-	mammillary body
MFG	-	middle frontal gyrus
MOrG	-	medial orbital gyrus
MOrG-WM	-	medial orbital gyrus white matter
MTG	-	middle temporal gyrus
MTG-WM	-	middle temporal gyrus white matter
O2	-	middle or lateral occipital gyrus
O2-WM	-	middle or lateral occipital gyrus white matter
O3	-	inferior occipital gyrus
O3-WM	-	inferior occipital gyrus white matter
O4	-	fourth occipital gyrus (part of the fusiform gyrus)
O5	-	lyngual gyrus

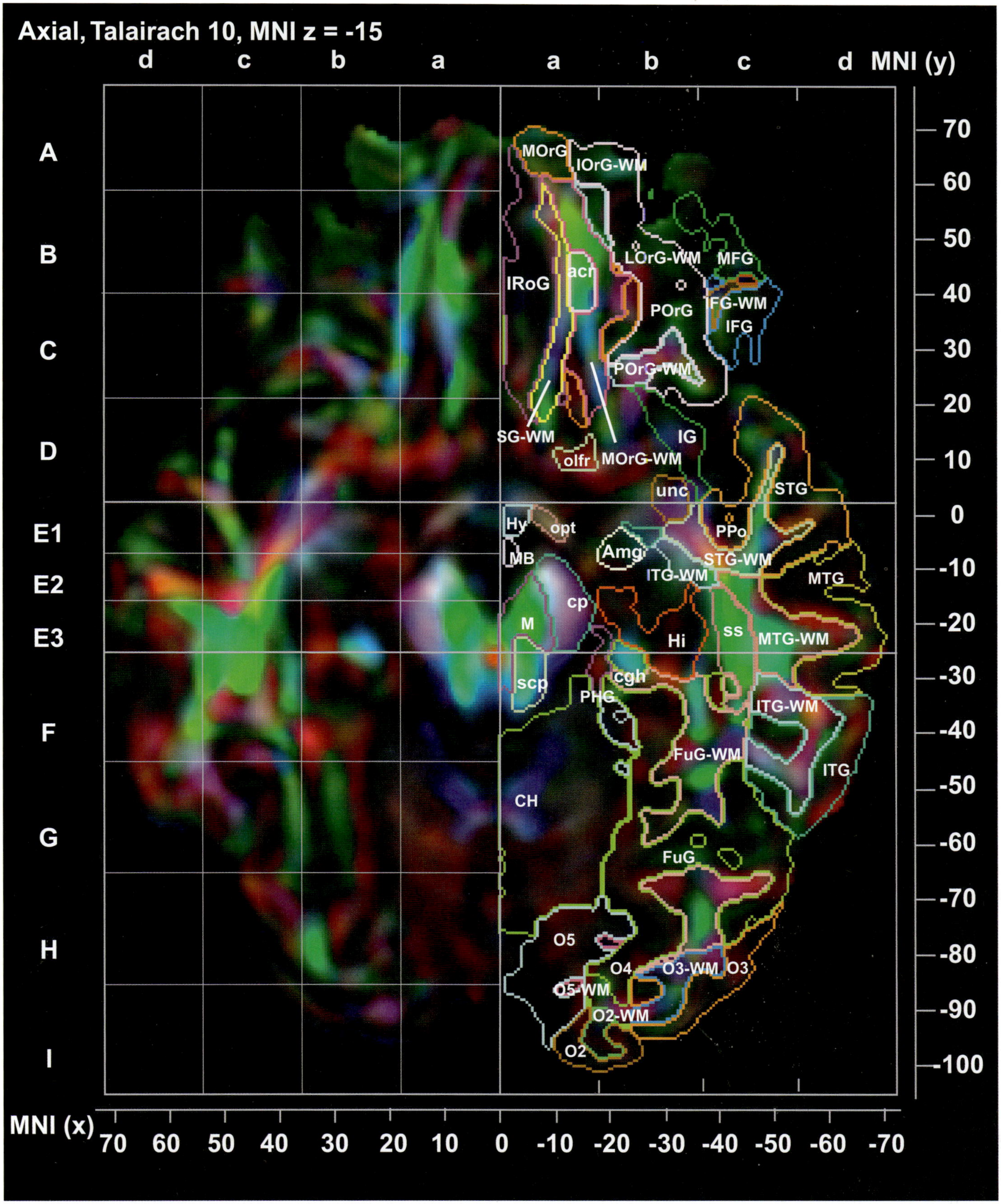

O5-WM	-	lyngual gyrus white matter
olfr	-	olfactory radiation
opt	-	optic tract
PHG	-	parahippocampal gyrus
POrG	-	posterior orbital gyrus
POrG-WM	-	posterior orbital gyrus white matter
PPo	-	planum polare
scp	-	superior cerebellar peduncle
SG-WM	-	straight gyrus white matter
ss	-	sagittal stratum (include inferior longitidinal fasciculus and inferior fronto-occipital fasciculus)
STG	-	superior temporal gyrus
STG-WM	-	superior temporal gyrus white matter
unc	-	uncinate fasciculus

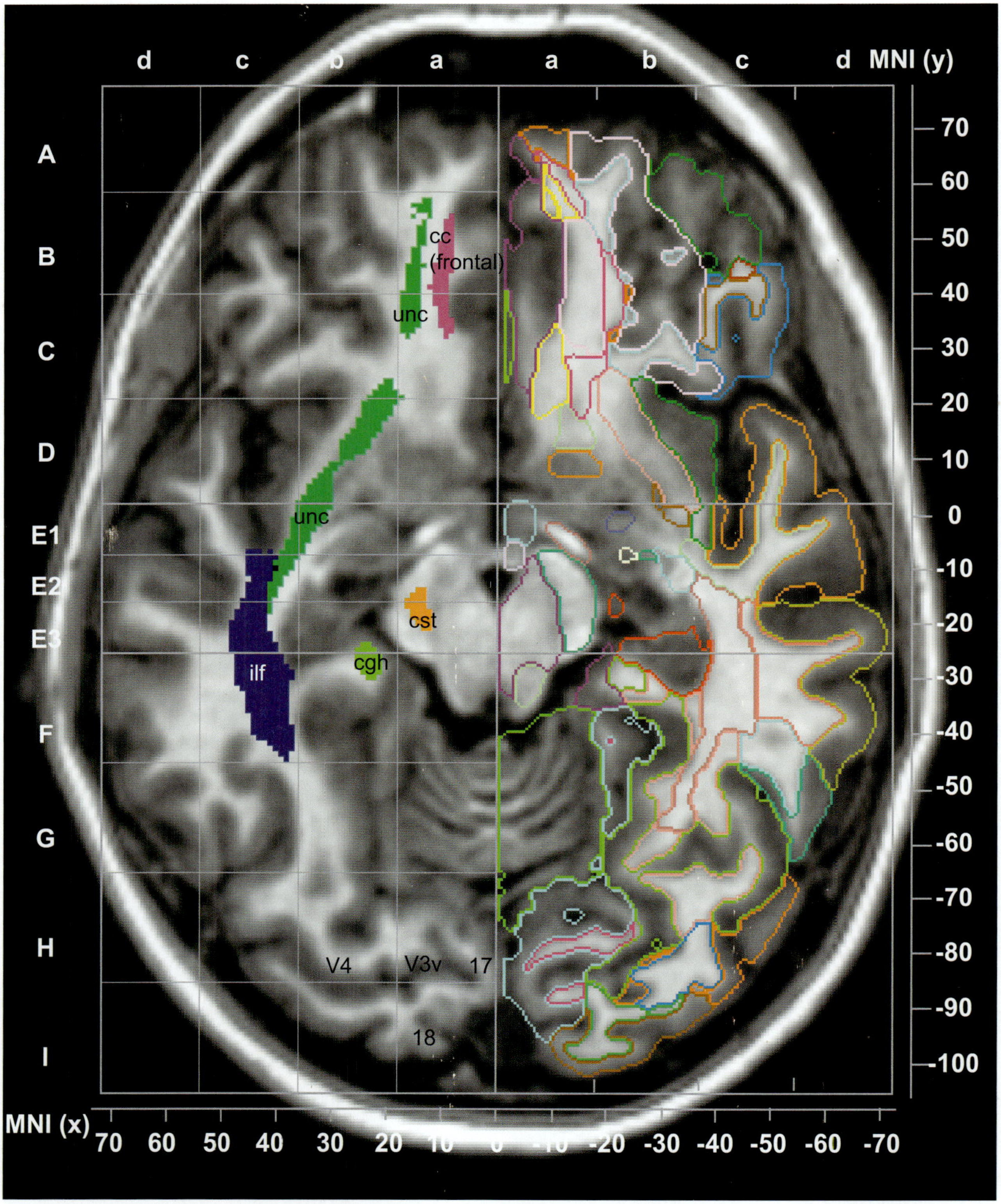

Ac - accumbens nucleus
acr - anterior corona radiata
al - ansa lenticularis
Amg - amygdala
cc (frontal) - corpus callosum (connecting bilateral frontal areas)
CG - cingulate gyrus
cgh - cingulum (hippocampal part)
CH - cerebellar hemisphere
cp - cerebral peduncle
cst - corticospinal tract
FuG - fusiform gyrus
FuG-WM - fusiform gyrus white matter
Hi - hippocampus
Hy - hypothalamus
IFG - inferior frontal gyrus
IFG-WM - inferior frontal gyrus white matter
ifo - inferior fronto-occipital fasciculus
IG - insular gyrus
ilf - inferior longitudinal fasciculus
IOrG - intermediate orbital gyrus
IRoG - inferior rostral gyrus
ITG - inferior temporal gyrus
ITG-WM - inferior temporal gyrus white matter
LOrG - lateral orbital gyrus
LOrG-WM - lateral orbital gyrus white matter
M - midbrain
MB - mammillary body
MFG - middle frontal gyrus
MOrG - medial orbital gyrus
MOrG-WM - medial orbital gyrus white matter
MTG - middle temporal gyrus
MTG-WM - middle temporal gyrus white matter
O2 - middle or lateral occipital gyrus
O2-WM - middle or lateral occipital gyrus white matter
O3 - inferior occipital gyrus
O3-WM - inferior occipital gyrus white matter
O4 - fourth occipital gyrus (part of the fusiform gyrus)
O5 - lyngual gyrus
O5-WM - lyngual gyrus white matter
olfr - olfactory radiation

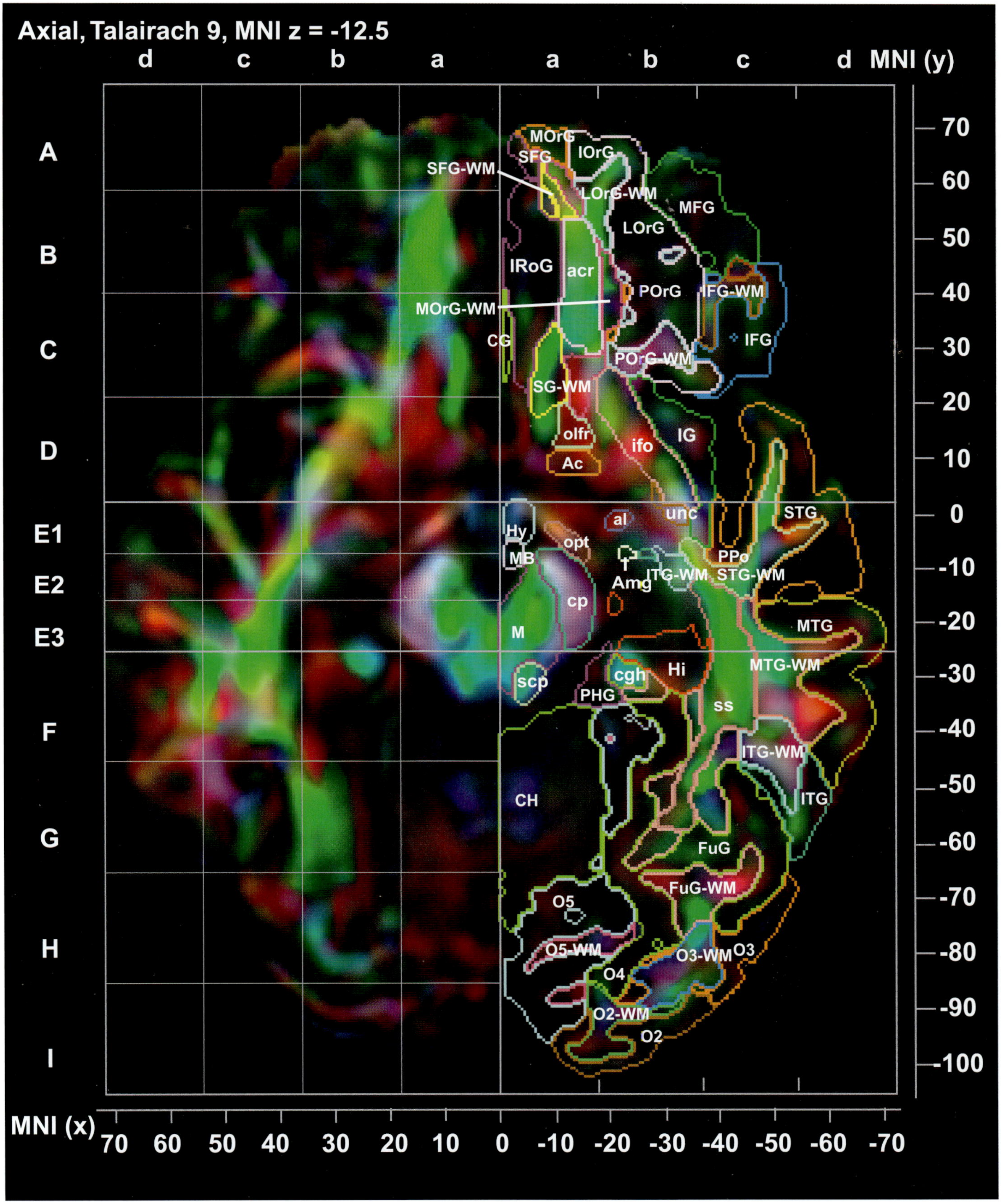

opt	-	optic tract
PHG	-	parahippocampal gyrus
POrG	-	posterior orbital gyrus
POrG-WM	-	posterior orbital gyrus white matter
PPo	-	planum polare
scp	-	superior cerebellar peduncle
SFG	-	superior frontal gyrus
SFG-WM	-	superior frontal gyrus white matter
SG-WM	-	straight gyrus white matter
ss	-	sagittal stratum (include inferior longitidinal fasciculus and inferior fronto-occipital fasciculus)
STG	-	superior temporal gyrus
STG-WM	-	superior temporal gyrus white matter
unc	-	uncinate fasciculus

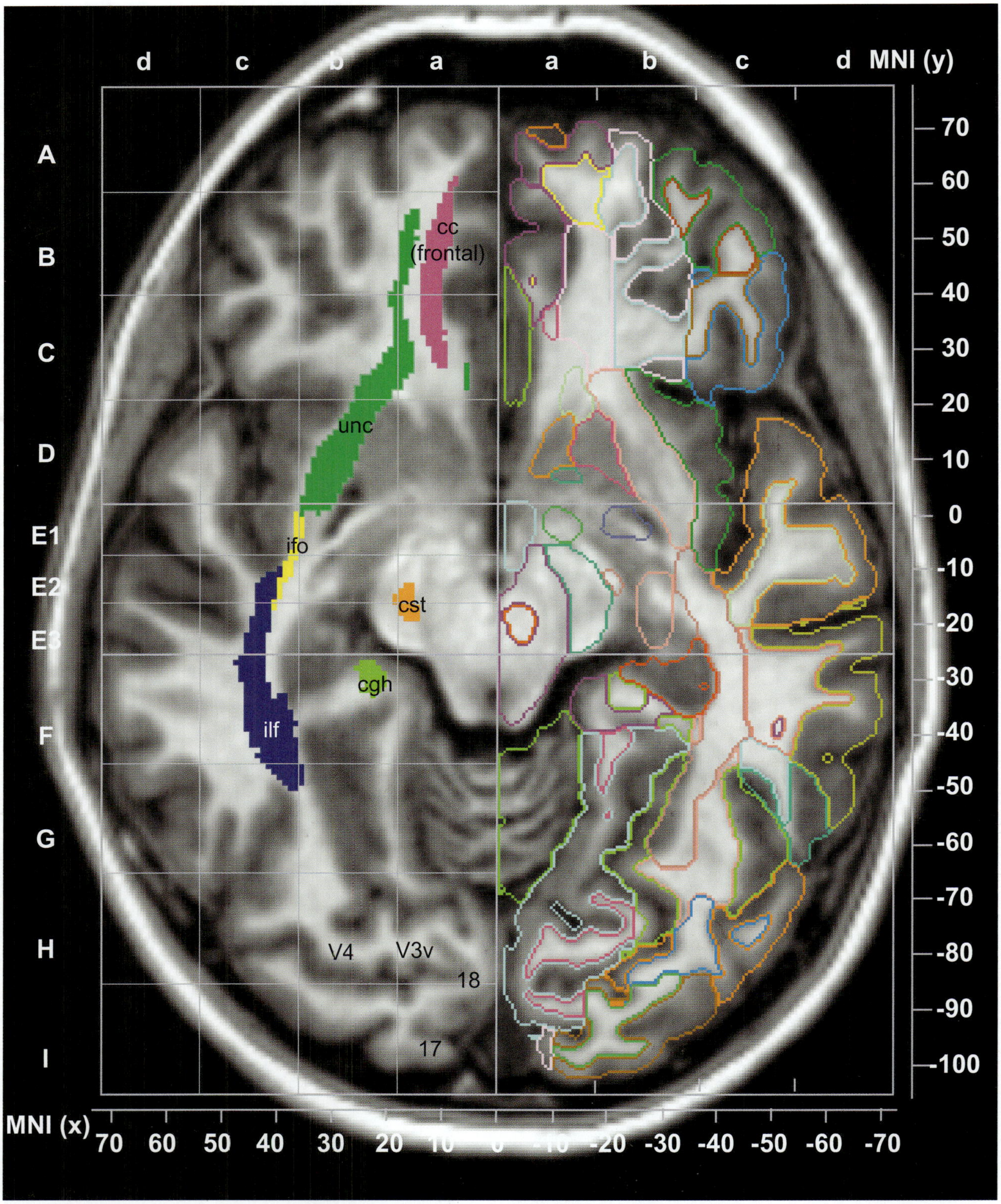

Ac - accumbens nucleus
acr - anterior corona radiata
al - ansa lenticularis
B - basal nucleus of Meynert
cc (frontal) - corpus callosum (connecting bilateral frontal areas)
CG - cingulate gyrus
cgc - cingulum (cingulate gyrus part)
cgh - cingulum (hippocampal part)
CH - cerebellar hemisphere
cp - cerebral peduncle
cst - corticospinal tract
FuG - fusiform gyrus
FuG-WM - fusiform gyrus white matter
fx/st - fornix (crus) / stria terminalis (cannot be resolved with current resolution)
Hi - hippocampus
Hy - hypothalamus
IFG - inferior frontal gyrus
IFG-WM - inferior frontal gyrus white matter
ifo - inferior fronto-occipital fasciculus
IG - insular gyrus
ilf - inferior longitudinal fasciculus
IOrG - intermediate orbital gyrus
IOrG-WM - intermediate orbital gyrus white matter
IRoG - inferior rostral gyrus
ITG - inferior temporal gyrus
ITG-WM - inferior temporal gyrus white matter
lenf - lenticular fasciculus
LOrG - lateral orbital gyrus
LOrG-WM - lateral orbital gyrus white matter
M - midbrain
MFG - middle frontal gyrus
MFG-WM - middle frontal gyrus white matter
MTG - middle temporal gyrus
MTG-WM - middle temporal gyrus white matter
O2 - middle or lateral occipital gyrus
O2-WM - middle or lateral occipital gyrus white matter
O3 - inferior occipital gyrus
O3-WM - inferior occipital gyrus white matter
O5 - lyngual gyrus
O5-WM - lyngual gyrus white matter

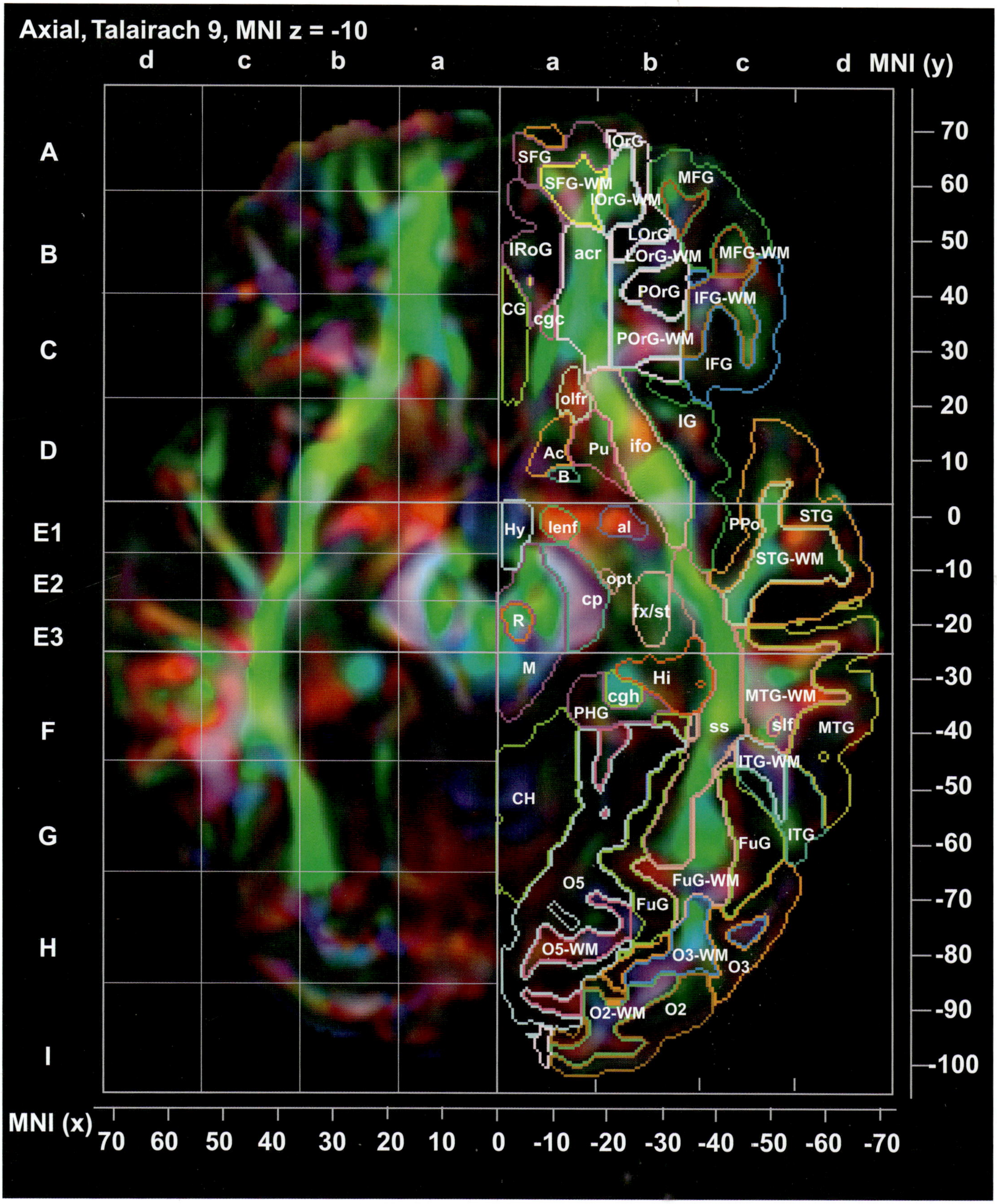

olfr - olfactory radiation
opt - optic tract
PHG - parahippocampal gyrus
POrG - posterior orbital gyrus
POrG-WM - posterior orbital gyrus white matter

PPo - planum polare
Pu - putamen
R - red nucleus
SFG - superior frontal gyrus
SFG-WM - superior frontal gyrus white matter
slf - superior longitudinal fasciculus

ss - sagittal stratum (include inferior longitidinal fasciculus and inferior fronto-occipital fasciculus)
STG - superior temporal gyrus
STG-WM - superior temporal gyrus white matter
unc - uncinate fasciculus

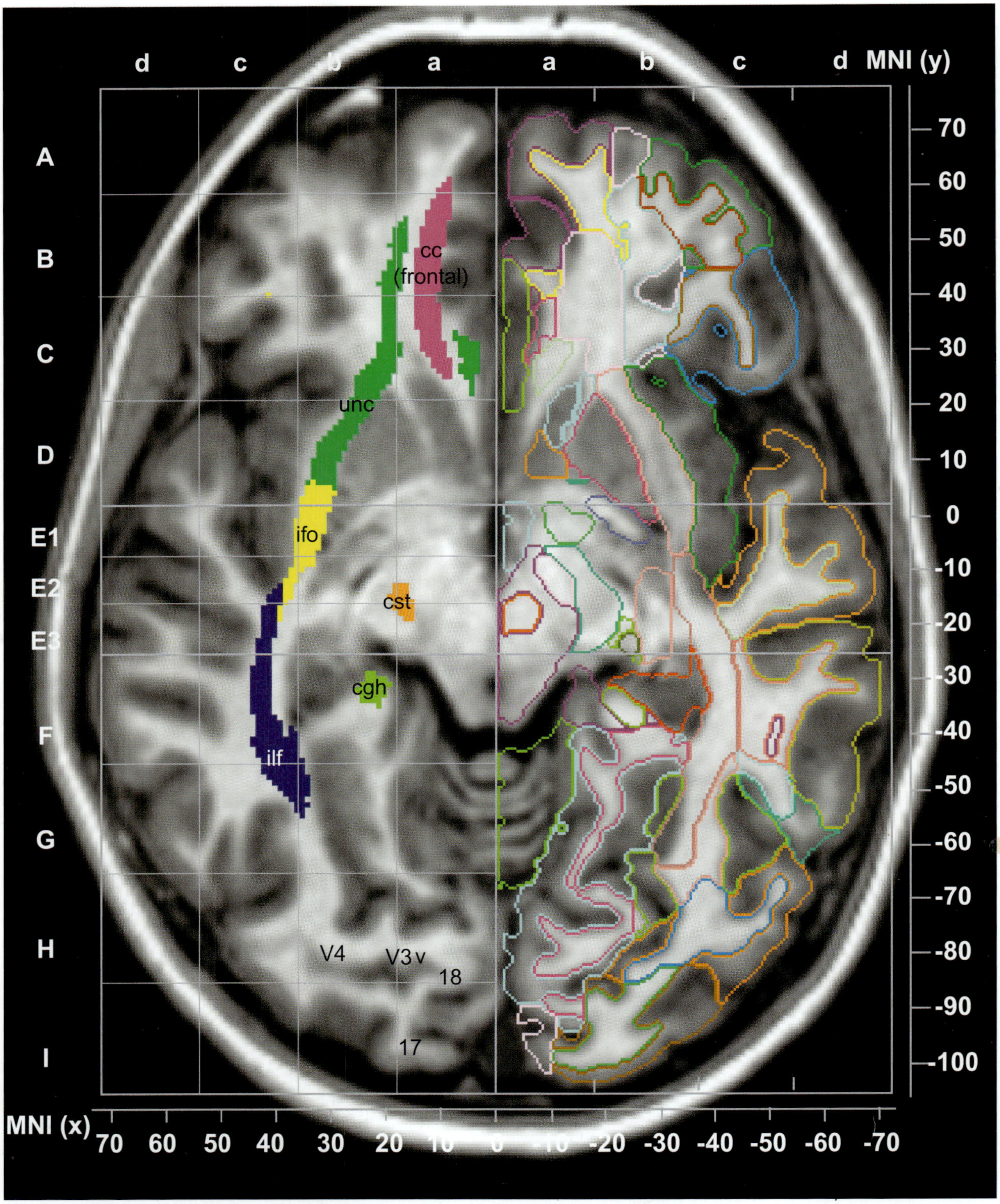

Ac - accumbens nucleus
acr - anterior corona radiata
al - ansa lenticularis
cc (frontal) - corpus callosum (connecting bilateral frontal areas)
CG - cingulate gyrus
cgc - cingulum (cingulate gyrus part)
cgh - cingulum (hippocampal part)
CH - cerebellar hemisphere
cp - cerebral peduncle
cst - corticospinal tract
Cun - cuneus
FuG - fusiform gyrus
FuG-WM - fusiform gyrus white matter
fx/st - fornix (crus) / stria terminalis (cannot be resolved with current resolution)
gcc - genu of the corpus callosum
Hi - hippocampus
Hy - hypothalamus
IFG - inferior frontal gyrus
IFG-WM - inferior frontal gyrus white matter
ifo - inferior fronto-occipital fasciculus
IG - insular gyrus
ilf - inferior longitudinal fasciculus
IOrG - intermediate orbital gyrus
IRoG - inferior rostral gyrus
ITG - inferior temporal gyrus
ITG-WM - inferior temporal gyrus white matter
lenf - lenticular fasciculus
LG - lateral geniculate nucleus (thalamus)
LOrG-WM - lateral orbital gyrus white matter
M - midbrain
MFG - middle frontal gyrus
MFG-WM - middle frontal gyrus white matter
MTG - middle temporal gyrus
MTG-WM - middle temporal gyrus white matter
O2 - middle or lateral occipital gyrus
O2-WM - middle or lateral occipital gyrus white matter
O3 - inferior occipital gyrus
O3-WM - inferior occipital gyrus white matter
O5 - lyngual gyrus
O5-WM - lyngual gyrus white matter

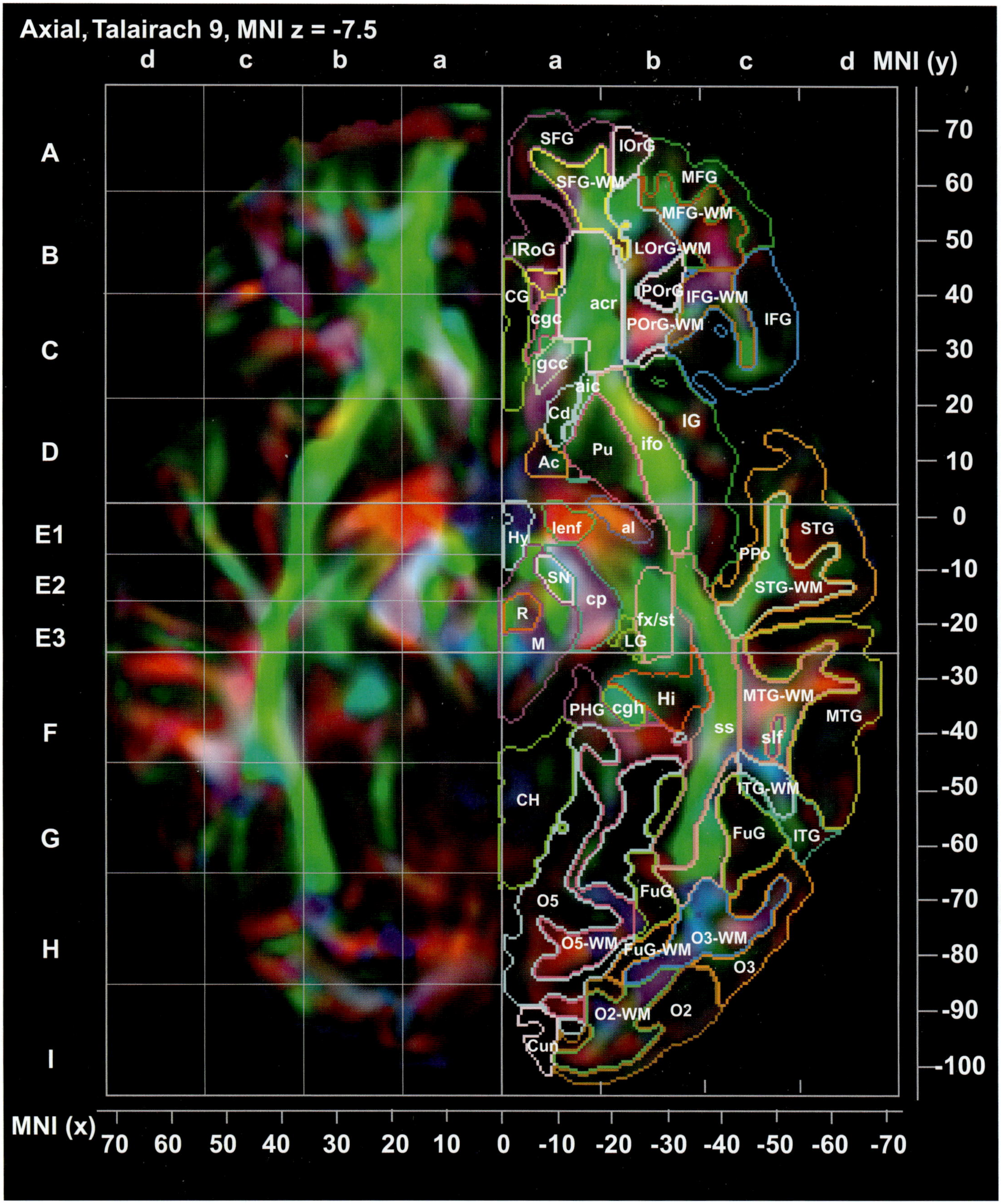

PHG	-	parahippocampal gyrus
POrG	-	posterior orbital gyrus
POrG-WM	-	posterior orbital gyrus white matter
PPo	-	planum polare
Pu	-	putamen
R	-	red nucleus
SFG	-	superior frontal gyrus
SFG-WM	-	superior frontal gyrus white matter
slf	-	superior longitudinal fasciculus
SN	-	substantia nigra
ss	-	sagittal stratum (include inferior longitidinal fasciculus and inferior fronto-occipital fasciculus)
STG	-	superior temporal gyrus
STG-WM	-	superior temporal gyrus white matter
unc	-	uncinate fasciculus

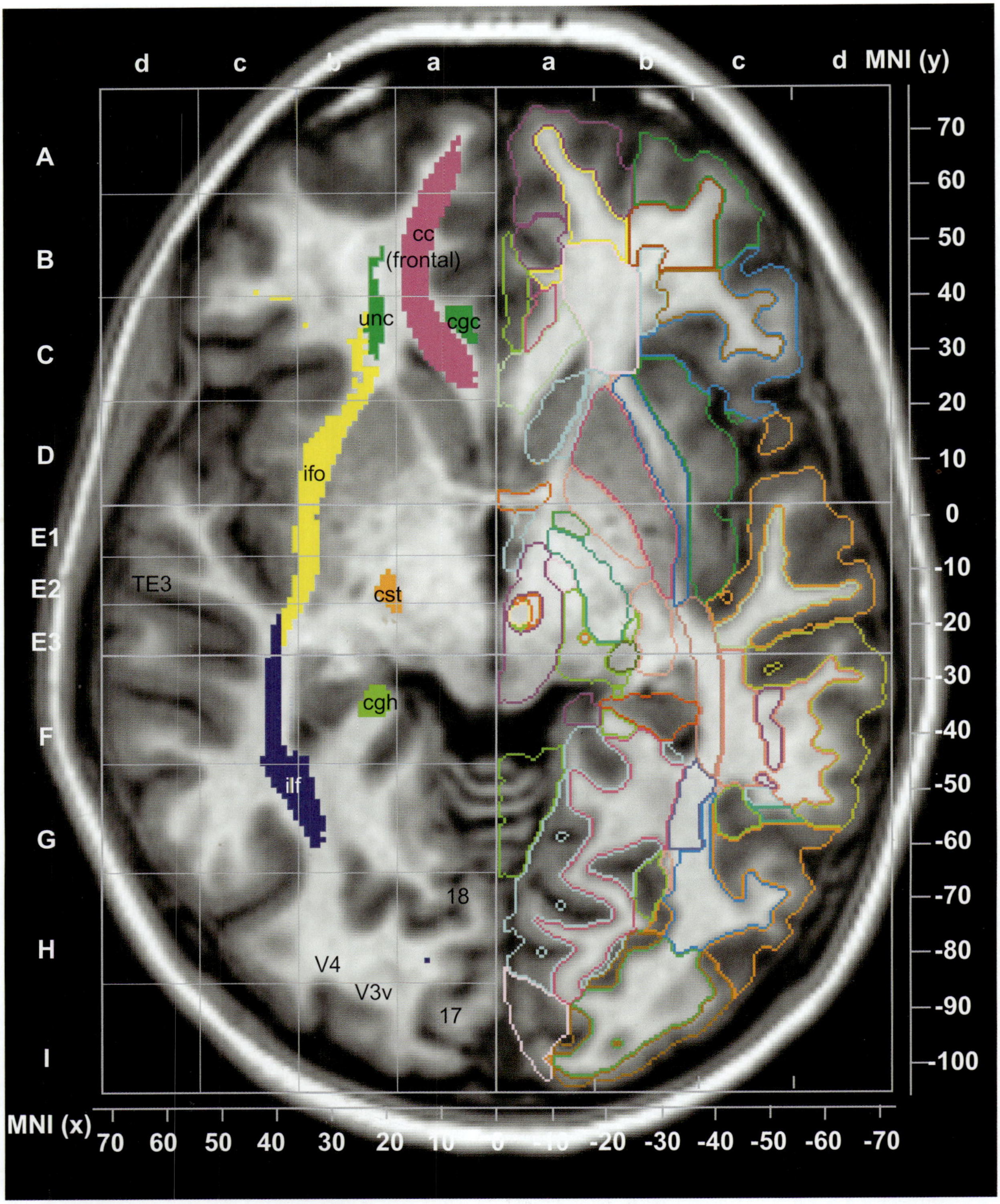

ac - anterior commissure
acr - anterior corona radiata
aic - anterior limb of the internal capsule
cc (frontal) - corpus callosum (connecting bilateral frontal areas)
Cd - caudate nucleus
CG - cingulate gyrus
cgc - cingulum (cingulate gyrus part)
cgh - cingulum (hippocampal part)
CH - cerebellar hemisphere
cp - cerebral peduncle
cst - corticospinal tract
Cun - cuneus
ec/ex/Cl - external capsule / extreme capsule / claustrum (cannot be resolved with current resolution)
FuG - fusiform gyrus
FuG-WM - fusiform gyrus white matter
fx/st - fornix (crus) / stria terminalis (cannot be resolved with current resolution)
gcc - genu of the corpus callosum
GP - globus pallidus
Hi - hippocampus
Hy - hypothalamus
IFG - inferior frontal gyrus
IFG-WM - inferior frontal gyrus white matter
ifo - inferior fronto-occipital fasciculus
IG - insular gyrus
ilf - inferior longitudinal fasciculus
IRoG - inferior rostral gyrus
ITG - inferior temporal gyrus
ITG-WM - inferior temporal gyrus white matter
lenf - lenticular fasciculus
LG - lateral geniculate nucleus (thalamus)
M - midbrain
MFG - middle frontal gyrus
MFG-WM - middle frontal gyrus white matter
MTG - middle temporal gyrus
MTG-WM - middle temporal gyrus white matter
O2 - middle or lateral occipital gyrus
O2-WM - middle or lateral occipital gyrus white matter
O3 - inferior occipital gyrus
O3-WM - inferior occipital gyrus white matter
O5 - lyngual gyrus
O5-WM - lyngual gyrus white matter

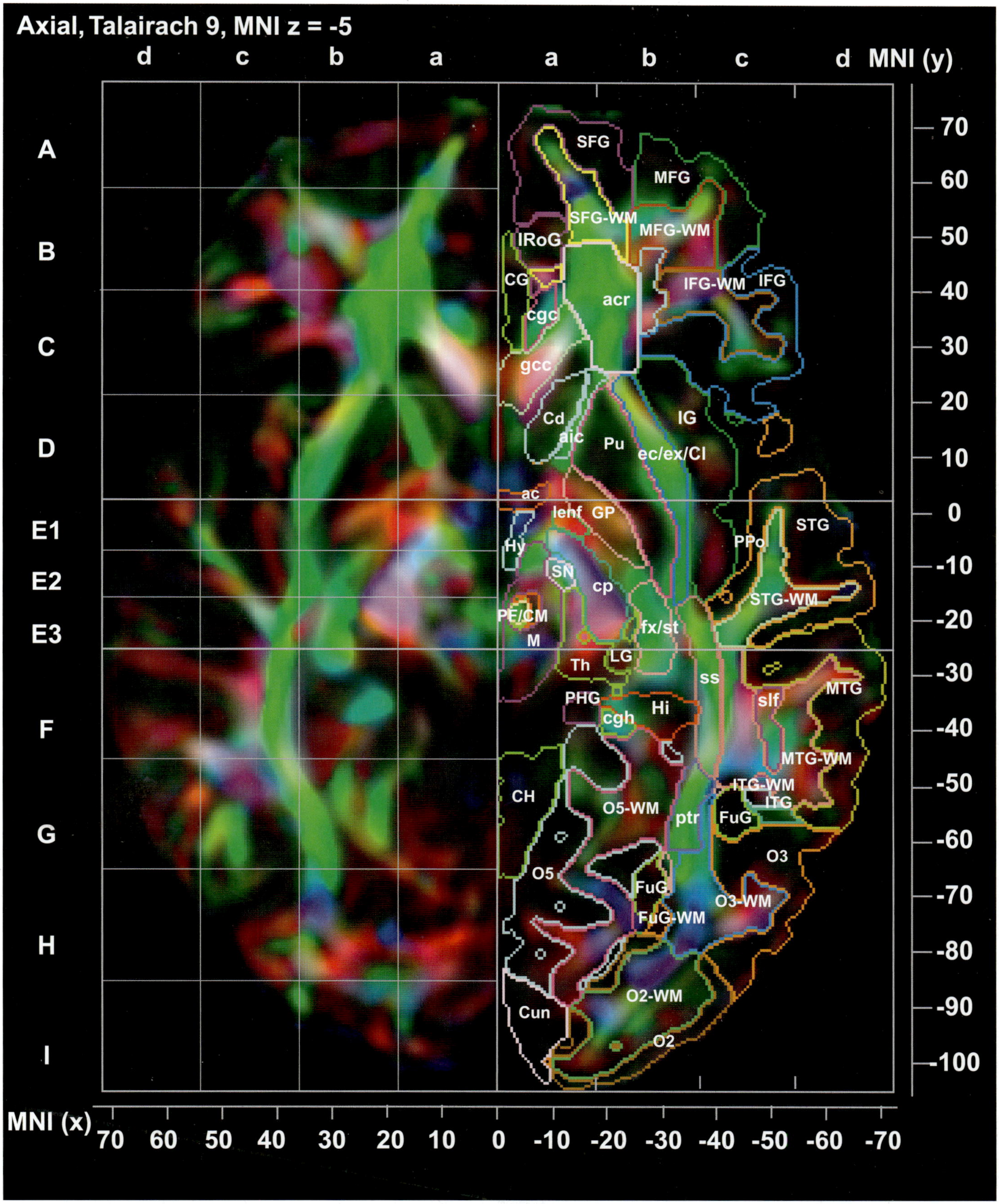

PF/CM - parafascicular nucleus / centromedian nucleus (thalamus)
PHG - parahippocampal gyrus
PPo - planum polare
ptr - posterior thalamic radiation (include the optic radiation)
Pu - putamen
SFG - superior frontal gyrus
SFG-WM - superior frontal gyrus white matter
slf - superior longitudinal fasciculus
SN - substantia nigra
ss - sagittal stratum (include inferior longitidinal fasciculus and inferior fronto-occipital fasciculus)
STG - superior temporal gyrus
STG-WM - superior temporal gyrus white matter
Th - thalamus
unc - uncinate fasciculus

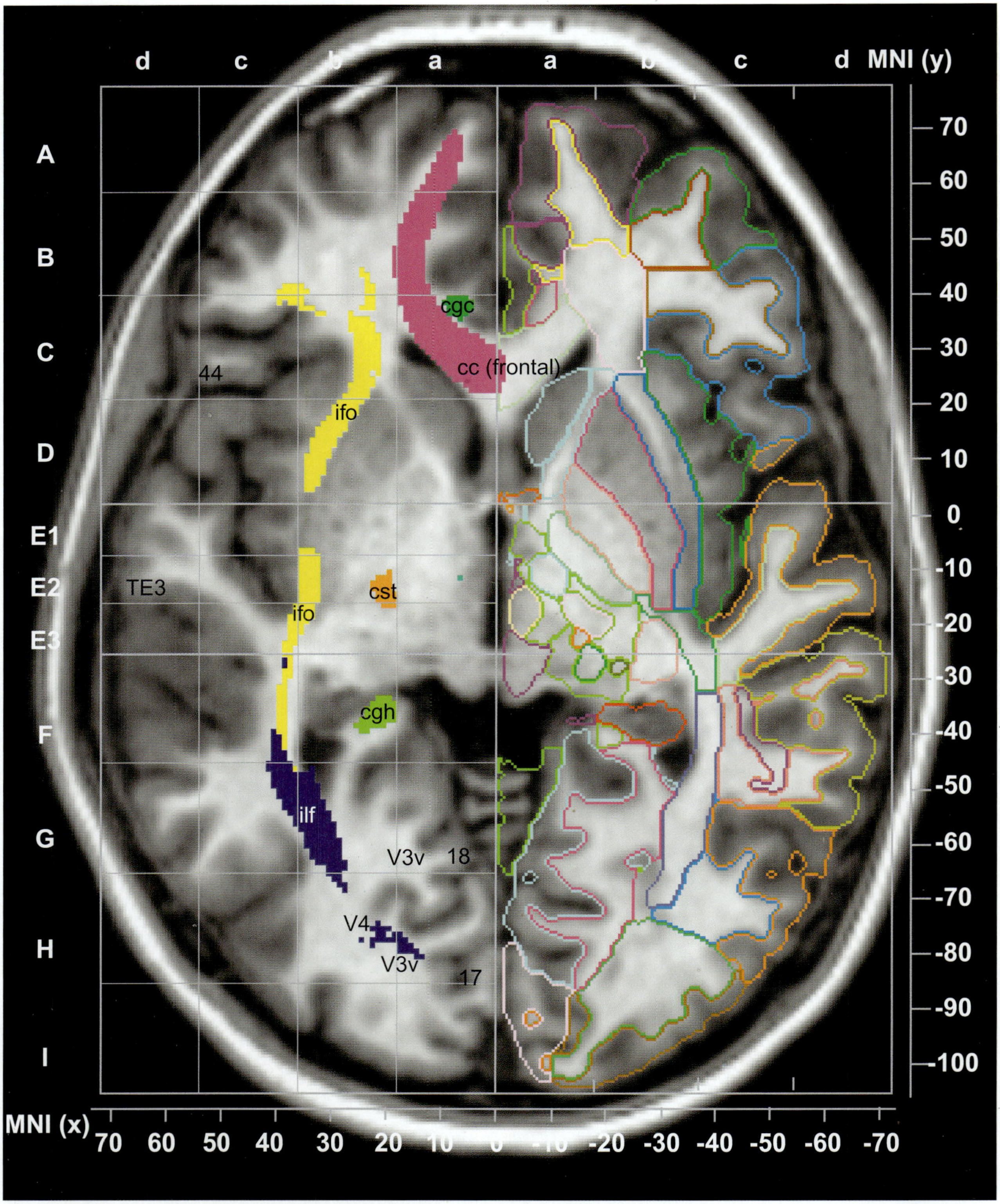

ac - anterior commissure
acr - anterior corona radiata
aic - anterior limb of the internal capsule
cc (frontal) - corpus callosum (connecting bilateral frontal areas)
Cd - caudate nucleus
CG - cingulate gyrus
cgc - cingulum (cingulate gyrus part)
cgh - cingulum (hippocampal part)
CH - cerebellar hemisphere
cst - corticospinal tract
Cun - cuneus
ec/ex/Cl - external capsule / extreme capsule / claustrum (cannot be resolved with current resolution)
Envelope - nuclei of the internal medullary lamina (thalamus)
fx/st - fornix (crus) / stria terminalis (can not be resolved with current resolution)
gcc - genu of the corpus callosum
GP - globus pallidus
Hi - hippocampus
IFG - inferior frontal gyrus
IFG-WM - inferior frontal gyrus white matter
ifo - inferior fronto-occipital fasciculus
IG - insular gyrus
ilf - inferior longitudinal fasciculus
LG - lateral geniculate nucleus (thalamus)
M - midbrain
MFG - middle frontal gyrus
MFG-WM - middle frontal gyrus white matter
MTG - middle temporal gyrus
MTG-WM - middle temporal gyrus white matter
O2 - middle or lateral occipital gyrus
O2-WM - middle or lateral occipital gyrus white matter
O3 - inferior occipital gyrus
O3-WM - inferior occipital gyrus white matter
O5 - lyngual gyrus
O5-WM - lyngual gyrus white matter
PF/CM - parafascicular nucleus / centromedian nucleus (thalamus)
PHG - parahippocampal gyrus

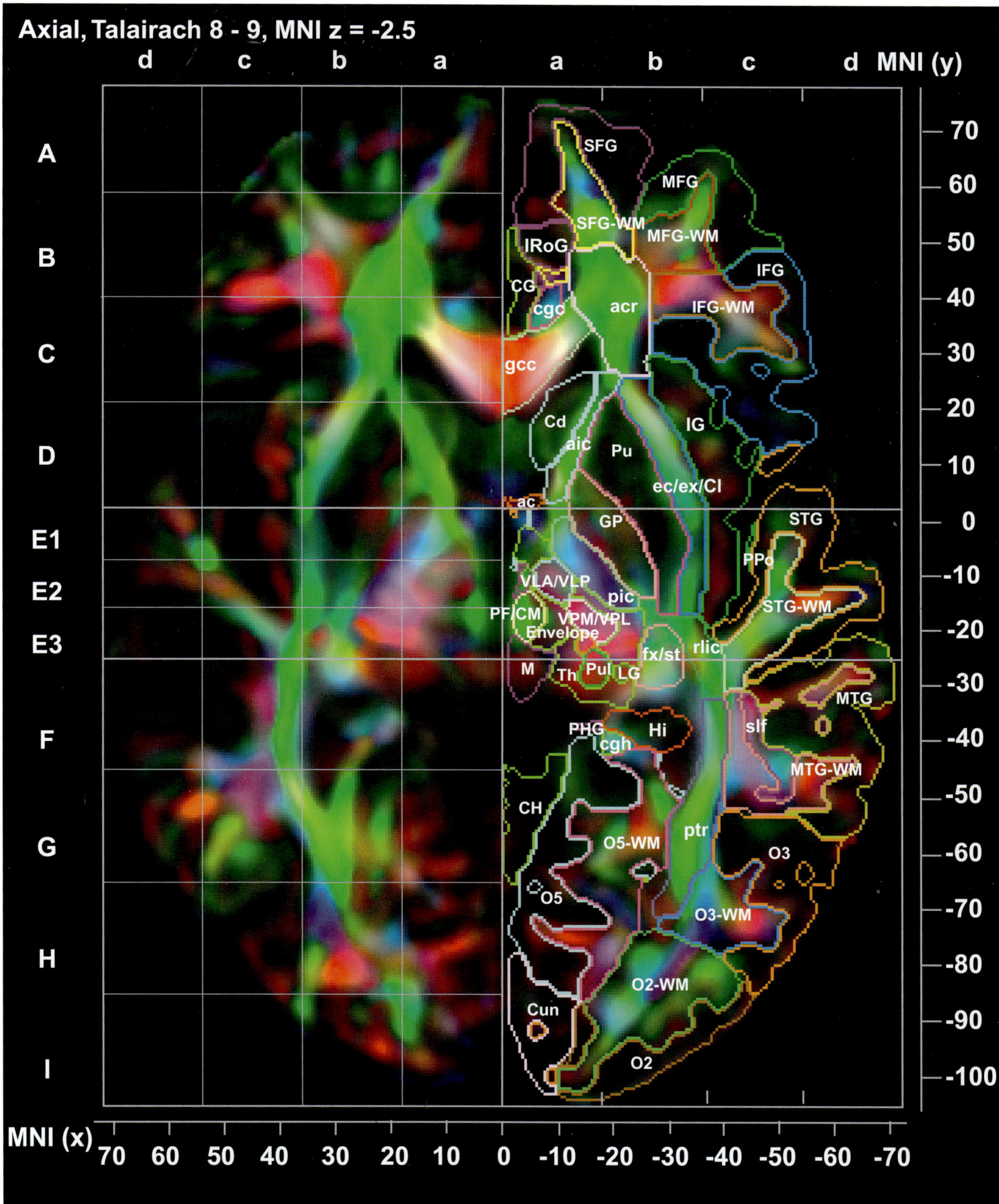

pic - posterior limb of the internal capsule
PPo - planum polare
ptr - posterior thalamic radiation (include the optic radiation)
Pu - putamen
rlic - retrolenticular part of the internal capsule

SFG - superior frontal gyrus
SFG-WM - superior frontal gyrus white matter
slf - superior longitudinal fasciculus
STG - superior temporal gyrus
STG-WM - superior temporal gyrus white matter

Th - thalamus
VLA/VLP - anterior and posterior ventrolateral thalamic nuclei
VPM/VPL - ventroposterior medialis / ventroposterior lateralis (thalamus)

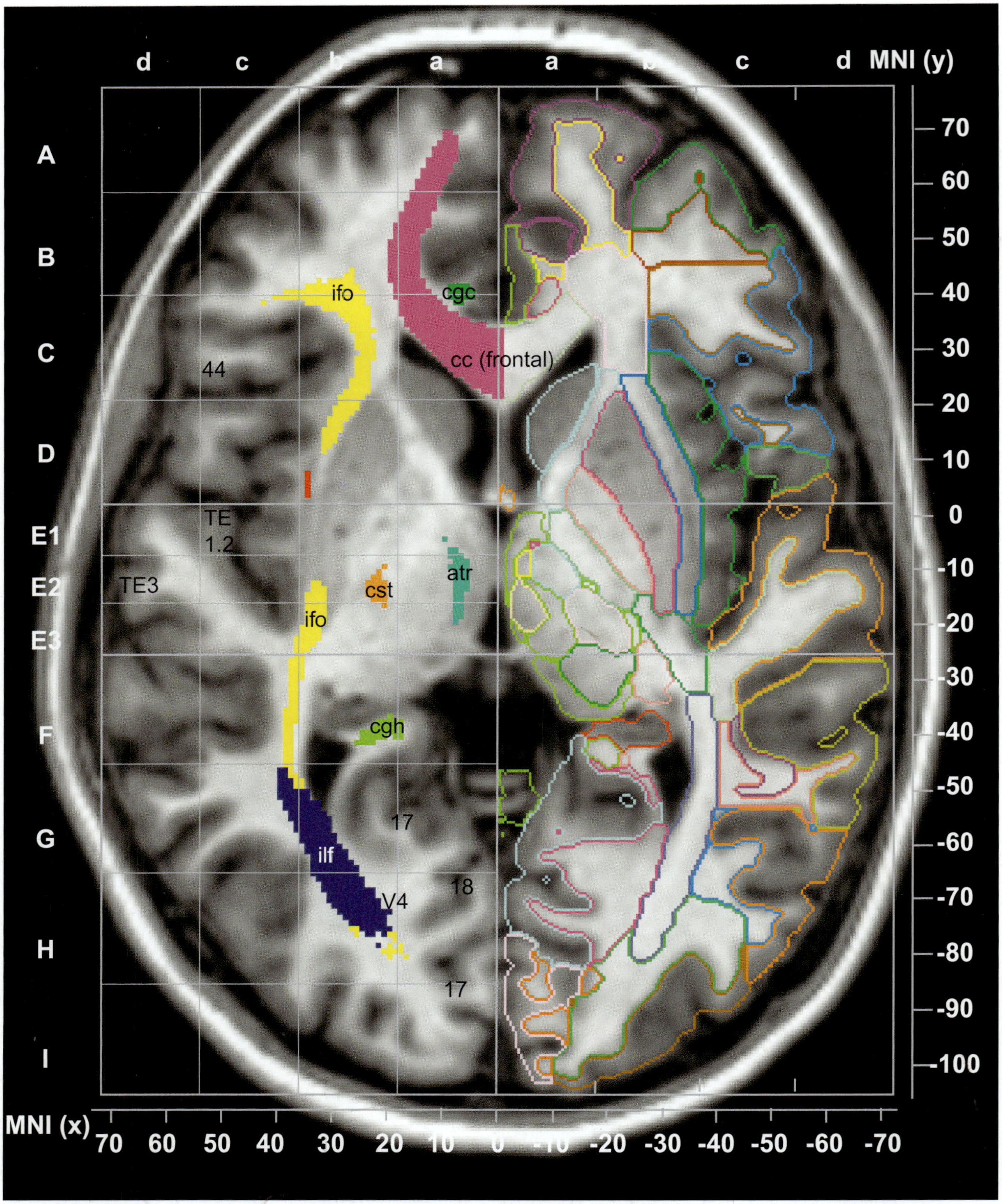

acr - anterior corona radiata
aic - anterior limb of the internal capsule
atr - anterior thalamic radiation
cc (frontal) - corpus callosum (connecting bilateral frontal areas)
Cd - caudate nucleus
CG - cingulate gyrus
cgc - cingulum (cingulate gyrus part)
cgh - cingulum (hippocampal part)
CH - cerebellar hemisphere
cst - corticospinal tract
Cun - cuneus
Cun-WM - cuneus white matter
ec/ex/Cl - external capsule / extreme capsule / claustrum (cannot be resolved with current resolution)
Envelope - nuclei of the internal medullary lamina (thalamus)
fx - fornix (column and body)
fx/st - fornix (crus) / stria terminalis (can not be resolved with current resolution)
gcc - genu of the corpus callosum
GP - globus pallidus
Hi - hippocampus
IFG - inferior frontal gyrus
IFG-WM - inferior frontal gyrus white matter
ifo - inferior fronto-occipital fasciculus
IG - insular gyrus
ilf - inferior longitudinal fasciculus
IRoG - inferior rostral gyrus
MFG - middle frontal gyrus
MFG-WM - middle frontal gyrus white matter
MTG - middle temporal gyrus
MTG-WM - middle temporal gyrus white matter
O2 - middle or lateral occipital gyrus
O2-WM - middle or lateral occipital gyrus white matter
O3 - inferior occipital gyrus
O3-WM - inferior occipital gyrus white matter
O5 - lyngual gyrus
O5-WM - lyngual gyrus white matter

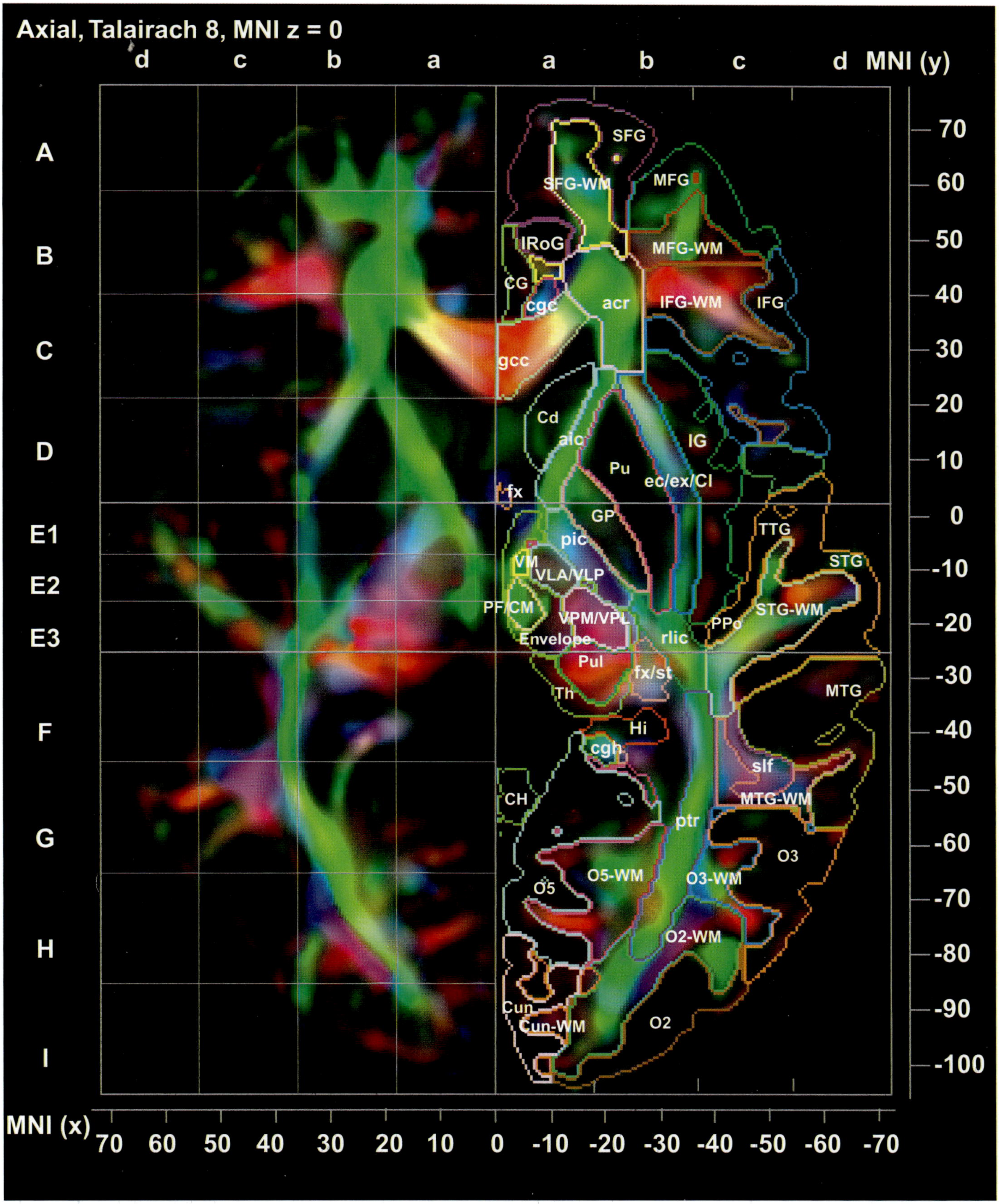

PF/CM - parafascicular nucleus / centromedian nucleus (thalamus)
pic - posterior limb of the internal capsule
PPo - planum polare
ptr - posterior thalamic radiation (include the optic radiation)
Pu - putamen
Pul - pulvinar (thalamus)
rlic - retrolenticular part of the internal capsule
SFG - superior frontal gyrus
SFG-WM - superior frontal gyrus white matter
slf - superior longitudinal fasciculus
STG - superior temporal gyrus
STG-WM - superior temporal gyrus white matter
Th - thalamus
VLA/VLP - anterior and posterior ventrolateral thalamic nuclei
VM - ventromedial thalamic nucleus
VPM/VPL - ventroposterior medialis / ventroposterior lateralis (thalamus)

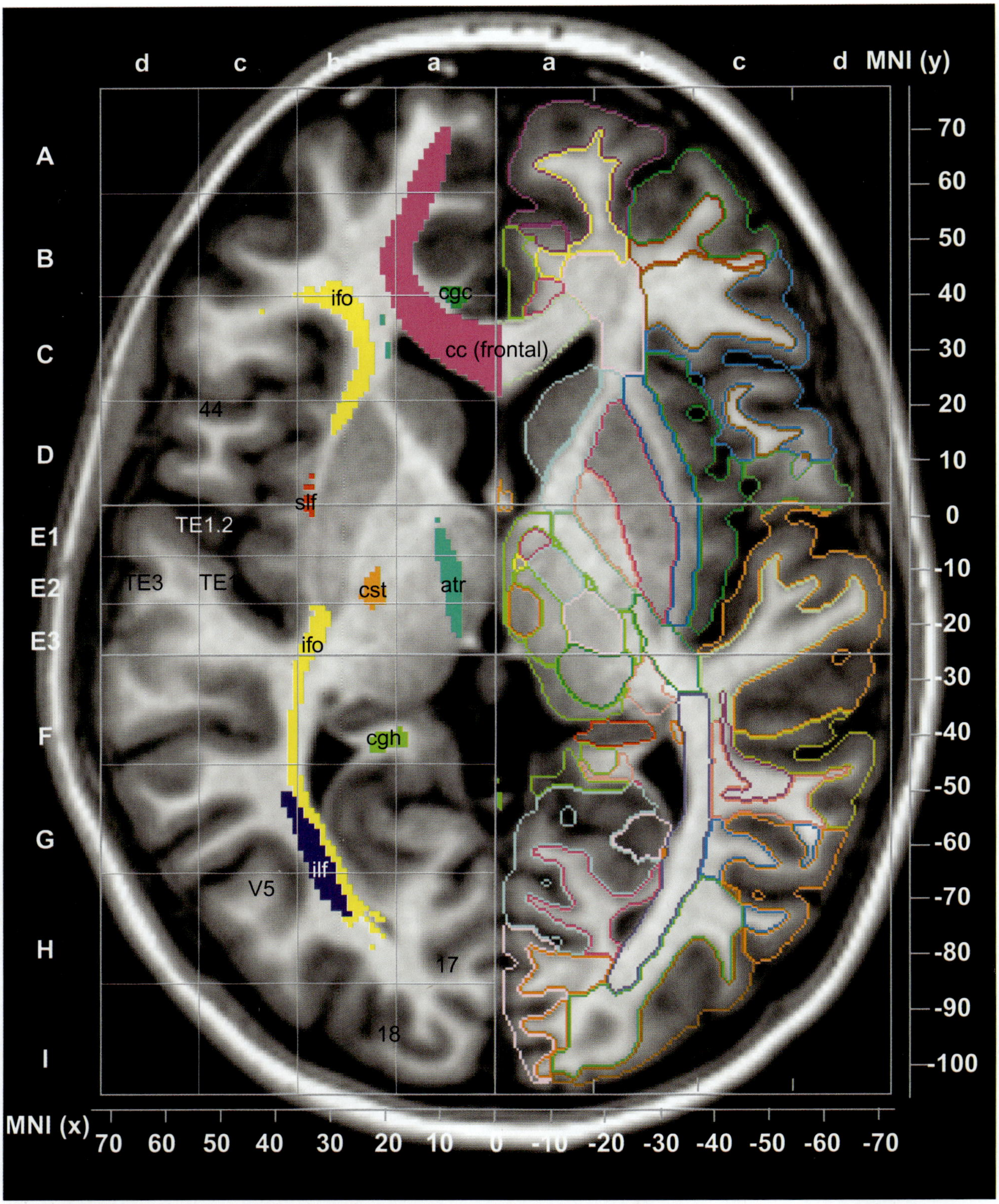

acr - anterior corona radiata
aic - anterior limb of the internal capsule
atr - anterior thalamic radiation
cc (frontal) - corpus callosum (connecting bilateral frontal areas)
Cd - caudate nucleus
CG - cingulate gyrus
cgc - cingulum (cingulate gyrus part)
cgh - cingulum (hippocampal part)
cst - corticospinal tract
Cun - cuneus
Cun-WM - cuneus white matter
ec/ex/Cl - external capsule / extreme capsule / claustrum (cannot be resolved with current resolution)

Envelope - nuclei of the internal medullary lamina (thalamus)
fx - fornix (column and body)
fx/st - fornix (crus) / stria terminalis (cannot be resolved with current resolution)
gcc - genu of the corpus callosum
GP - globus pallidus
Hi - hippocampus
IFG - inferior frontal gyrus
IFG-WM - inferior frontal gyrus white matter
ifo - inferior fronto-occipital fasciculus
IG - insular gyrus
ilf - inferior longitudinal fasciculus
IRoG - inferior rostral gyrus

MD - mediodorsal thalamic nucleus
MFG - middle frontal gyrus
MFG-WM - middle frontal gyrus white matter
MTG - middle temporal gyrus
MTG-WM - middle temporal gyrus white matter
O2 - middle or lateral occipital gyrus
O2-WM - middle or lateral occipital gyrus white matter
O3 - inferior occipital gyrus
O3-WM - inferior occipital gyrus white matter
O5 - lyngual gyrus
O5-WM - lyngual gyrus white matter
pic - posterior limb of the internal capsule
PrG - precentral gyrus

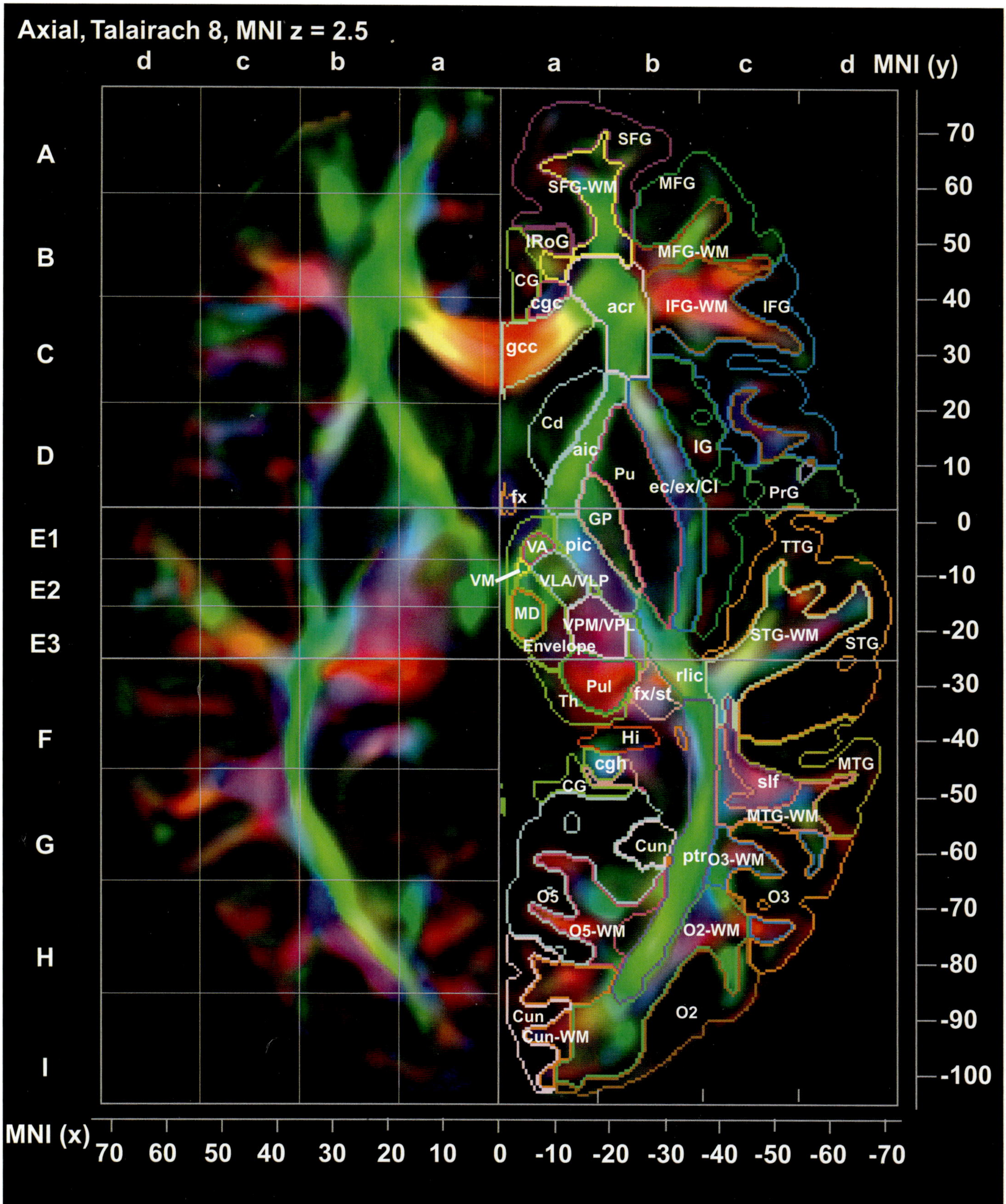

ptr - posterior thalamic radiation (include the optic radiation)
Pu - putamen
Pul - pulvinar (thalamus)
rlic - retrolenticular part of the internal capsule
SFG - superior frontal gyrus
SFG-WM - superior frontal gyrus white matter
slf - superior longitudinal fasciculus
STG - superior temporal gyrus
STG-WM - superior temporal gyrus white matter
Th - thalamus
TTG - transverse temporal gyrus
VA - ventroanterior thalamic nucleus
VLA/VLP - anterior and posterior ventrolateral thalamic nuclei
VM - ventromedial thalamic nucleus
VPM/VPL - ventroposterior medialis / ventroposterior lateralis (thalamus)

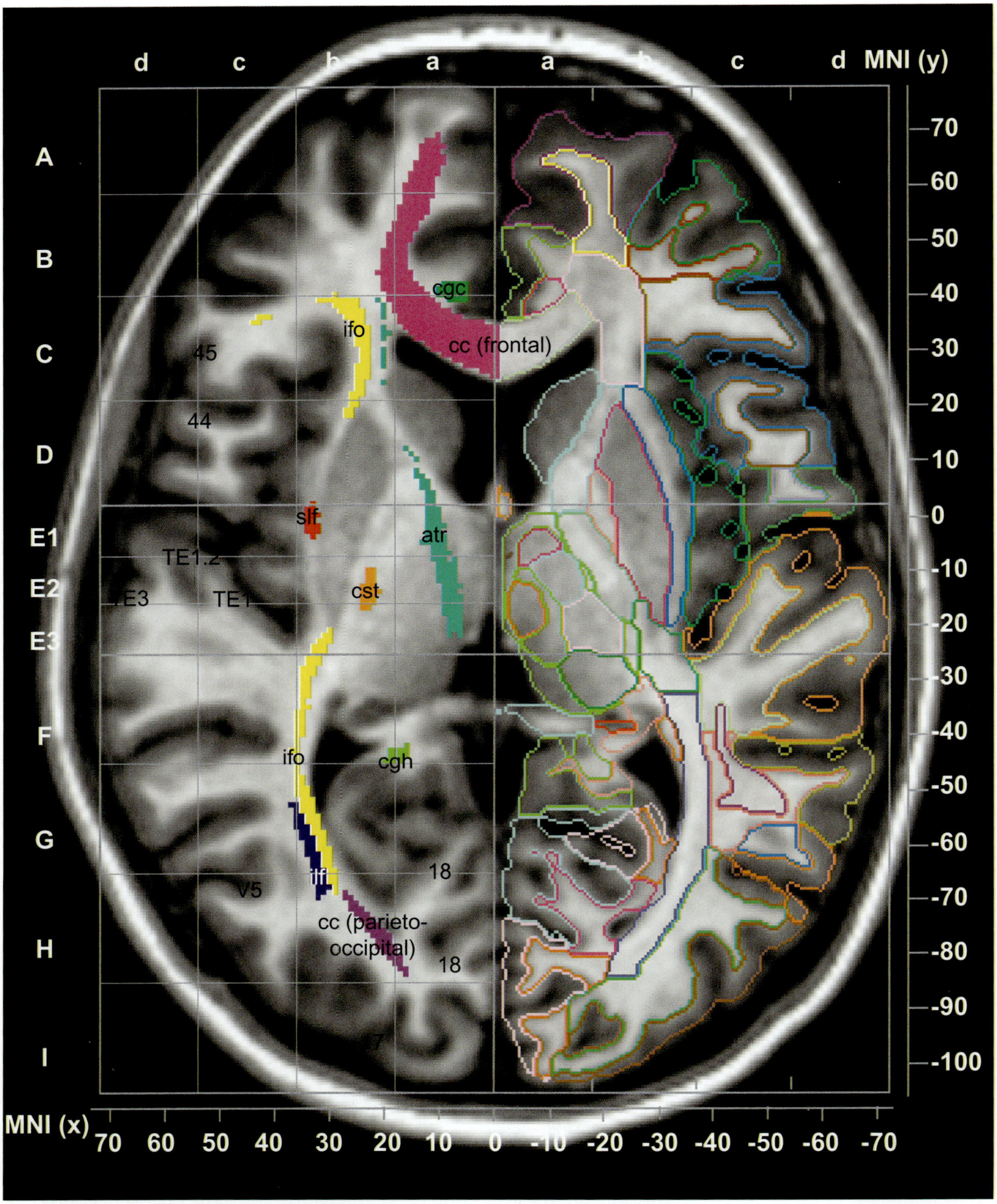

acr - anterior corona radiata
aic - anterior limb of the internal capsule
atr - anterior thalamic radiation
cc (frontal) - corpus callosum (connecting bilateral frontal areas)
cc (parieto-occipital) - corpus callosum (connecting bilateral parieto-occipital areas)
Cd - caudate nucleus
CG - cingulate gyrus
cgc - cingulum (cingulate gyrus part)
cgh - cingulum (hippocampal part)
CG-WM - cingulate gyrus white matter
cst - corticospinal tract
Cun - cuneus
Cun-WM - cuneus white matter

ec/ex/Cl - external capsule / extreme capsule / claustrum (cannnot be resolved with current resolution)
Envelope - nuclei of the internal medullary lamina (thalamus)
fx - fornix (column and body)
fx/st - fornix (crus) / stria terminalis (cannot be resolved with current resolution)
gcc - genu of the corpus callosum
GP - globus pallidus
Hi - hippocampus
IFG - inferior frontal gyrus
IFG-WM - inferior frontal gyrus white matter
ifo - inferior fronto-occipital fasciculus
IG - insular gyrus

ilf - inferior longitudinal fasciculus
MD - mediodorsal thalamic nucleus
MFG - middle frontal gyrus
MFG-WM - middle frontal gyrus white matter
MTG - middle temporal gyrus
MTG-WM - middle temporal gyrus white matter
O2 - middle or lateral occipital gyrus
O2-WM - middle or lateral occipital gyrus white matter
O3 - inferior occipital gyrus
O3-WM - inferior occipital gyrus white matter
O5 - lyngual gyrus
O5-WM - lyngual gyrus white matter
pic - posterior limb of the internal capsule
PrG - precentral gyrus
PrG-WM - precentral gyrus white matter

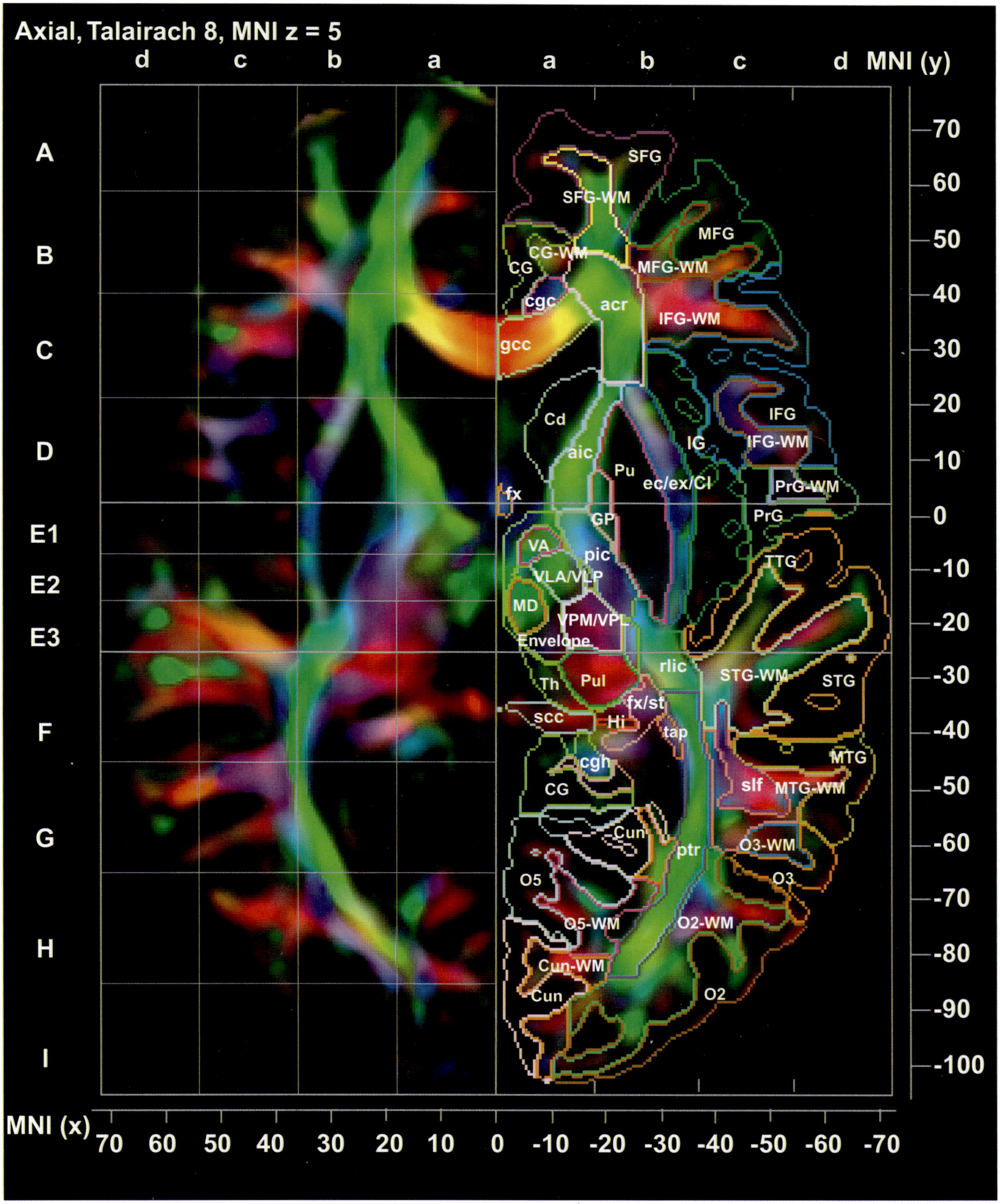

ptr - posterior thalamic radiation (include the optic radiation)
Pu - putamen
Pul - pulvinar (thalamus)
rlic - retrolenticular part of the internal capsule
scc - splenium of the corpus callosum

SFG - superior frontal gyrus
SFG-WM - superior frontal gyrus white matter
slf - superior longitudinal fasciculus
STG - superior temporal gyrus
STG-WM - superior temporal gyrus white matter
tap - tapetum of the corpus callosum

Th - thalamus
TTG - transverse temporal gyrus
VA - ventroanterior thalamic nucleus
VLA/VLP - anterior and posterior ventrolateral thalamic nuclei
VPM/VPL - ventroposterior medialis / ventroposterior lateralis (thalamus)

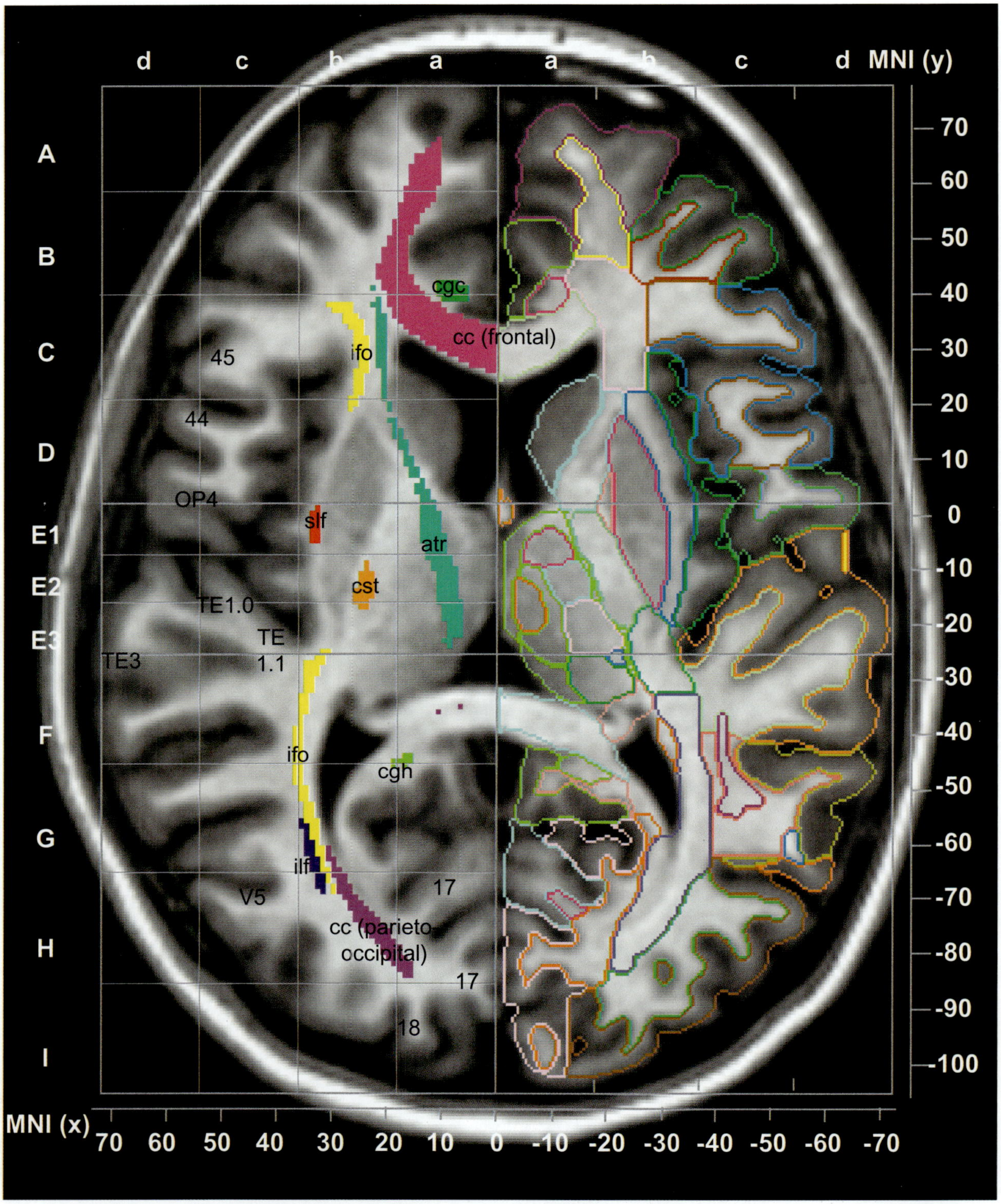

acr - anterior corona radiata
aic - anterior limb of the internal capsule
atr - anterior thalamic radiation
cc (frontal) - corpus callosum (connecting bilateral frontal areas)
cc (parieto-occipital) - corpus callosum (connecting bilateral parieto-occipital areas)
Cd - caudate nucleus
CG - cingulate gyrus
cgc - cingulum (cingulate gyrus part)
cgh - cingulum (hippocampal part)
CG-WM - cingulate gyrus white matter
cst - corticospinal tract
Cun - cuneus
Cun-WM - cuneus white matter
ec/ex/Cl - external capsule / extreme capsule / claustrum (cannot be resolved with current resolution)
Envelope - nuclei of the internal medullary lamina (thalamus)
fx - fornix (column and body)
fx/st - fornix (crus) / stria terminalis (cannot be resolved with current resolution)
gcc - genu of the corpus callosum
GP - globus pallidus
IFG - inferior frontal gyrus
IFG-WM - inferior frontal gyrus white matter
ifo - inferior fronto-occipital fasciculus
IG - insular gyrus
ilf - inferior longitudinal fasciculus
MD - mediodorsal thalamic nucleus
MFG - middle frontal gyrus
MFG-WM - middle frontal gyrus white matter
MTG - middle temporal gyrus
MTG-WM - middle temporal gyrus white matter
O2 - middle or lateral occipital gyrus
O2-WM - middle or lateral occipital gyrus white matter
O3 - inferior occipital gyrus
O3-WM - inferior occipital gyrus white matter
O5 - lyngual gyrus
O5-WM - lyngual gyrus white matter
pic - posterior limb of the internal capsule
PrG - precentral gyrus
PrG-WM - precentral gyrus white matter

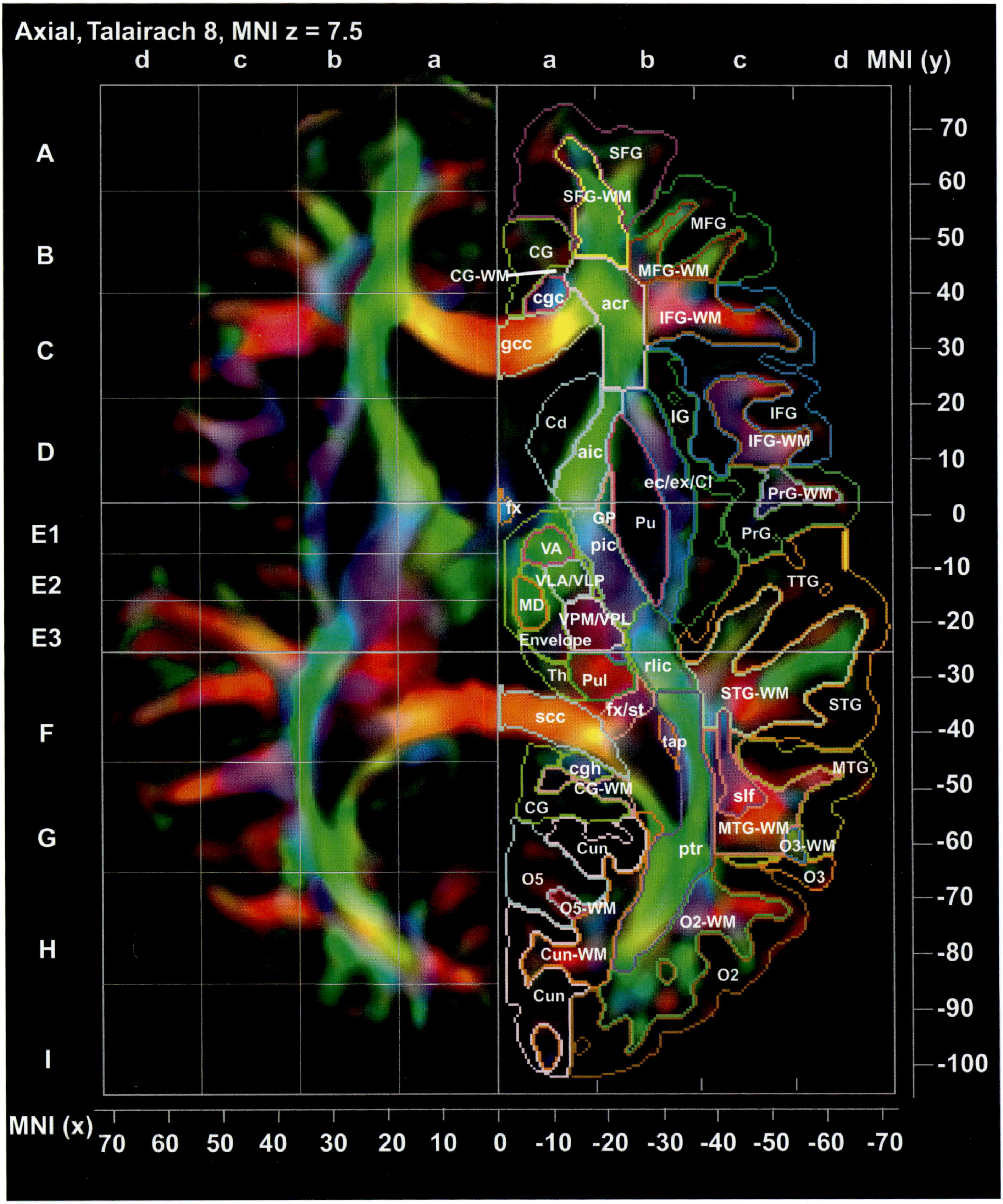

ptr - posterior thalamic radiation (include the optic radiation)
Pu - putamen
Pul - pulvinar (thalamus)
rlic - retrolenticular part of the internal capsule
scc - splenium of the corpus callosum
SFG - superior frontal gyrus
SFG-WM - superior frontal gyrus white matter
slf - superior longitudinal fasciculus
STG - superior temporal gyrus
STG-WM - superior temporal gyrus white matter
tap - tapetum of the corpus callosum
Th - thalamus
TTG - transverse temporal gyrus
VA - ventroanterior thalamic nucleus
VLA/VLP - anterior and posterior ventrolateral thalamic nuclei
VPM/VPL - ventroposterior medialis / ventroposterior lateralis (thalamus)

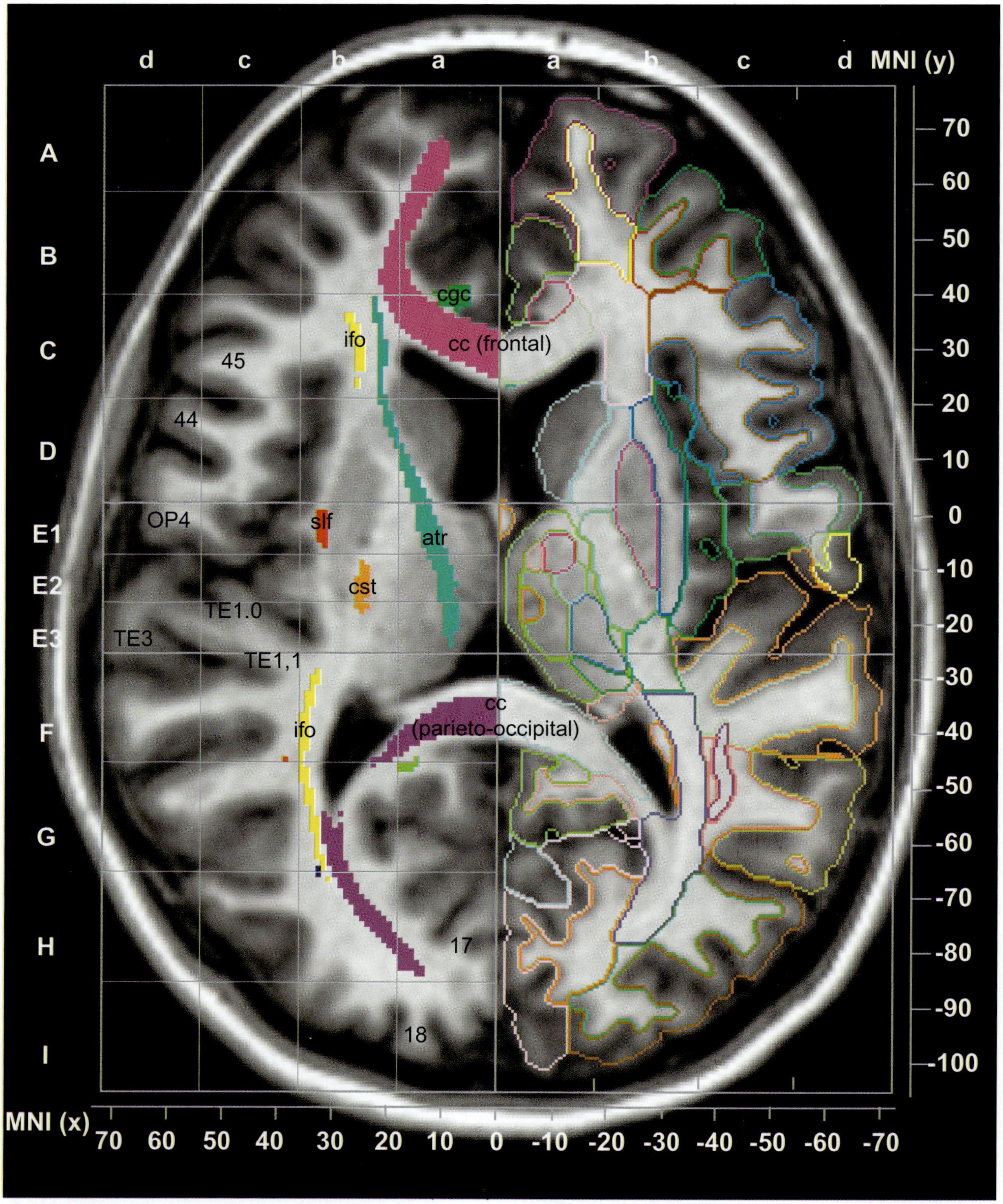

acr - anterior corona radiata
Ath - anterior nuclear group (thalamus)
atr - anterior thalamic radiation
cc (frontal) - corpus callosum (connecting bilateral frontal areas)
cc (parieto-occipital) - corpus callosum (connecting bilateral parieto-occipital areas)
Cd - caudate nucleus
CG - cingulate gyrus
cgc - cingulum (cingulate gyrus part)
cgh - cingulum (hippocampal part)
CG-WM - cingulate gyrus white matter
cst - corticospinal tract
Cun - cuneus
Cun-WM - cuneus white matter
ec/ex/Cl - external capsule / extreme capsule / claustrum (cannot be resolved with current resolution)
Envelope - nuclei of the internal medullary lamina (thalamus)
fx - fornix (column and body)
fx/st - fornix (crus) / stria terminalis (cannot be resolved with current resolution)
gcc - genu of the corpus callosum
IFG - inferior frontal gyrus
IFG-WM - inferior frontal gyrus white matter
ifo - inferior fronto-occipital fasciculus
IG - insular gyrus
MD - mediodorsal thalamic nucleus
MFG - middle frontal gyrus
MFG-WM - middle frontal gyrus white matter
MTG - middle temporal gyrus
MTG-WM - middle temporal gyrus white matter
O2 - middle or lateral occipital gyrus
O2-WM - middle or lateral occipital gyrus white matter
O3 - inferior occipital gyrus
O5 - lyngual gyrus
PoG - postcentral gyrus
PrG - precentral gyrus
PrG-WM - precentral gyrus white matter
ptr - posterior thalamic radiation (include the optic radiation)
Pu - putamen
Pul - pulvinar (thalamus)

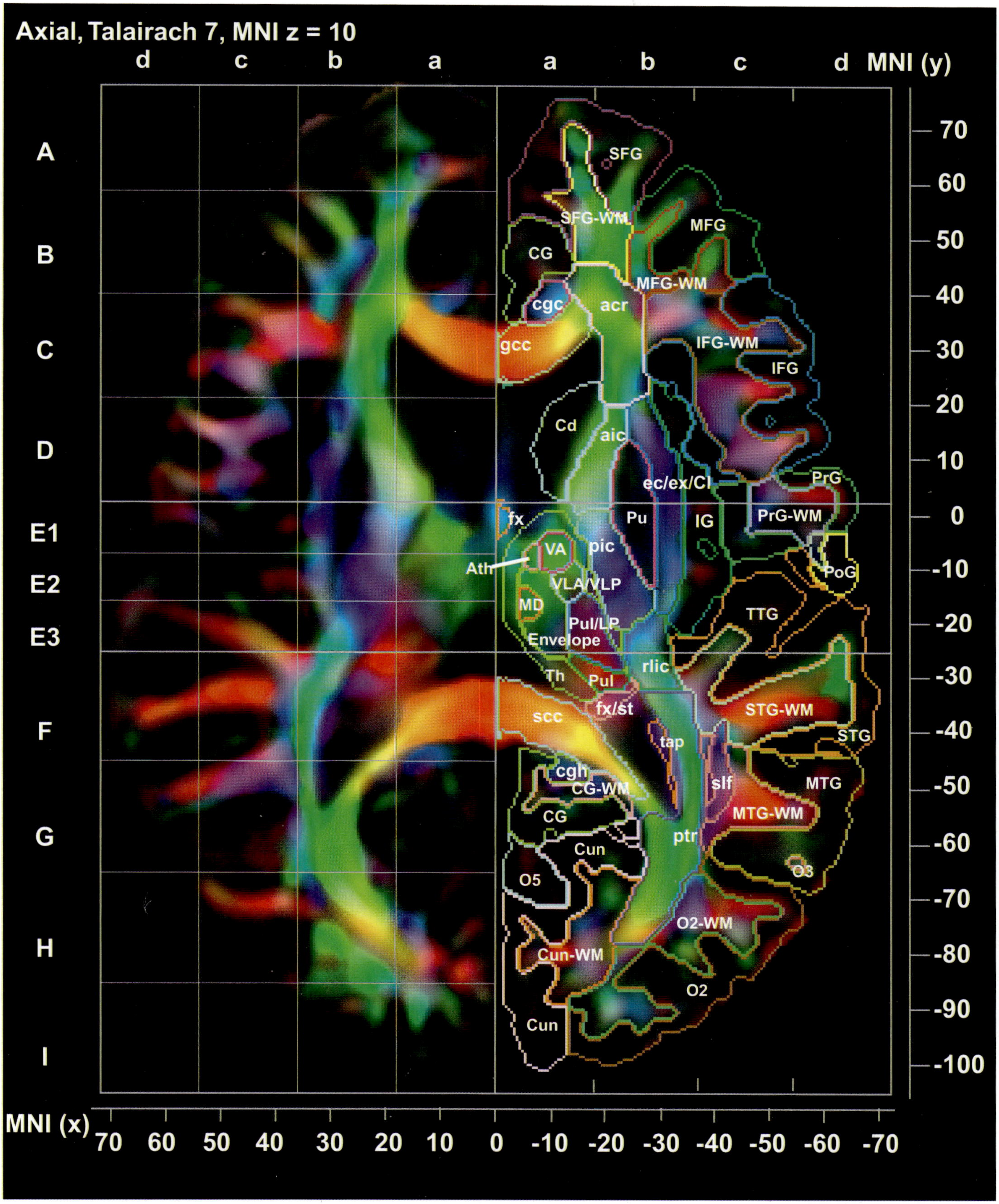

Pul/LP - pulvinar / lateral posterior thalamic nucleus
scc - splenium of the corpus callosum
SFG - superior frontal gyrus
SFG-WM - superior frontal gyrus white matter
slf - superior longitudinal fasciculus
STG - superior temporal gyrus
STG-WM - superior temporal gyrus white matter
tap - tapetum of the corpus callosum
Th - thalamus
TTG - transverse temporal gyrus
VA - ventroanterior thalamic nucleus
VLA/VLP - anterior and posterior ventrolateral thalamic nuclei

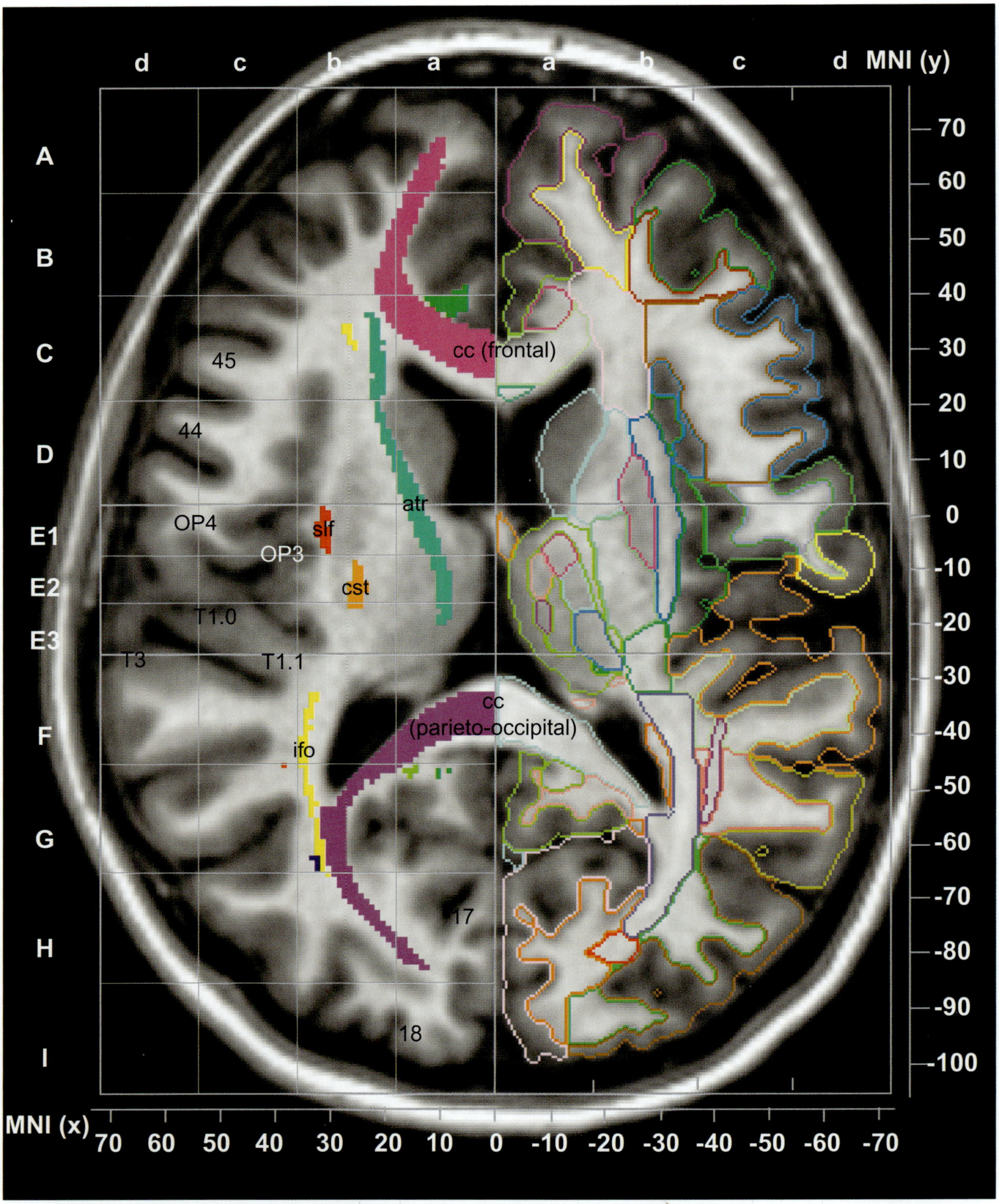

acr - anterior corona radiata
aic - anterior limb of the internal capsule
Ath - anterior nuclear group (thalamus)
atr - anterior thalamic radiation
cc (frontal) - corpus callosum (connecting bilateral frontal areas)
cc (parieto-occipital) - corpus callosum (connecting bilateral parieto-occipital areas)
Cd - caudate nucleus
CG - cingulate gyrus
cgc - cingulum (cingulate gyrus part)
cgh - cingulum (hippocampal part)
CG-WM - cingulate gyrus white matter
cst - corticospinal tract
Cun - cuneus
Cun-WM - cuneus white matter
ec/ex/Cl - external capsule / extreme capsule / claustrum (cannot be resolved with current resolution)
Envelope - nuclei of the internal medullary lamina (thalamus)
fx - fornix (column and body)
fx/st - fornix (crus) / stria terminalis (cannot be resolved with current resolution)
gcc - genu of the corpus callosum
IFG - inferior frontal gyrus
IFG-WM - inferior frontal gyrus white matter
ifo - inferior fronto-occipital fasciculus
IG - insular gyrus
LD - laterodorsal thalamic nucleus
MFG - middle frontal gyrus
MFG-WM - middle frontal gyrus white matter
MTG - middle temporal gyrus
MTG-WM - middle temporal gyrus white matter
O1-WM - superior occipital gyrus white matter
O2 - middle or lateral occipital gyrus
O2-WM - middle or lateral occipital gyrus white matter
pic - posterior limb of the internal capsule
PoG - postcentral gyrus
PoG-WM - postcentral gyrus white matter
PrG - precentral gyrus
PrG-WM - precentral gyrus white matter
PTe - planum temporale

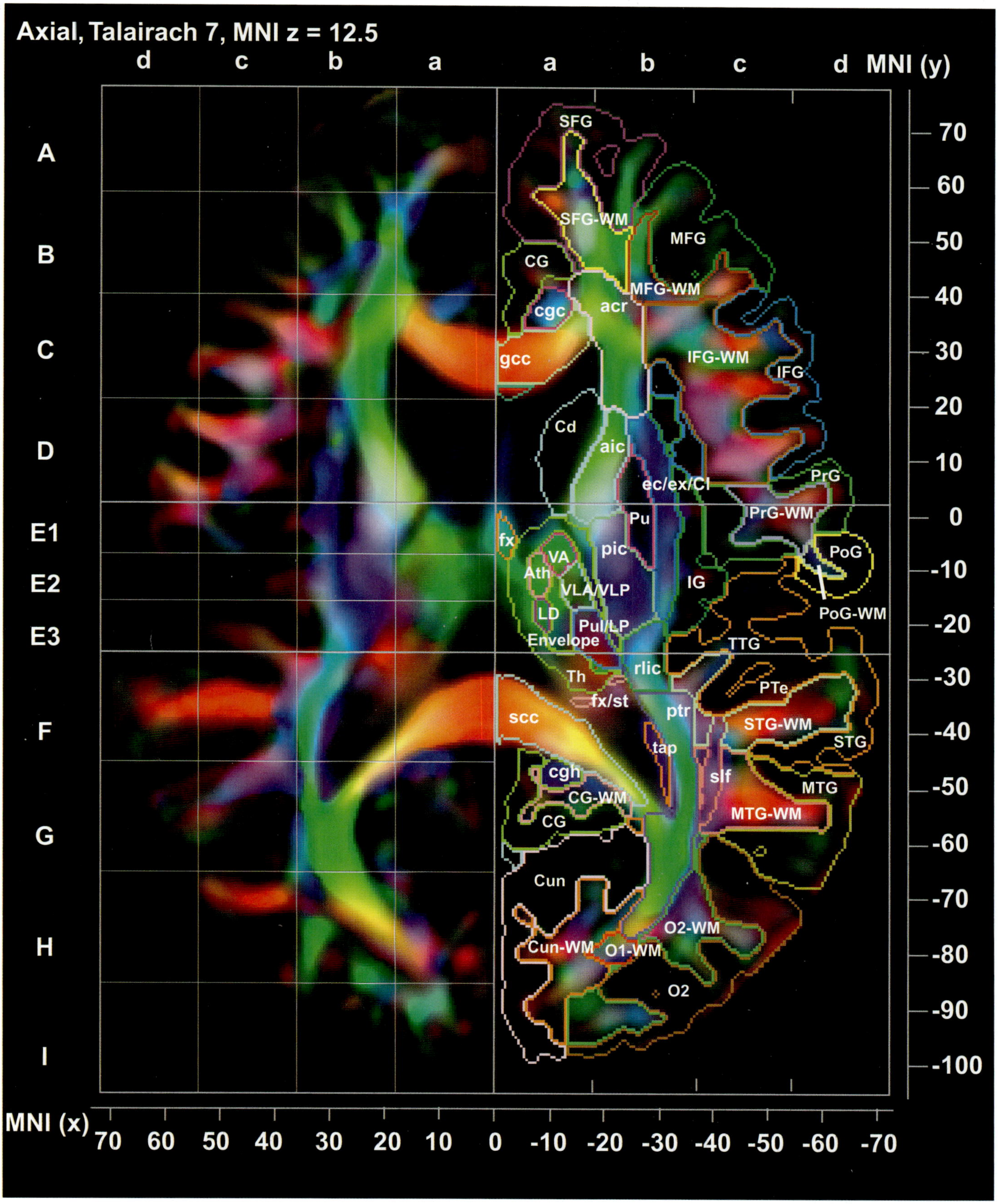

ptr - posterior thalamic radiation (include the optic radiation)
Pu - putamen
Pul/LP - pulvinar / lateral posterior thalamic nucleus
rlic - retrolenticular part of the internal capsule
scc - splenium of the corpus callosum
SFG - superior frontal gyrus
SFG-WM - superior frontal gyrus white matter
slf - superior longitudinal fasciculus
STG - superior temporal gyrus
STG-WM - superior temporal gyrus white matter
tap - tapetum of the corpus callosum
Th - thalamus
TTG - transverse temporal gyrus
VA - ventroanterior thalamic nucleus
VLA/VLP - anterior and posterior ventrolateral thalamic nuclei

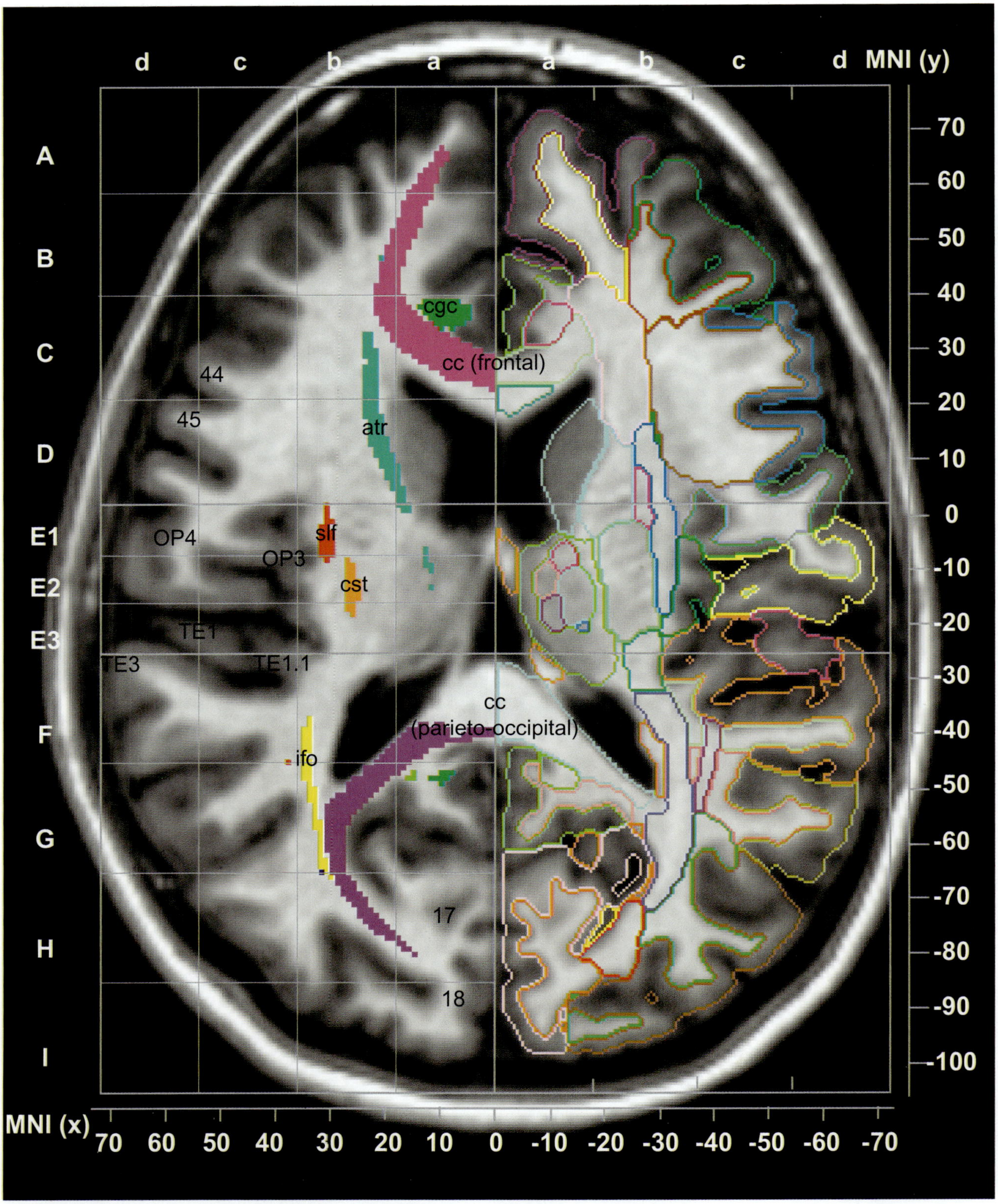

acr	- anterior corona radiata	CG-WM	- cingulate gyrus white matter	LD	- laterodorsal thalamic nucleus
aic	- anterior limb of the internal capsule	cst	- corticospinal tract	MFG	- middle frontal gyrus
Ath	- anterior nuclear group (thalamus)	Cun	- cuneus	MFG-WM	- middle frontal gyrus white matter
atr	- anterior thalamic radiation	Cun-WM	- cuneus white matter	MTG	- middle temporal gyrus
bcc	- body of the corpus callosum	ec/ex/Cl	- external capsule / extreme capsule / claustrum (cannot be resolved with current resolution)	MTG-WM	- middle temporal gyrus white matter
cc (frontal)	- corpus callosum (connecting bilateral frontal areas)			O1	- superior occipital gyrus
				O1-WM	- superior occipital gyrus white matter
cc (parieto-occipital)	- corpus callosum (connecting bilateral parieto-occipital areas)	fx	- fornix (column and body)	O2	- middle or lateral occipital gyrus
		gcc	- genu of the corpus callosum	O2-WM	- middle or lateral occipital gyrus white matter
Cd	- caudate nucleus	IFG	- inferior frontal gyrus	pic	- posterior limb of the internal capsule
CG	- cingulate gyrus	IFG-WM	- inferior frontal gyrus white matter	PoG	- postcentral gyrus
cgc	- cingulum (cingulate gyrus part)	ifo	- inferior fronto-occipital fasciculus	PoG-WM	- postcentral gyrus white matter
cgh	- cingulum (hippocampal part)	IG	- insular gyrus	PrG	- precentral gyrus

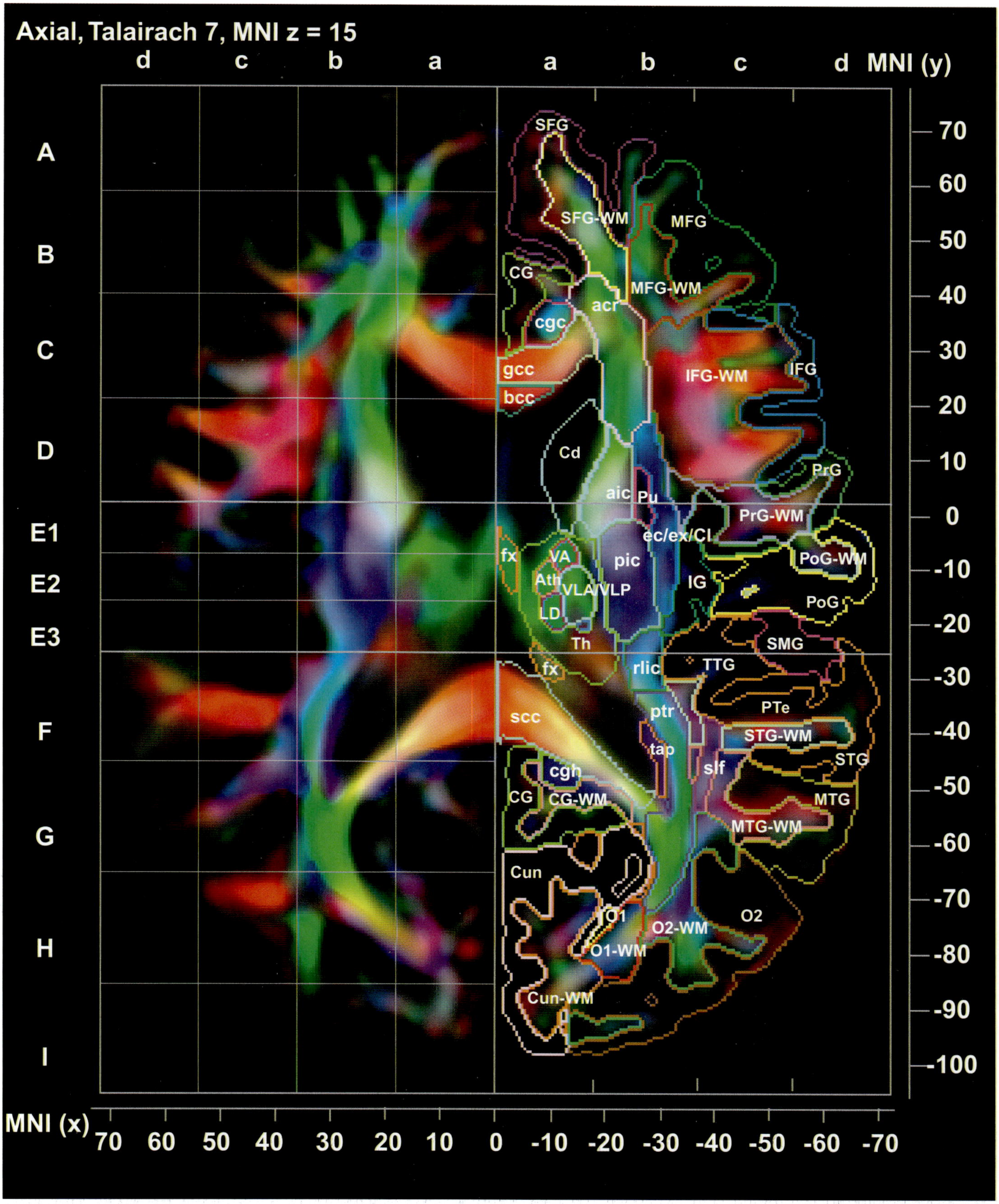

PrG-WM - precentral gyrus white matter
PTe - planum temporale
ptr - posterior thalamic radiation (include the optic radiation)
Pu - putamen
rlic - retrolenticular part of the internal capsule
scc - splenium of the corpus callosum
SFG - superior frontal gyrus
SFG-WM - superior frontal gyrus white matter
slf - superior longitudinal fasciculus
SMG - supramarginal gyrus
STG - superior temporal gyrus
STG-WM - superior temporal gyrus white matter
tap - tapetum of the corpus callosum
Th - thalamus
TTG - transverse temporal gyrus
VA - ventroanterior thalamic nucleus
VLA/VLP - anterior and posterior ventrolateral thalamic nuclei

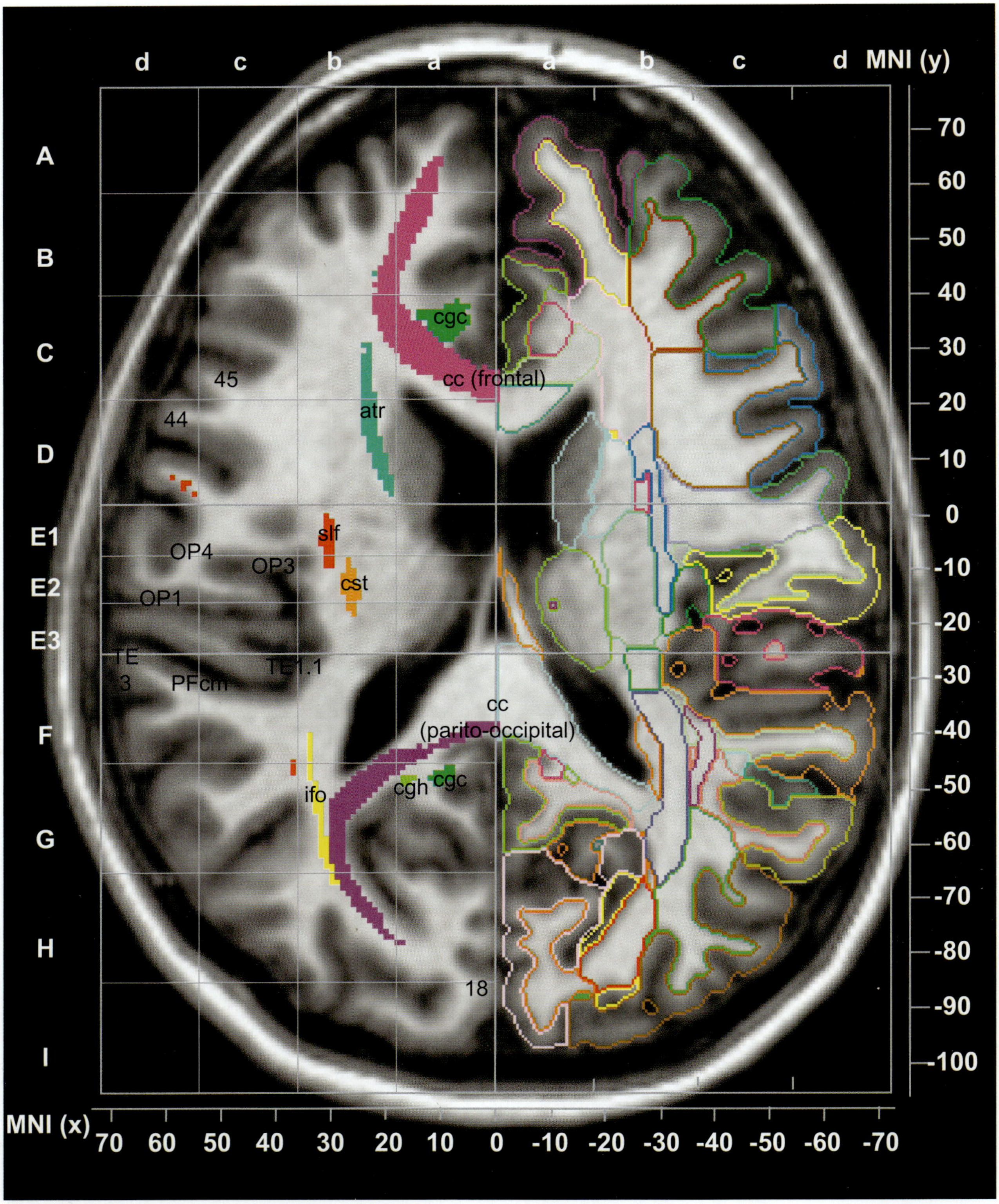

acr - anterior corona radiata
aic - anterior limb of the internal capsule
AnG - angular gyrus
atr - anterior thalamic radiation
bcc - body of the corpus callosum
cc (frontal) - corpus callosum (connecting bilateral frontal areas)
cc (parieto-occipital) - corpus callosum (connecting bilateral parieto-occipital areas)
Cd - caudate nucleus
CG - cingulate gyrus
cgc - cingulum (cingulate gyrus part)
cgh - cingulum (hippocampal part)
CG-WM - cingulate gyrus white matter
cst - corticospinal tract
Cun - cuneus
Cun-WM - cuneus white matter
ec/ex/Cl - external capsule / extreme capsule / claustrum (cannot be resolved with current resolution)
fx - fornix (column and body)
gcc - genu of the corpus callosum
IFG - inferior frontal gyrus
IFG-WM - inferior frontal gyrus white matter
ifo - inferior fronto-occipital fasciculus
IG - insular gyrus
LD - laterodorsal thalamic nucleus
MFG - middle frontal gyrus
MFG-WM - middle frontal gyrus white matter
MTG - middle temporal gyrus
MTG-WM - middle temporal gyrus white matter
O1 - superior occipital gyrus
O1-WM - superior occipital gyrus white matter
O2 - middle or lateral occipital gyrus
O2-WM - middle or lateral occipital gyrus white matter
PCun - precuneus
pic - posterior limb of the internal capsule
PoG - postcentral gyrus
PoG-WM - postcentral gyrus white matter
PrG - precentral gyrus
PrG-WM - precentral gyrus white matter

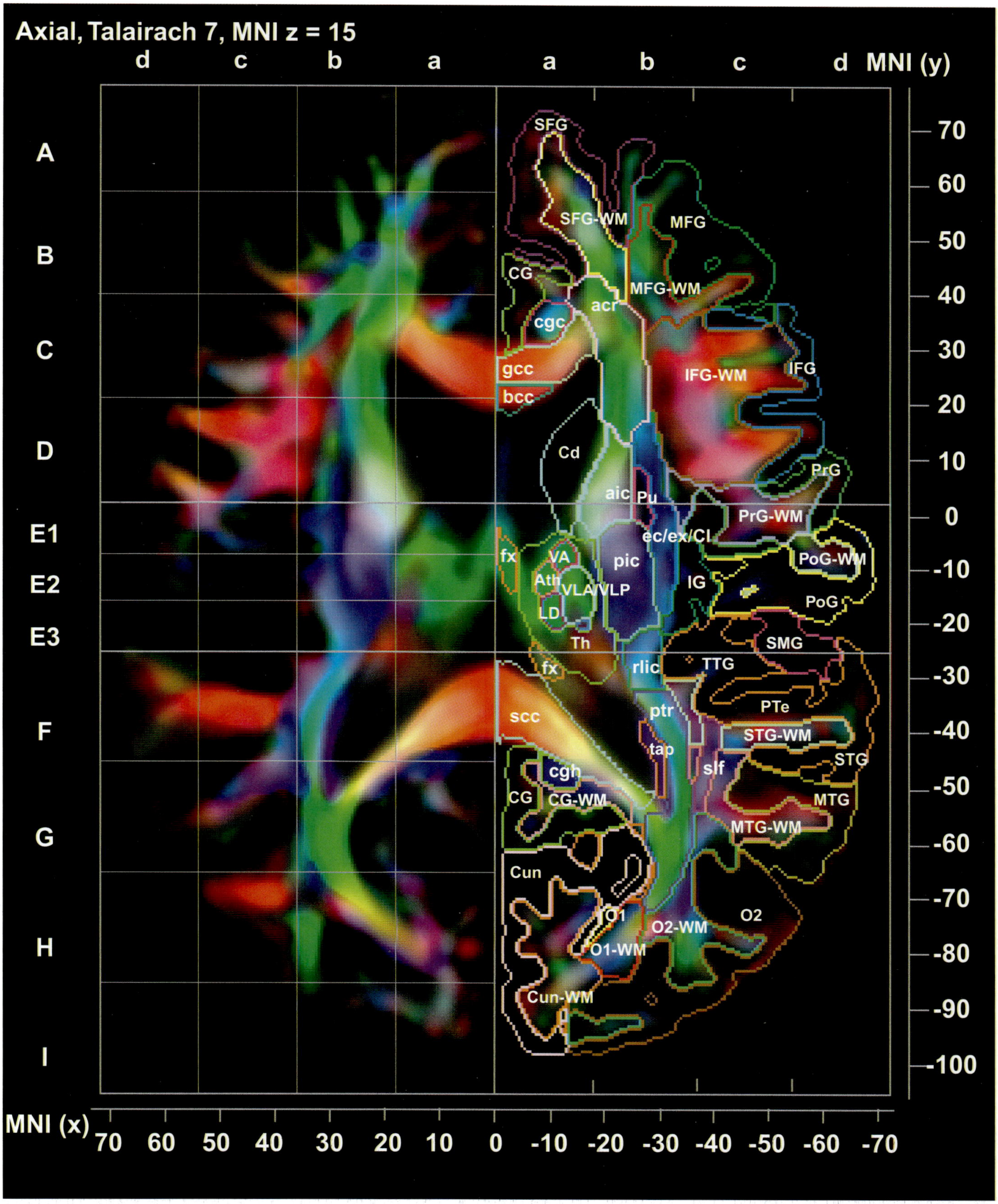

PrG-WM - precentral gyrus white matter
PTe - planum temporale
ptr - posterior thalamic radiation (include the optic radiation)
Pu - putamen
rlic - retrolenticular part of the internal capsule
scc - splenium of the corpus callosum
SFG - superior frontal gyrus
SFG-WM - superior frontal gyrus white matter
slf - superior longitudinal fasciculus
SMG - supramarginal gyrus
STG - superior temporal gyrus
STG-WM - superior temporal gyrus white matter
tap - tapetum of the corpus callosum
Th - thalamus
TTG - transverse temporal gyrus
VA - ventroanterior thalamic nucleus
VLA/VLP - anterior and posterior ventrolateral thalamic nuclei

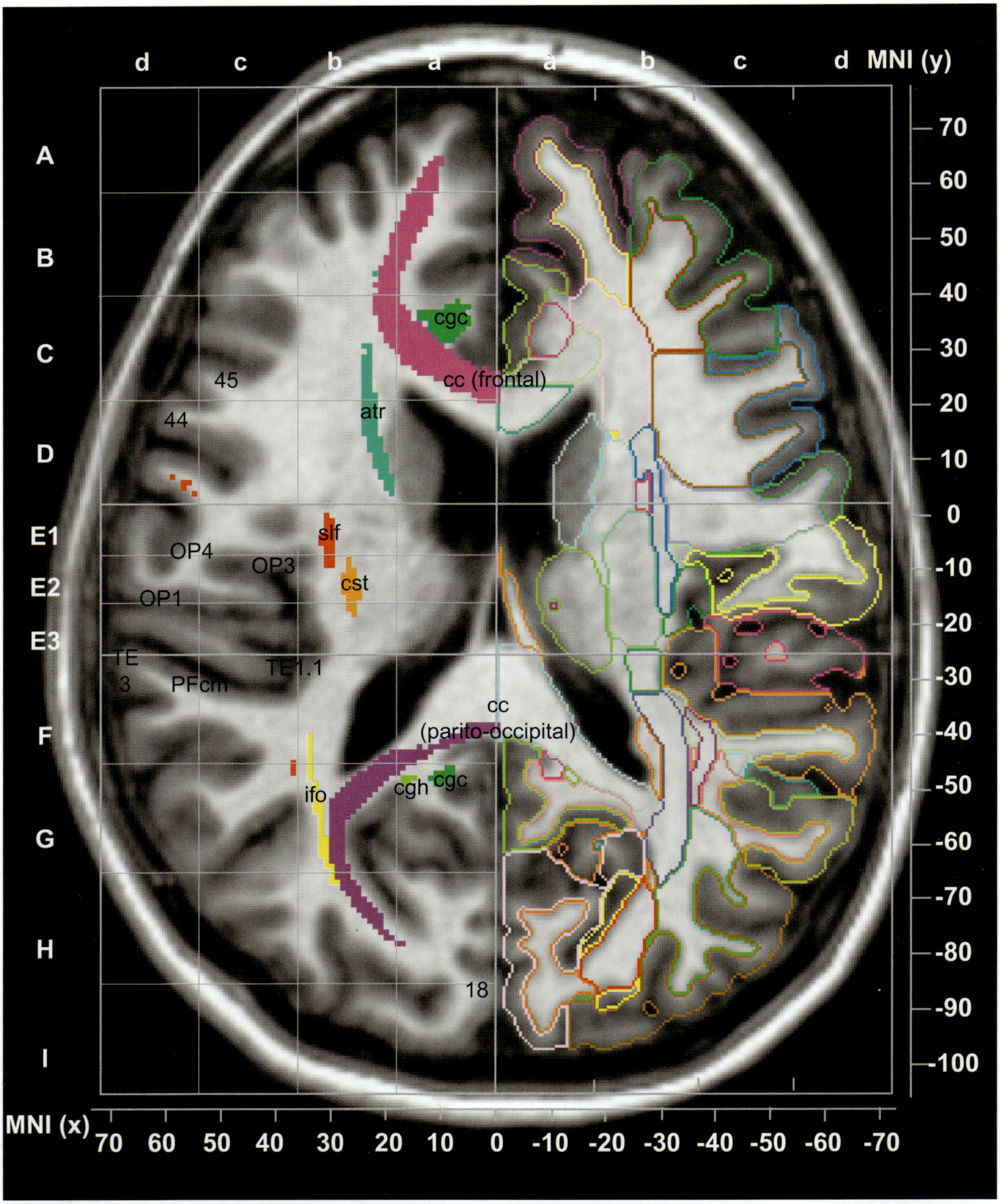

acr - anterior corona radiata
aic - anterior limb of the internal capsule
AnG - angular gyrus
atr - anterior thalamic radiation
bcc - body of the corpus callosum
cc (frontal) - corpus callosum (connecting bilateral frontal areas)
cc (parieto-occipital) - corpus callosum (connecting bilateral parieto-occipital areas)
Cd - caudate nucleus
CG - cingulate gyrus
cgc - cingulum (cingulate gyrus part)
cgh - cingulum (hippocampal part)
CG-WM - cingulate gyrus white matter
cst - corticospinal tract
Cun - cuneus
Cun-WM - cuneus white matter
ec/ex/Cl - external capsule / extreme capsule / claustrum (cannot be resolved with current resolution)
fx - fornix (column and body)
gcc - genu of the corpus callosum
IFG - inferior frontal gyrus
IFG-WM - inferior frontal gyrus white matter
ifo - inferior fronto-occipital fasciculus
IG - insular gyrus
LD - laterodorsal thalamic nucleus
MFG - middle frontal gyrus
MFG-WM - middle frontal gyrus white matter
MTG - middle temporal gyrus
MTG-WM - middle temporal gyrus white matter
O1 - superior occipital gyrus
O1-WM - superior occipital gyrus white matter
O2 - middle or lateral occipital gyrus
O2-WM - middle or lateral occipital gyrus white matter
PCun - precuneus
pic - posterior limb of the internal capsule
PoG - postcentral gyrus
PoG-WM - postcentral gyrus white matter
PrG - precentral gyrus
PrG-WM - precentral gyrus white matter

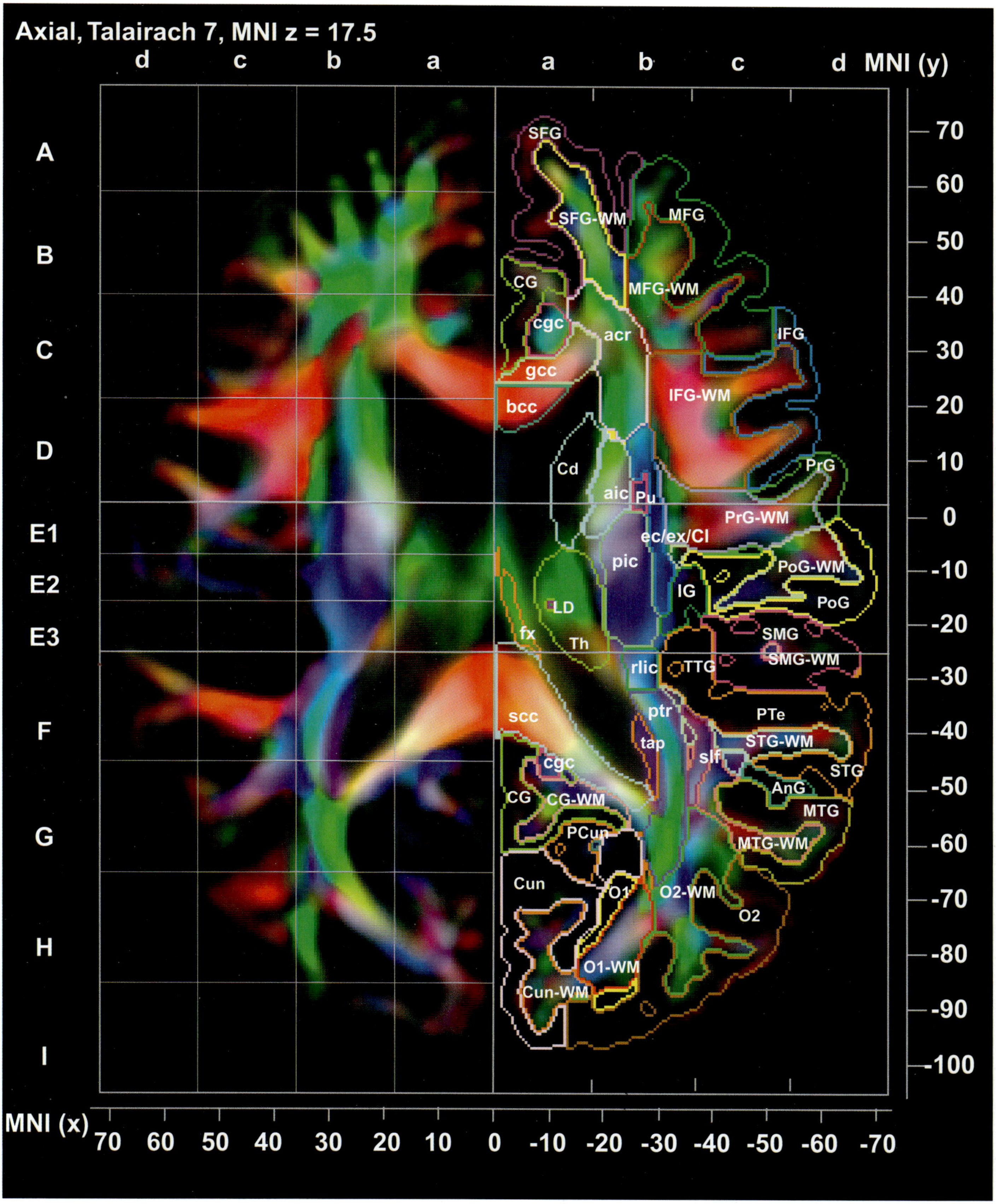

PTe - planum temporale
ptr - posterior thalamic radiation (include the optic radiation)
Pu - putamen
rlic - retrolenticular part of the internal capsule
scc - splenium of the corpus callosum
SFG - superior frontal gyrus
SFG-WM - superior frontal gyrus white matter
slf - superior longitudinal fasciculus
SMG - supramarginal gyrus
SMG-WM - supramarginal gyrus white matter
STG - superior temporal gyrus
STG-WM - superior temporal gyrus white matter
tap - tapetum of the corpus callosum
Th - thalamus
TTG - transverse temporal gyrus

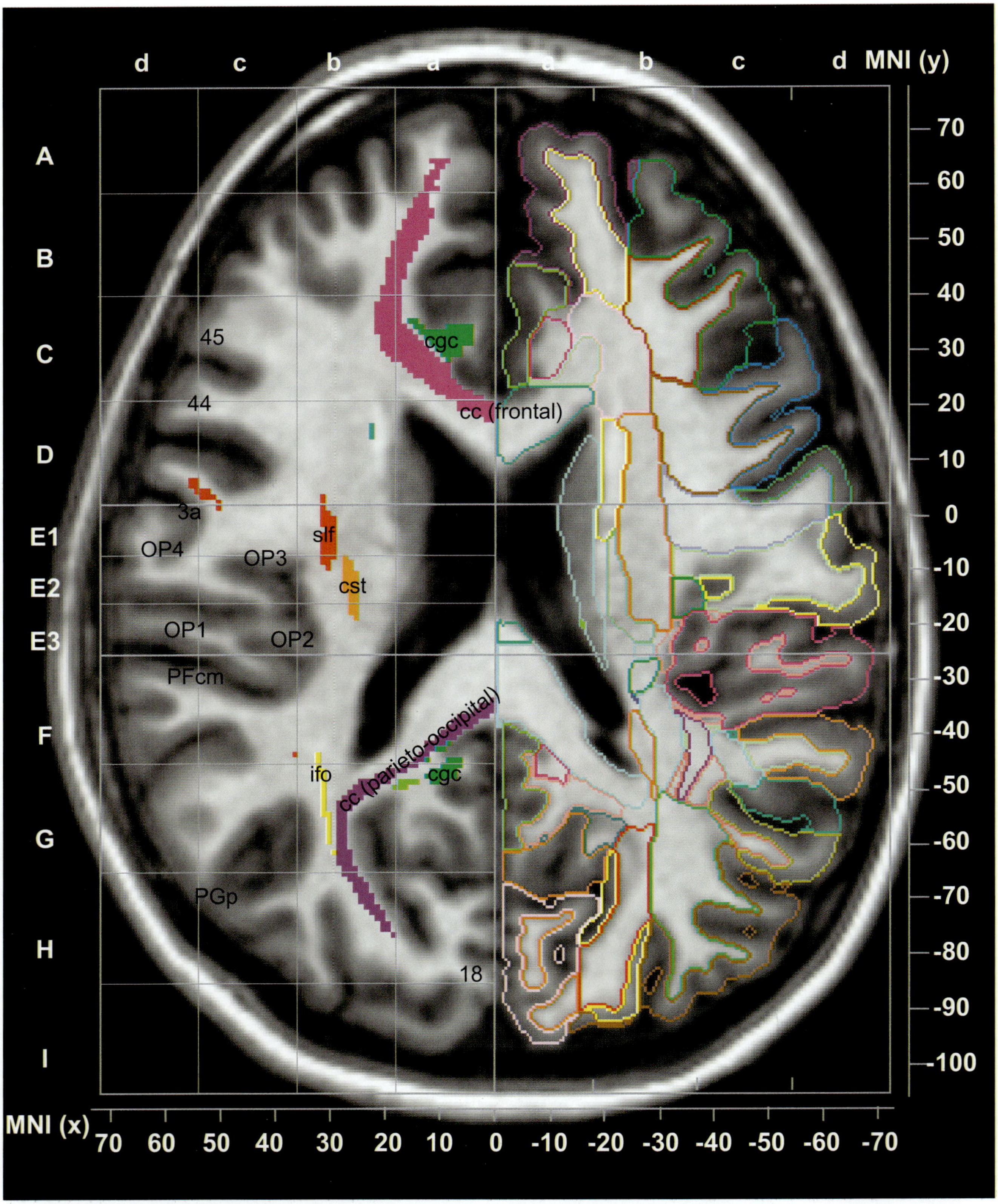

acr - anterior corona radiata
AnG - angular gyrus
AnG-WM - angular gyrus white matter
bcc - body of the corpus callosum
cc (frontal) - corpus callosum (connecting bilateral frontal areas)
cc (parieto-occipital) - corpus callosum (connecting bilateral parieto-occipital areas)
Cd - caudate nucleus
CG - cingulate gyrus
cgc - cingulum (cingulate gyrus part)
CG-WM - cingulate gyrus white matter
cst - corticospinal tract
Cun - cuneus
Cun-WM - cuneus white matter
gcc - genu of the corpus callosum
IFG - inferior frontal gyrus
IFG-WM - inferior frontal gyrus white matter
ifo - inferior fronto-occipital fasciculus
IG - insular gyrus
MFG - middle frontal gyrus
MFG-WM - middle frontal gyrus white matter
MTG - middle temporal gyrus
MTG-WM - middle temporal gyrus white matter
O1 - superior occipital gyrus
O1-WM - superior occipital gyrus white matter
O2 - middle or lateral occipital gyrus
O2-WM - middle or lateral occipital gyrus white matter
pcr - posterior corona radiata
PCun - precuneus
PCun-WM - precuneus white matter
pic - posterior limb of the internal capsule

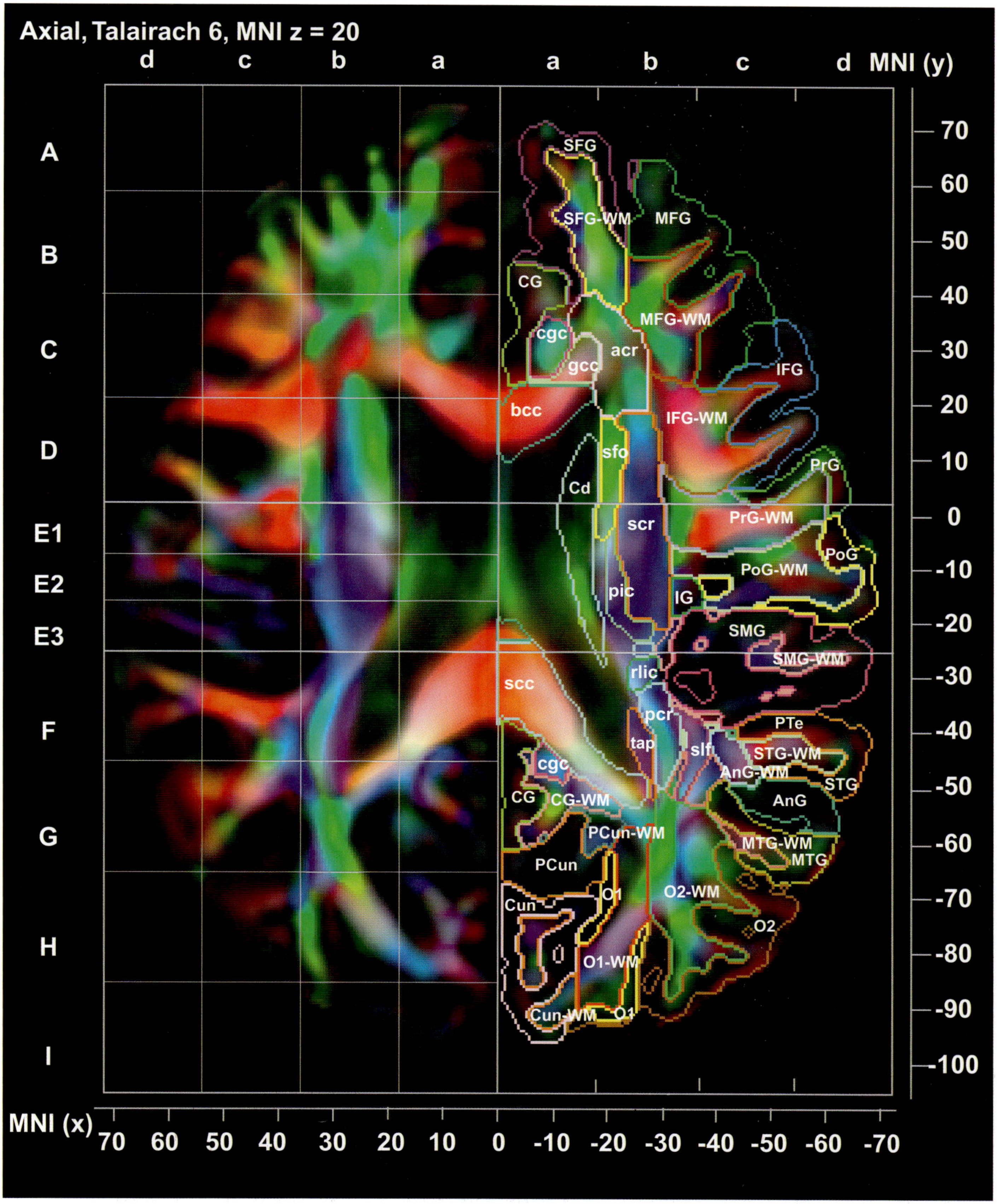

PoG	- postcentral gyrus	scc	- splenium of the corpus callosum	SMG	- supramarginal gyrus
PoG-WM	- postcentral gyrus white matter	scr	- superior corona radiata	SMG-WM	- supramarginal gyrus white matter
PrG	- precentral gyrus	SFG	- superior frontal gyrus	STG	- superior temporal gyrus
PrG-WM	- precentral gyrus white matter	SFG-WM	- superior frontal gyrus white matter	STG-WM	- superior temporal gyrus white matter
PTe	- planum temporale	sfo	- Superior fronto-occipital fasciculus	tap	- tapetum of the corpus callosum
rlic	- retrolenticular part of the internal capsule	slf	- superior longitudinal fasciculus		

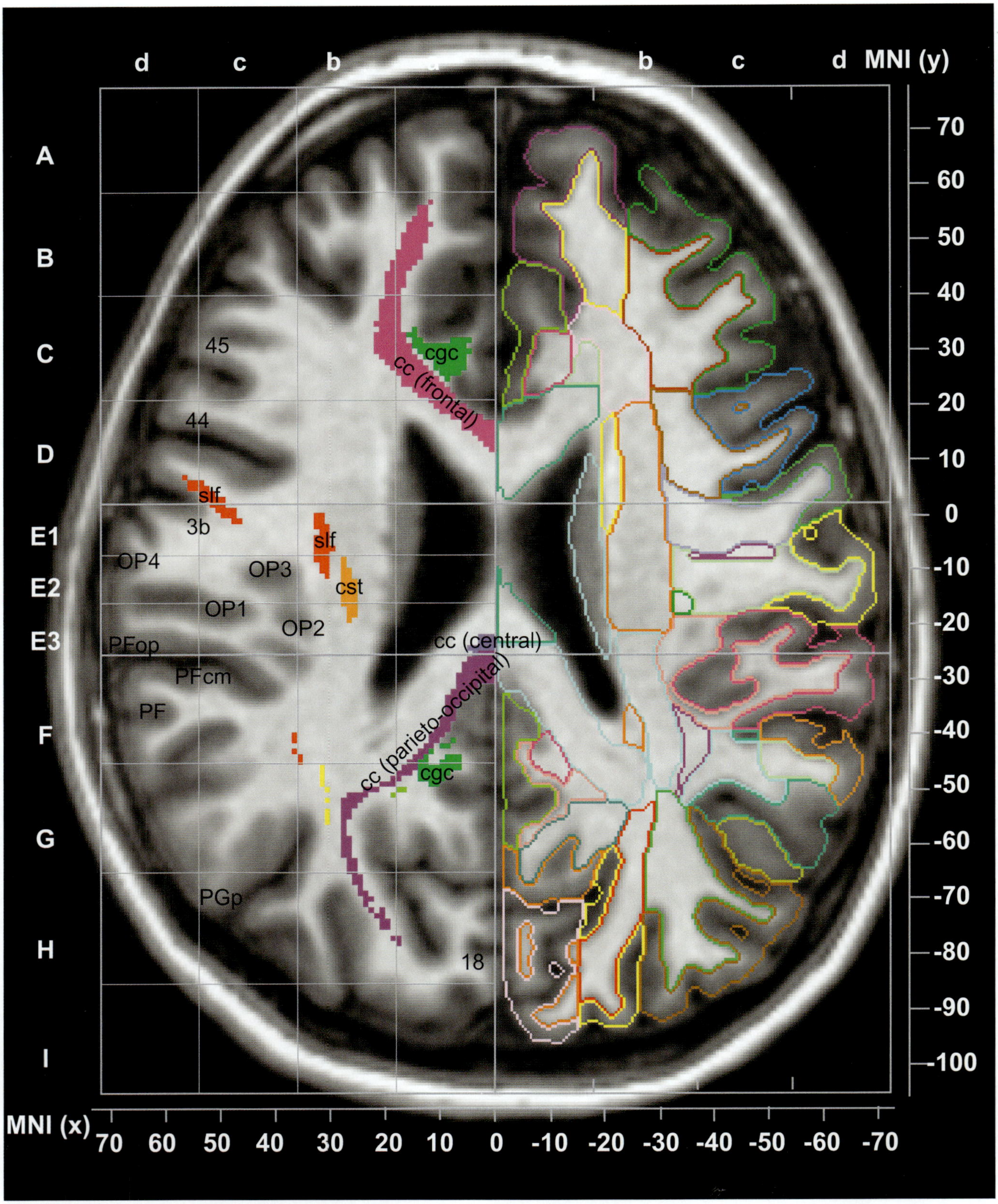

acr - anterior corona radiata
AnG - angular gyrus
AnG-WM - angular gyrus white matter
bcc - body of the corpus callosum
cc (central) - corpus callosum (connecting bilateral central areas)
cc (frontal) - corpus callosum (connecting bilateral frontal areas)
cc (parieto-occipital) - corpus callosum (connecting bilateral parieto-occipital areas)
Cd - caudate nucleus
CG - cingulate gyrus
cgc - cingulum (cingulate gyrus part)
CG-WM - cingulate gyrus white matter
cst - corticospinal tract
Cun - cuneus
Cun-WM - cuneus white matter
gcc - genu of the corpus callosum
IFG - inferior frontal gyrus
IFG-WM - inferior frontal gyrus white matter
IG - insular gyrus
MFG - middle frontal gyrus
MFG-WM - middle frontal gyrus white matter
MTG - middle temporal gyrus
O1 - superior occipital gyrus
O1-WM - superior occipital gyrus white matter
O2 - middle or lateral occipital gyrus
O2-WM - middle or lateral occipital gyrus white matter
pcr - posterior corona radiata
PCun - precuneus
PCun-WM - precuneus white matter
PoG - postcentral gyrus

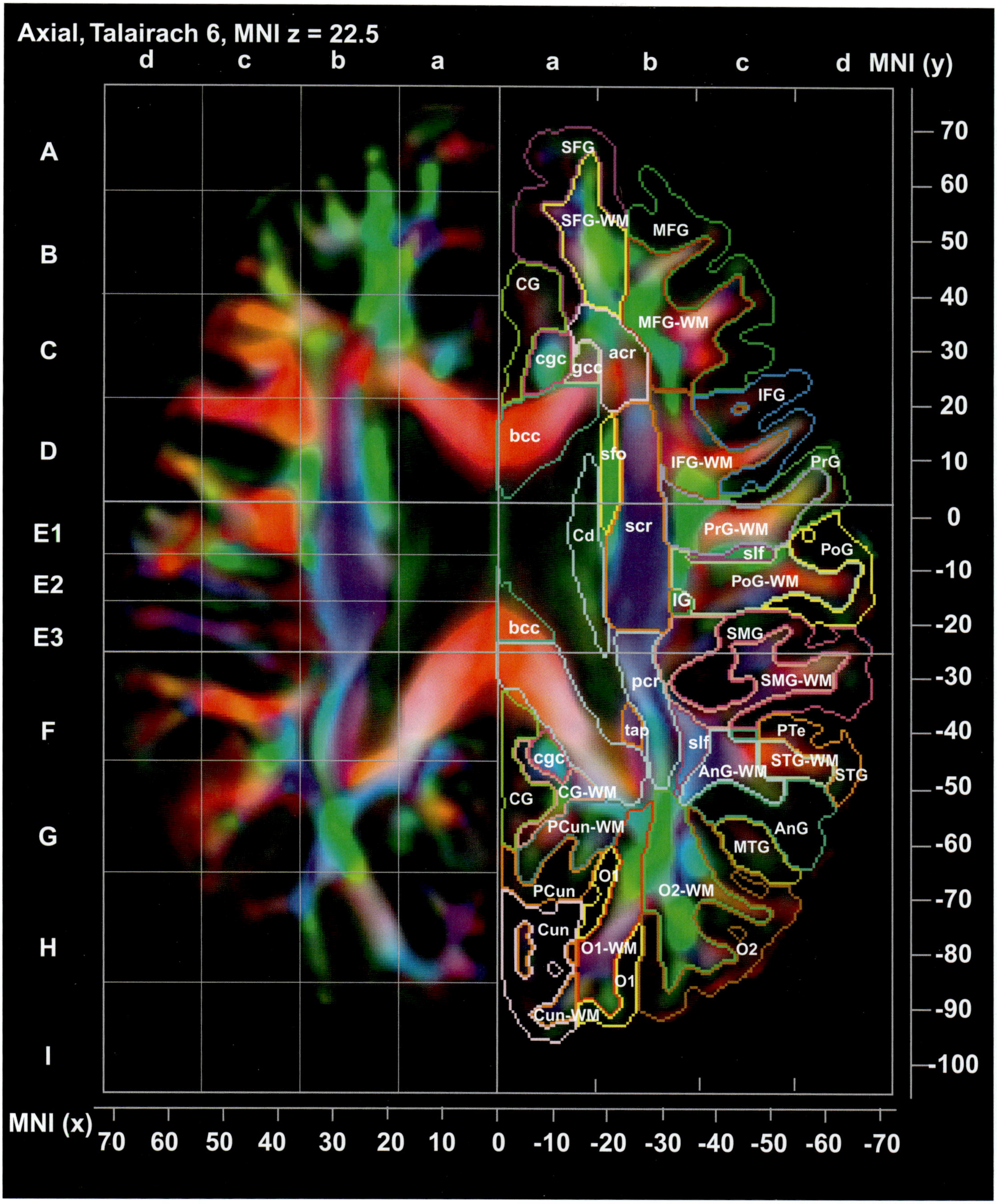

PoG-WM - postcentral gyrus white matter
PrG - precentral gyrus
PrG-WM - precentral gyrus white matter
PTe - planum temporale
scr - superior corona radiata
SFG - superior frontal gyrus
SFG-WM - superior frontal gyrus white matter
sfo - Superior fronto-occipital fasciculus
slf - superior longitudinal fasciculus
SMG - supramarginal gyrus
SMG-WM - supramarginal gyrus white matter
STG - superior temporal gyrus
STG-WM - superior temporal gyrus white matter
tap - tapetum of the corpus callosum

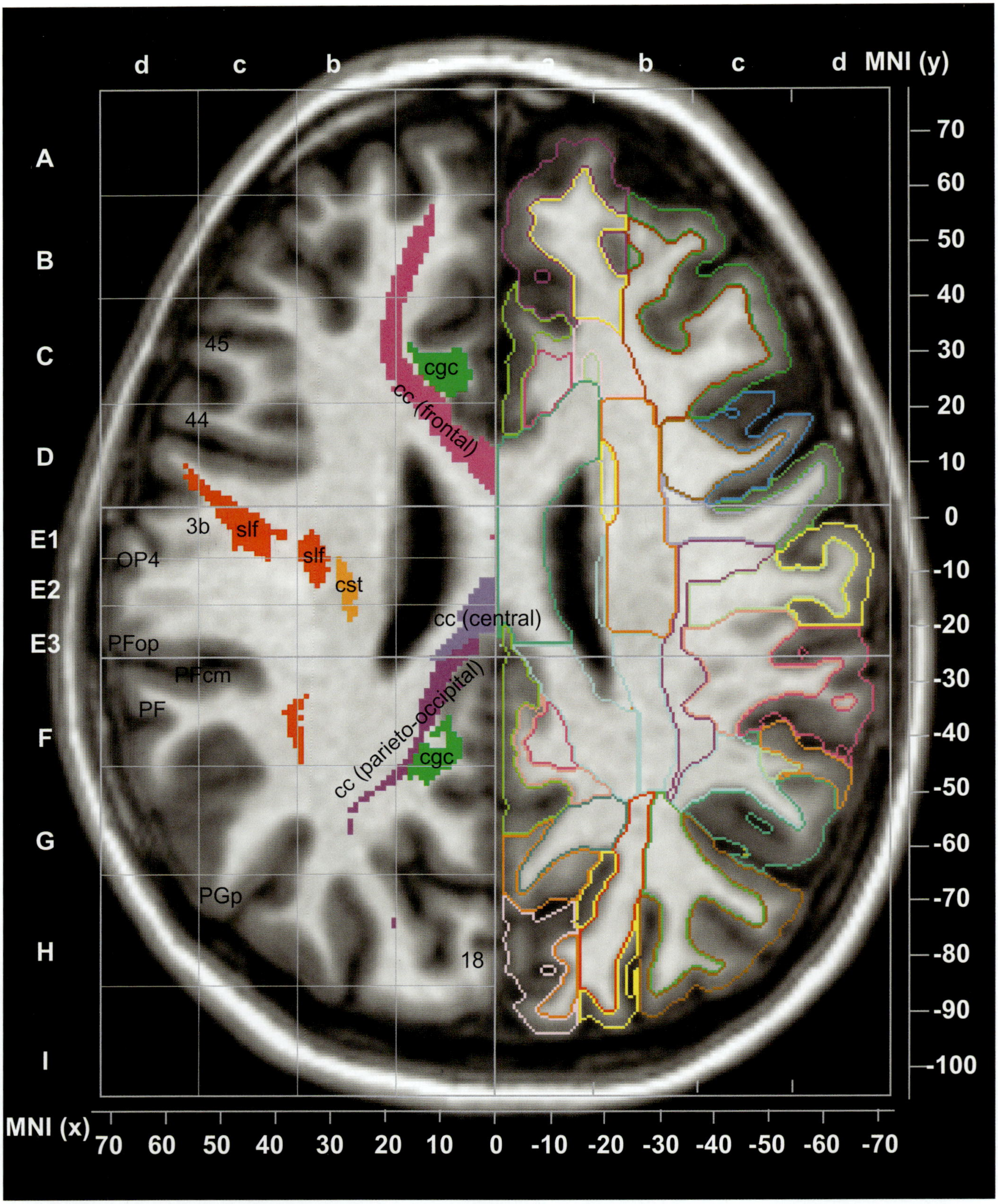

acr - anterior corona radiata
AnG - angular gyrus
AnG-WM - angular gyrus white matter
bcc - body of the corpus callosum
cc (central) - corpus callosum (connecting bilateral central areas)
cc (frontal) - corpus callosum (connecting bilateral frontal areas)
cc (parieto-occipital) - corpus callosum (connecting bilateral parieto-occipital areas)
Cd - caudate nucleus
CG - cingulate gyrus
cgc - cingulum (cingulate gyrus part)
CG-WM - cingulate gyrus white matter
cst - corticospinal tract
Cun - cuneus
Cun-WM - cuneus white matter
IFG - inferior frontal gyrus
IFG-WM - inferior frontal gyrus white matter
MFG - middle frontal gyrus
MFG-WM - middle frontal gyrus white matter
O1 - superior occipital gyrus
O1-WM - superior occipital gyrus white matter
O2 - middle or lateral occipital gyrus
O2-WM - middle or lateral occipital gyrus white matter

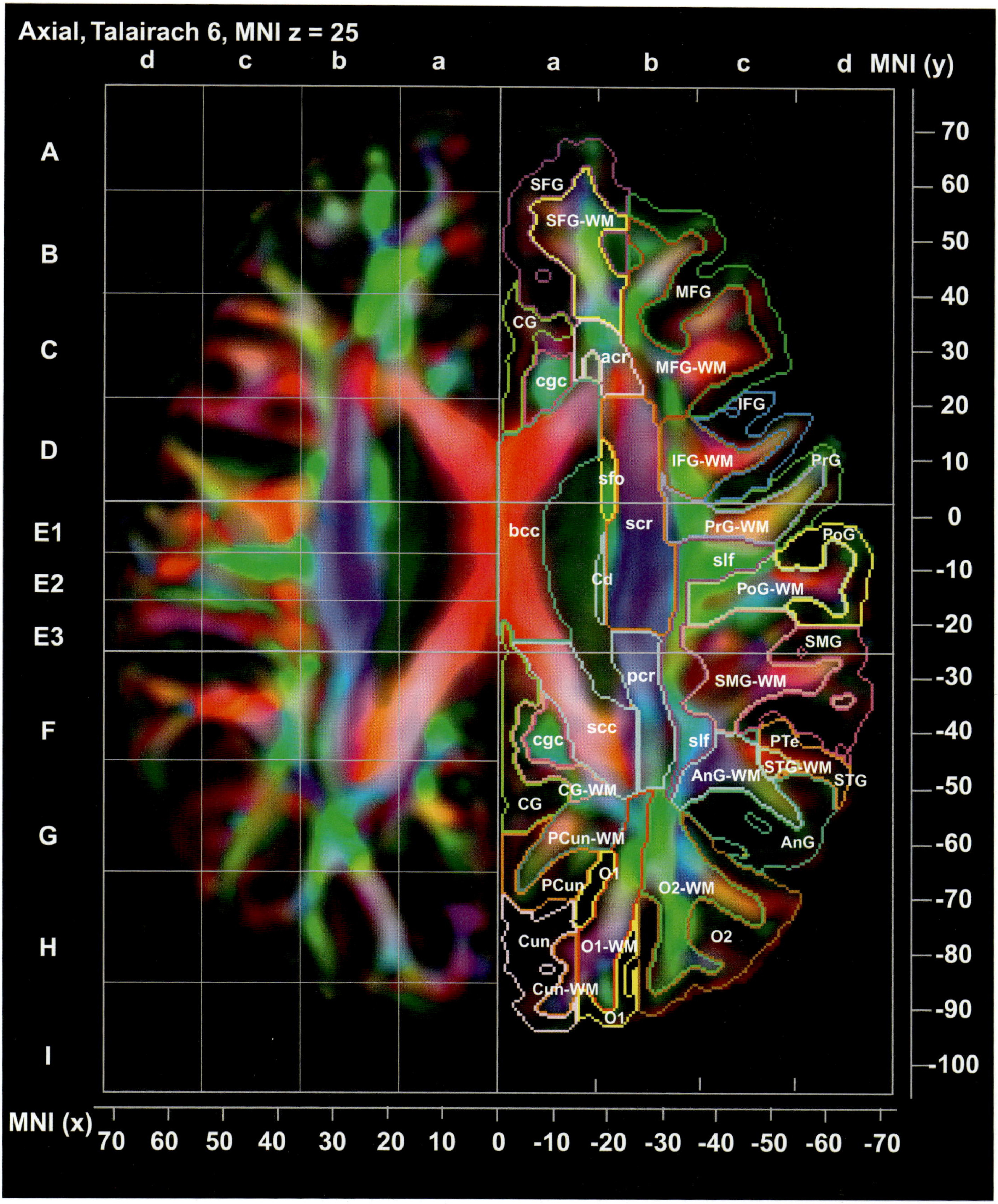

pcr	-	posterior corona radiata
PCun	-	precuneus
PCun-WM	-	precuneus white matter
PoG	-	postcentral gyrus
PoG-WM	-	postcentral gyrus white matter
PrG	-	precentral gyrus
PrG-WM	-	precentral gyrus white matter
PTe	-	planum temporale
scc	-	splenium of the corpus callosum
scr	-	superior corona radiata
SFG	-	superior frontal gyrus
SFG-WM	-	superior frontal gyrus white matter
sfo	-	superior fronto-occipital fasciculus
slf	-	superior longitudinal fasciculus
SMG	-	supramarginal gyrus
SMG-WM	-	supramarginal gyrus white matter
STG	-	superior temporal gyrus
STG-WM	-	superior temporal gyrus white matter

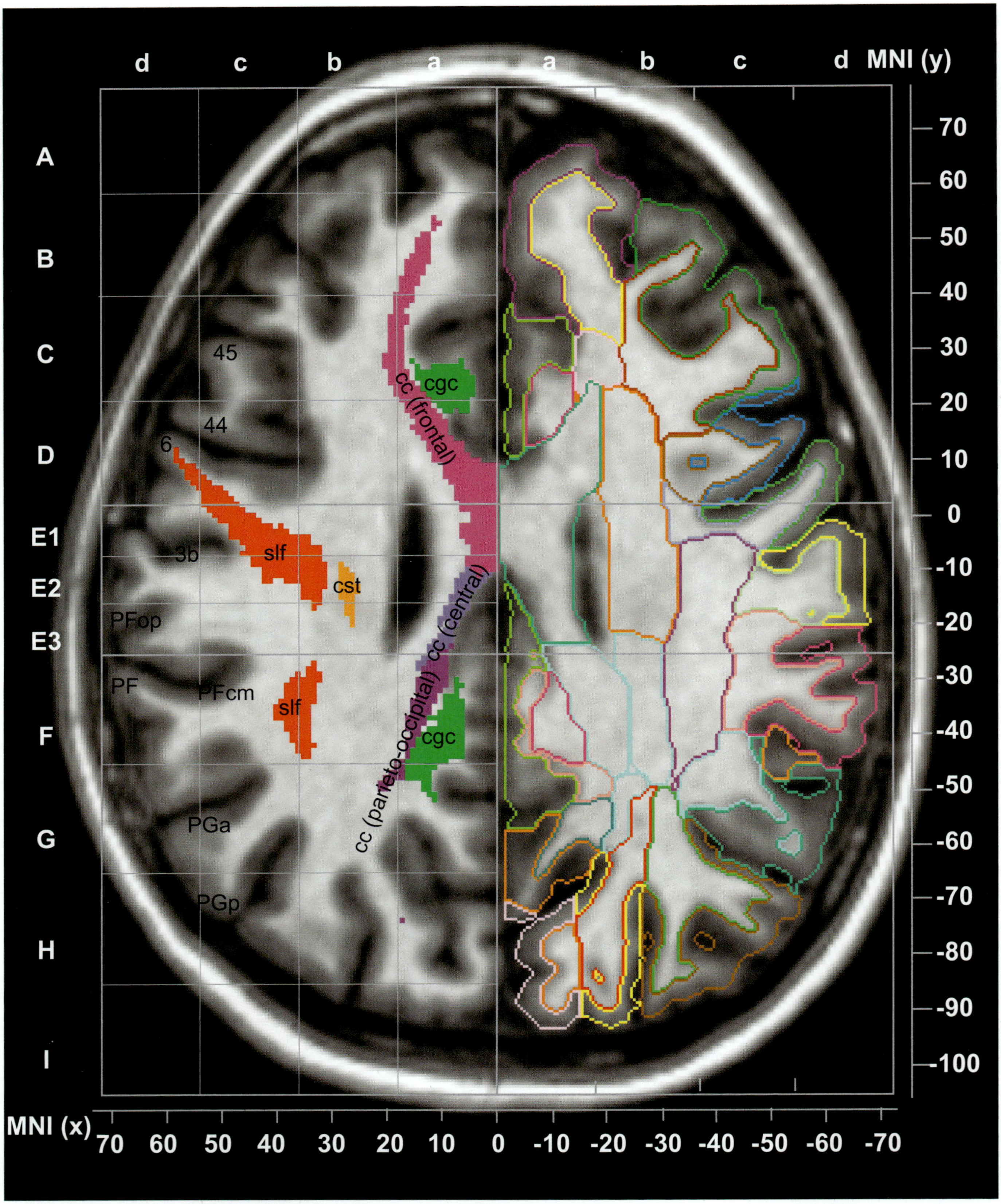

acr - anterior corona radiata
AnG - angular gyrus
AnG-WM - angular gyrus white matter
bcc - body of the corpus callosum
cc (central) - corpus callosum (connecting bilateral central areas)
cc (frontal) - corpus callosum (connecting bilateral frontal areas)
cc (parieto-occipital) - corpus callosum (connecting bilateral parieto-occipital areas)
CG - cingulate gyrus
cgc - cingulum (cingulate gyrus part)
CG-WM - cingulate gyrus white matter
cst - corticospinal tract
Cun - cuneus
Cun-WM - cuneus white matter
IFG - inferior frontal gyrus
IFG-WM - inferior frontal gyrus white matter
MFG - middle frontal gyrus
MFG-WM - middle frontal gyrus white matter
O1 - superior occipital gyrus
O1-WM - superior occipital gyrus white matter
O2 - middle or lateral occipital gyrus
O2-WM - middle or lateral occipital gyrus white matter

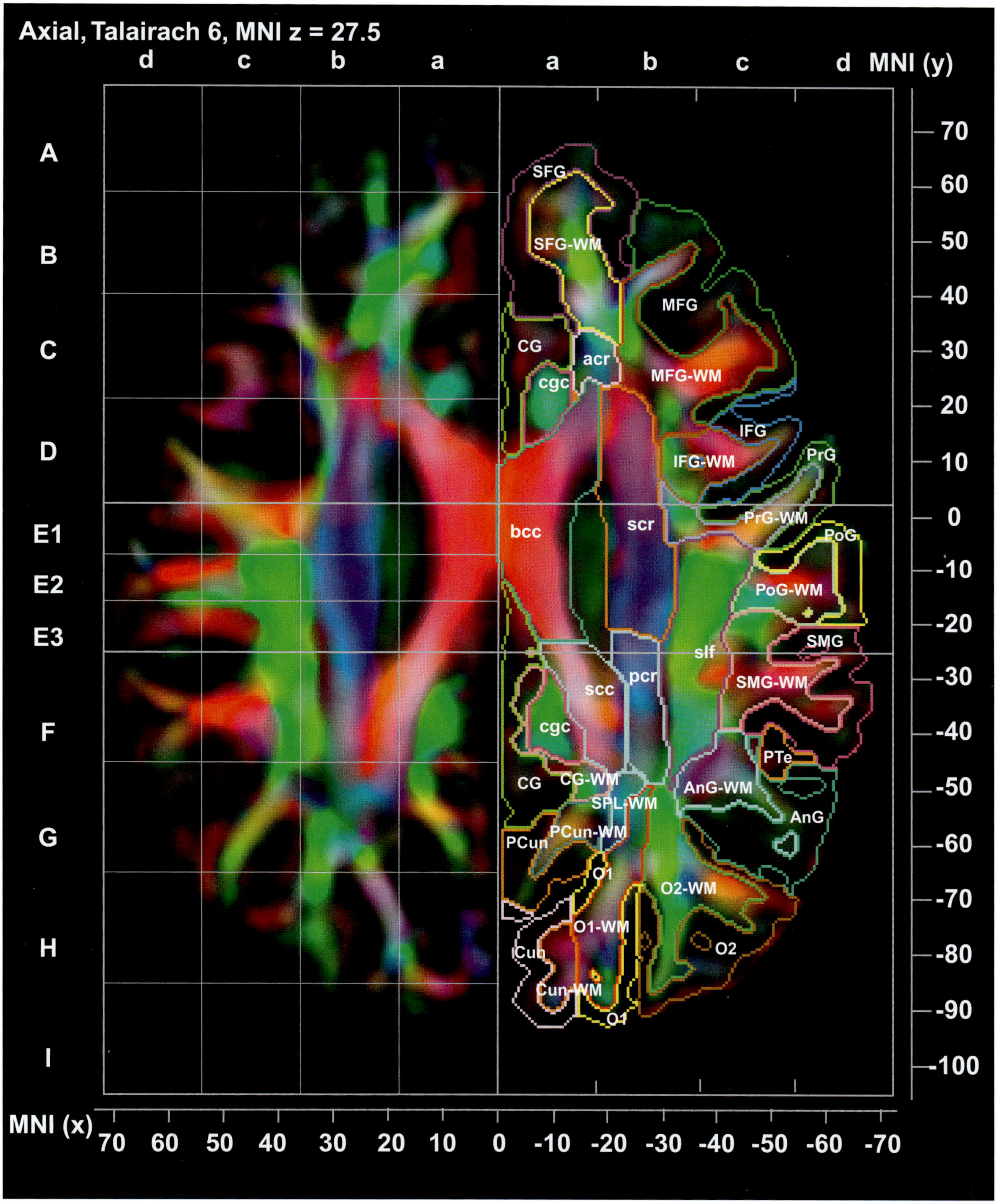

pcr - posterior corona radiata
PCun - precuneus
PCun-WM - precuneus white matter
PoG - postcentral gyrus
PoG-WM - postcentral gyrus white matter
PrG - precentral gyrus
PrG-WM - precentral gyrus white matter
PTe - planum temporale
scc - splenium of the corpus callosum
scr - superior corona radiata
SFG - superior frontal gyrus
SFG-WM - superior frontal gyrus white matter
slf - superior longitudinal fasciculus
SMG - supramarginal gyrus
SMG-WM - supramarginal gyrus white matter
SPL-WM - superior parietal lobule white matter

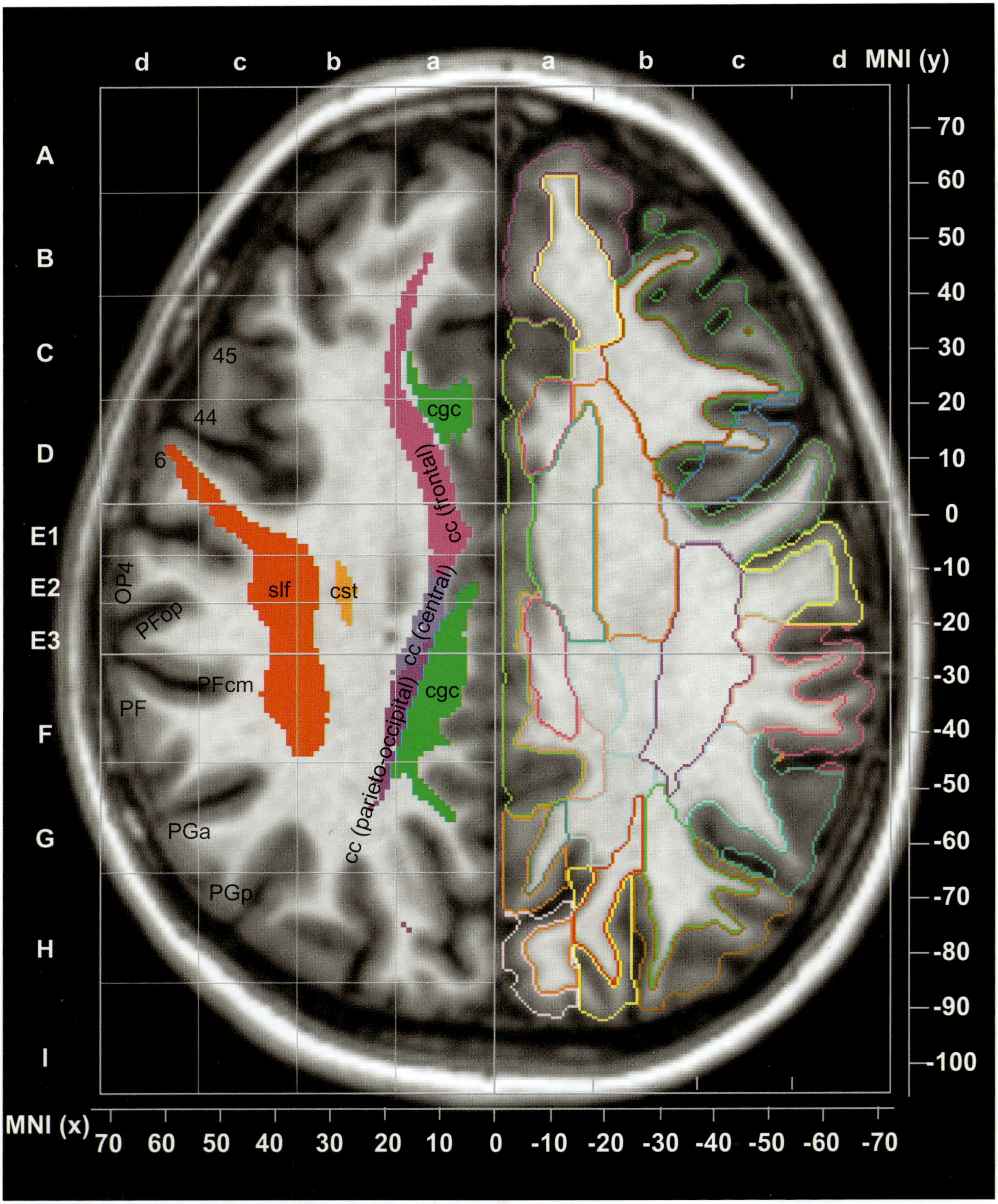

acr - anterior corona radiata
AnG - angular gyrus
AnG-WM - angular gyrus white matter
bcc - body of the corpus callosum
cc (central) - corpus callosum (connecting bilateral central areas)
cc (frontal) - corpus callosum (connecting bilateral frontal areas)
cc (parieto-occipital) - corpus callosum (connecting bilateral parieto-occipital areas)
CG - cingulate gyrus
cgc - cingulum (cingulate gyrus part)
CG-WM - cingulate gyrus white matter
cst - corticospinal tract
Cun - cuneus
Cun-WM - cuneus white matter
IFG - inferior frontal gyrus
IFG-WM - inferior frontal gyrus white matter
MFG - middle frontal gyrus
MFG-WM - middle frontal gyrus white matter

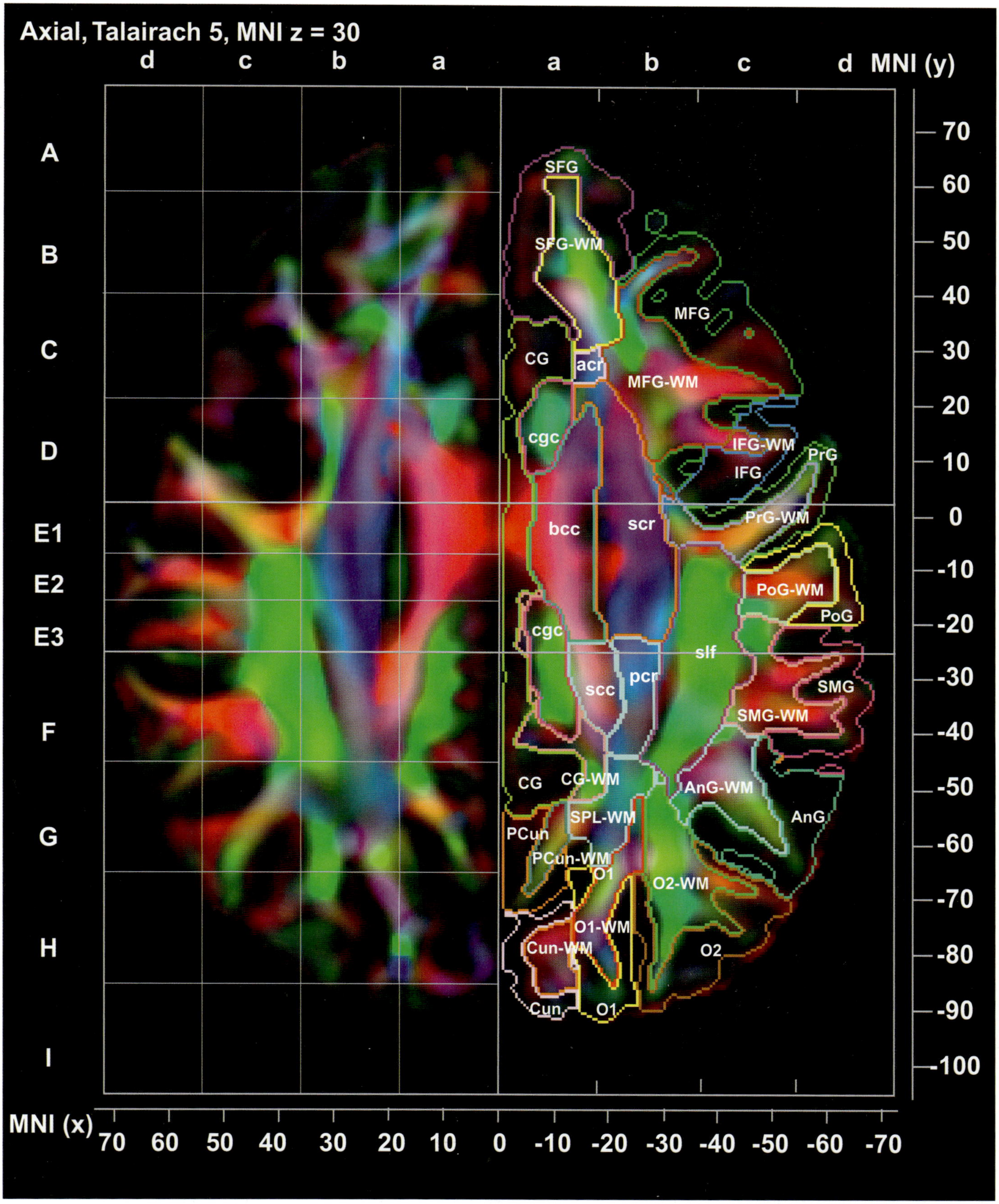

O1	- superior occipital gyrus	PoG	- postcentral gyrus	SFG	- superior frontal gyrus
O1-WM	- superior occipital gyrus white matter	PoG-WM	- postcentral gyrus white matter	SFG-WM	- superior frontal gyrus white matter
O2	- middle or lateral occipital gyrus	PrG	- precentral gyrus	slf	- superior longitudinal fasciculus
O2-WM	- middle or lateral occipital gyrus white matter	PrG-WM	- precentral gyrus white matter	SMG	- supramarginal gyrus
pcr	- posterior corona radiata	scc	- splenium of the corpus callosum	SMG-WM	- supramarginal gyrus white matter
PCun	- precuneus	scr	- superior corona radiata	SPL-WM	- superior parietal lobule white matter
PCun-WM	- precuneus white matter				

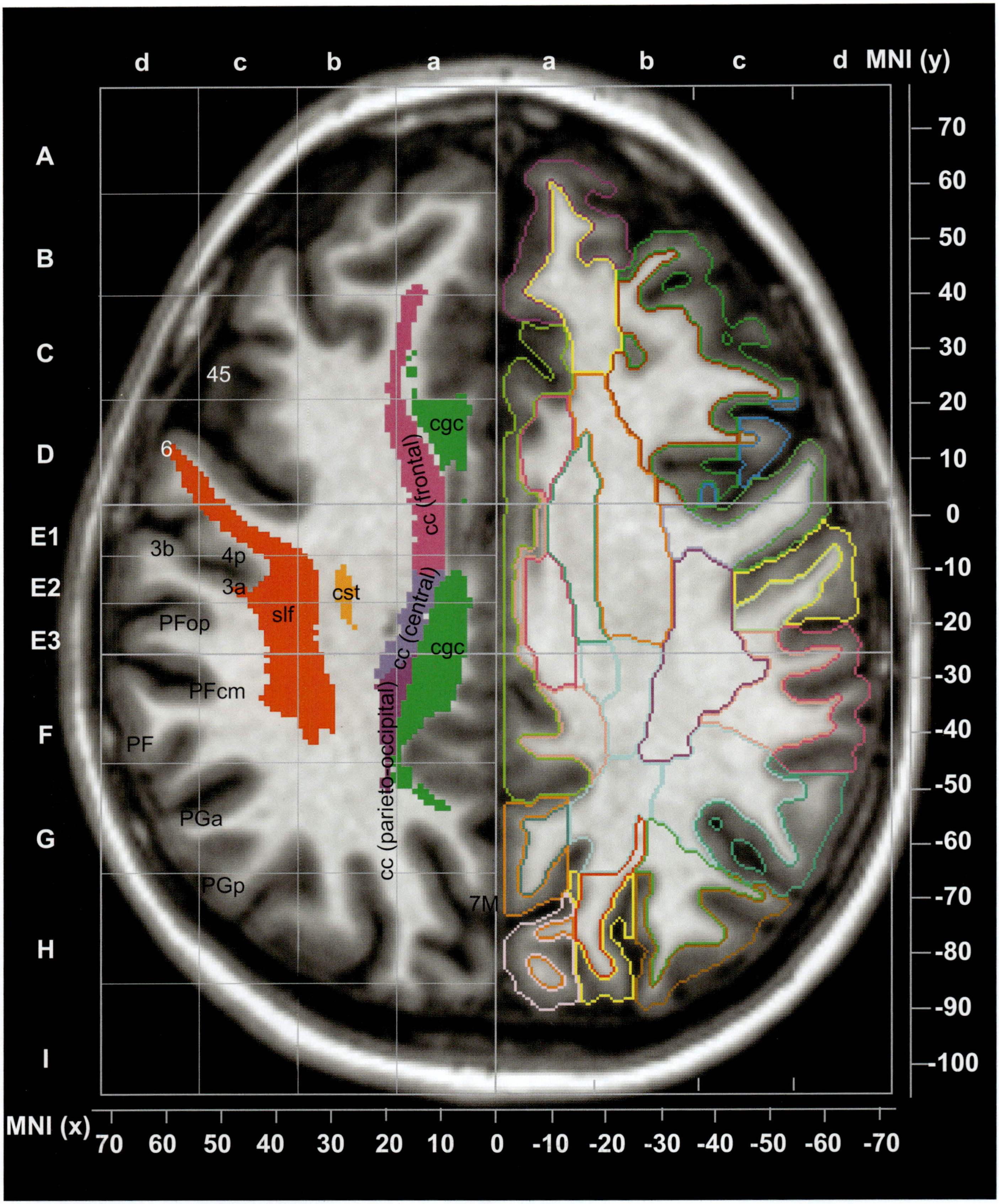

AnG - angular gyrus
AnG-WM - angular gyrus white matter
bcc - body of the corpus callosum
cc (central) - corpus callosum (connecting bilateral central areas)
cc (frontal) - corpus callosum (connecting bilateral frontal areas)
cc (parieto-occipital) - corpus callosum (connecting bilateral parieto-occipital areas)
CG - cingulate gyrus
cgc - cingulum (cingulate gyrus part)
CG-WM - cingulate gyrus white matter
cst - corticospinal tract
Cun - cuneus
Cun-WM - cuneus white matter
IFG - inferior frontal gyrus
MFG - middle frontal gyrus
MFG-WM - middle frontal gyrus white matter
O1 - superior occipital gyrus

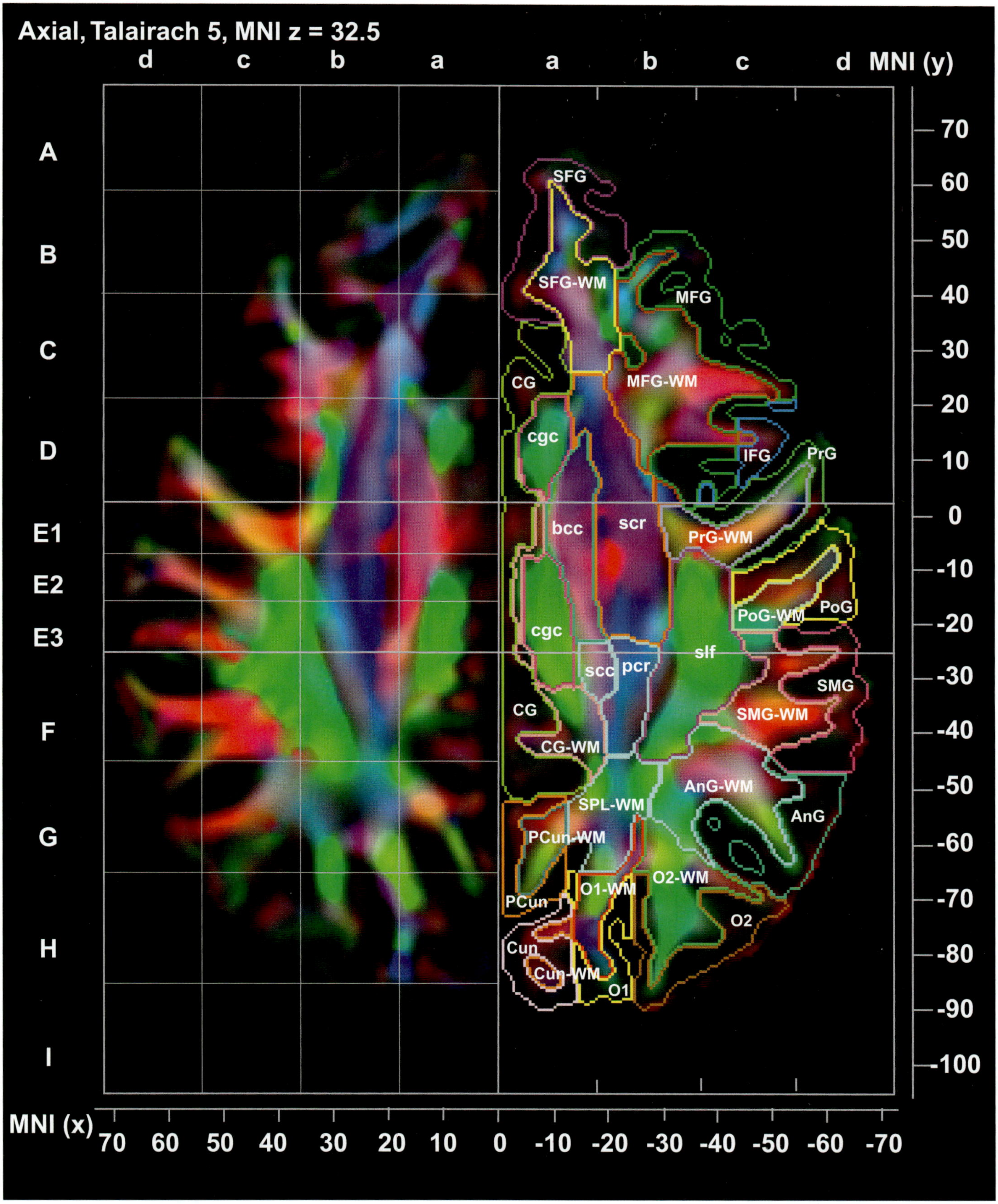

O1-WM - superior occipital gyrus white matter
O2 - middle or lateral occipital gyrus
O2-WM - middle or lateral occipital gyrus white matter
pcr - posterior corona radiata
PCun - precuneus
PCun-WM - precuneus white matter
PoG - postcentral gyrus
PoG-WM - postcentral gyrus white matter
PrG - precentral gyrus
PrG-WM - precentral gyrus white matter
scc - splenium of the corpus callosum
scr - superior corona radiata
SFG - superior frontal gyrus
SFG-WM - superior frontal gyrus white matter
slf - superior longitudinal fasciculus
SMG - supramarginal gyrus
SMG-WM - supramarginal gyrus white matter
SPL-WM - superior parietal lobule white matter

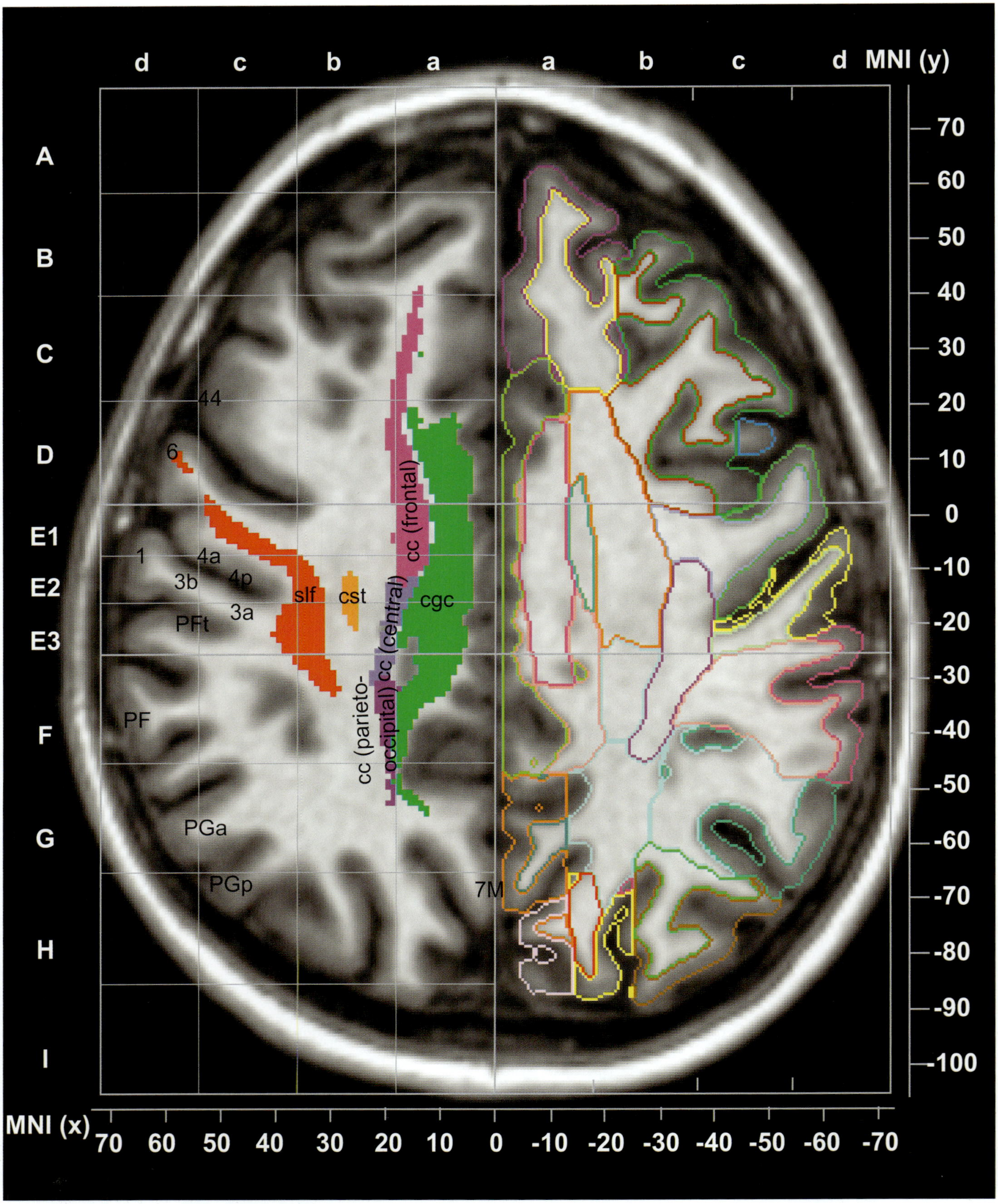

AnG - angular gyrus
AnG-WM - angular gyrus white matter
bcc - body of the corpus callosum
cc (central) - corpus callosum (connecting bilateral central areas)
cc (frontal) - corpus callosum (connecting bilateral frontal areas)
cc (parieto-occipital) - corpus callosum (connecting bilateral parieto-occipital areas)
CG - cingulate gyrus
cgc - cingulum (cingulate gyrus part)
CG-WM - cingulate gyrus white matter
cst - corticospinal tract
Cun - cuneus
Cun-WM - cuneus white matter
IFG - inferior frontal gyrus
MFG - middle frontal gyrus
MFG-WM - middle frontal gyrus white matter
O1 - superior occipital gyrus

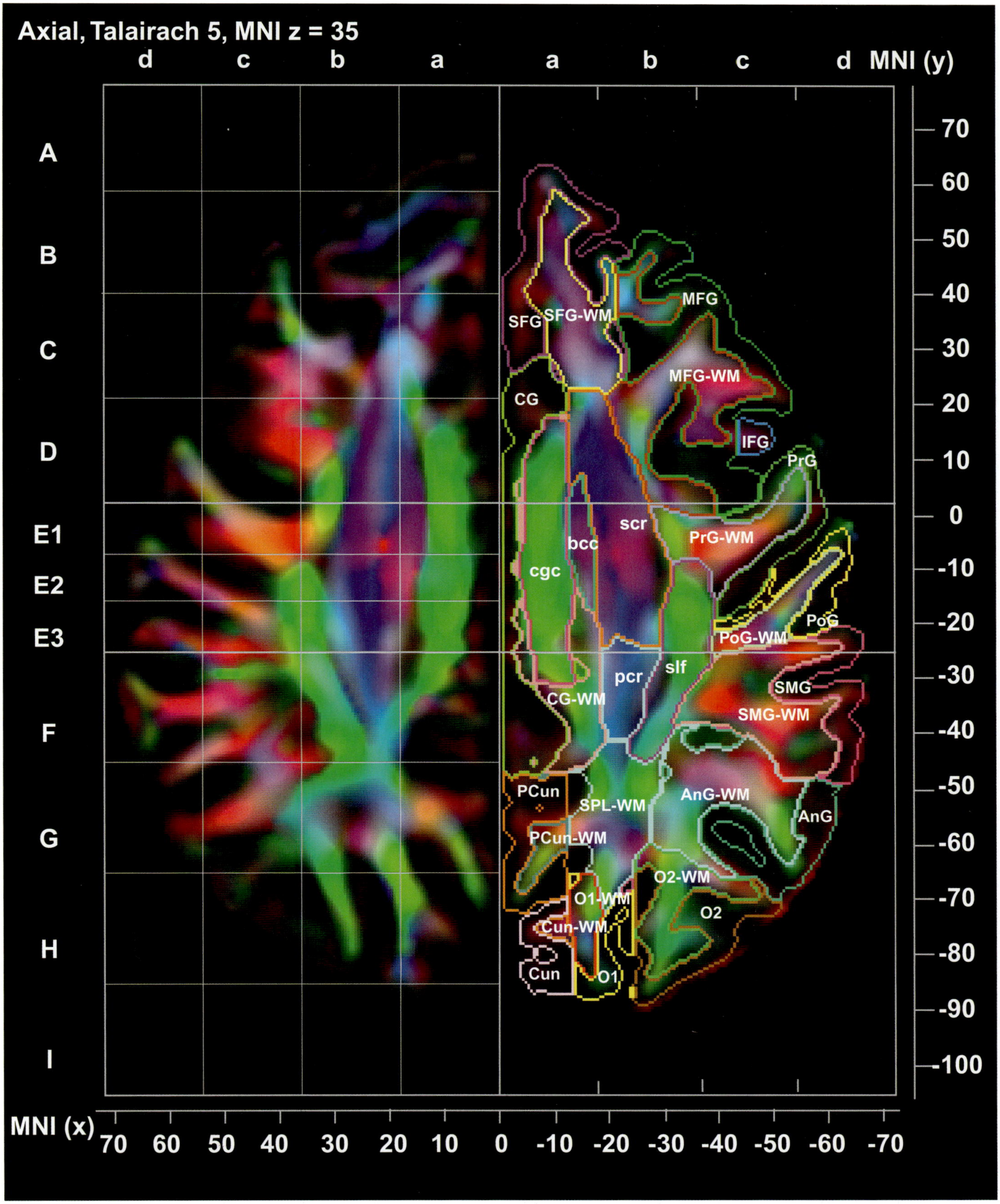

O1-WM	-	superior occipital gyrus white matter
O2	-	middle or lateral occipital gyrus
O2-WM	-	middle or lateral occipital gyrus white matter
pcr	-	posterior corona radiata
PCun	-	precuneus
PCun-WM	-	precuneus white matter
PoG	-	postcentral gyrus
PoG-WM	-	postcentral gyrus white matter
PrG	-	precentral gyrus
PrG-WM	-	precentral gyrus white matter
scr	-	superior corona radiata
SFG	-	superior frontal gyrus
SFG-WM	-	superior frontal gyrus white matter
slf	-	superior longitudinal fasciculus
SMG	-	supramarginal gyrus
SMG-WM	-	supramarginal gyrus white matter
SPL-WM	-	superior parietal lobule white matter

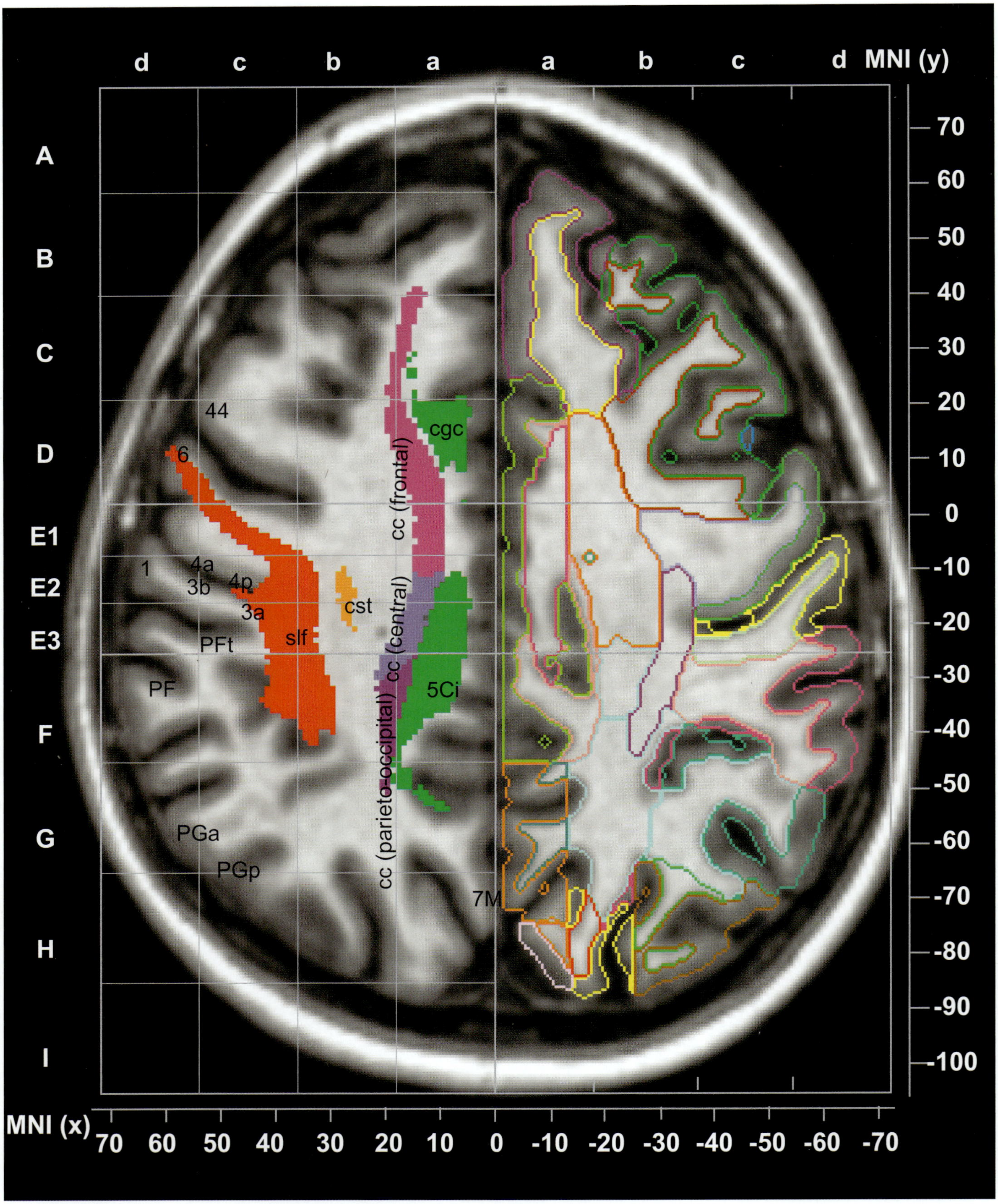

AnG - angular gyrus
AnG-WM - angular gyrus white matter
cc (central) - corpus callosum (connecting bilateral central areas)
cc (frontal) - corpus callosum (connecting bilateral frontal areas)
cc (parieto-occipital) - corpus callosum (connecting bilateral parieto-occipital areas)
CG - cingulate gyrus
cgc - cingulum (cingulate gyrus part)
CG-WM - cingulate gyrus white matter
cst - corticospinal tract
Cun - cuneus
MFG - middle frontal gyrus
MFG-WM - middle frontal gyrus white matter
O1 - superior occipital gyrus
O1-WM - superior occipital gyrus white matter
O2 - middle or lateral occipital gyrus

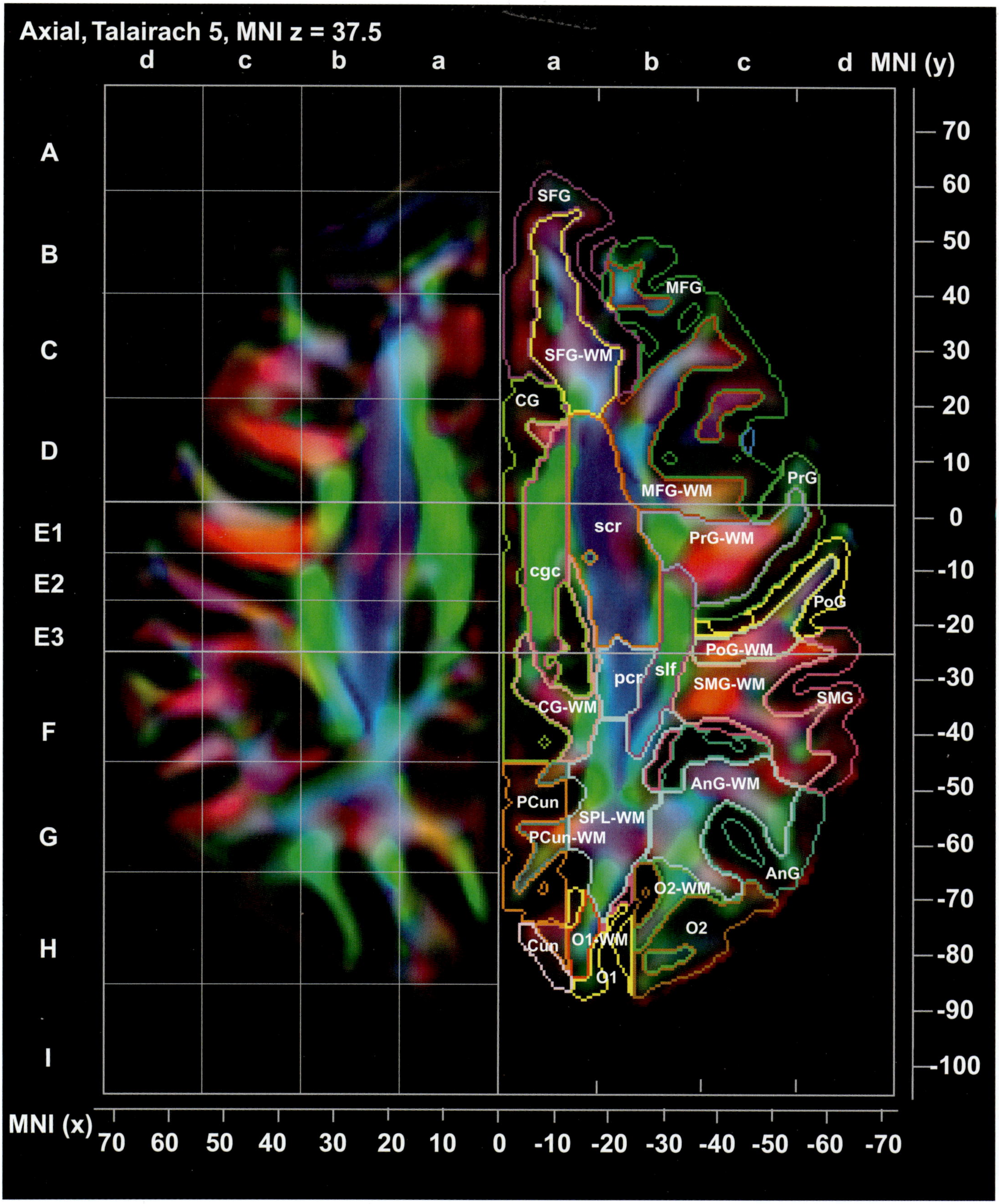

O2-WM - middle or lateral occipital gyrus white matter
pcr - posterior corona radiata
PCun - precuneus
PCun-WM - precuneus white matter
PoG - postcentral gyrus
PoG-WM - postcentral gyrus white matter
PrG - precentral gyrus
PrG-WM - precentral gyrus white matter
scr - superior corona radiata
SFG - superior frontal gyrus
SFG-WM - superior frontal gyrus white matter
slf - superior longitudinal fasciculus
SMG - supramarginal gyrus
SMG-WM - supramarginal gyrus white matter
SPL-WM - superior parietal lobule white matter

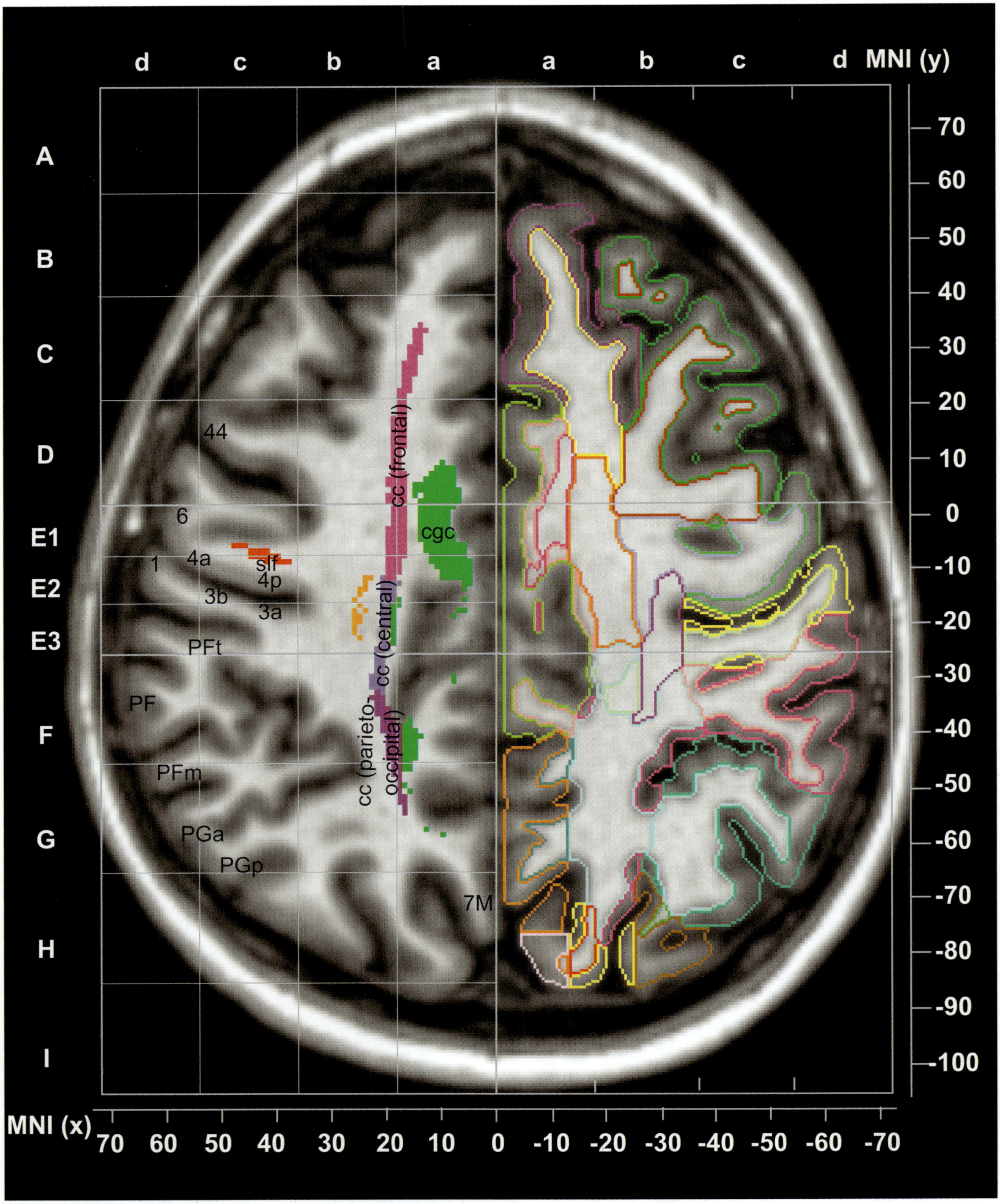

AnG - angular gyrus
AnG-WM - angular gyrus white matter
cc (central) - corpus callosum (connecting bilateral central areas)
cc (frontal) - corpus callosum (connecting bilateral frontal areas)
cc (parieto-occipital) - corpus callosum (connecting bilateral parieto-occipital areas)
CG - cingulate gyrus
cgc - cingulum (cingulate gyrus part)
CG-WM - cingulate gyrus white matter
Cun - cuneus
MFG - middle frontal gyrus
MFG-WM - middle frontal gyrus white matter
O1 - superior occipital gyrus
O1-WM - superior occipital gyrus white matter
O2 - middle or lateral occipital gyrus
pcr - posterior corona radiata

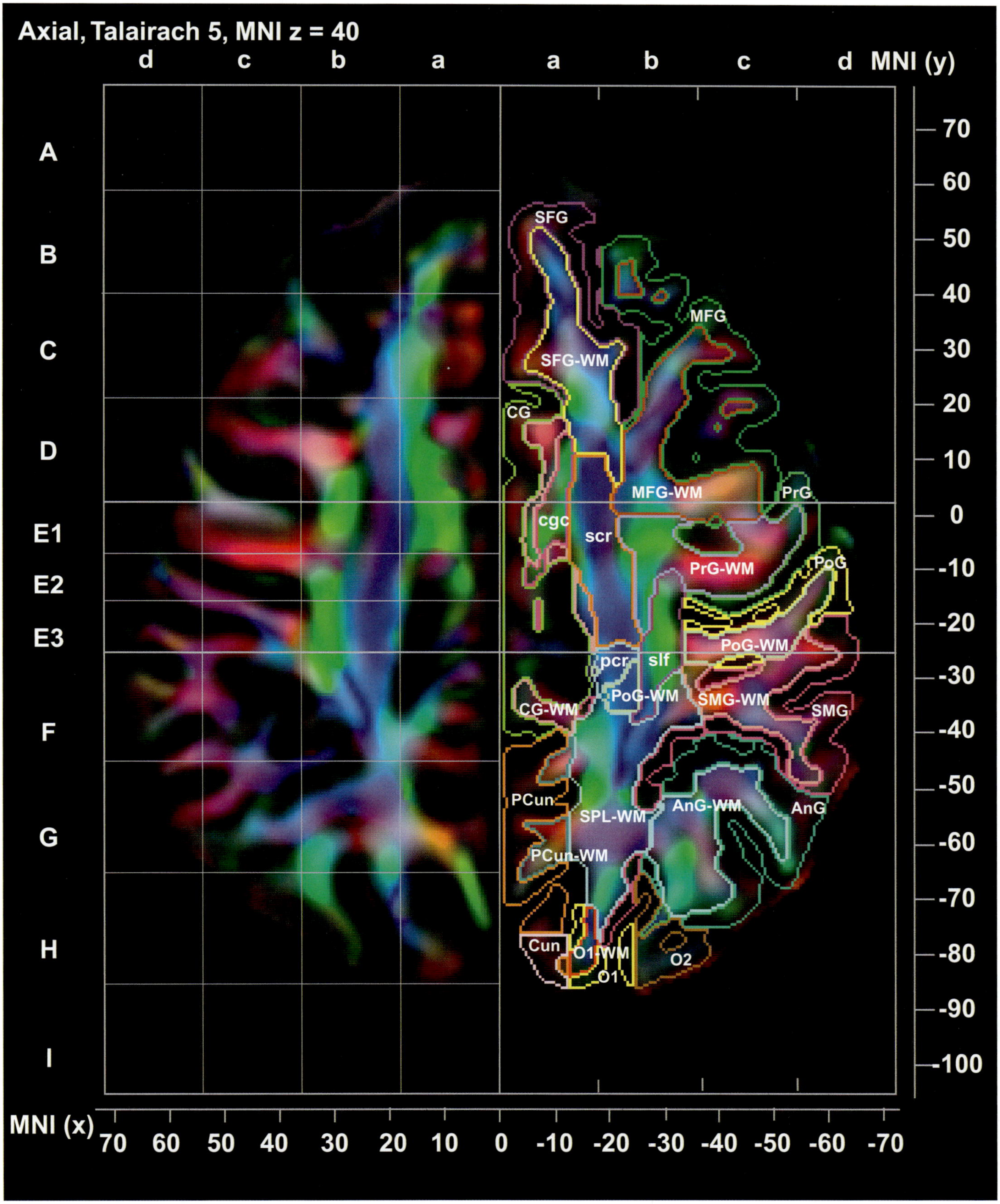

PCun - precuneus
PCun-WM - precuneus white matter
PoG - postcentral gyrus
PoG-WM - postcentral gyrus white matter
PrG - precentral gyrus
PrG-WM - precentral gyrus white matter
scr - superior corona radiata
SFG - superior frontal gyrus
SFG-WM - superior frontal gyrus white matter
slf - superior longitudinal fasciculus
SMG - supramarginal gyrus
SMG-WM - supramarginal gyrus white matter
SPL-WM - superior parietal lobule white matter

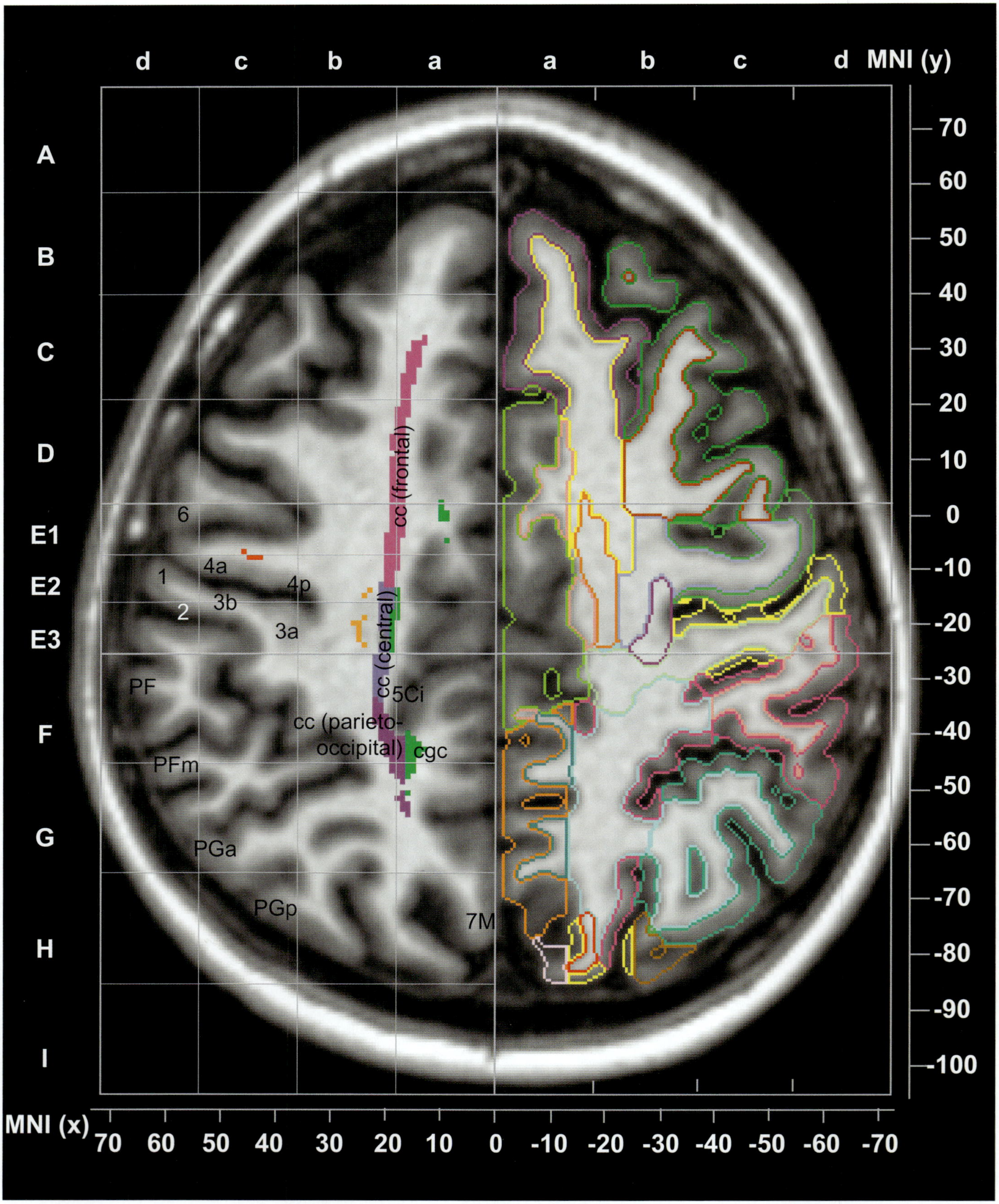

AnG	- angular gyrus	cc (parieto-occipital)	- corpus callosum (connecting bilateral parieto-occipital areas)	MFG	- middle frontal gyrus
AnG-WM	- angular gyrus white matter	CG	- cingulate gyrus	MFG-WM	- middle frontal gyrus white matter
cc (central)	- corpus callosum (connecting bilateral central areas)	cgc	- cingulum (cingulate gyrus part)	O1	- superior occipital gyrus
cc (frontal)	- corpus callosum (connecting bilateral frontal areas)	CG-WM	- cingulate gyrus white matter	O1-WM	- superior occipital gyrus white matter
		Cun	- cuneus	O2	- middle or lateral occipital gyrus
				PCun	- precuneus

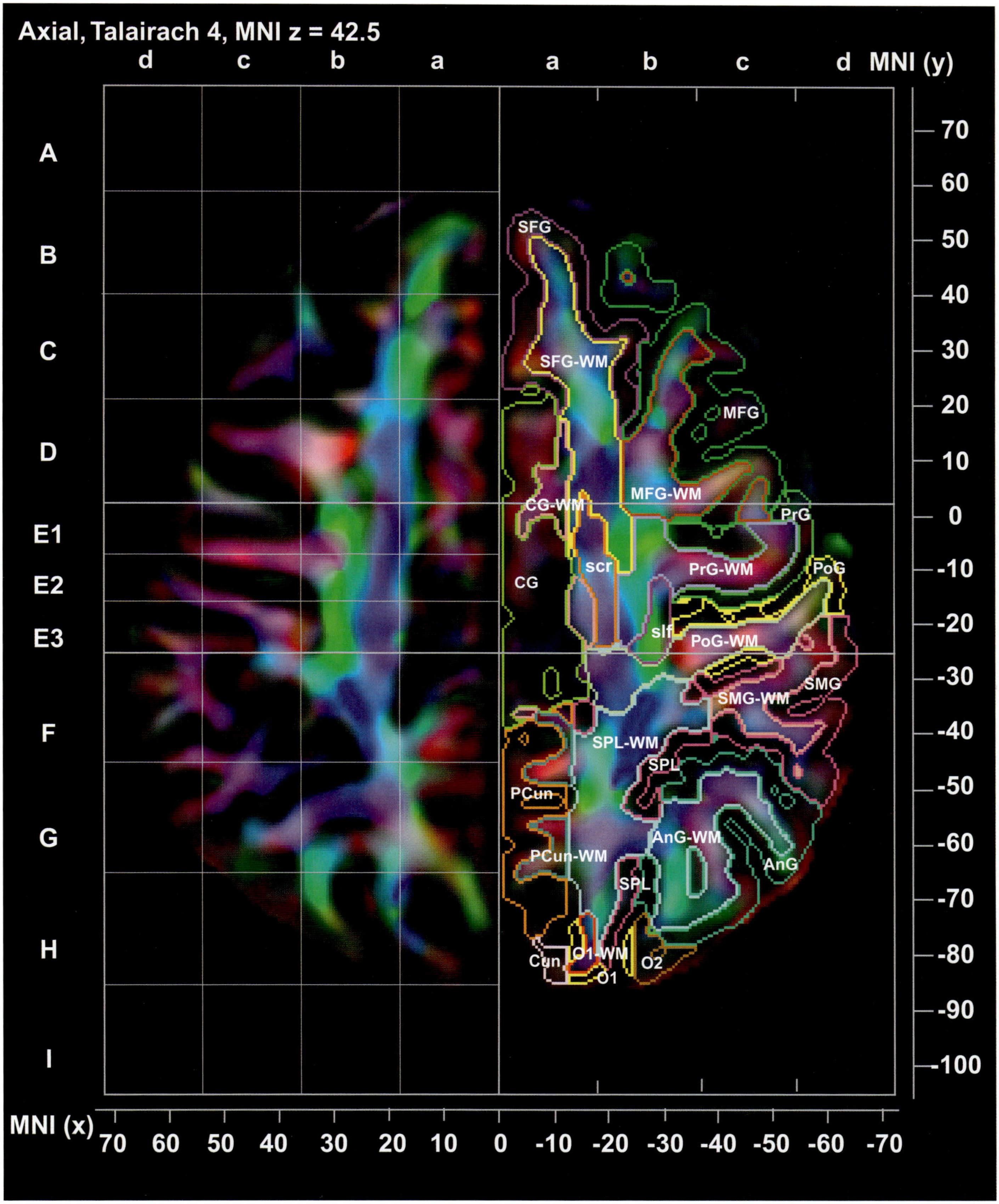

PCun-WM - precuneus white matter
PoG - postcentral gyrus
PoG-WM - postcentral gyrus white matter
PrG - precentral gyrus
PrG-WM - precentral gyrus white matter
scr - superior corona radiata
SFG - superior frontal gyrus
SFG-WM - superior frontal gyrus white matter
slf - superior longitudinal fasciculus
SMG - supramarginal gyrus
SMG-WM - supramarginal gyrus white matter
SPL - superior parietal lobule
SPL-WM - superior parietal lobule white matter

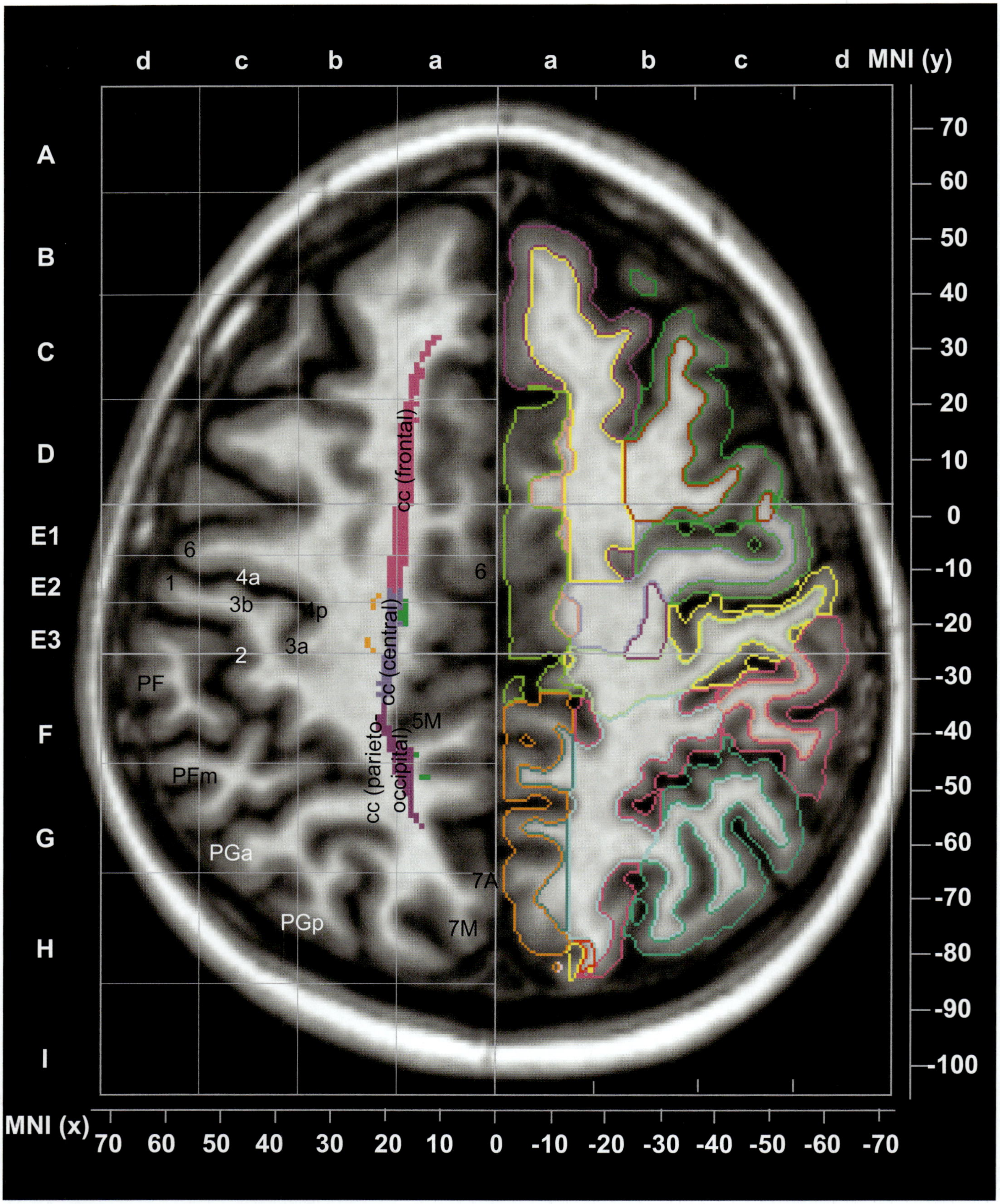

AnG - angular gyrus
AnG-WM - angular gyrus white matter
cc (central) - corpus callosum (connecting bilateral central areas)
cc (frontal) - corpus callosum (connecting bilateral frontal areas)
cc (parieto-occipital) - corpus callosum (connecting bilateral parieto-occipital areas)
CG - cingulate gyrus
CG-WM - cingulate gyrus white matter
MFG - middle frontal gyrus
MFG-WM - middle frontal gyrus white matter
O1 - superior occipital gyrus
PCun - precuneus
PCun-WM - precuneus white matter
PoG - postcentral gyrus
PoG-WM - postcentral gyrus white matter
PrG - precentral gyrus

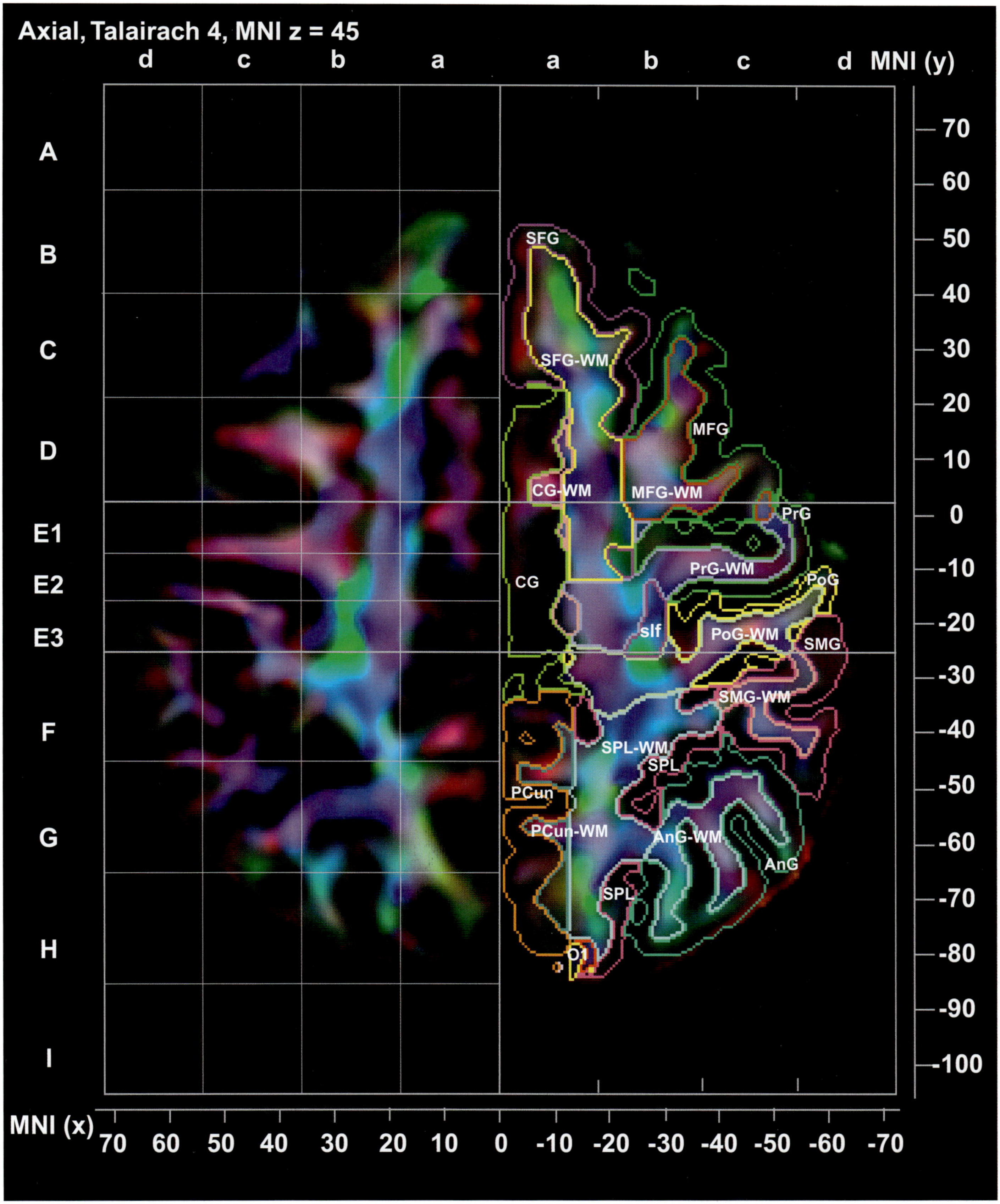

PrG-WM - precentral gyrus white matter
scr - superior corona radiata
SFG - superior frontal gyrus
SFG-WM - superior frontal gyrus white matter
slf - superior longitudinal fasciculus
SMG - supramarginal gyrus
SMG-WM - supramarginal gyrus white matter
SPL - superior parietal lobule
SPL-WM - superior parietal lobule white matter

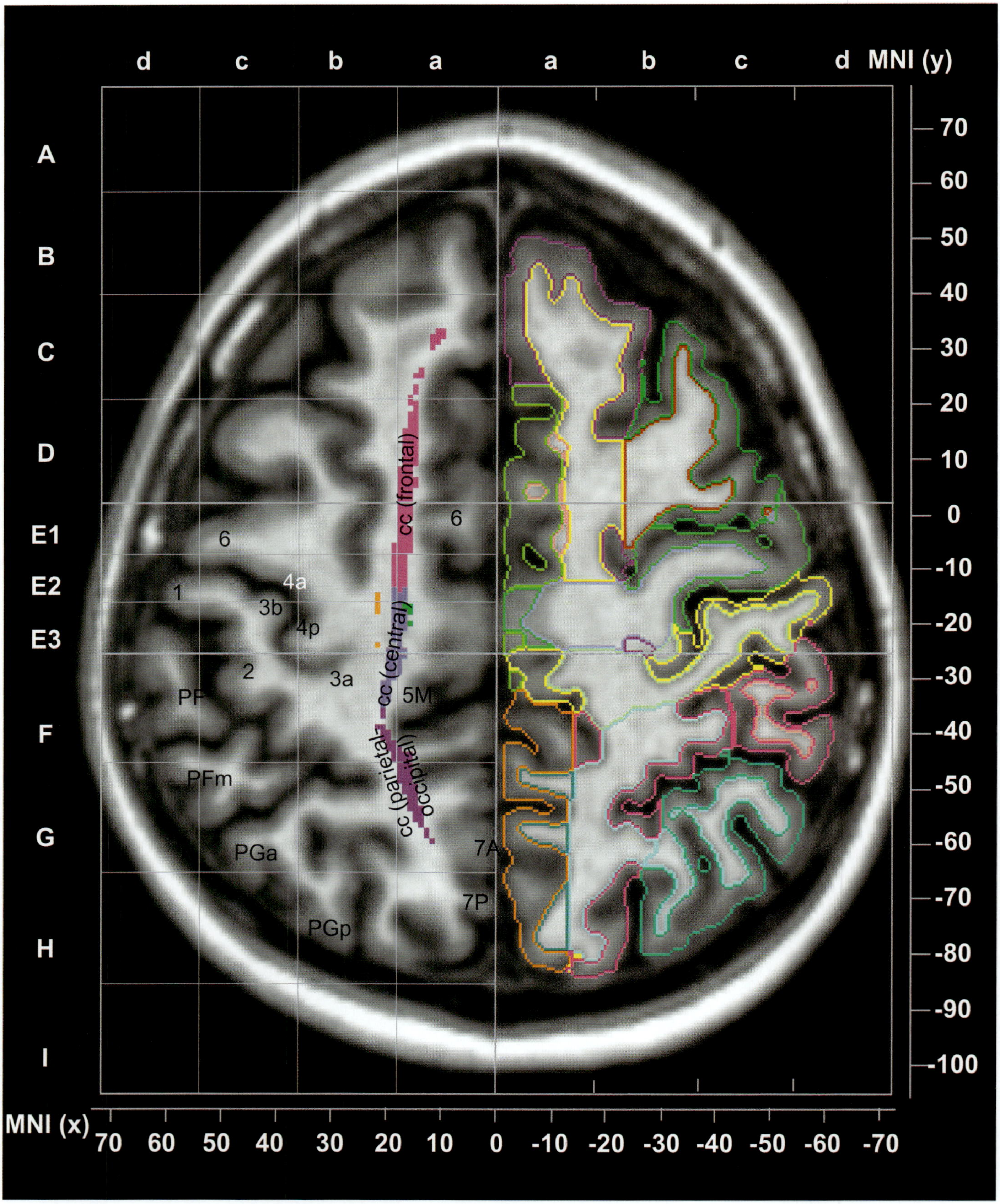

AnG - angular gyrus
AnG-WM - angular gyrus white matter
cc (central) - corpus callosum (connecting bilateral central areas)
cc (frontal) - corpus callosum (connecting bilateral frontal areas)
cc (parieto-occipital) - corpus callosum (connecting bilateral parieto-occipital areas)
CG - cingulate gyrus
CG-WM - cingulate gyrus white matter
MFG - middle frontal gyrus
MFG-WM - middle frontal gyrus white matter
PCun - precuneus
PCun-WM - precuneus white matter
PoG - postcentral gyrus
PoG-WM - postcentral gyrus white matter
PrG - precentral gyrus
PrG-WM - precentral gyrus white matter

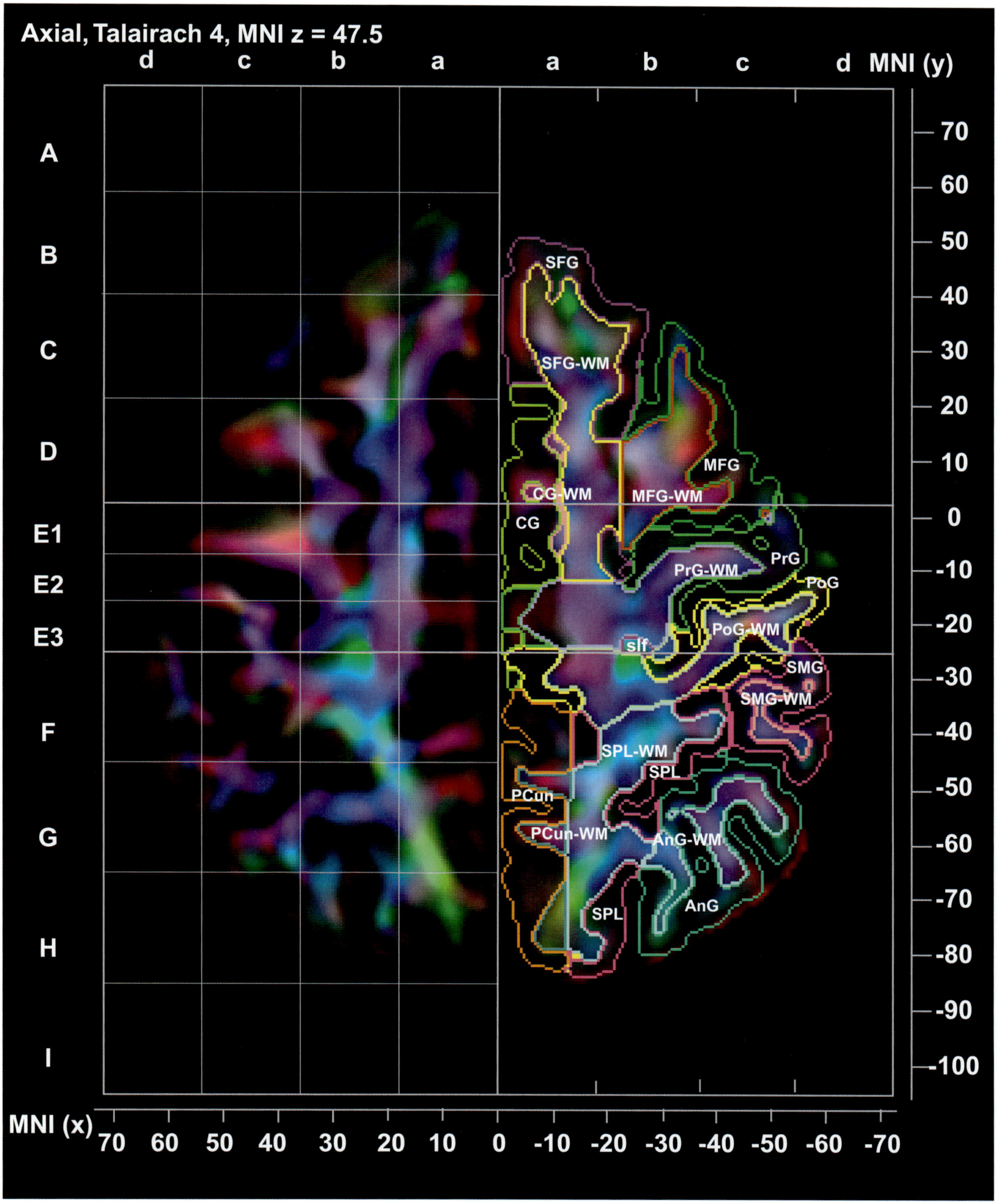

SFG - superior frontal gyrus
SFG-WM - superior frontal gyrus white matter
slf - superior longitudinal fasciculus
SMG - supramarginal gyrus
SMG-WM - supramarginal gyrus white matter
SPL - superior parietal lobule
SPL-WM - superior parietal lobule white matter

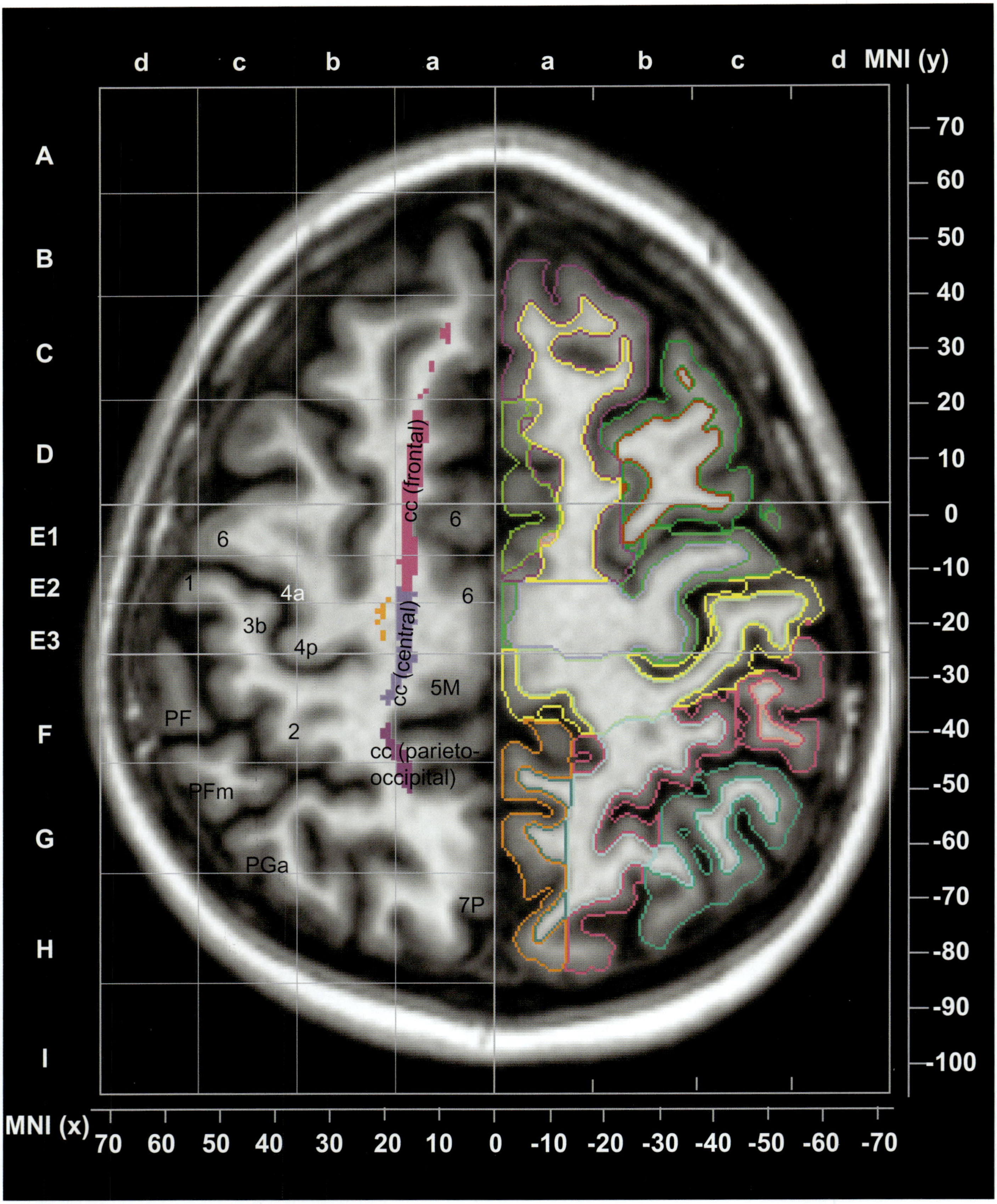

AnG - angular gyrus
AnG-WM - angular gyrus white matter
cc (central) - corpus callosum (connecting bilateral central areas)
cc (frontal) - corpus callosum (connecting bilateral frontal areas)
cc (parieto-occipital) - corpus callosum (connecting bilateral parieto-occipital areas)
CG - cingulate gyrus
MFG - middle frontal gyrus
MFG-WM - middle frontal gyrus white matter
PCun - precuneus
PCun-WM - precuneus white matter
PoG - postcentral gyrus
PoG-WM - postcentral gyrus white matter
PrG - precentral gyrus
PrG-WM - precentral gyrus white matter
SFG - superior frontal gyrus

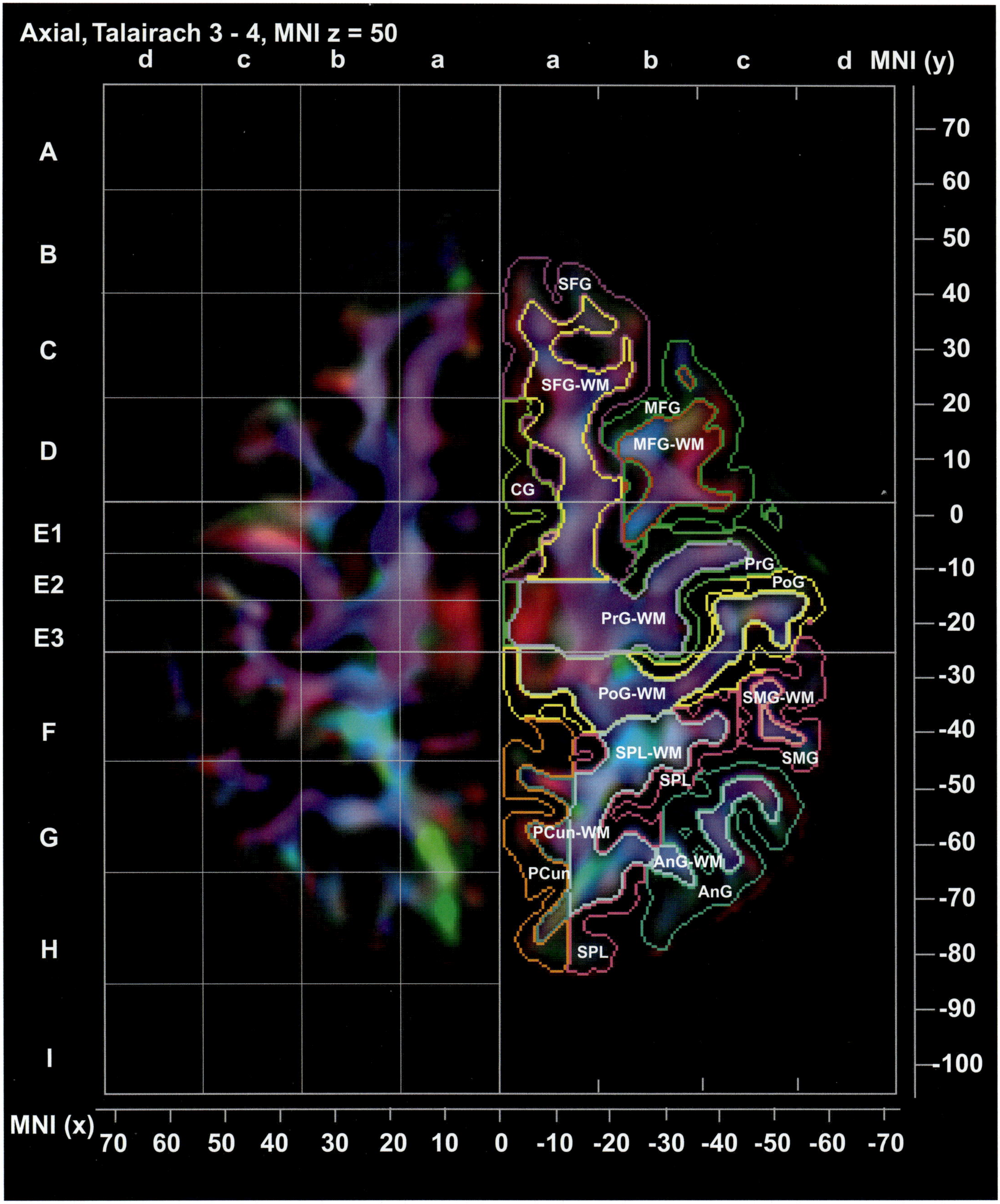

SFG-WM - superior frontal gyrus white matter
SMG - supramarginal gyrus
SMG-WM - supramarginal gyrus white matter
SPL - superior parietal lobule
SPL-WM - superior parietal lobule white matter

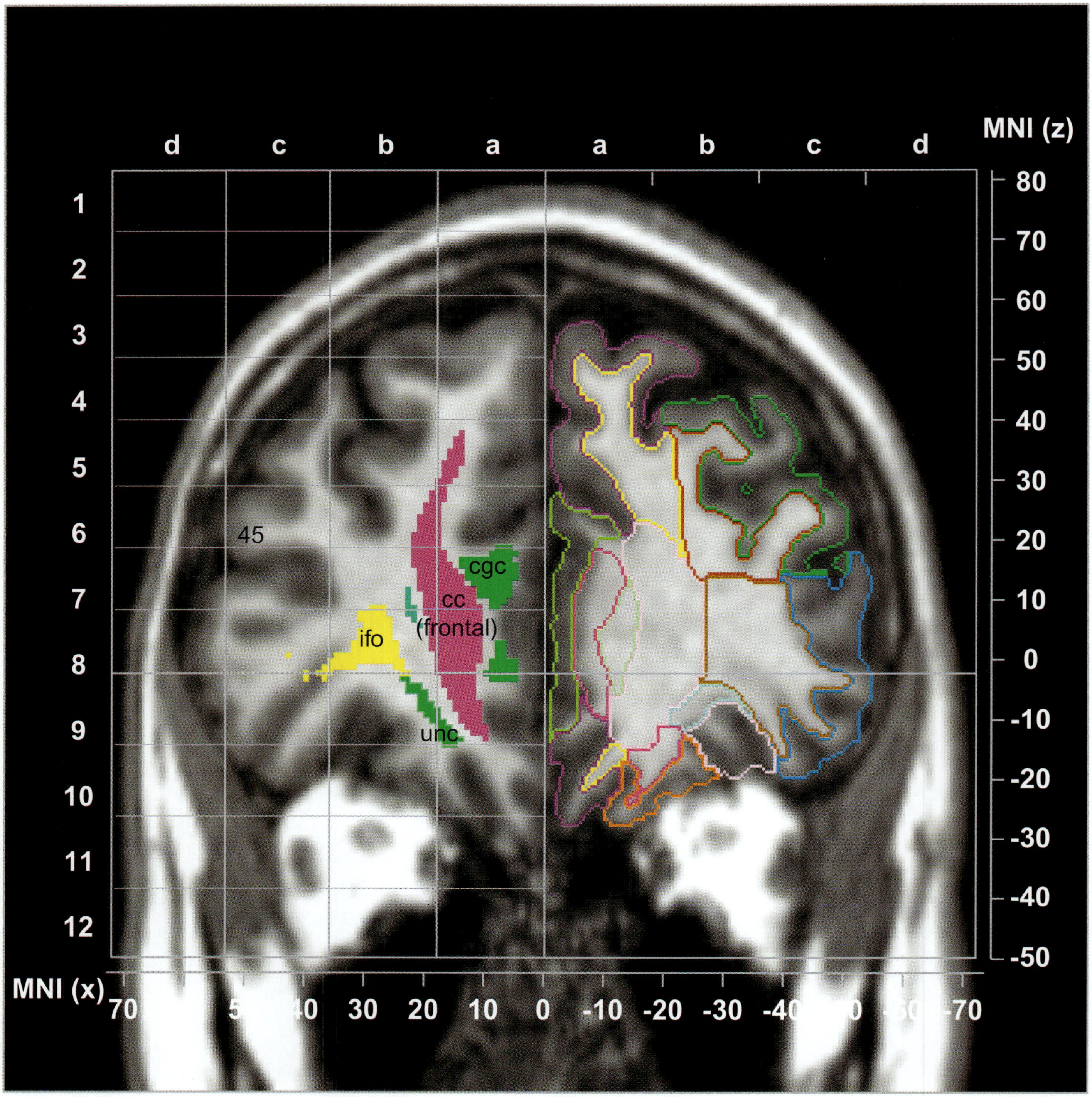

cc (frontal) - corpus callosum (connecting bilateral frontal areas)
cgc - cingulum (cingulate gyrus part)
ifo - inferior fronto-occipital fasciculus
unc - uncinate fasciculus

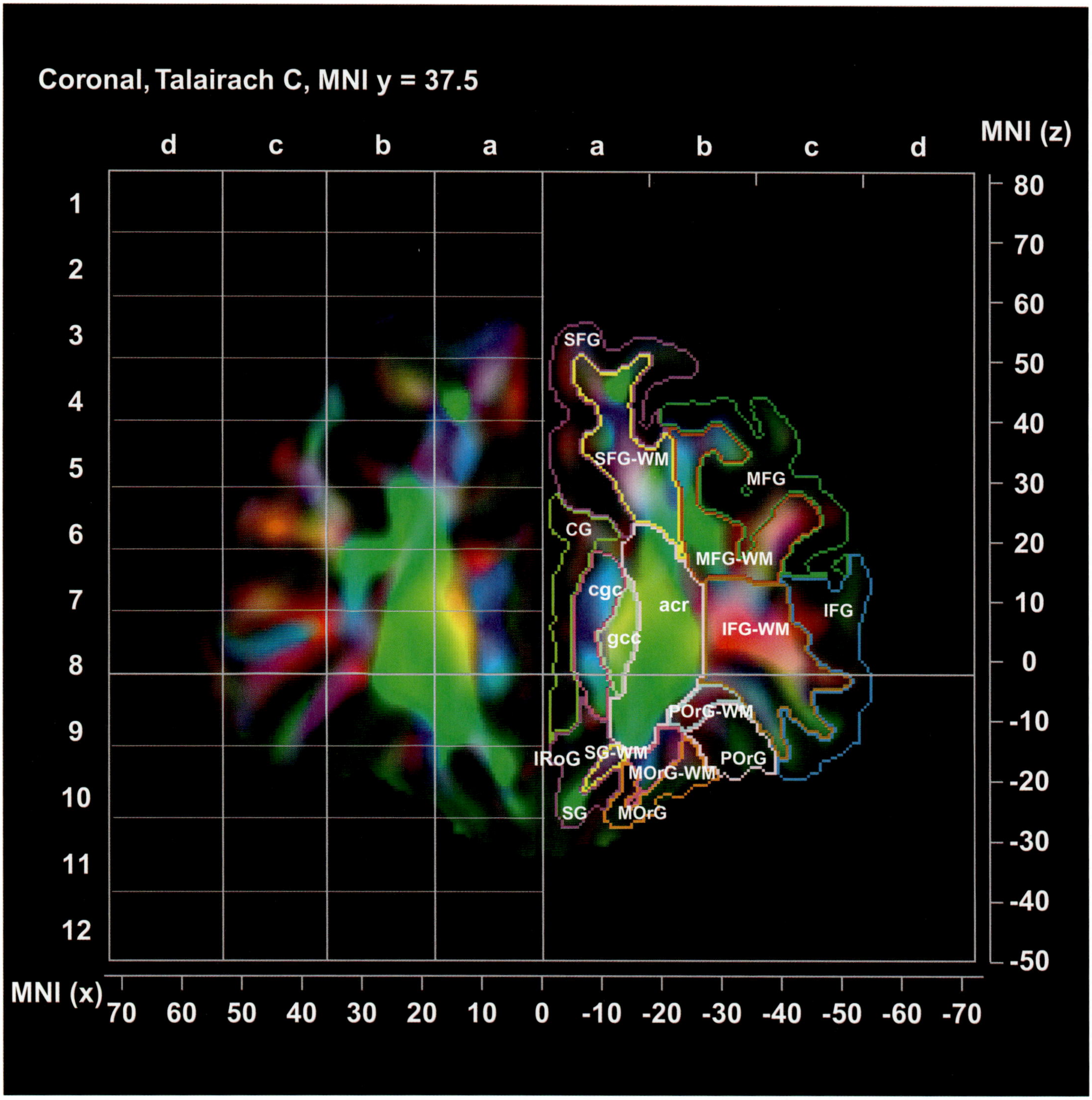

acr - anterior corona radiata
CG - cingulate gyrus
gcc - genu of the corpus callosum
IFG - inferior frontal gyrus
IFG-WM - inferior frontal gyrus white matter
IRoG - inferior rostral gyrus
MFG - middle frontal gyrus
MFG-WM - middle frontal gyrus white matter
MOrG - medial orbital gyrus
MOrG-WM - medial orbital gyrus white matter
POrG - posterior orbital gyrus
POrG-WM - posterior orbital gyrus white matter
SFG - superior frontal gyrus
SFG-WM - superior frontal gyrus white matter
SG - straight gyrus
SG-WM - straight gyrus white matter

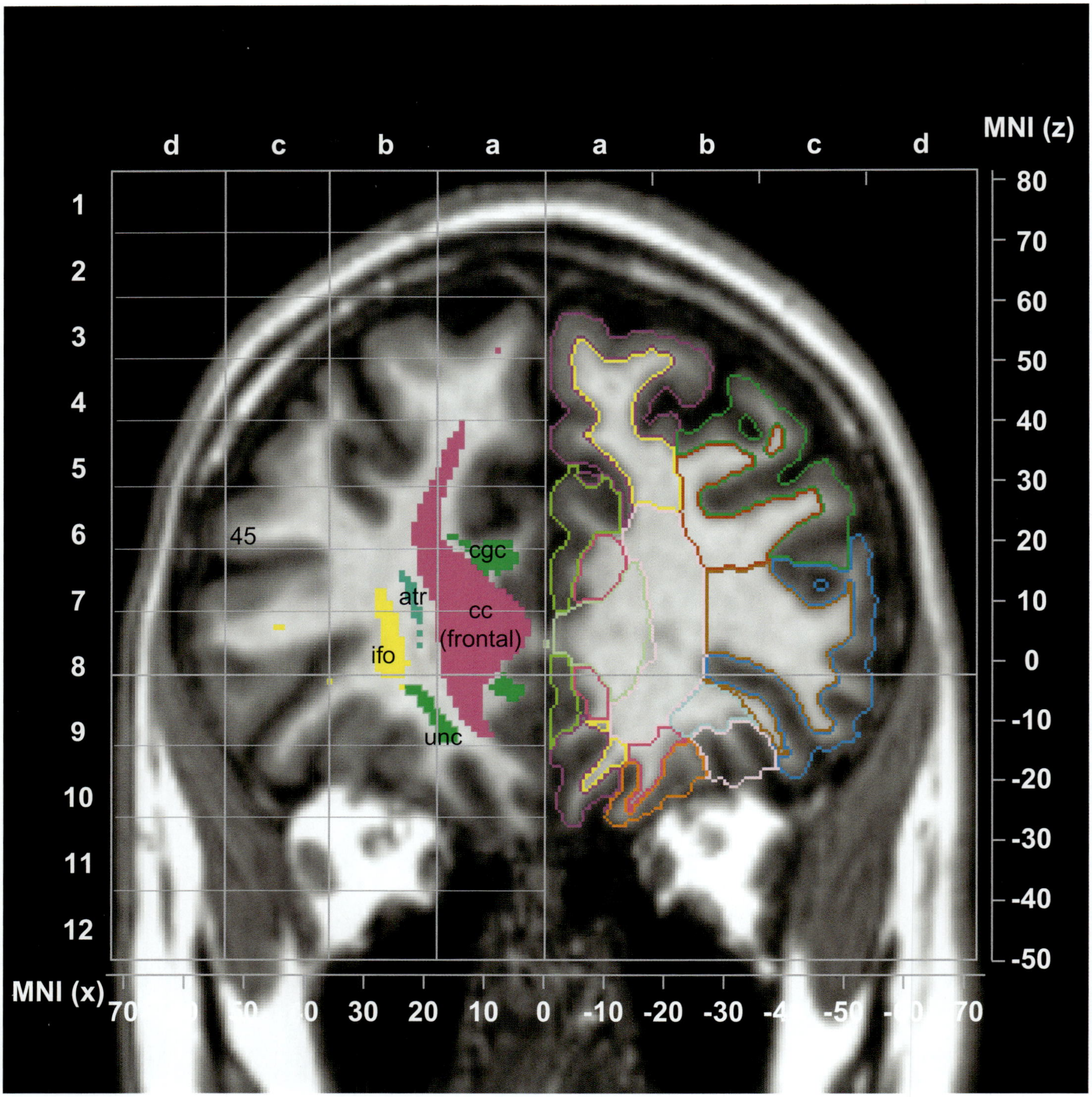

atr - anterior thalamic radiation
cc (frontal) - corpus callosum (connecting bilateral frontal areas)
cgc - cingulum (cingulate gyrus part)
ifo - inferior fronto-occipital fasciculus
unc - uncinate fasciculus

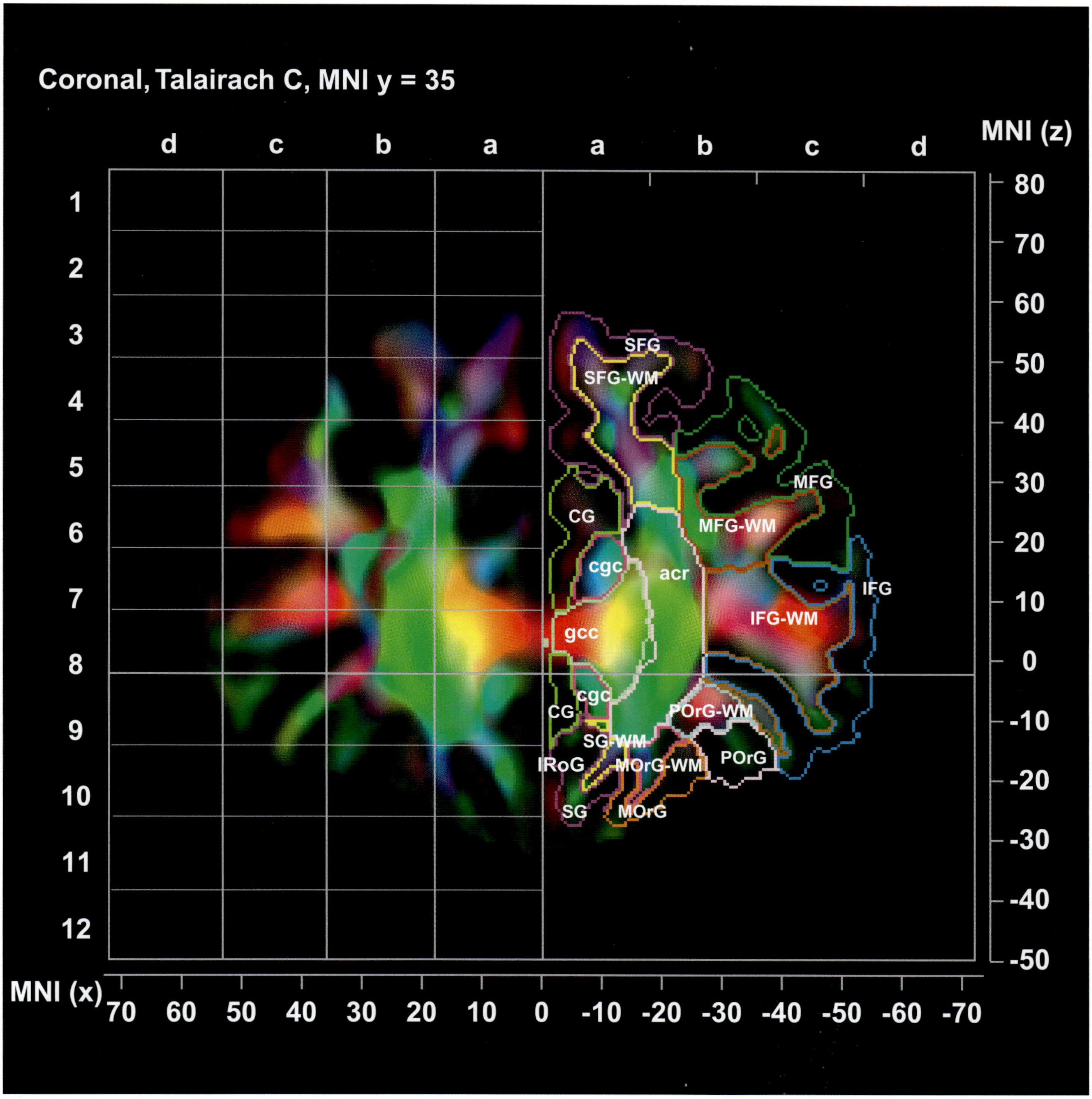

acr	-	anterior corona radiata
CG	-	cingulate gyrus
cgc	-	cingulum (cingulate gyrus part)
gcc	-	genu of the corpus callosum
IFG	-	inferior frontal gyrus
IFG-WM	-	inferior frontal gyrus white matter
IRoG	-	inferior rostral gyrus
MFG	-	middle frontal gyrus
MFG-WM	-	middle frontal gyrus white matter
MOrG	-	medial orbital gyrus
MOrG-WM	-	medial orbital gyrus white matter
POrG	-	posterior orbital gyrus
POrG-WM	-	posterior orbital gyrus white matter
SFG	-	superior frontal gyrus
SFG-WM	-	superior frontal gyrus white matter
SG	-	straight gyrus
SG-WM	-	straight gyrus white matter

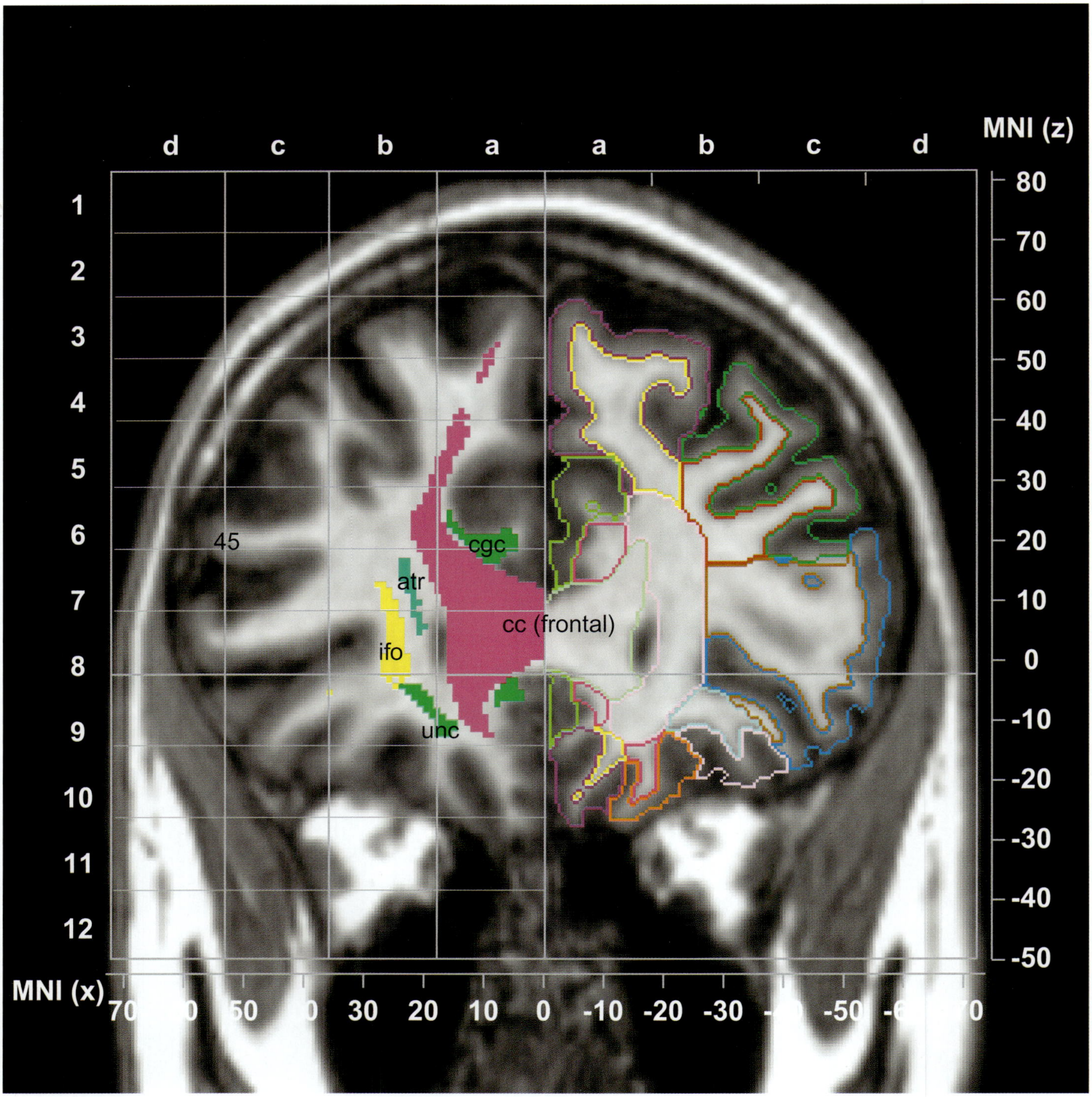

atr - anterior thalamic radiation
cc (frontal) - corpus callosum (connecting bilateral frontal areas)
cgc - cingulum (cingulate gyrus part)
ifo - inferior fronto-occipital fasciculus
unc - uncinate fasciculus

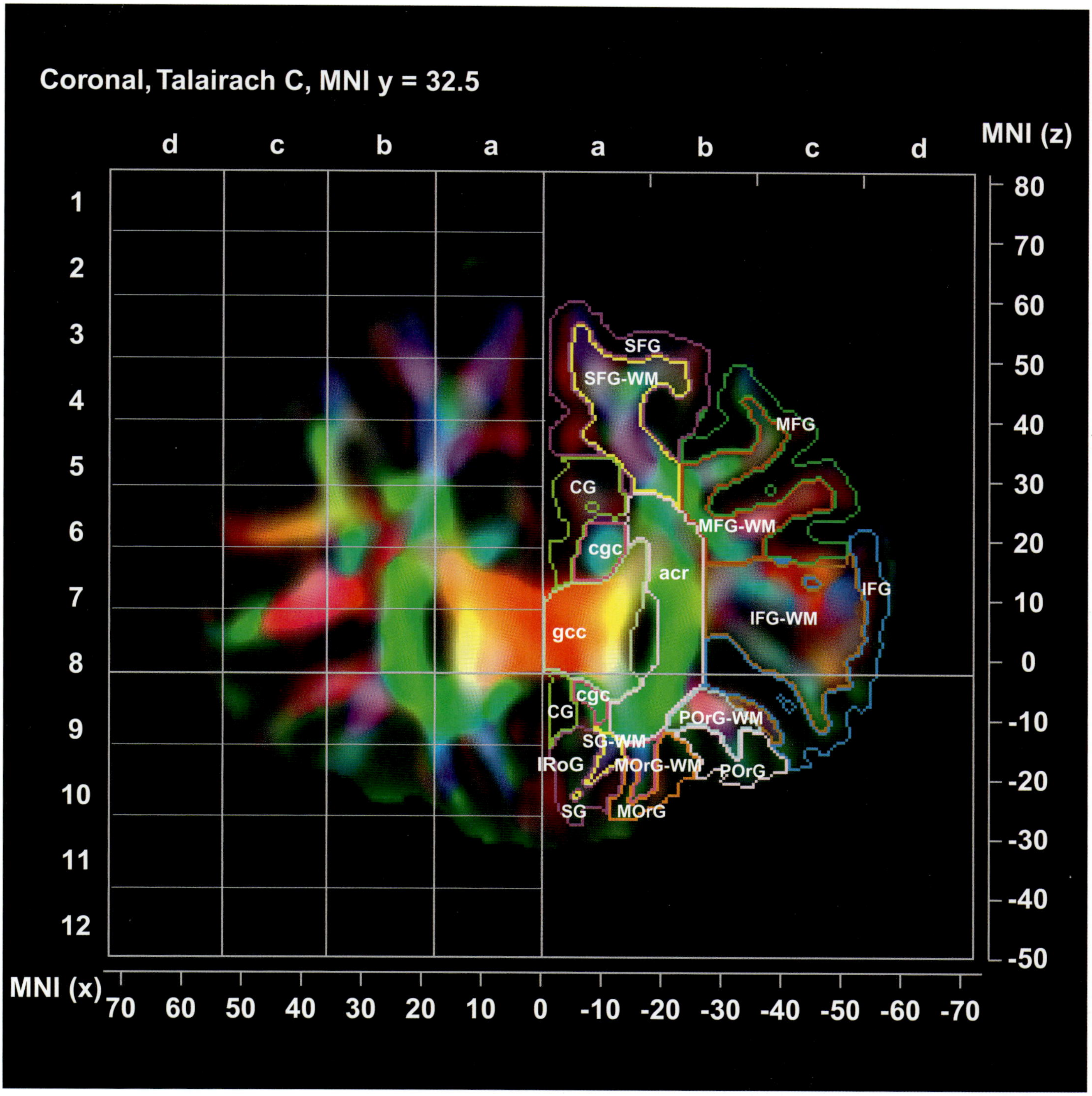

acr - anterior corona radiata
CG - cingulate gyrus
cgc - cingulum (cingulate gyrus part)
gcc - genu of the corpus callosum
IFG - inferior frontal gyrus
IFG-WM - inferior frontal gyrus white matter
IRoG - inferior rostral gyrus
MFG - middle frontal gyrus
MFG-WM - middle frontal gyrus white matter
MOrG - medial orbital gyrus
MOrG-WM - medial orbital gyrus white matter
POrG - posterior orbital gyrus
POrG-WM - posterior orbital gyrus white matter
SFG - superior frontal gyrus
SFG-WM - superior frontal gyrus white matter
SG - straight gyrus
SG-WM - straight gyrus white matter

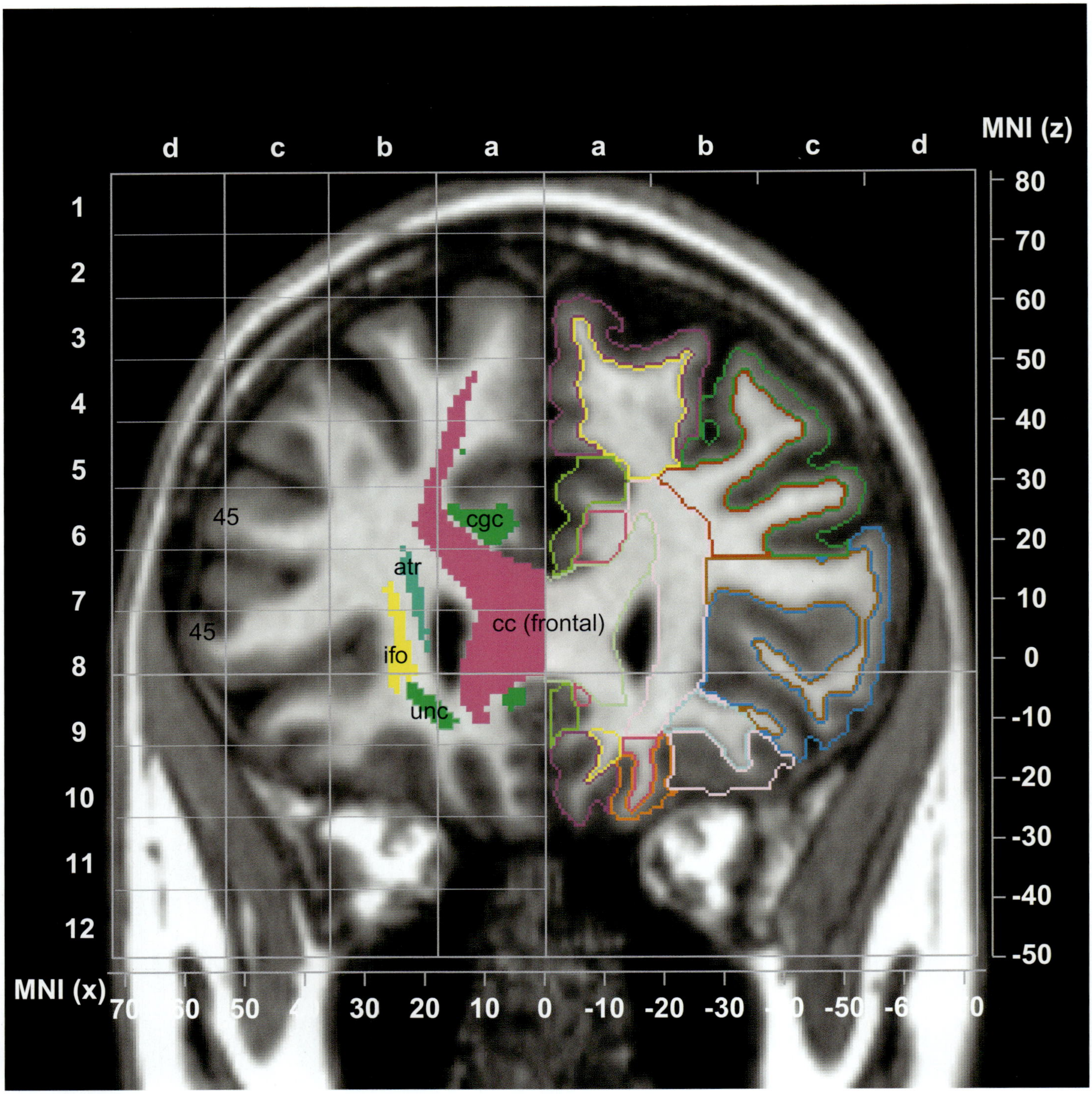

atr - anterior thalamic radiation
cc (frontal) - corpus callosum (connecting bilateral frontal areas)
cgc - cingulum (cingulate gyrus part)
ifo - inferior fronto-occipital fasciculus
unc - uncinate fasciculus

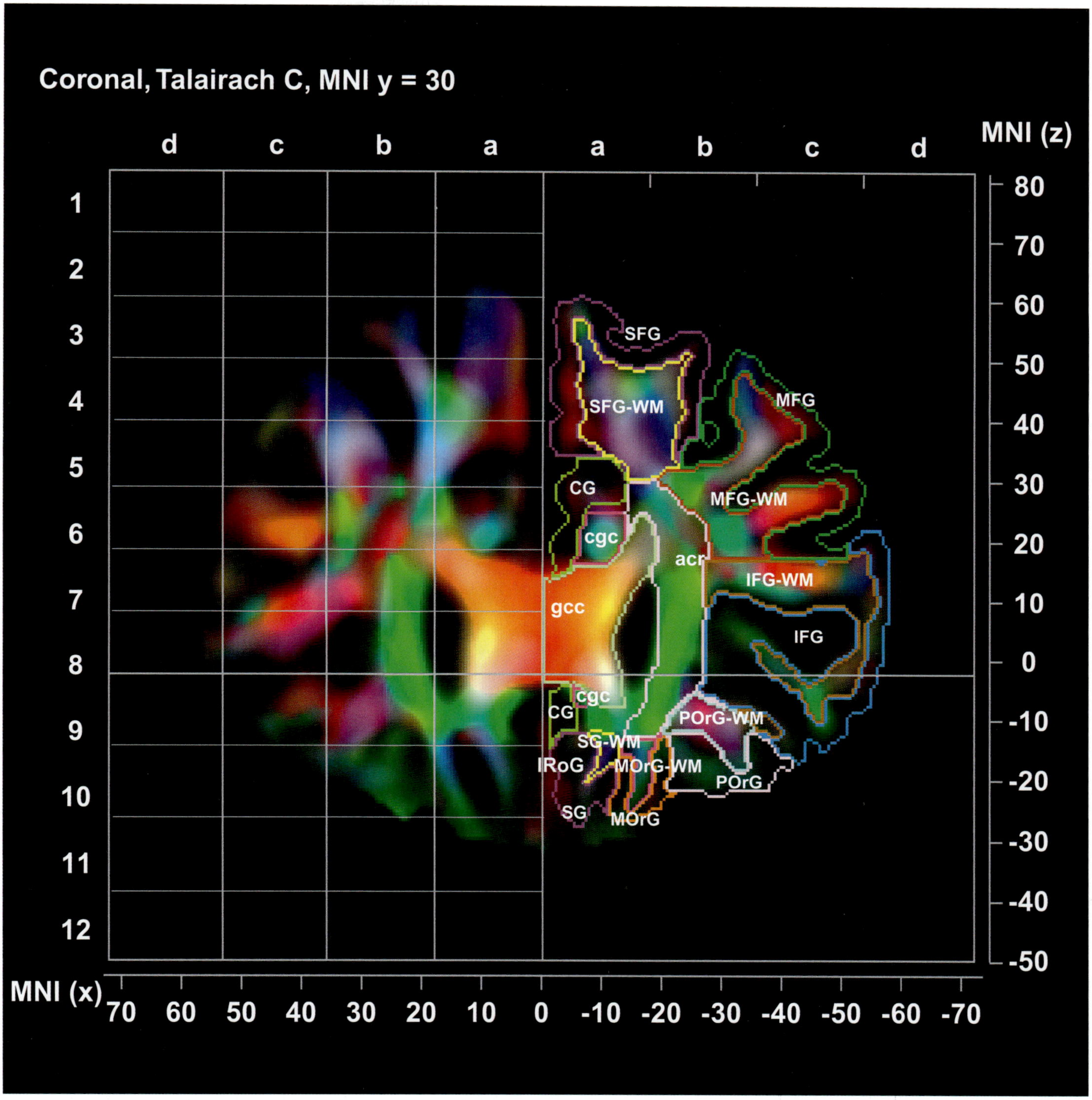

acr	-	anterior corona radiata
CG	-	cingulate gyrus
cgc	-	cingulum (cingulate gyrus part)
gcc	-	genu of the corpus callosum
IFG	-	inferior frontal gyrus
IFG-WM	-	inferior frontal gyrus white matter
IRoG	-	inferior rostral gyrus
MFG	-	middle frontal gyrus
MFG-WM	-	middle frontal gyrus white matter
MOrG	-	medial orbital gyrus
MOrG-WM	-	medial orbital gyrus white matter
POrG	-	posterior orbital gyrus
POrG-WM	-	posterior orbital gyrus white matter
SFG	-	superior frontal gyrus
SFG-WM	-	superior frontal gyrus white matter
SG	-	straight gyrus
SG-WM	-	straight gyrus white matter

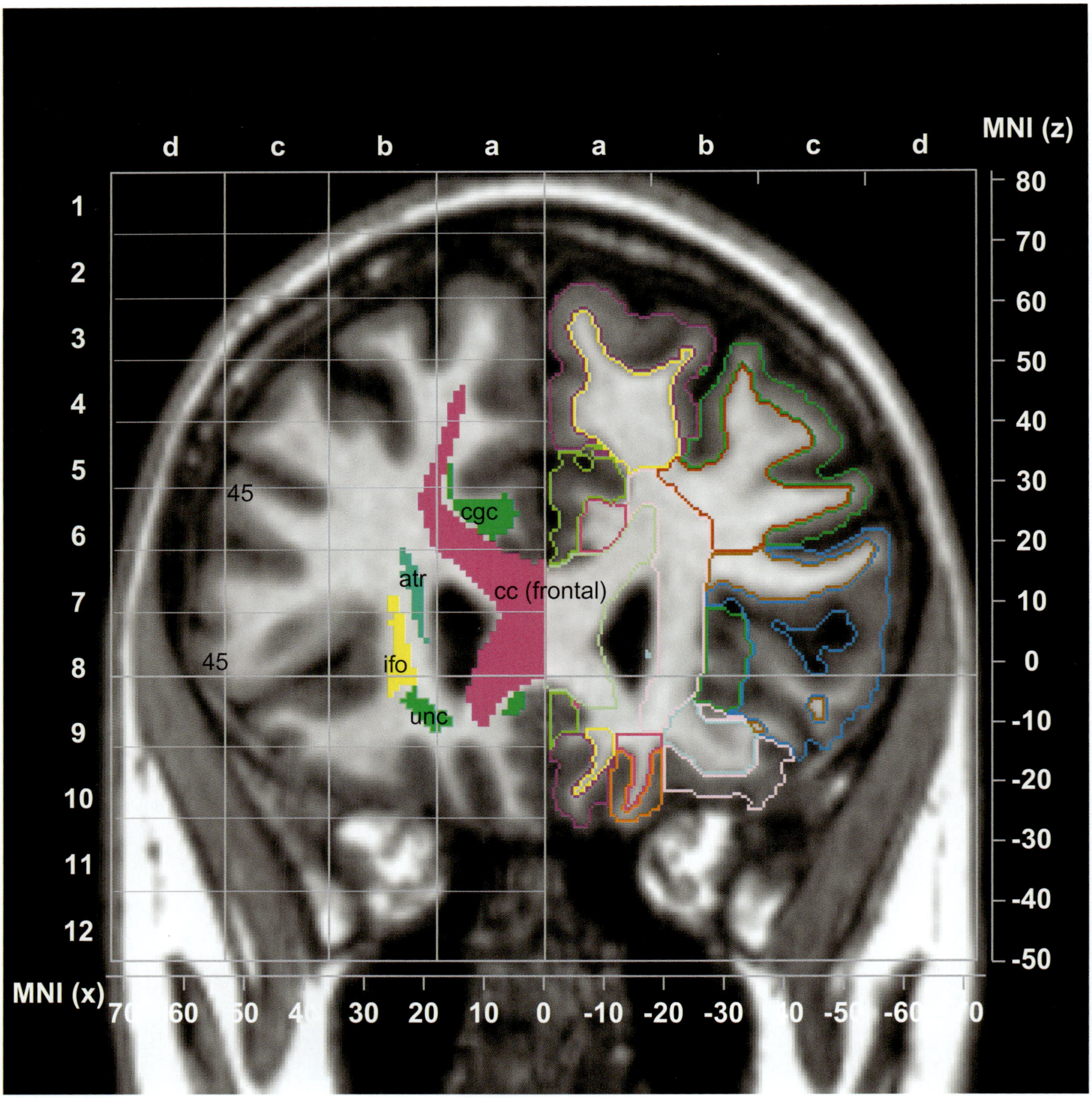

atr	-	anterior thalamic radiation	cgc	-	cingulum (cingulate gyrus part)
cc (frontal)	-	corpus callosum (connecting bilateral frontal areas)	ifo	-	inferior fronto-occipital fasciculus
			unc	-	uncinate fasciculus

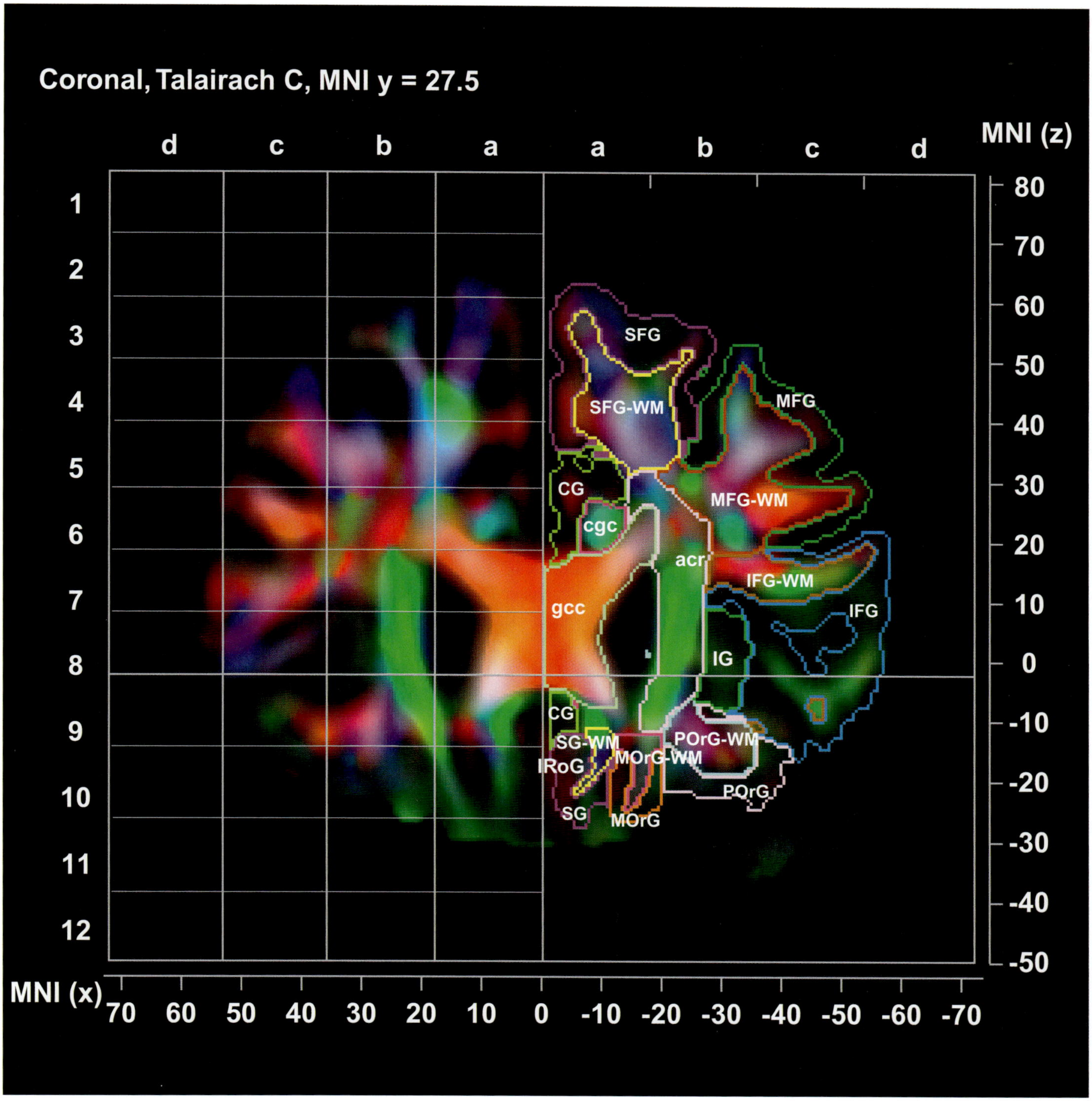

acr	-	anterior corona radiata
CG	-	cingulate gyrus
cgc	-	cingulum (cingulate gyrus part)
gcc	-	genu of the corpus callosum
IFG	-	inferior frontal gyrus
IFG-WM	-	inferior frontal gyrus white matter
IRoG	-	inferior rostral gyrus
MFG	-	middle frontal gyrus
MFG-WM	-	middle frontal gyrus white matter
MOrG	-	medial orbital gyrus
MOrG-WM	-	medial orbital gyrus white matter
POrG	-	posterior orbital gyrus
POrG-WM	-	posterior orbital gyrus white matter
SFG	-	superior frontal gyrus
SFG-WM	-	superior frontal gyrus white matter
SG	-	straight gyrus
SG-WM	-	straight gyrus white matter

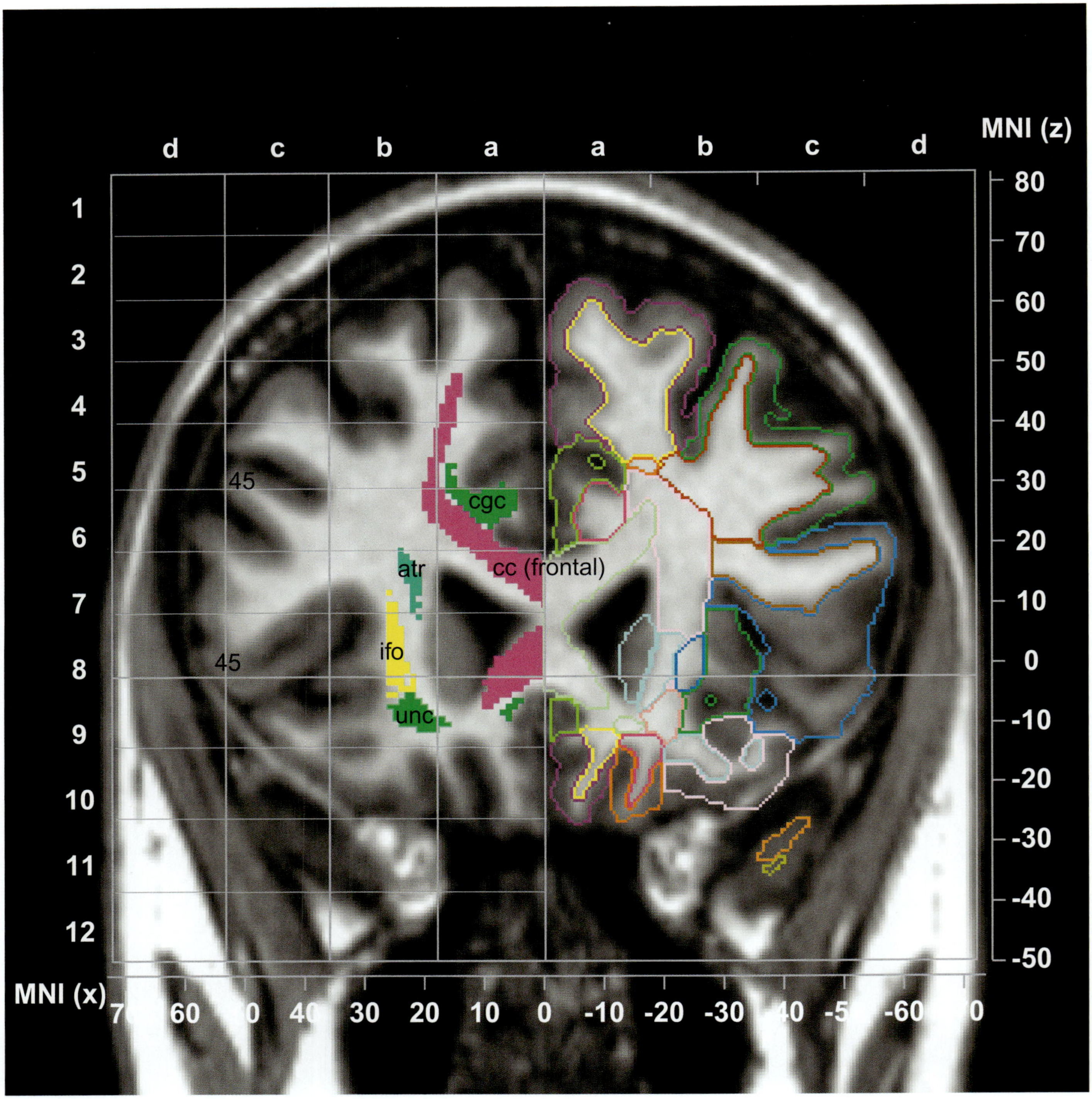

atr	-	anterior thalamic radiation	cgc	-	cingulum (cingulate gyrus part)
cc (frontal)	-	corpus callosum (connecting bilateral frontal areas)	ifo	-	inferior fronto-occipital fasciculus
			unc	-	uncinate fasciculus

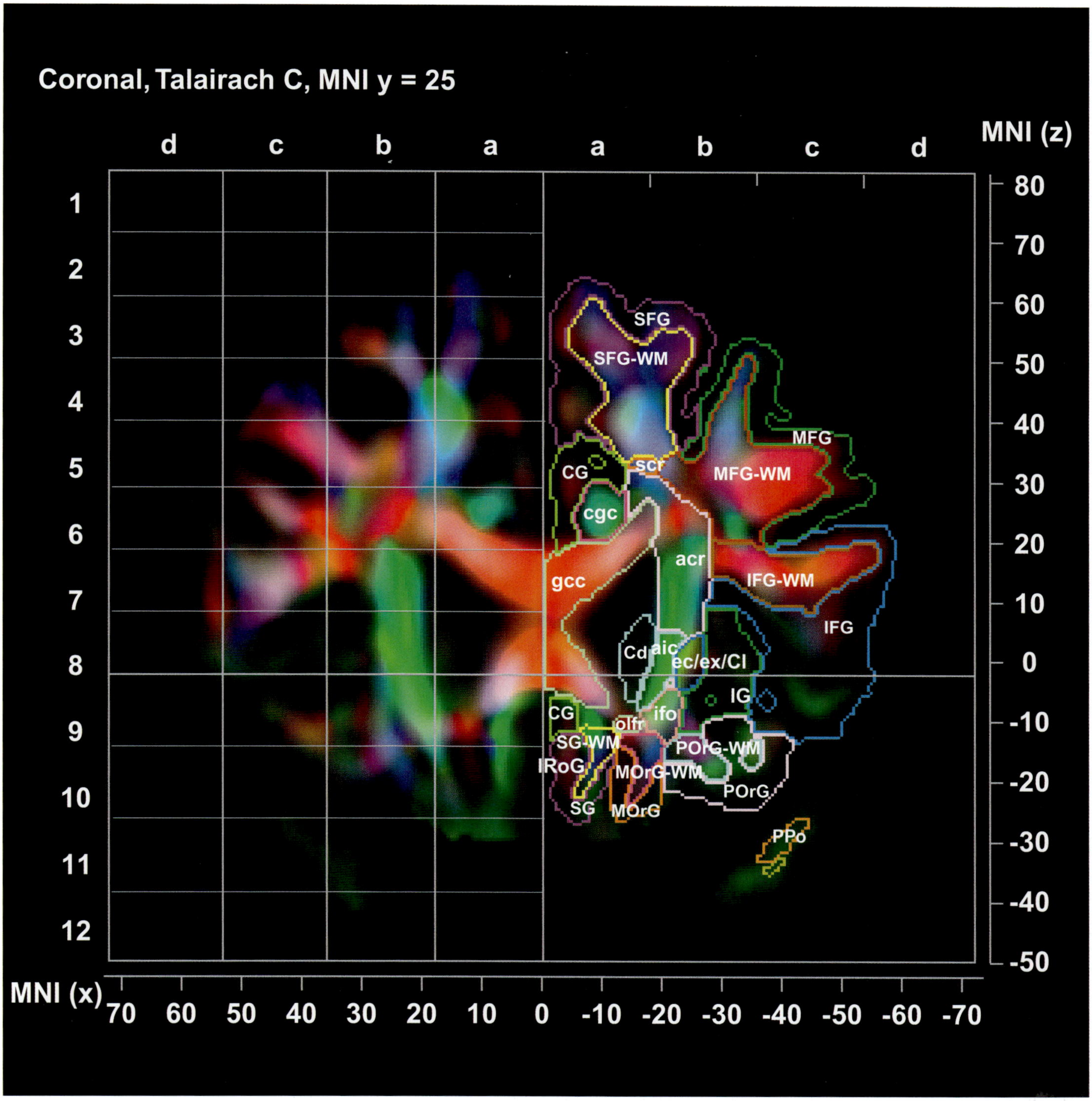

acr	-	anterior corona radiata
aic	-	anterior limb of the internal capsule
Cd	-	caudate nucleus
CG	-	cingulate gyrus
cgc	-	cingulum (cingulate gyrus part)
ec/ex/Cl	-	external capsule / extreme capsule / claustrum (cannot be resolved with current resolution)
gcc	-	genu of the corpus callosum
IFG	-	inferior frontal gyrus
IFG-WM	-	inferior frontal gyrus white matter
ifo	-	inferior fronto-occipital fasciculus
IG	-	insular gyrus
IRoG	-	inferior rostral gyrus
MFG	-	middle frontal gyrus
MFG-WM	-	middle frontal gyrus white matter
MOrG	-	medial orbital gyrus
MOrG-WM	-	medial orbital gyrus white matter
olfr	-	olfactory radiation
POrG	-	posterior orbital gyrus
POrG-WM	-	posterior orbital gyrus white matter
PPo	-	planum polare
scr	-	superior corona radiata
SFG	-	superior frontal gyrus
SFG-WM	-	superior frontal gyrus white matter
SG	-	straight gyrus
SG-WM	-	straight gyrus white matter

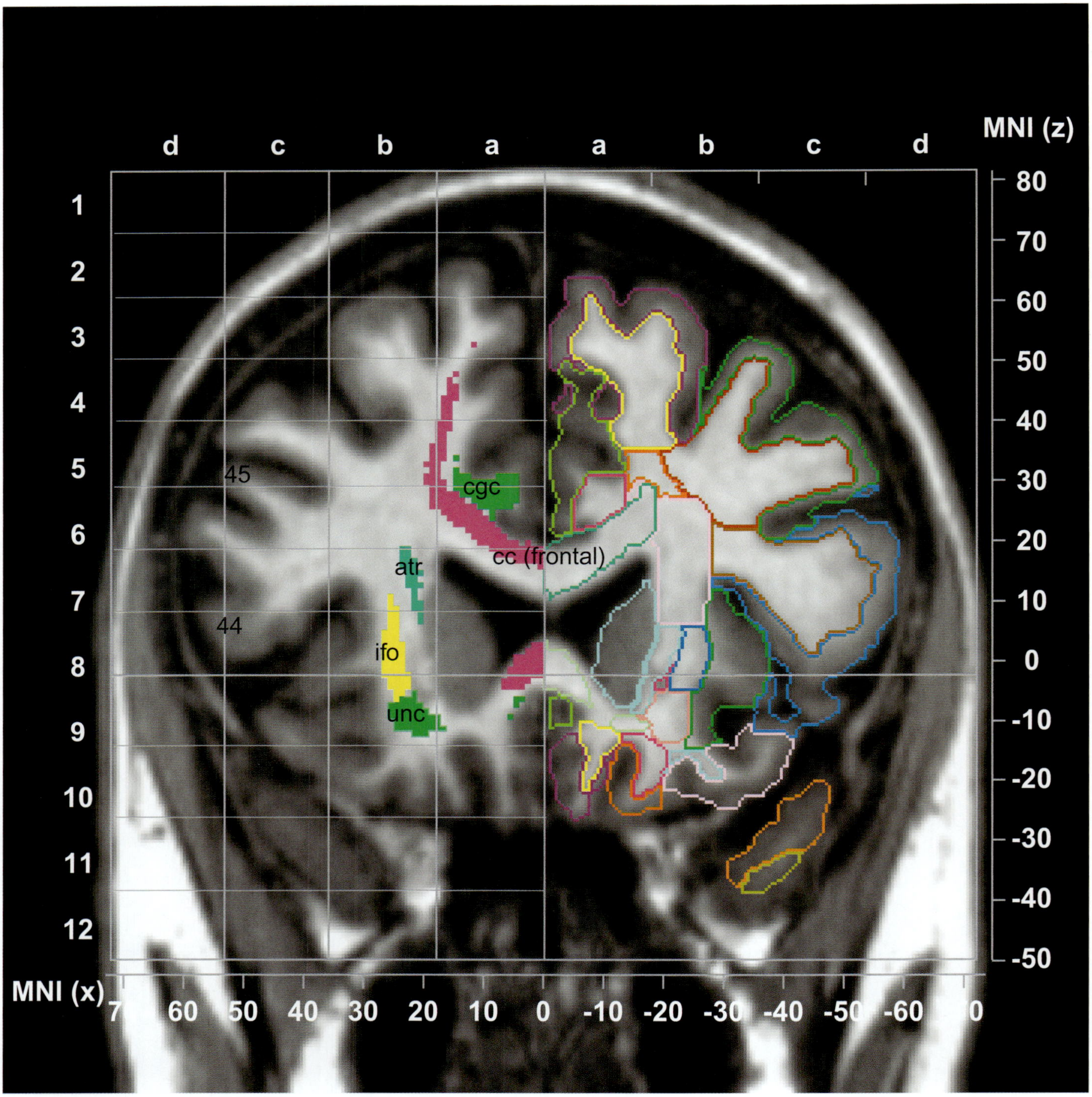

atr - anterior thalamic radiation
cc (frontal) - corpus callosum (connecting bilateral frontal areas)
cgc - cingulum (cingulate gyrus part)
ifo - inferior fronto-occipital fasciculus
unc - uncinate fasciculus

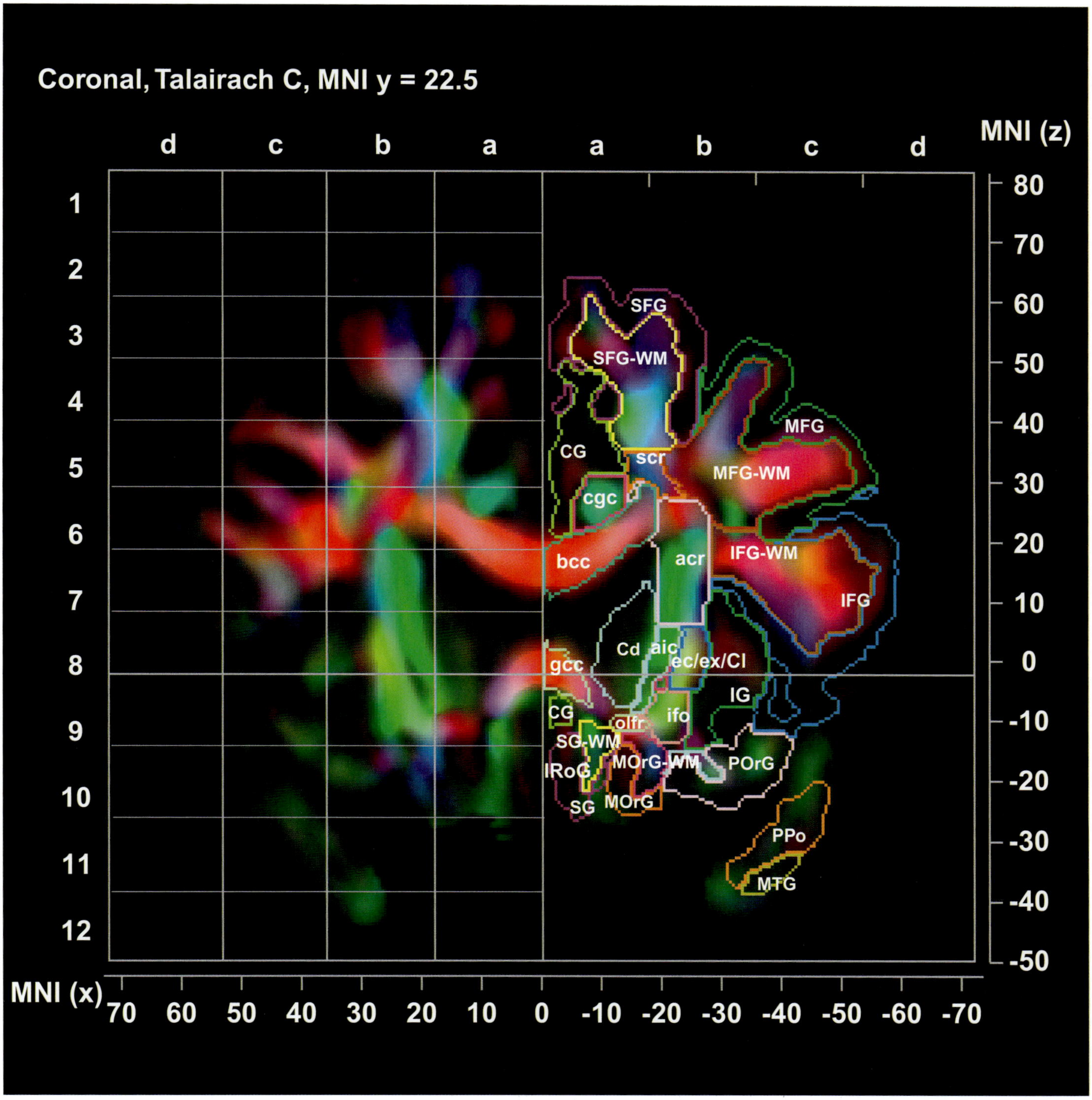

acr - anterior corona radiata
aic - anterior limb of the internal capsule
bcc - body of the corpus callosum
Cd - caudate nucleus
CG - cingulate gyrus
cgc - cingulum (cingulate gyrus part)
ec/ex/Cl - external capsule / extreme capsule / claustrum (cannot be resolved with current resolution)
gcc - genu of the corpus callosum
IFG - inferior frontal gyrus
IFG-WM - inferior frontal gyrus white matter
ifo - inferior fronto-occipital fasciculus
IG - insular gyrus
IRoG - inferior rostral gyrus
MFG - middle frontal gyrus
MFG-WM - middle frontal gyrus white matter
MOrG - medial orbital gyrus
MOrG-WM - medial orbital gyrus white matter
MTG - middle temporal gyrus
olfr - olfactory radiation
POrG - posterior orbital gyrus
PPo - planum polare
scr - superior corona radiata
SFG - superior frontal gyrus
SFG-WM - superior frontal gyrus white matter
SG - straight gyrus
SG-WM - straight gyrus white matter

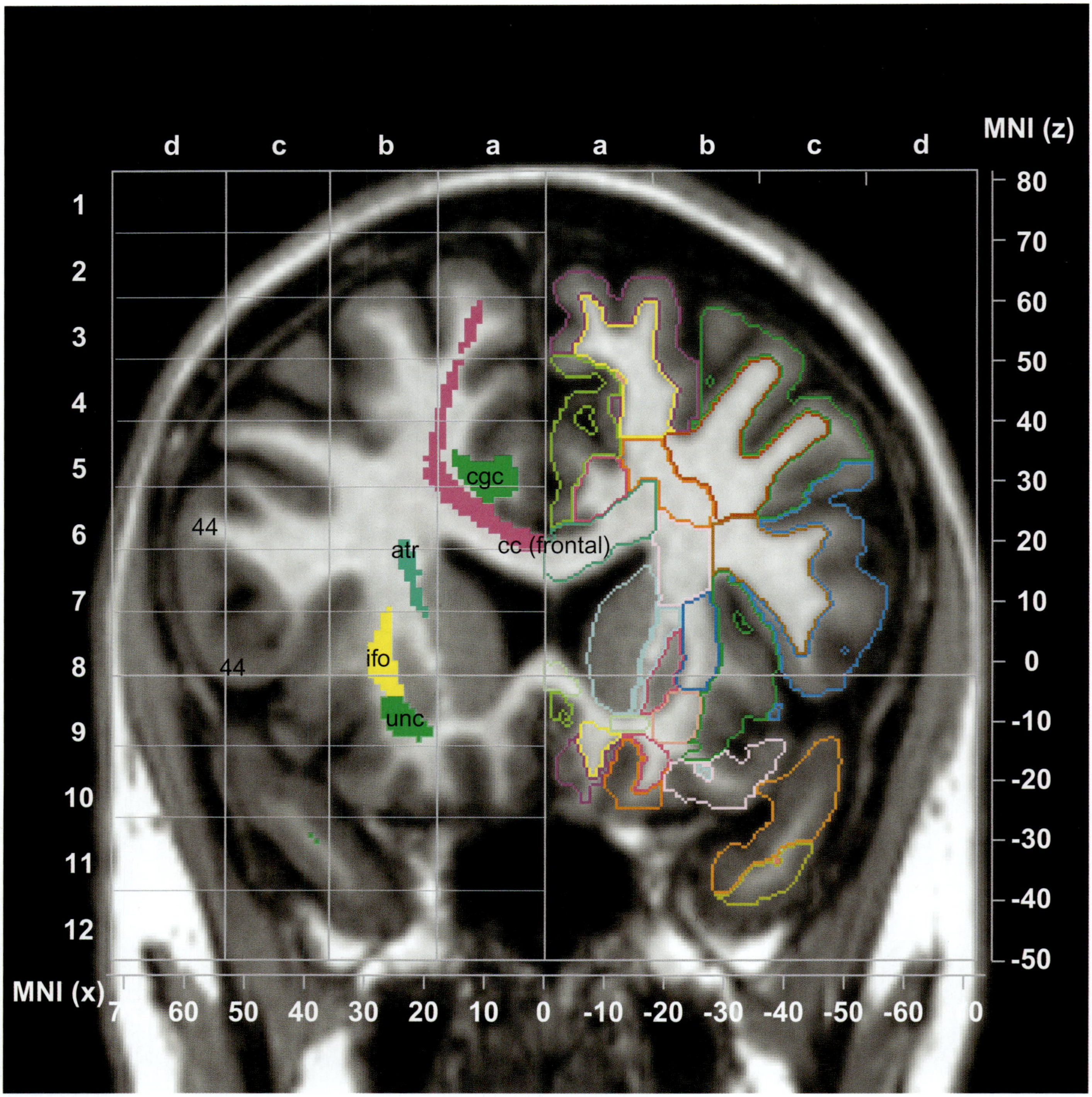

atr	-	anterior thalamic radiation	cgc	-	cingulum (cingulate gyrus part)
cc (frontal)	-	corpus callosum (connecting bilateral frontal areas)	ifo	-	inferior fronto-occipital fasciculus
			unc	-	uncinate fasciculus

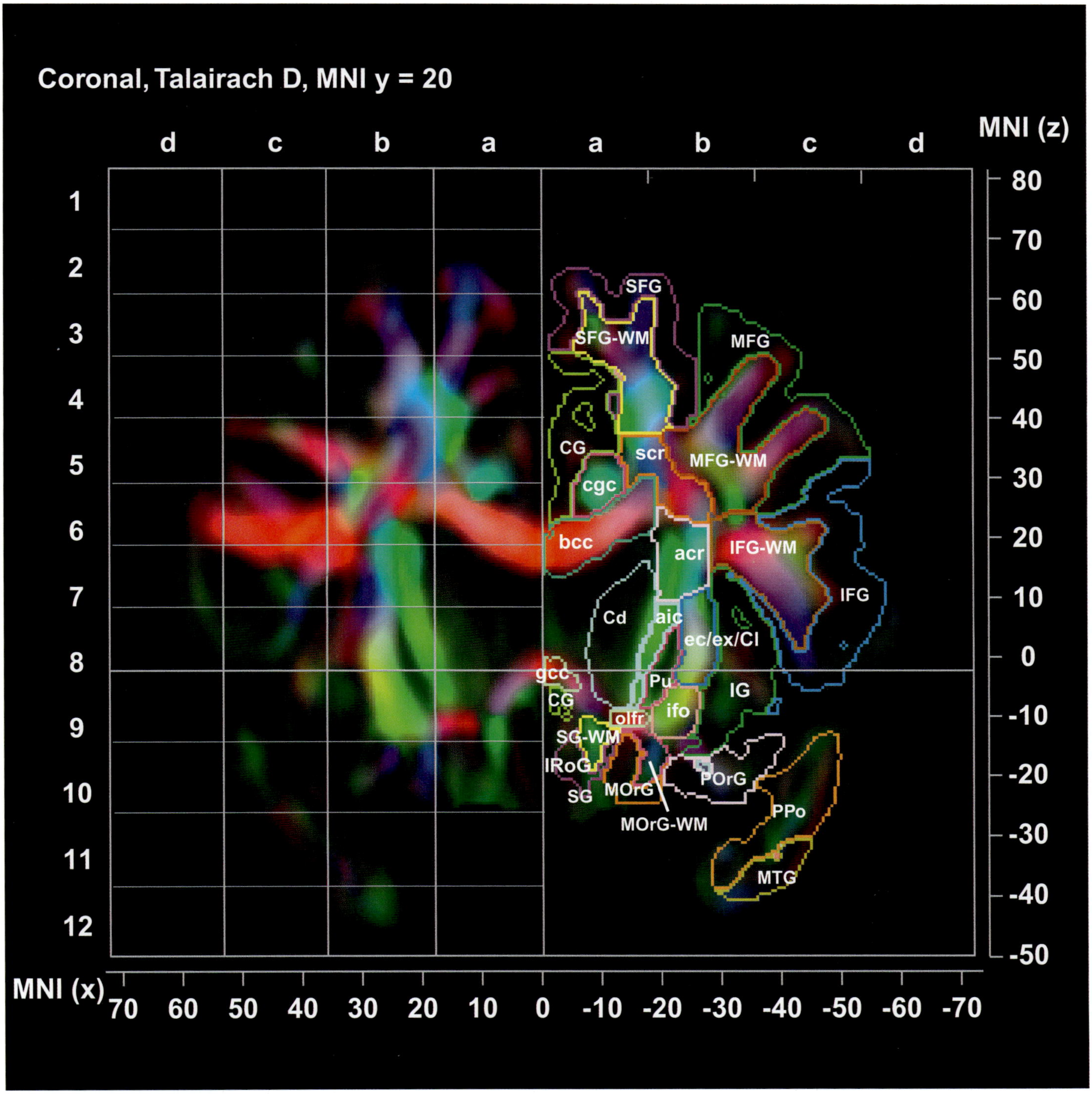

acr	-	anterior corona radiata
aic	-	anterior limb of the internal capsule
bcc	-	body of the corpus callosum
Cd	-	caudate nucleus
CG	-	cingulate gyrus
cgc	-	cingulum (cingulate gyrus part)
ec/ex/Cl	-	external capsule / extreme capsule / claustrum (cannot be resolved with current resolution)
gcc	-	genu of the corpus callosum
IFG	-	inferior frontal gyrus
IFG-WM	-	inferior frontal gyrus white matter
ifo	-	inferior fronto-occipital fasciculus
IG	-	insular gyrus
IRoG	-	inferior rostral gyrus
MFG	-	middle frontal gyrus
MFG-WM	-	middle frontal gyrus white matter
MOrG	-	medial orbital gyrus
MOrG-WM	-	medial orbital gyrus white matter
MTG	-	middle temporal gyrus
olfr	-	olfactory radiation
POrG	-	posterior orbital gyrus
PPo	-	planum polare
Pu	-	putamen
scr	-	superior corona radiata
SFG	-	superior frontal gyrus
SFG-WM	-	superior frontal gyrus white matter
SG	-	straight gyrus
SG-WM	-	straight gyrus white matter

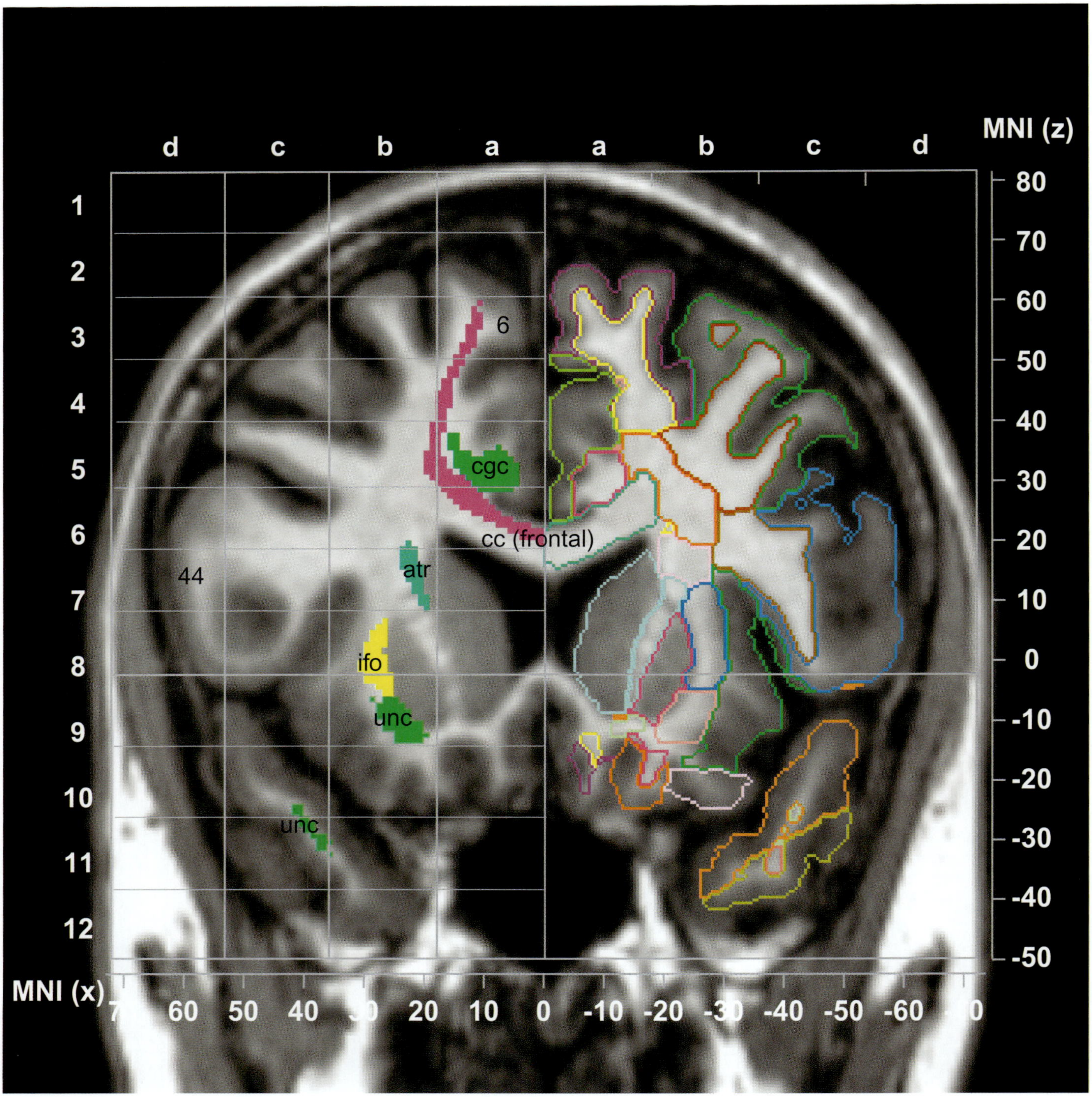

atr - anterior thalamic radiation
cc (frontal) - corpus callosum (connecting bilateral frontal areas)
cgc - cingulum (cingulate gyrus part)
ifo - inferior fronto-occipital fasciculus
unc - uncinate fasciculus

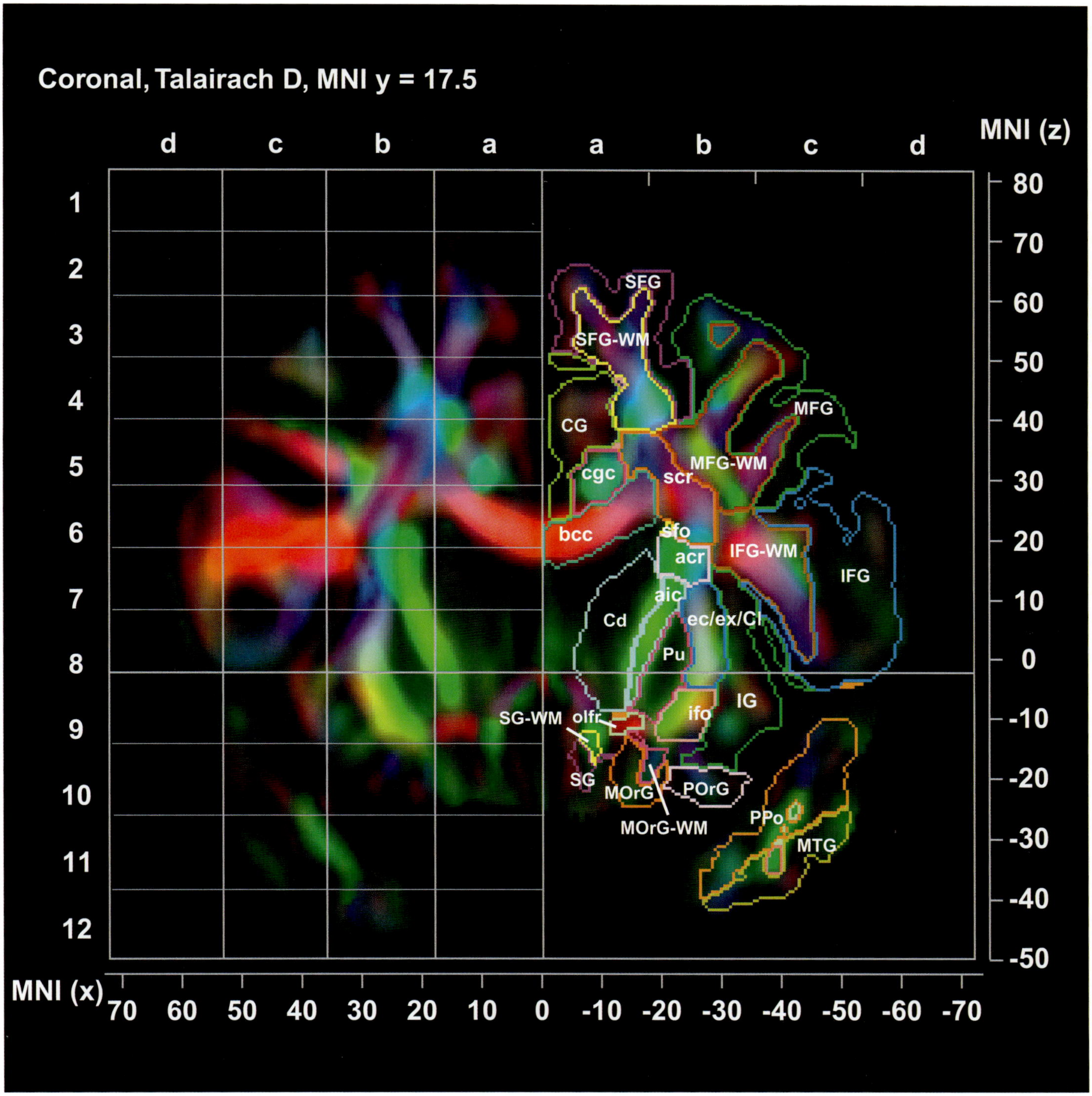

acr - anterior corona radiata
aic - anterior limb of the internal capsule
bcc - body of the corpus callosum
Cd - caudate nucleus
CG - cingulate gyrus
cgc - cingulum (cingulate gyrus part)
ec/ex/Cl - external capsule / extreme capsule / claustrum (cannot be resolved with current resolution)
IFG - inferior frontal gyrus
IFG-WM - inferior frontal gyrus white matter
ifo - inferior fronto-occipital fasciculus
IG - insular gyrus
MFG - middle frontal gyrus
MFG-WM - middle frontal gyrus white matter
MOrG - medial orbital gyrus
MOrG-WM - medial orbital gyrus white matter
MTG - middle temporal gyrus
olfr - olfactory radiation
POrG - posterior orbital gyrus
PPo - planum polare
Pu - putamen
scr - superior corona radiata
SFG - superior frontal gyrus
SFG-WM - superior frontal gyrus white matter
sfo - superior fronto-occipital fasciculus
SG - straight gyrus
SG-WM - straight gyrus white matter

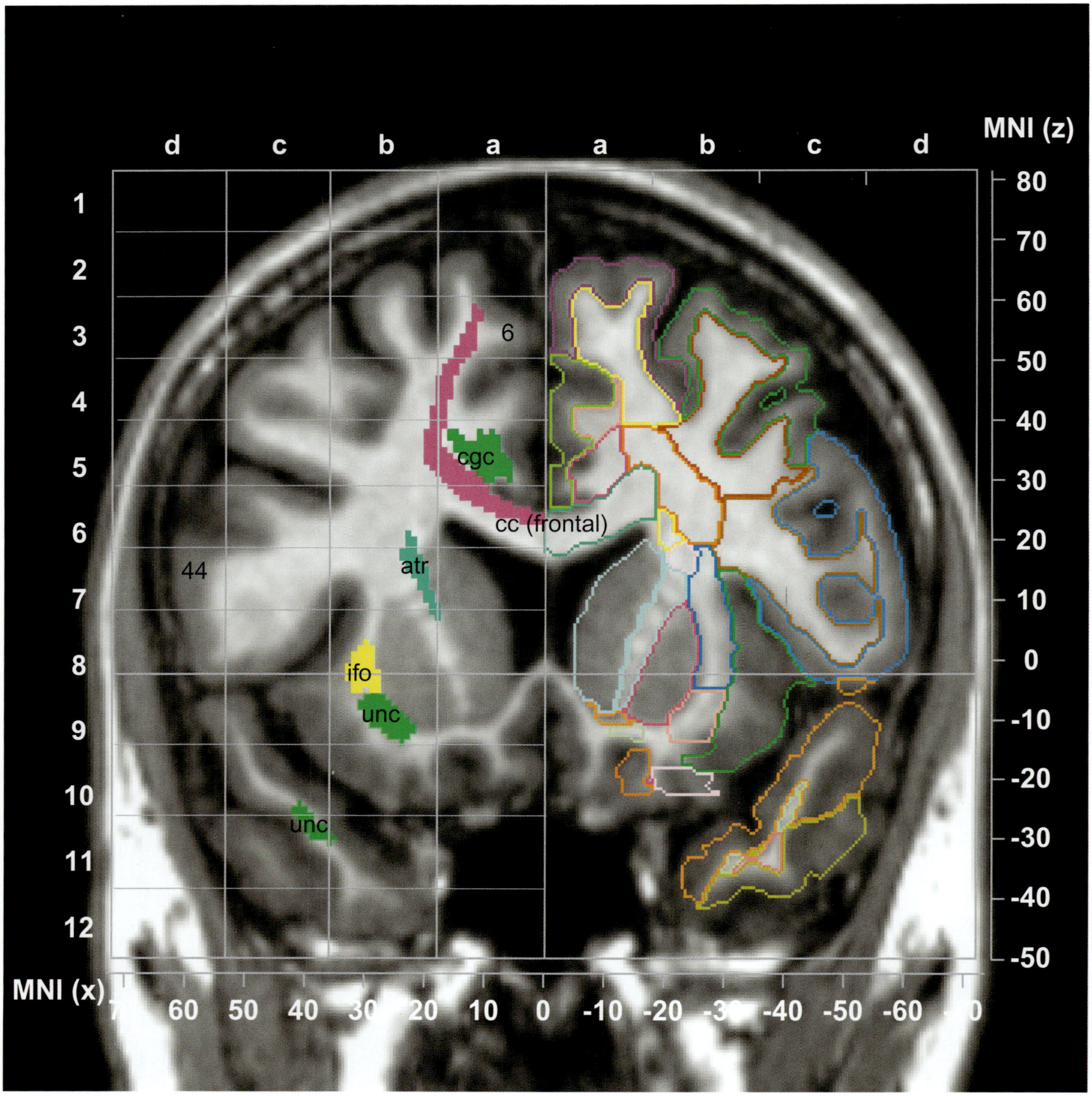

atr	-	anterior thalamic radiation	cgc	-	cingulum (cingulate gyrus part)
cc (frontal)	-	corpus callosum (connecting bilateral frontal areas)	ifo	-	inferior fronto-occipital fasciculus
			unc	-	uncinate fasciculus

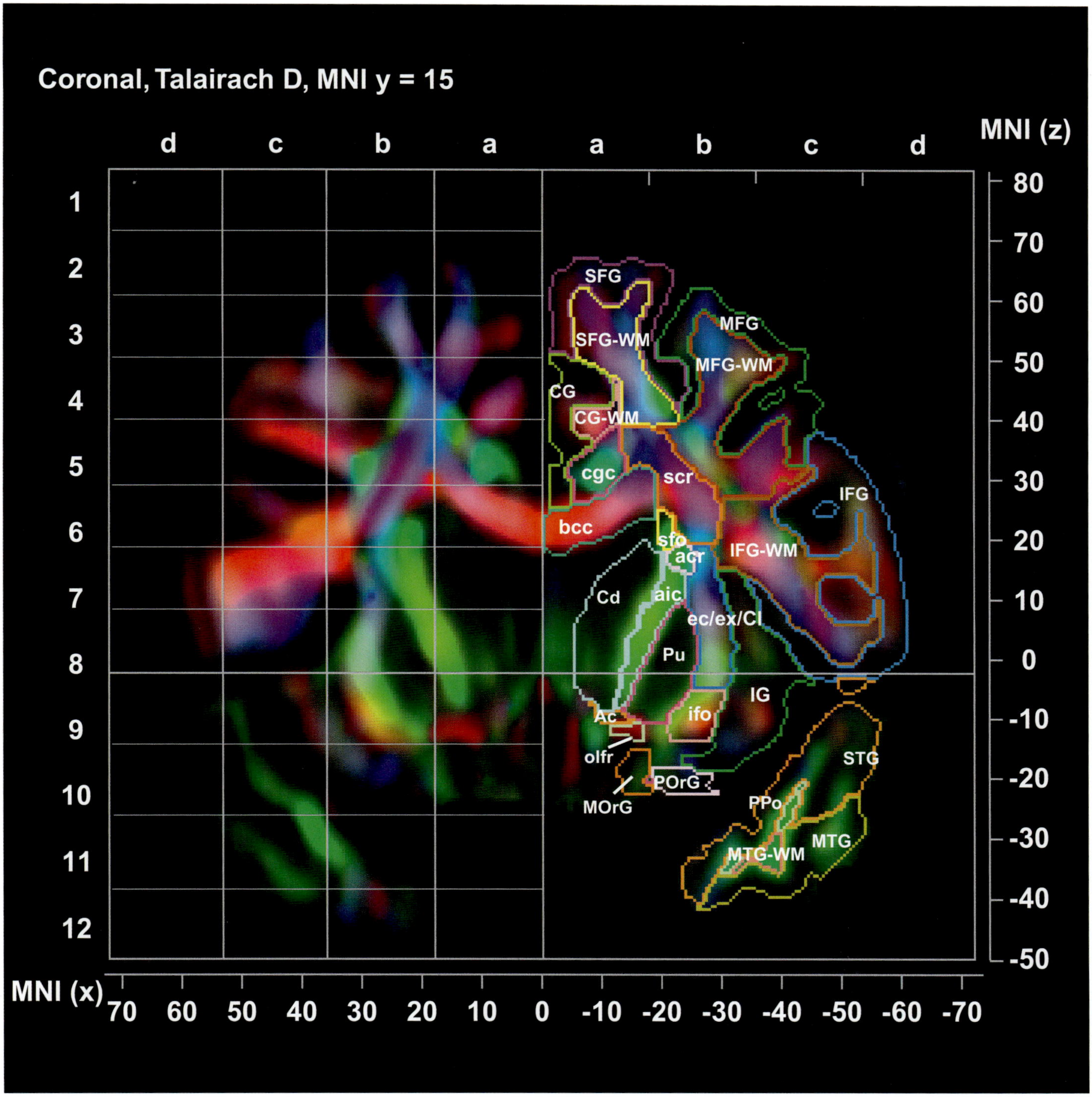

Ac	-	accumbens nucleus
acr	-	anterior corona radiata
aic	-	anterior limb of the internal capsule
bcc	-	body of the corpus callosum
Cd	-	caudate nucleus
CG	-	cingulate gyrus
cgc	-	cingulum (cingulate gyrus part)
CG-WM	-	cingulate gyrus white matter
ec/ex/Cl	-	external capsule / extreme capsule / claustrum (cannot be resolved with current resolution)
IFG	-	inferior frontal gyrus
IFG-WM	-	inferior frontal gyrus white matter
ifo	-	inferior fronto-occipital fasciculus
IG	-	insular gyrus
MFG	-	middle frontal gyrus
MFG-WM	-	middle frontal gyrus white matter
MOrG	-	medial orbital gyrus
MTG	-	middle temporal gyrus
MTG-WM	-	middle temporal gyrus white matter
olfr	-	olfactory radiation
POrG	-	posterior orbital gyrus
PPo	-	planum polare
Pu	-	putamen
scr	-	superior corona radiata
SFG	-	superior frontal gyrus
SFG-WM	-	superior frontal gyrus white matter
sfo	-	superior fronto-occipital fasciculus
STG	-	superior temporal gyrus

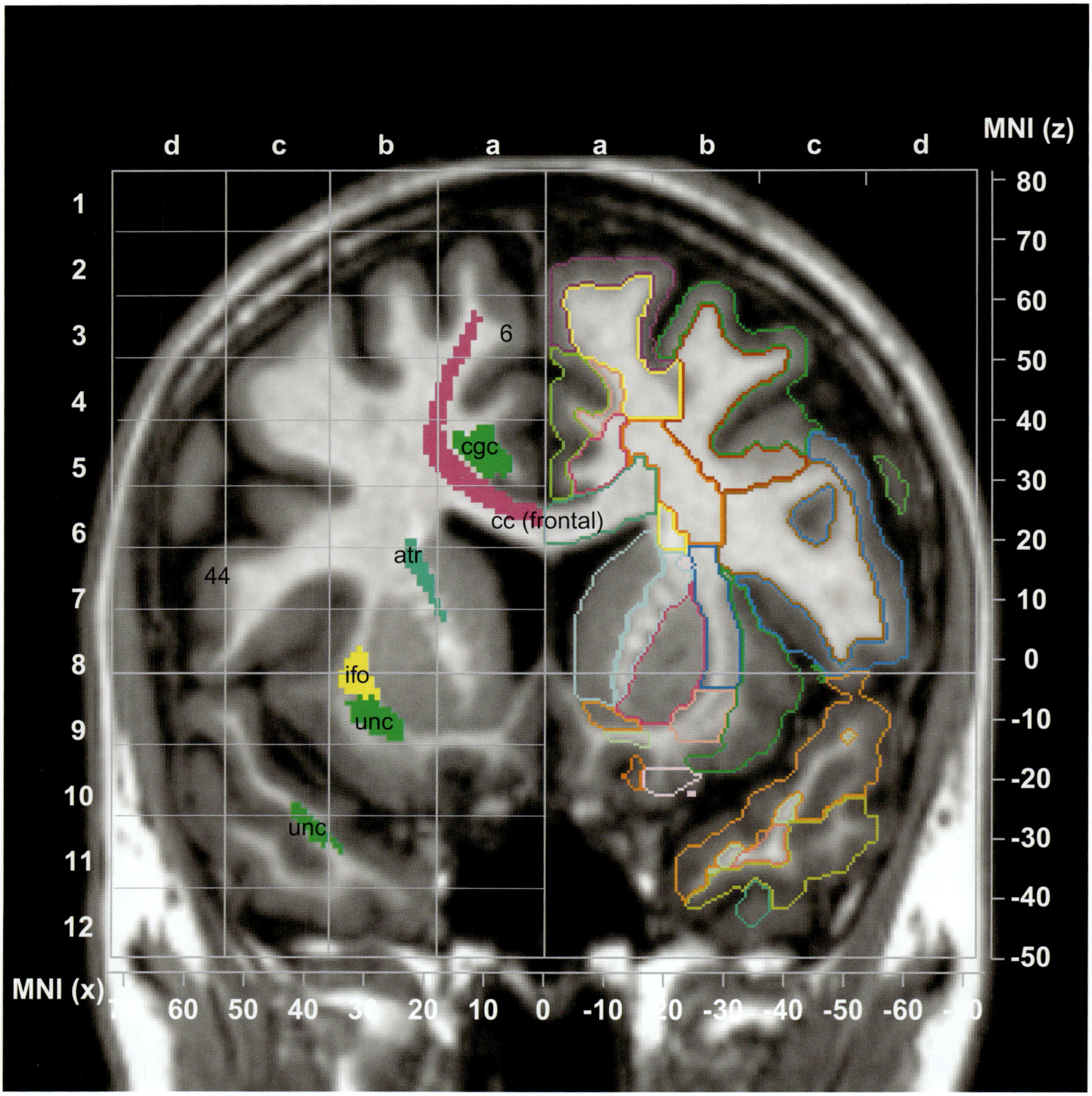

atr	-	anterior thalamic radiation
cc (frontal)	-	corpus callosum (connecting bilateral frontal areas)
cgc	-	cingulum (cingulate gyrus part)
ifo	-	inferior fronto-occipital fasciculus
unc	-	uncinate fasciculus

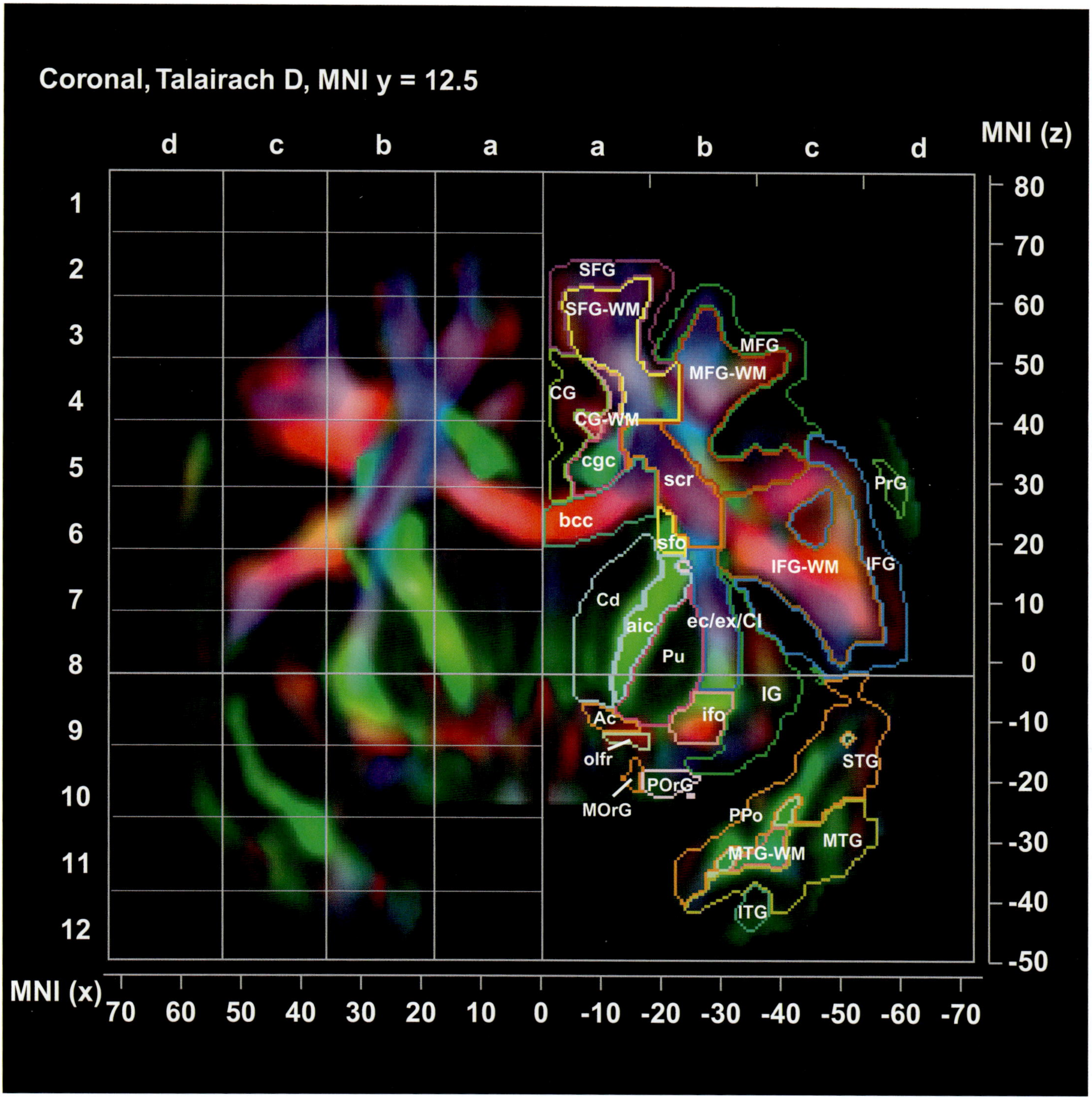

Ac - accumbens nucleus
aic - anterior limb of the internal capsule
bcc - body of the corpus callosum
Cd - caudate nucleus
CG - cingulate gyrus
cgc - cingulum (cingulate gyrus part)
CG-WM - cingulate gyrus white matter
ec/ex/Cl - external capsule / extreme capsule / claustrum (cannot be resolved with current resolution)
IFG - inferior frontal gyrus
IFG-WM - inferior frontal gyrus white matter
ifo - inferior fronto-occipital fasciculus
IG - insular gyrus
ITG - inferior temporal gyrus
MFG - middle frontal gyrus
MFG-WM - middle frontal gyrus white matter
MOrG - medial orbital gyrus
MTG - middle temporal gyrus
MTG-WM - middle temporal gyrus white matter
olfr - olfactory radiation
POrG - posterior orbital gyrus
PPo - planum polare
PrG - precentral gyrus
Pu - putamen
scr - superior corona radiata
SFG - superior frontal gyrus
SFG-WM - superior frontal gyrus white matter
sfo - superior fronto-occipital fasciculus
STG - superior temporal gyrus

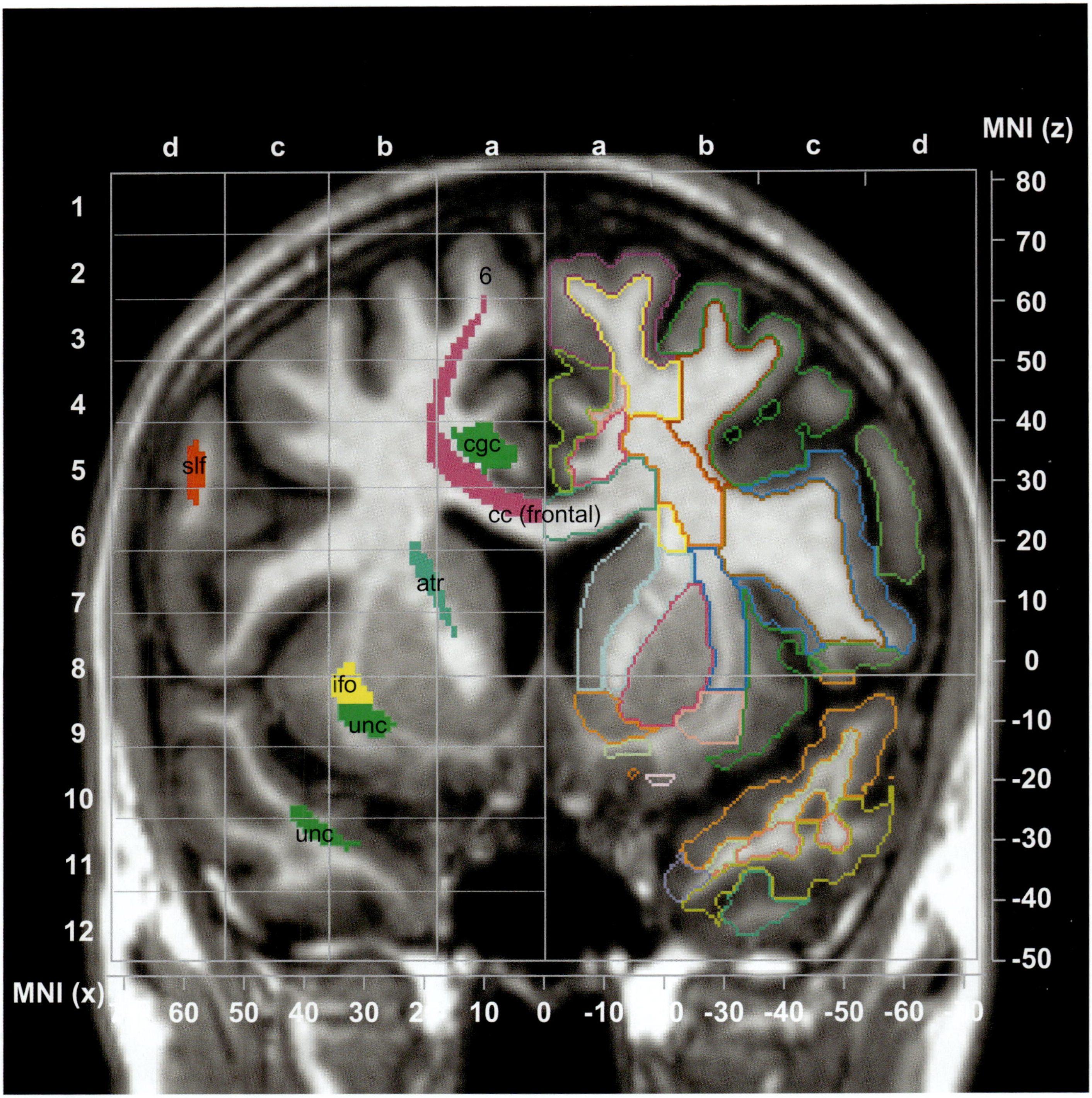

atr - anterior thalamic radiation
cc (frontal) - corpus callosum (connecting bilateral frontal areas)
cgc - cingulum (cingulate gyrus part)
ifo - inferior fronto-occipital fasciculus
slf - superior longitudinal fasciculus
unc - uncinate fasciculus

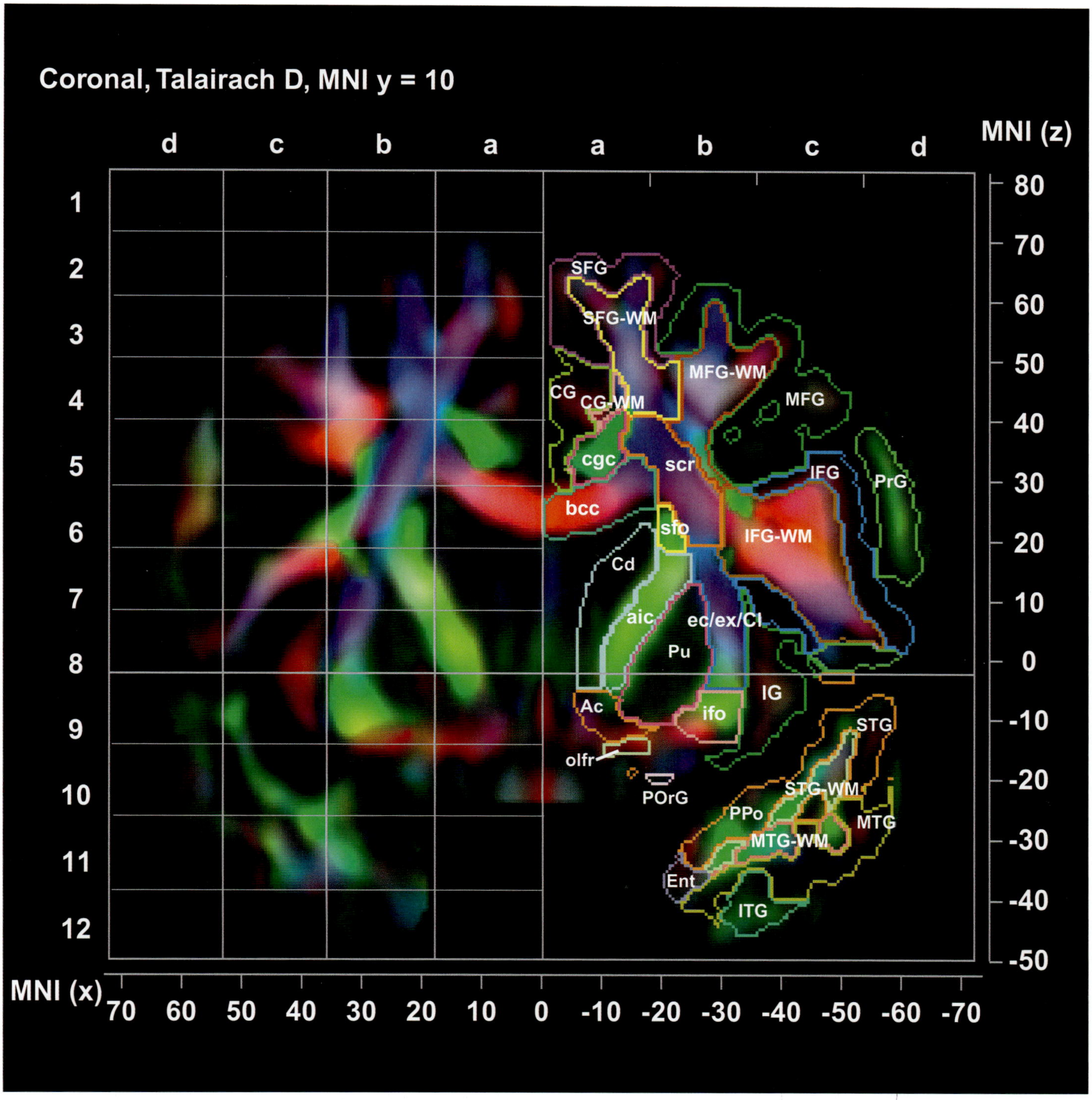

Ac - accumbens nucleus
aic - anterior limb of the internal capsule
bcc - body of the corpus callosum
Cd - caudate nucleus
CG - cingulate gyrus
cgc - cingulum (cingulate gyrus part)
CG-WM - cingulate gyrus white matter
ec/ex/Cl - external capsule / extreme capsule / claustrum (cannot be resolved with current resolution)
Ent - entorhinal cortex

IFG - inferior frontal gyrus
IFG-WM - inferior frontal gyrus white matter
ifo - inferior fronto-occipital fasciculus
IG - insular gyrus
ITG - inferior temporal gyrus
MFG - middle frontal gyrus
MFG-WM - middle frontal gyrus white matter
MTG - middle temporal gyrus
MTG-WM - middle temporal gyrus white matter
olfr - olfactory radiation

POrG - posterior orbital gyrus
PPo - planum polare
PrG - precentral gyrus
Pu - putamen
scr - superior corona radiata
SFG - superior frontal gyrus
SFG-WM - superior frontal gyrus white matter
sfo - superior fronto-occipital fasciculus
STG - superior temporal gyrus
STG-WM - superior temporal gyrus white matter

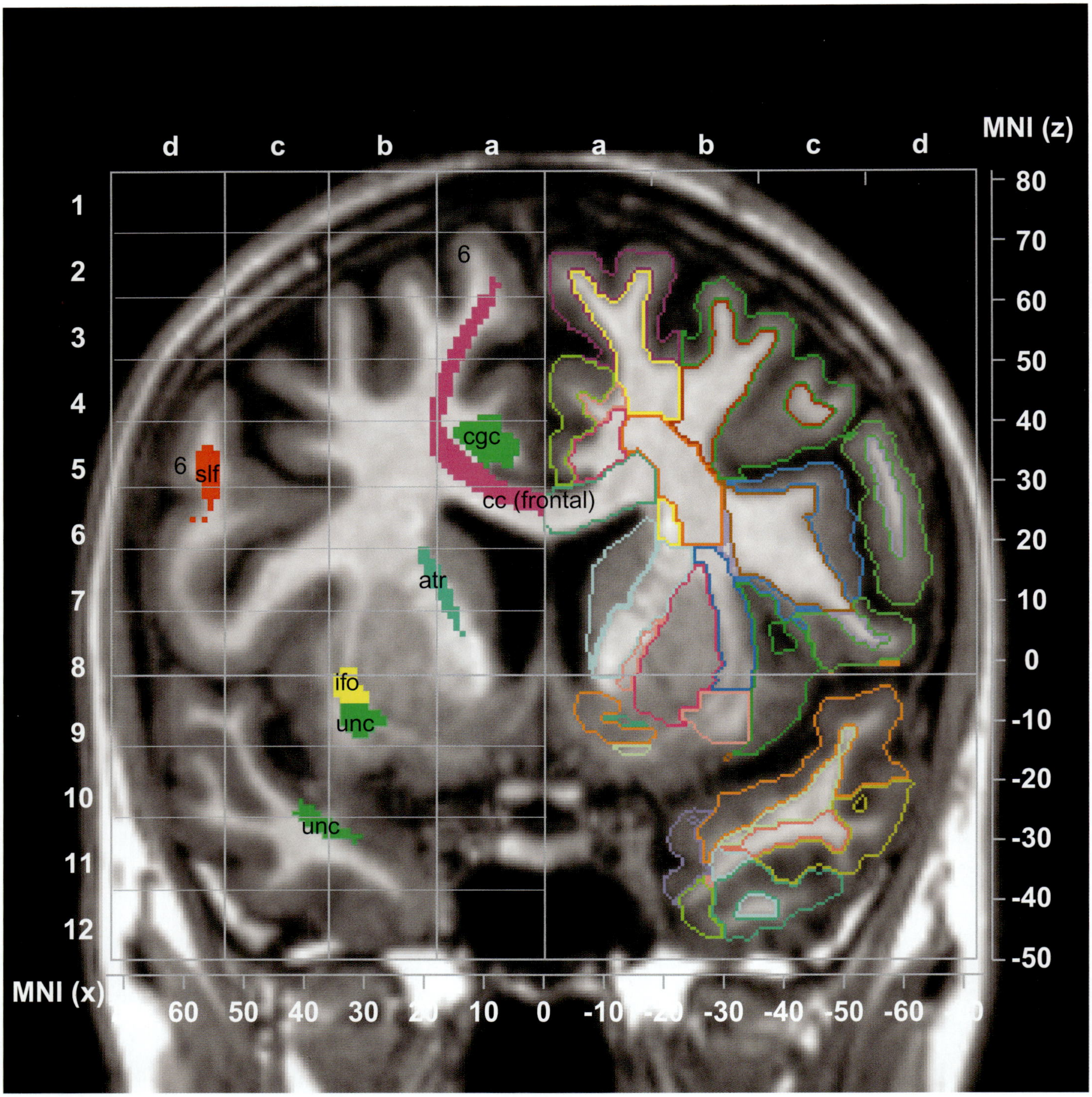

atr - anterior thalamic radiation
cc (frontal) - corpus callosum (connecting bilateral frontal areas)
cgc - cingulum (cingulate gyrus part)
ifo - inferior fronto-occipital fasciculus
slf - superior longitudinal fasciculus
unc - uncinate fasciculus

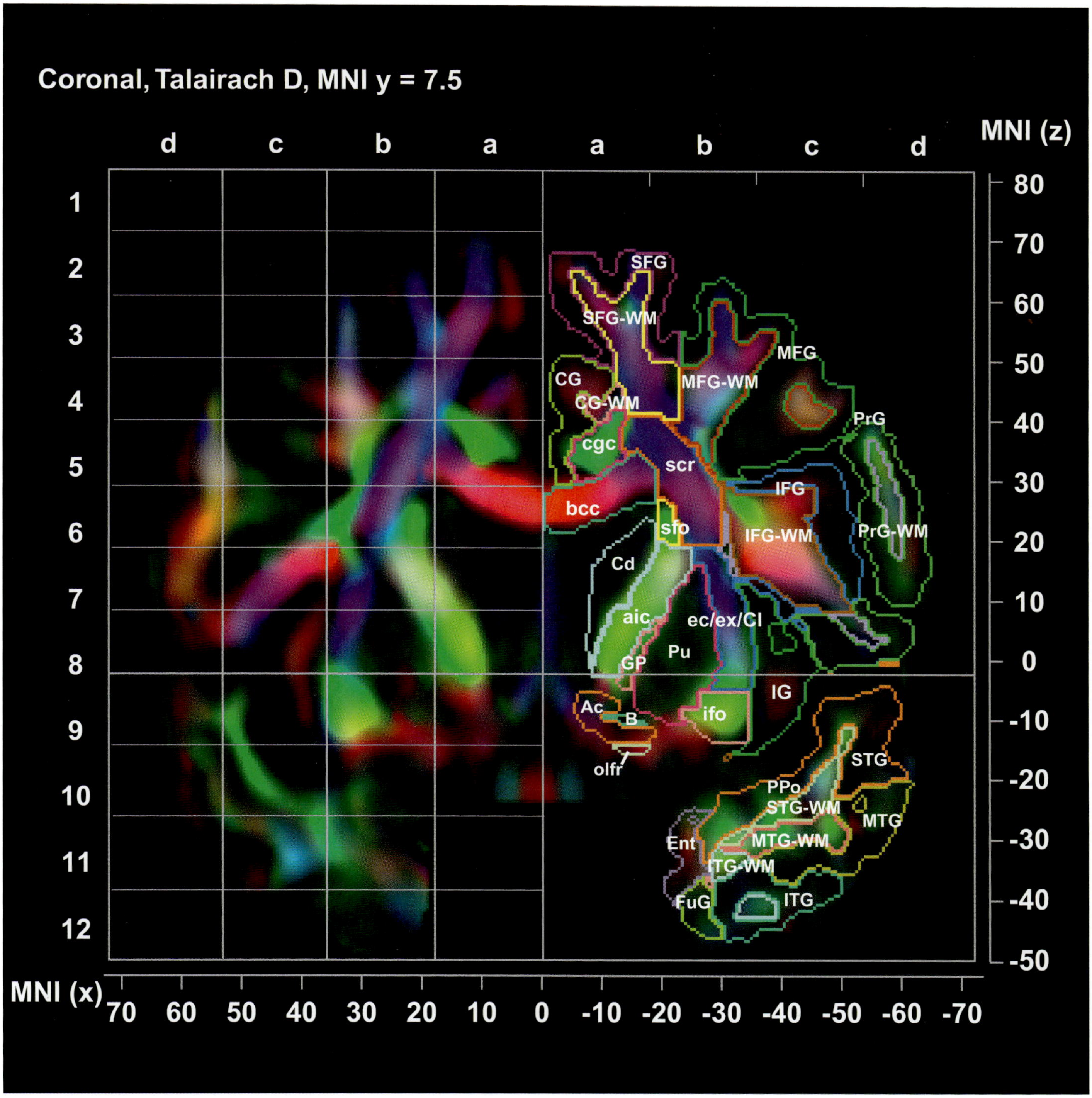

Ac - accumbens nucleus
aic - anterior limb of the internal capsule
B - basal nucleus of Meynert
bcc - body of the corpus callosum
Cd - caudate nucleus
CG - cingulate gyrus
cgc - cingulum (cingulate gyrus part)
CG-WM - cingulate gyrus white matter
ec/ex/Cl - external capsule / extreme capsule / claustrum (cannnot be resolved with current resolution)
Ent - entorhinal cortex
FuG - fusiform gyrus
GP - globus pallidus
IFG - inferior frontal gyrus
IFG-WM - inferior frontal gyrus white matter
ifo - inferior fronto-occipital fasciculus
IG - insular gyrus
ITG - inferior temporal gyrus
ITG-WM - inferior temporal gyrus white matter
MFG - middle frontal gyrus
MFG-WM - middle frontal gyrus white matter
MTG - middle temporal gyrus
MTG-WM - middle temporal gyrus white matter
olfr - olfactory radiation
PPo - planum polare
PrG - precentral gyrus
PrG-WM - precentral gyrus white matter
Pu - putamen
scr - superior corona radiata
SFG - superior frontal gyrus
SFG-WM - superior frontal gyrus white matter
sfo - superior fronto-occipital fasciculus
STG - superior temporal gyrus
STG-WM - superior temporal gyrus white matter

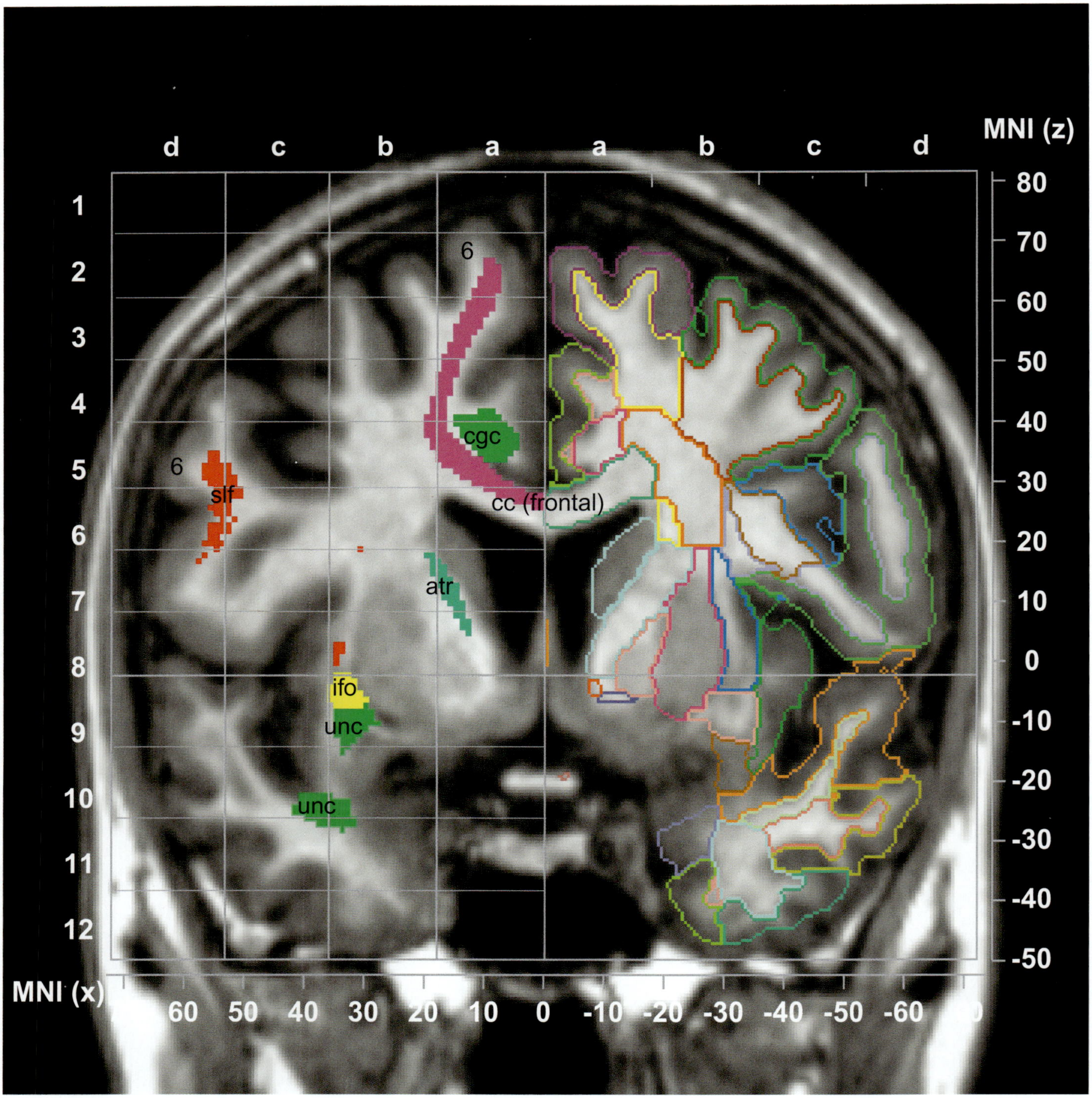

atr - anterior thalamic radiation	cgc - cingulum (cingulate gyrus part)	slf - superior longitudinal fasciculus
cc (frontal) - corpus callosum (connecting bilateral frontal areas)	ifo - inferior fronto-occipital fasciculus	unc - uncinate fasciculus

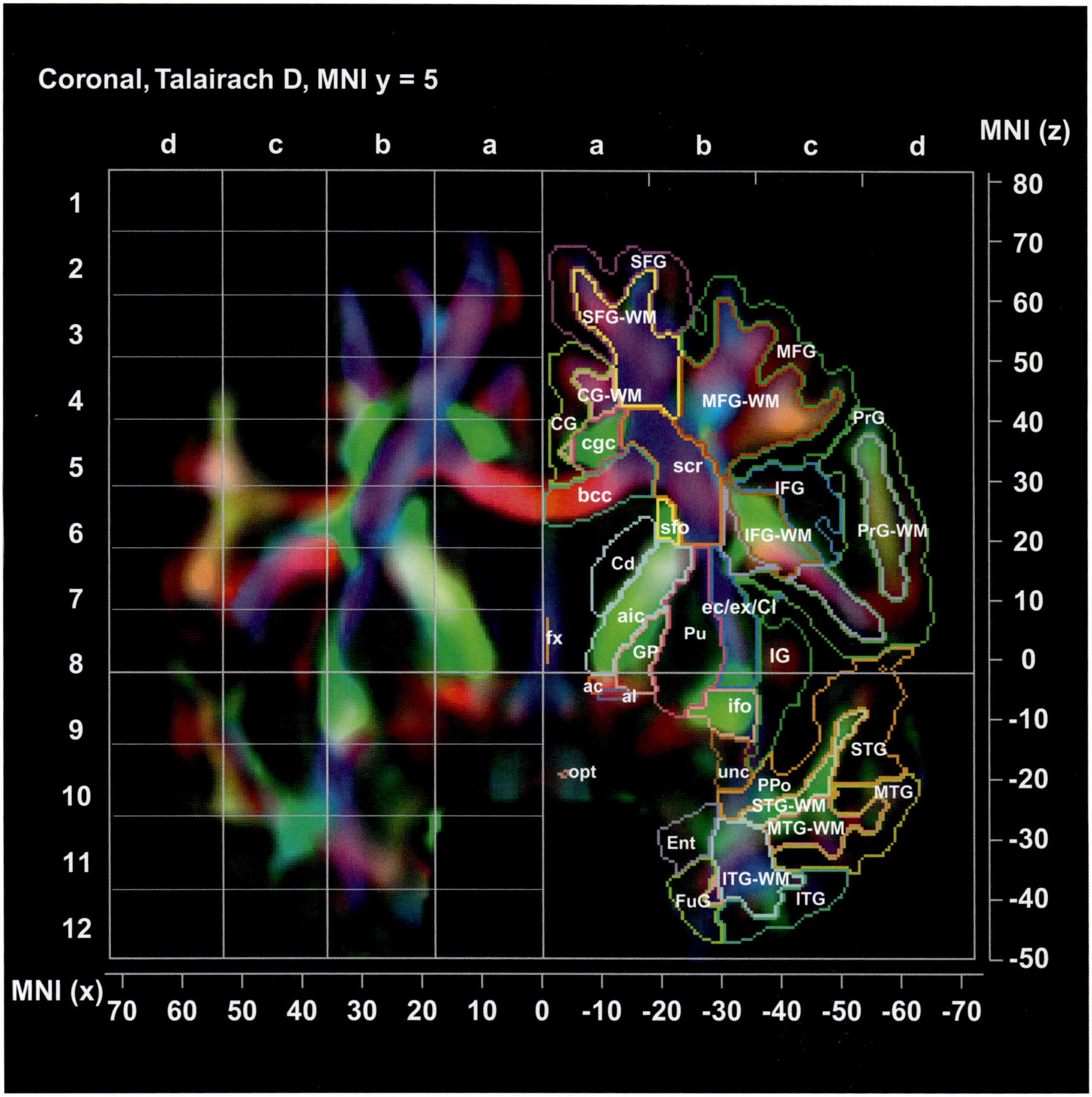

ac - anterior commissure
aic - anterior limb of the internal capsule
al - ansa lenticularis
bcc - body of the corpus callosum
Cd - caudate nucleus
CG - cingulate gyrus
cgc - cingulum (cingulate gyrus part)
CG-WM - cingulate gyrus white matter
ec/ex/Cl - external capsule / extreme capsule / claustrum (cannot be resolved with current resolution)
Ent - entorhinal cortex
FuG - fusiform gyrus
fx - fornix (column and body)
GP - globus pallidus
IFG - inferior frontal gyrus
IFG-WM - inferior frontal gyrus white matter
ifo - inferior fronto-occipital fasciculus
IG - insular gyrus
ITG - inferior temporal gyrus
ITG-WM - inferior temporal gyrus white matter
MFG - middle frontal gyrus
MFG-WM - middle frontal gyrus white matter
MTG - middle temporal gyrus
MTG-WM - middle temporal gyrus white matter
opt - optic tract
PPo - planum polare
PrG - precentral gyrus
PrG-WM - precentral gyrus white matter
Pu - putamen
scr - superior corona radiata
SFG - superior frontal gyrus
SFG-WM - superior frontal gyrus white matter
sfo - Superior fronto-occipital fasciculus
STG - superior temporal gyrus
STG-WM - superior temporal gyrus white matter
unc - uncinate fasciculus

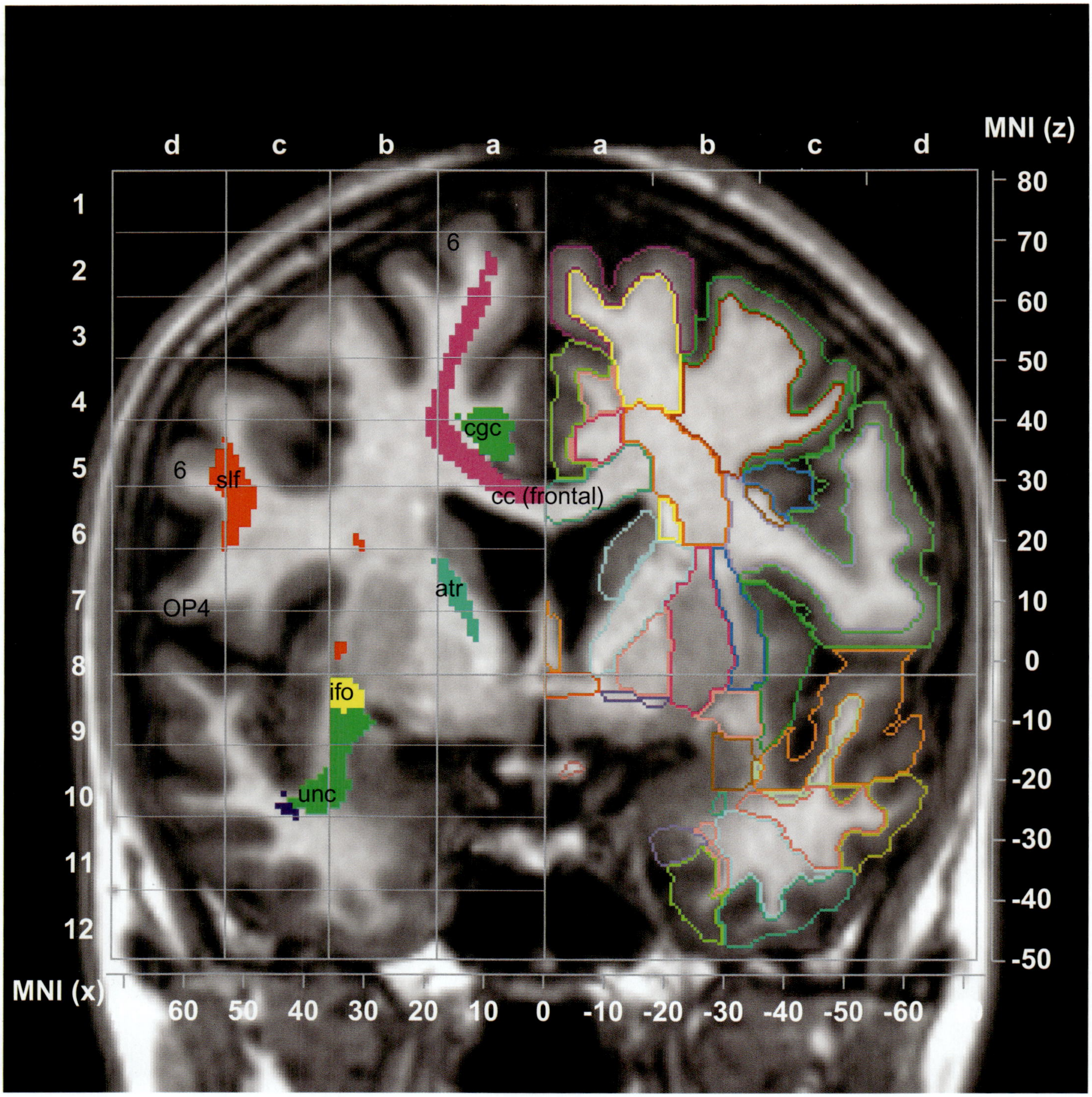

atr	-	anterior thalamic radiation
cc (frontal)	-	corpus callosum (connecting bilateral frontal areas)
cgc	-	cingulum (cingulate gyrus part)
ifo	-	inferior fronto-occipital fasciculus
slf	-	superior longitudinal fasciculus
unc	-	uncinate fasciculus

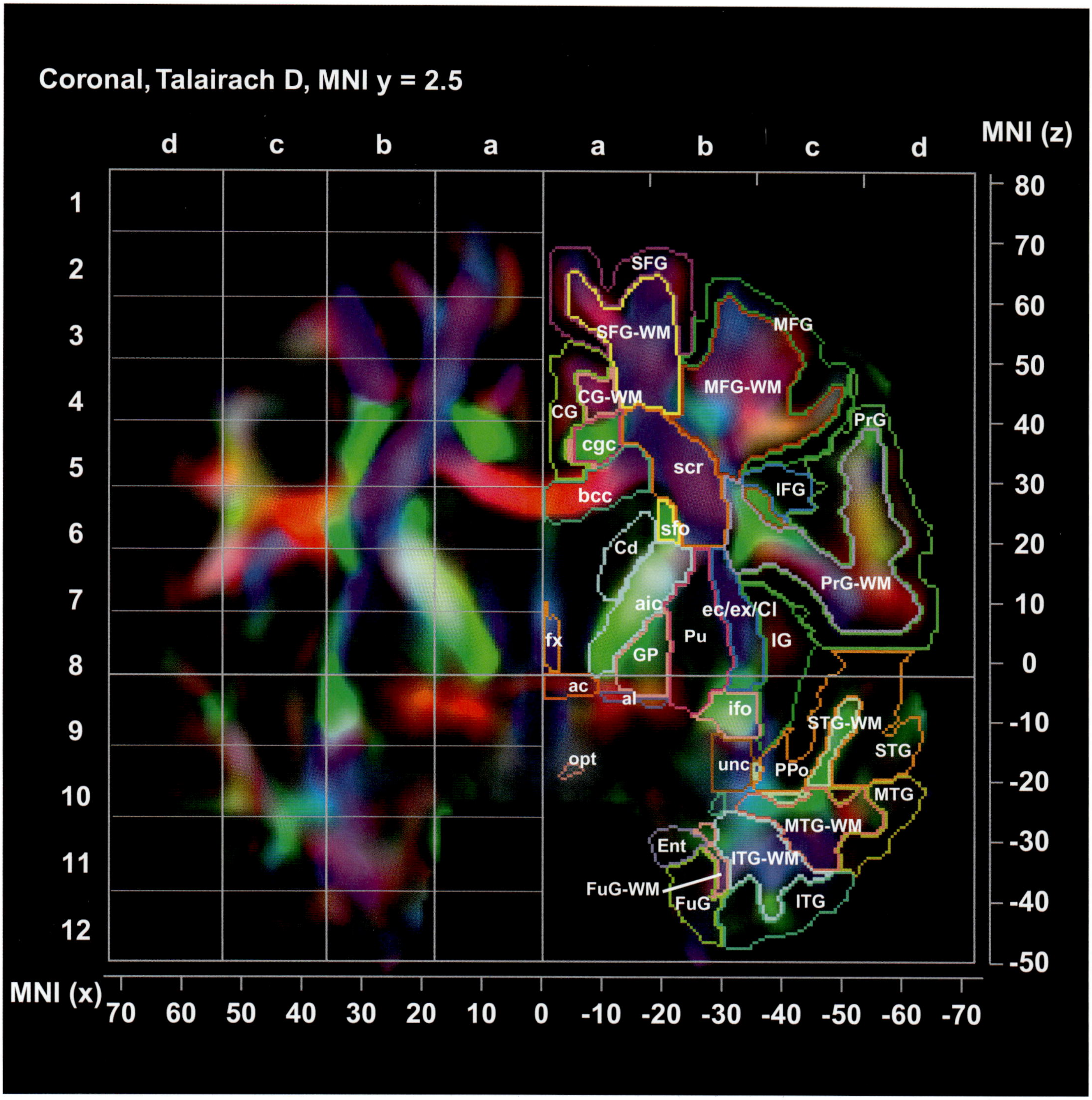

ac	anterior commissure	FuG-WM	fusiform gyrus white matter	opt	optic tract
aic	anterior limb of the internal capsule	fx	fornix (column and body)	PPo	planum polare
al	ansa lenticularis	GP	globus pallidus	PrG	precentral gyrus
bcc	body of the corpus callosum	IFG	inferior frontal gyrus	PrG-WM	precentral gyrus white matter
Cd	caudate nucleus	ifo	inferior fronto-occipital fasciculus	Pu	putamen
CG	cingulate gyrus	IG	insular gyrus	scr	superior corona radiata
cgc	cingulum (cingulate gyrus part)	ITG	inferior temporal gyrus	SFG	superior frontal gyrus
CG-WM	cingulate gyrus white matter	ITG-WM	inferior temporal gyrus white matter	SFG-WM	superior frontal gyrus white matter
ec/ex/Cl	external capsule / extreme capsule / claustrum (cannot be resolved with current resolution)	MFG	middle frontal gyrus	sfo	superior fronto-occipital fasciculus
Ent	entorhinal cortex	MFG-WM	middle frontal gyrus white matter	STG	superior temporal gyrus
FuG	fusiform gyrus	MTG	middle temporal gyrus	STG-WM	superior temporal gyrus white matter
		MTG-WM	middle temporal gyrus white matter	unc	uncinate fasciculus

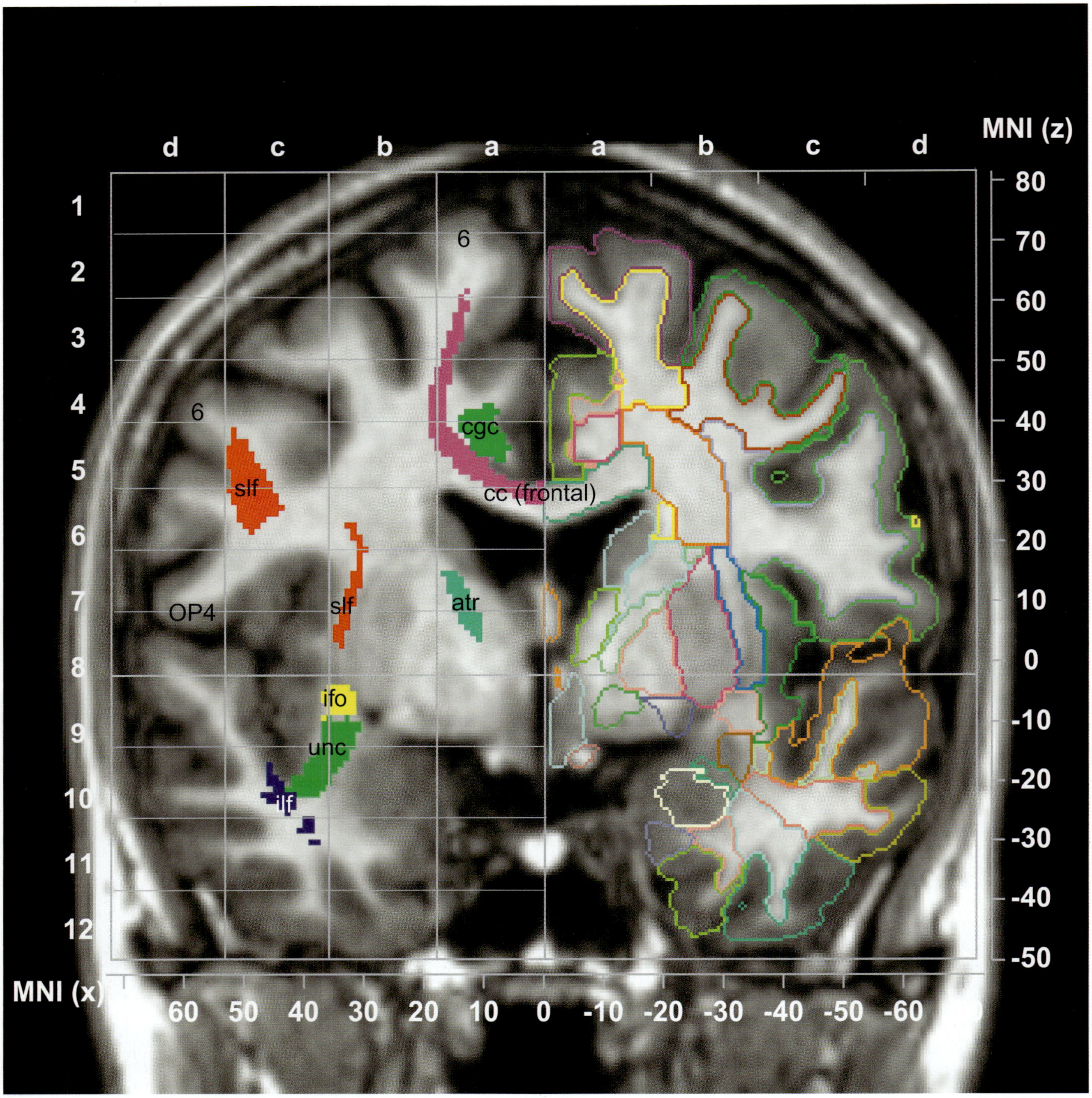

atr	-	anterior thalamic radiation
cc (frontal)	-	corpus callosum (connecting bilateral frontal areas)
cgc	-	cingulum (cingulate gyrus part)
ifo	-	inferior fronto-occipital fasciculus
ilf	-	inferior longitudinal fasciculus
slf	-	superior longitudinal fasciculus
unc	-	uncinate fasciculus

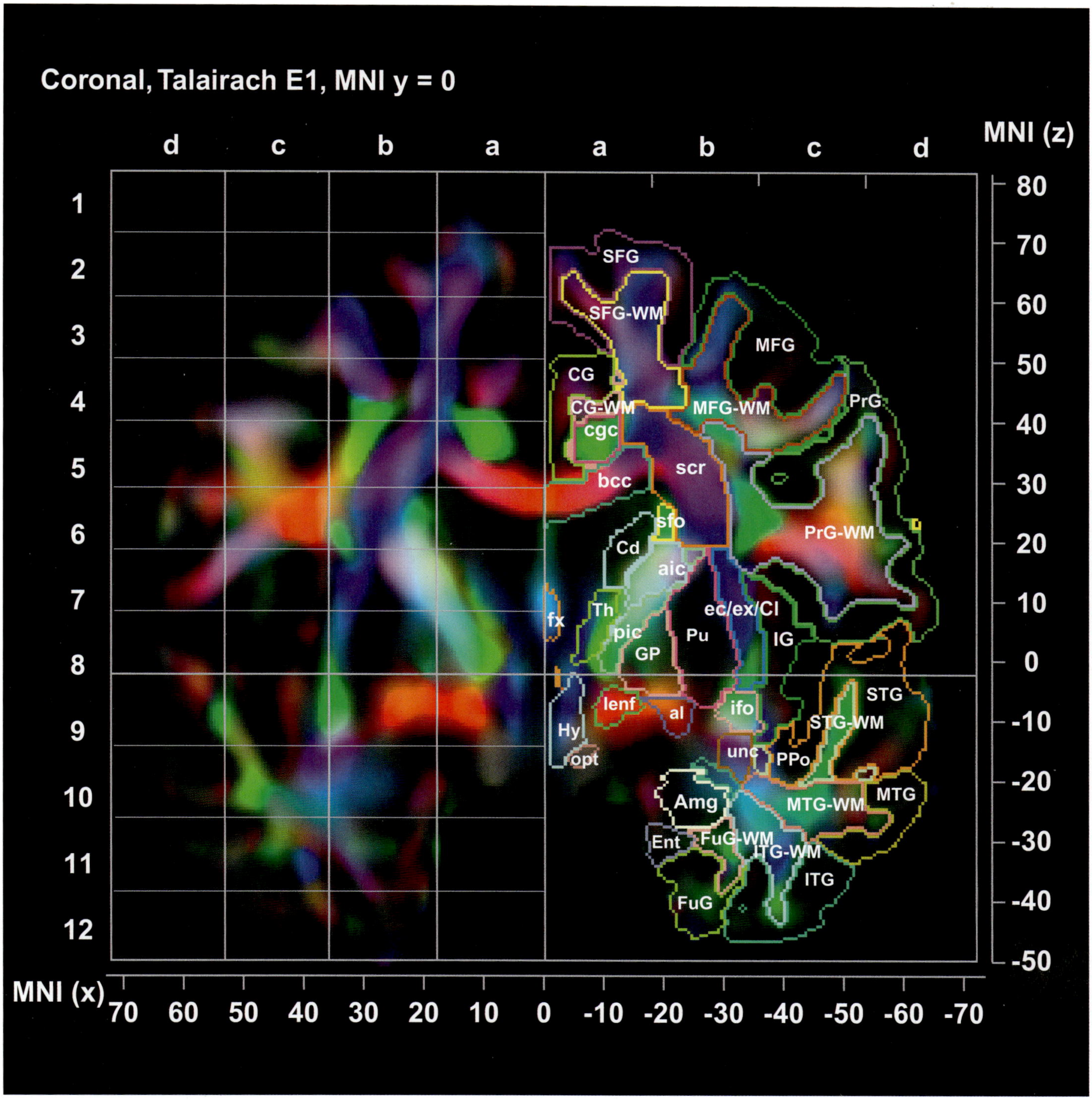

Amg - amygdala
aic - anterior limb of the internal capsule
al - ansa lenticularis
bcc - body of the corpus callosum
Cd - caudate nucleus
CG - cingulate gyrus
cgc - cingulum (cingulate gyrus part)
CG-WM - cingulate gyrus white matter
ec/ex/Cl - external capsule / extreme capsule / claustrum (cannot be resolved with current resolution)
Ent - entorhinal cortex
FuG - fusiform gyrus
FuG-WM - fusiform gyrus white matter
fx - fornix (column and body)
GP - globus pallidus
Hy - hypothalamus
ifo - inferior fronto-occipital fasciculus
IG - insular gyrus
ITG - inferior temporal gyrus
ITG-WM - inferior temporal gyrus white matter
lenf - lenticular fasciculus
MFG - middle frontal gyrus
MFG-WM - middle frontal gyrus white matter
MTG - middle temporal gyrus
MTG-WM - middle temporal gyrus white matter
opt - optic tract
pic - posterior limb of the internal capsule
PPo - planum polare
PrG - precentral gyrus
PrG-WM - precentral gyrus white matter
Pu - putamen
scr - superior corona radiata
SFG - superior frontal gyrus
SFG-WM - superior frontal gyrus white matter
sfo - Superior fronto-occipital fasciculus
STG - superior temporal gyrus
STG-WM - superior temporal gyrus white matter
Th - thalamus
unc - uncinate fasciculus

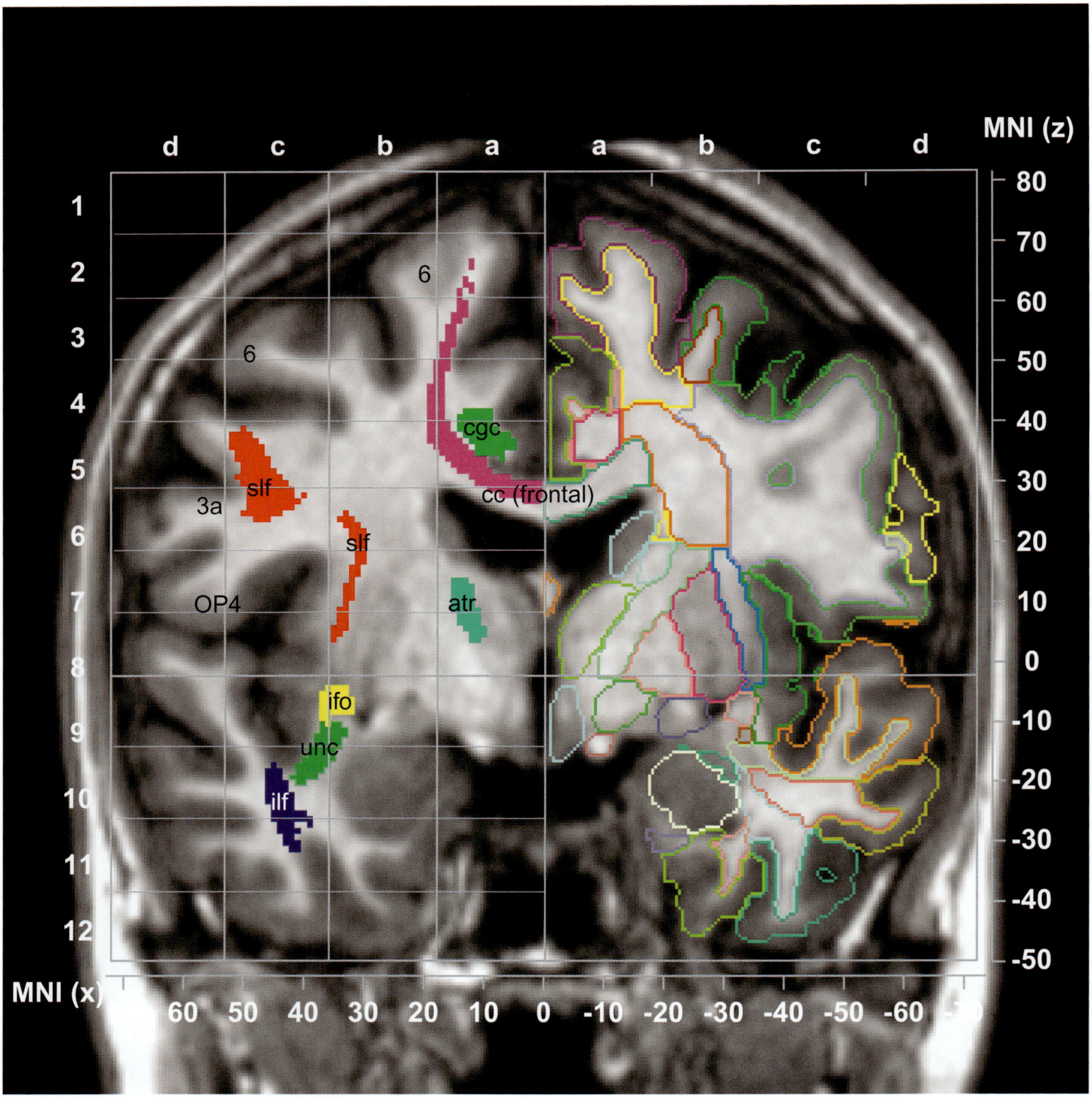

atr	-	anterior thalamic radiation	ifo	-	inferior fronto-occipital fasciculus
cc (frontal)	-	corpus callosum (connecting bilateral frontal areas)	ilf	-	inferior longitudinal fasciculus
			slf	-	superior longitudinal fasciculus
cgc	-	cingulum (cingulate gyrus part)	unc	-	uncinate fasciculus

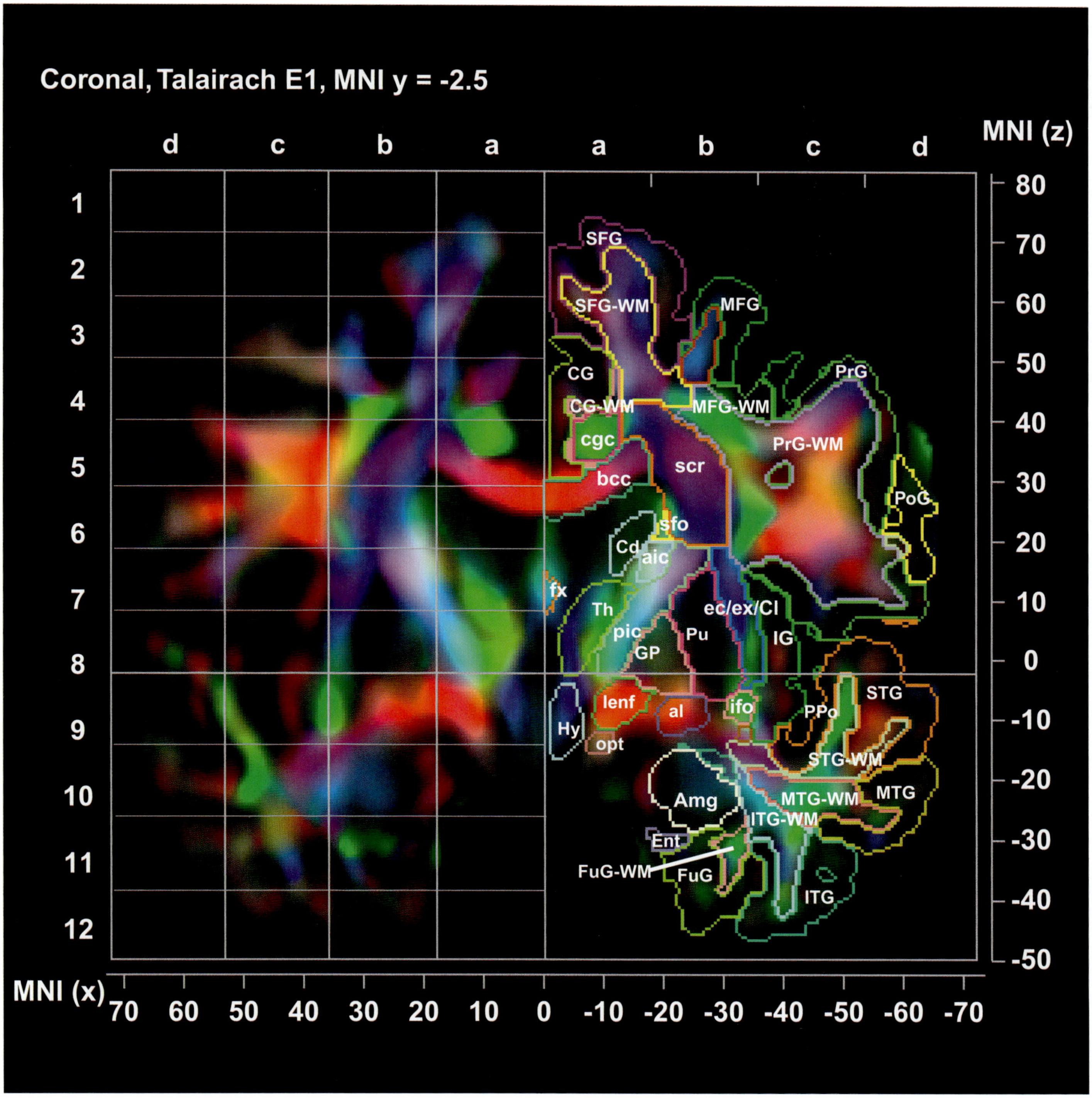

Amg - amygdala
aic - anterior limb of the internal capsule
al - ansa lenticularis
bcc - body of the corpus callosum
Cd - caudate nucleus
CG - cingulate gyrus
cgc - cingulum (cingulate gyrus part)
CG-WM - cingulate gyrus white matter
ec/ex/Cl - external capsule / extreme capsule / claustrum (cannot be resolved with current resolution)
Ent - entorhinal cortex
FuG - fusiform gyrus
FuG-WM - fusiform gyrus white matter
fx - fornix (column and body)
GP - globus pallidus
Hy - hypothalamus
ifo - inferior fronto-occipital fasciculus
IG - insular gyrus
ITG - inferior temporal gyrus
ITG-WM - inferior temporal gyrus white matter
lenf - lenticular fasciculus
MFG - middle frontal gyrus
MFG-WM - middle frontal gyrus white matter
MTG - middle temporal gyrus
MTG-WM - middle temporal gyrus white matter
opt - optic tract
pic - posterior limb of the internal capsule
PoG - postcentral gyrus
PPo - planum polare
PrG - precentral gyrus
PrG-WM - precentral gyrus white matter
Pu - putamen
scr - superior corona radiata
SFG - superior frontal gyrus
SFG-WM - superior frontal gyrus white matter
sfo - Superior fronto-occipital fasciculus
STG - superior temporal gyrus
STG-WM - superior temporal gyrus white matter
Th - thalamus

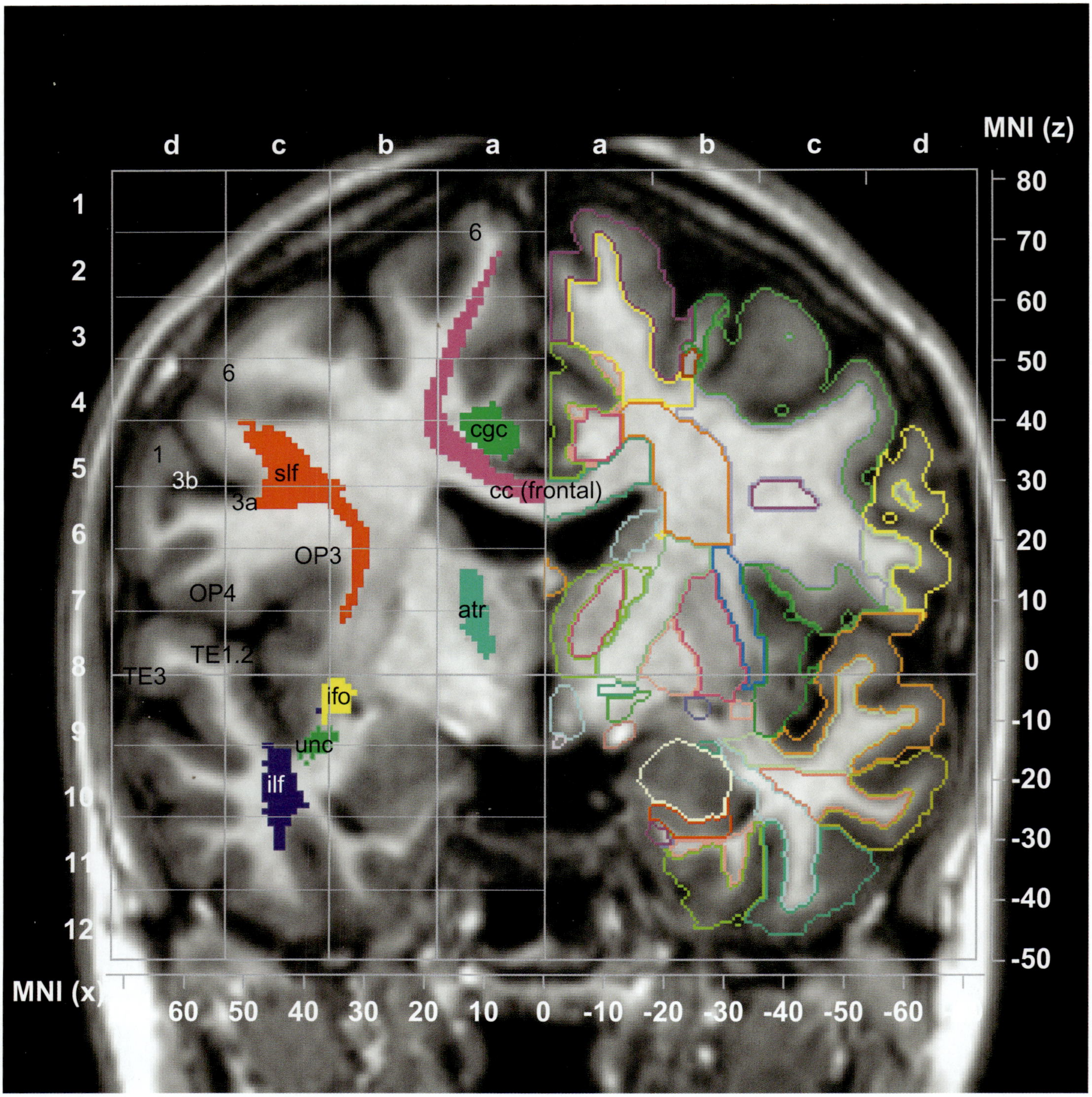

atr - anterior thalamic radiation
cc (frontal) - corpus callosum (connecting bilateral frontal areas)
cgc - cingulum (cingulate gyrus part)
ifo - inferior fronto-occipital fasciculus
ilf - inferior longitudinal fasciculus
slf - superior longitudinal fasciculus
unc - uncinate fasciculus

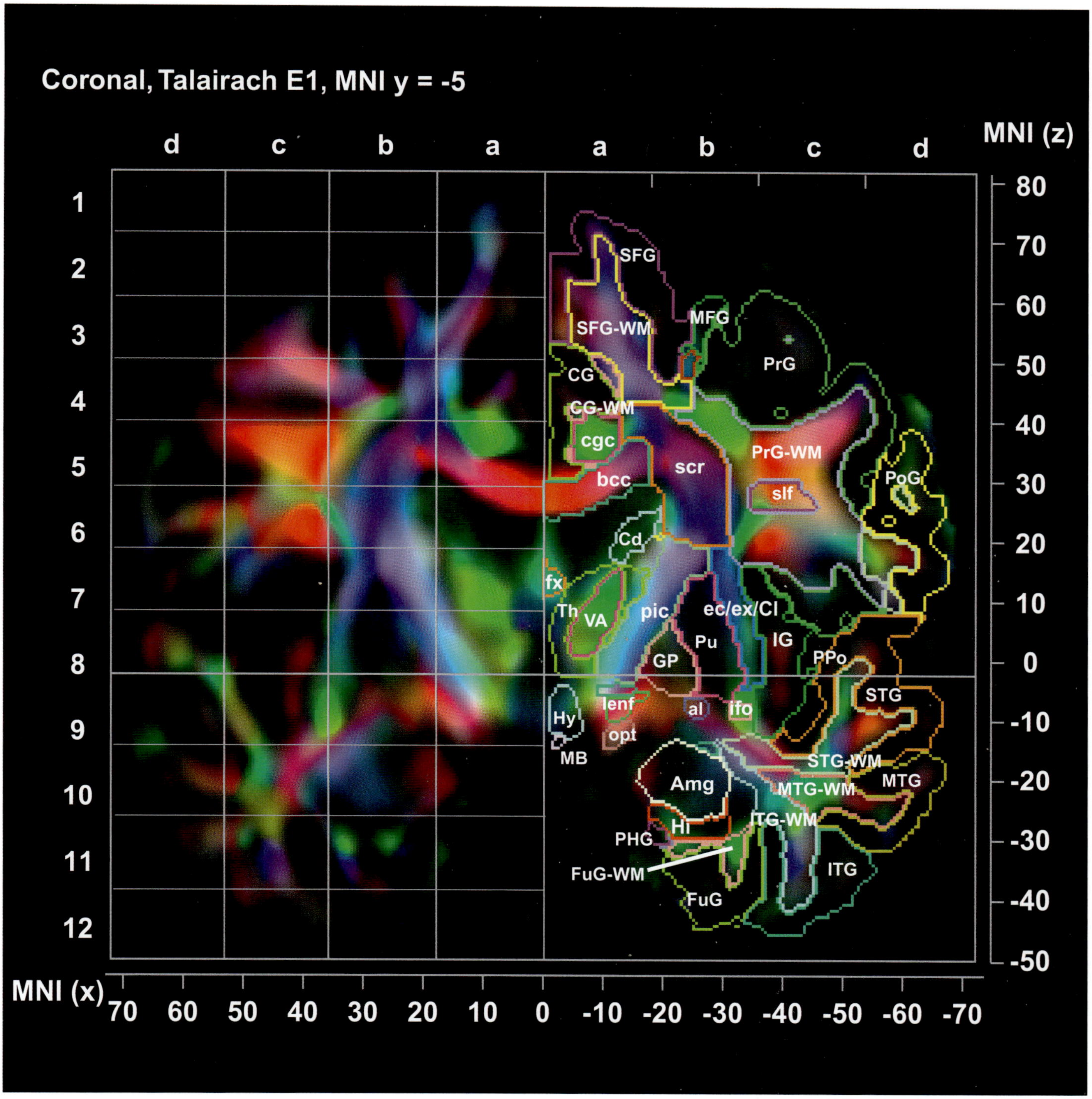

al - ansa lenticularis
Amg - amygdala
bcc - body of the corpus callosum
Cd - caudate nucleus
CG - cingulate gyrus
cgc - cingulum (cingulate gyrus part)
CG-WM - cingulate gyrus white matter
ec/ex/CI - external capsule / extreme capsule / claustrum (cannot be resolved with current resolution)
FuG - fusiform gyrus
FuG-WM - fusiform gyrus white matter
fx - fornix (column and body)
GP - globus pallidus
Hi - hippocampus
Hy - hypothalamus
ifo - inferior fronto-occipital fasciculus
IG - insular gyrus
ITG - inferior temporal gyrus
ITG-WM - inferior temporal gyrus white matter
lenf - lenticular fasciculus
MB - mammillary body
MFG - middle frontal gyrus
MTG - middle temporal gyrus
MTG-WM - middle temporal gyrus white matter
opt - optic tract
PHG - parahippocampal gyrus
pic - posterior limb of the internal capsule
PoG - postcentral gyrus
PPo - planum polare
PrG - precentral gyrus
PrG-WM - precentral gyrus white matter
Pu - putamen
scr - superior corona radiata
SFG - superior frontal gyrus
SFG-WM - superior frontal gyrus white matter
slf - superior longitudinal fasciculus
STG - superior temporal gyrus
STG-WM - superior temporal gyrus white matter
Th - thalamus
VA - ventroanterior thalamic nucleus

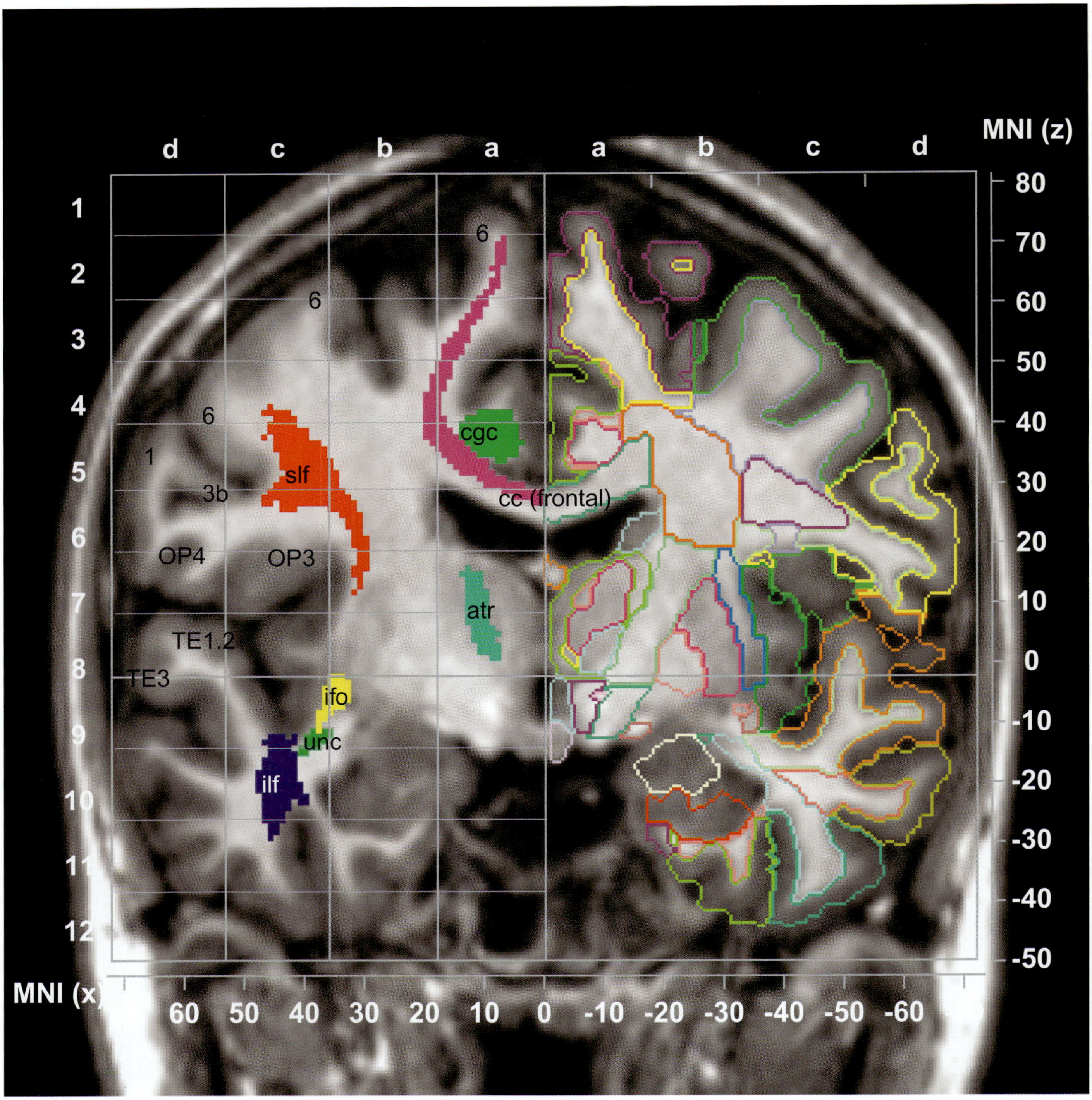

atr - anterior thalamic radiation
cc (frontal) - corpus callosum (connecting bilateral frontal areas)
cgc - cingulum (cingulate gyrus part)
ifo - inferior fronto-occipital fasciculus
ilf - inferior longitudinal fasciculus
slf - superior longitudinal fasciculus
unc - uncinate fasciculus

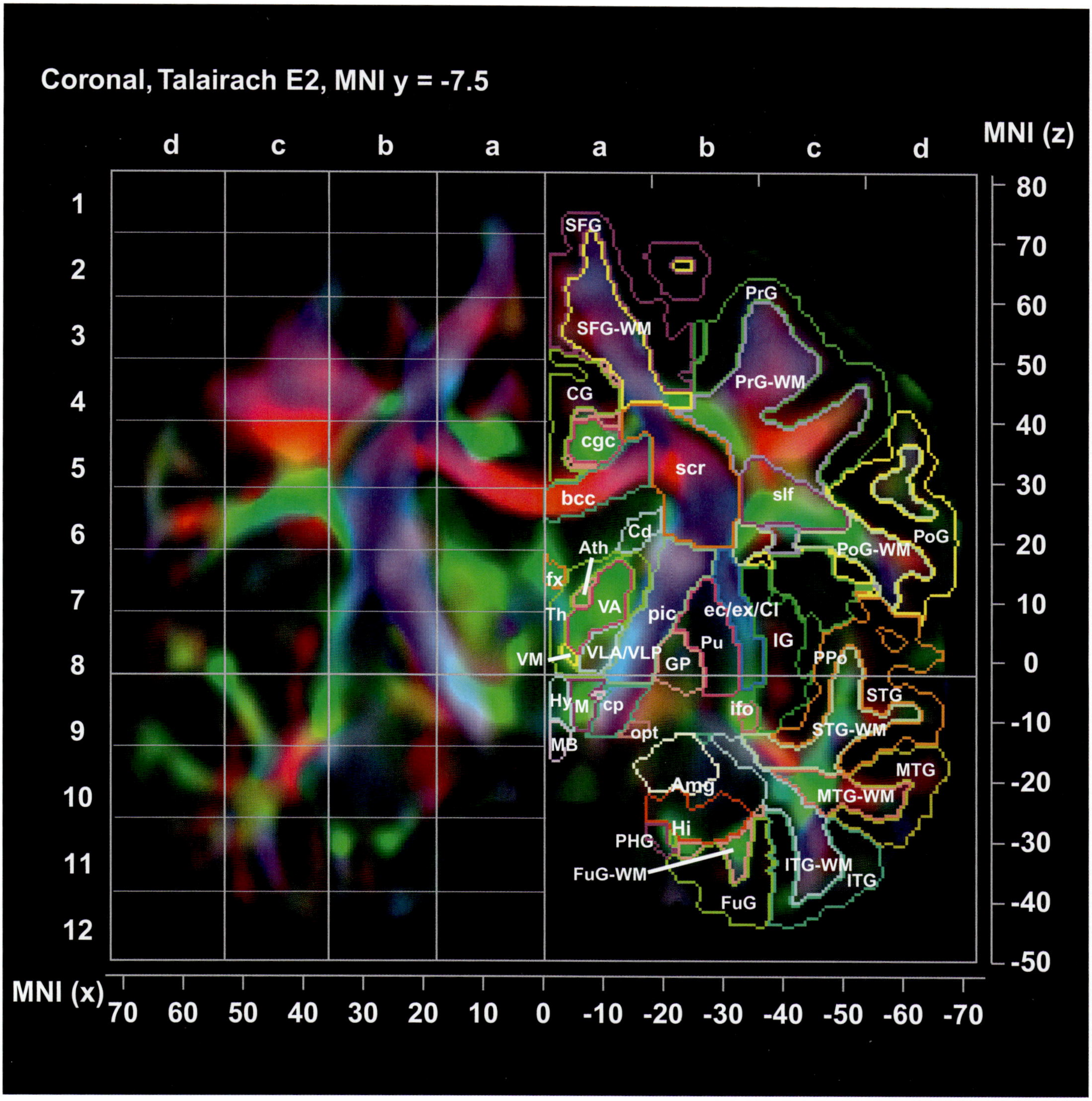

Amg - amygdala
Ath - anterior nuclear group (thalamus)
bcc - body of the corpus callosum
Cd - caudate nucleus
CG - cingulate gyrus
cgc - cingulum (cingulate gyrus part)
cp - cerebral peduncle
ec/ex/Cl - external capsule / extreme capsule / claustrum (cannot be resolved with current resolution)
FuG - fusiform gyrus
FuG-WM - fusiform gyrus white matter
fx - fornix (column and body)
GP - globus pallidus
Hi - hippocampus
Hy - hypothalamus
ifo - inferior fronto-occipital fasciculus
IG - insular gyrus
ITG - inferior temporal gyrus
ITG-WM - inferior temporal gyrus white matter
M - midbrain
MB - mammillary body
MTG - middle temporal gyrus
MTG-WM - middle temporal gyrus white matter
opt - optic tract
PHG - parahippocampal gyrus
pic - posterior limb of the internal capsule
PoG - postcentral gyrus
PoG-WM - postcentral gyrus white matter
PPo - planum polare
PrG - precentral gyrus
PrG-WM - precentral gyrus white matter
Pu - putamen
scr - superior corona radiata
SFG - superior frontal gyrus
SFG-WM - superior frontal gyrus white matter
slf - superior longitudinal fasciculus
STG - superior temporal gyrus
STG-WM - superior temporal gyrus white matter
Th - thalamus
VA - ventroanterior thalamic nucleus
VLA/VLP - anterior and posterior ventrolateral thalamic nuclei
VM - ventromedial thalamic nucleus

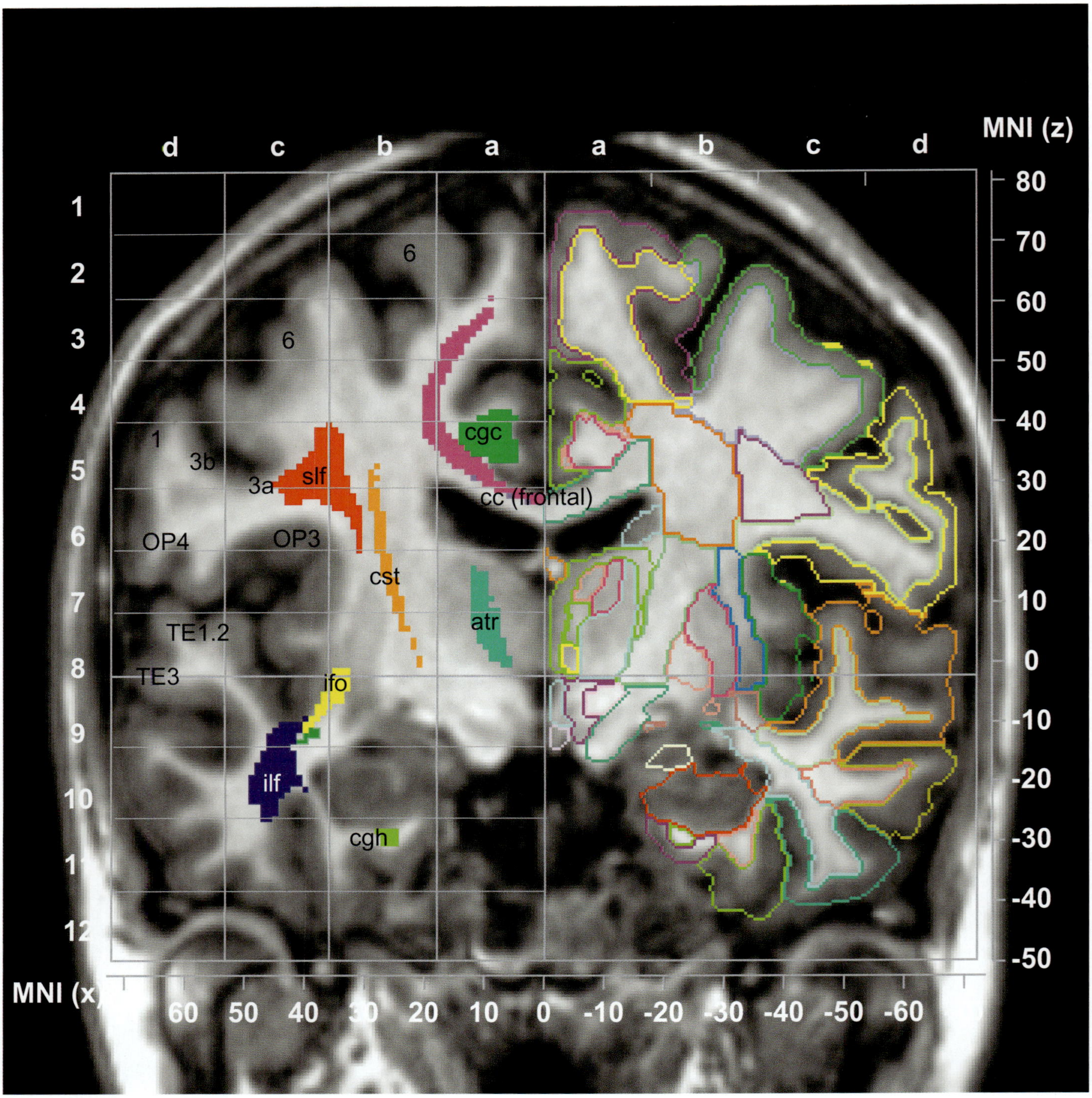

atr - anterior thalamic radiation
cc (frontal) - corpus callosum (connecting bilateral frontal areas)
cgc - cingulum (cingulate gyrus part)
cgh - cingulum (hippocampal part)
cst - corticospinal tract
ifo - inferior fronto-occipital fasciculus
ilf - inferior longitudinal fasciculus
slf - superior longitudinal fasciculus

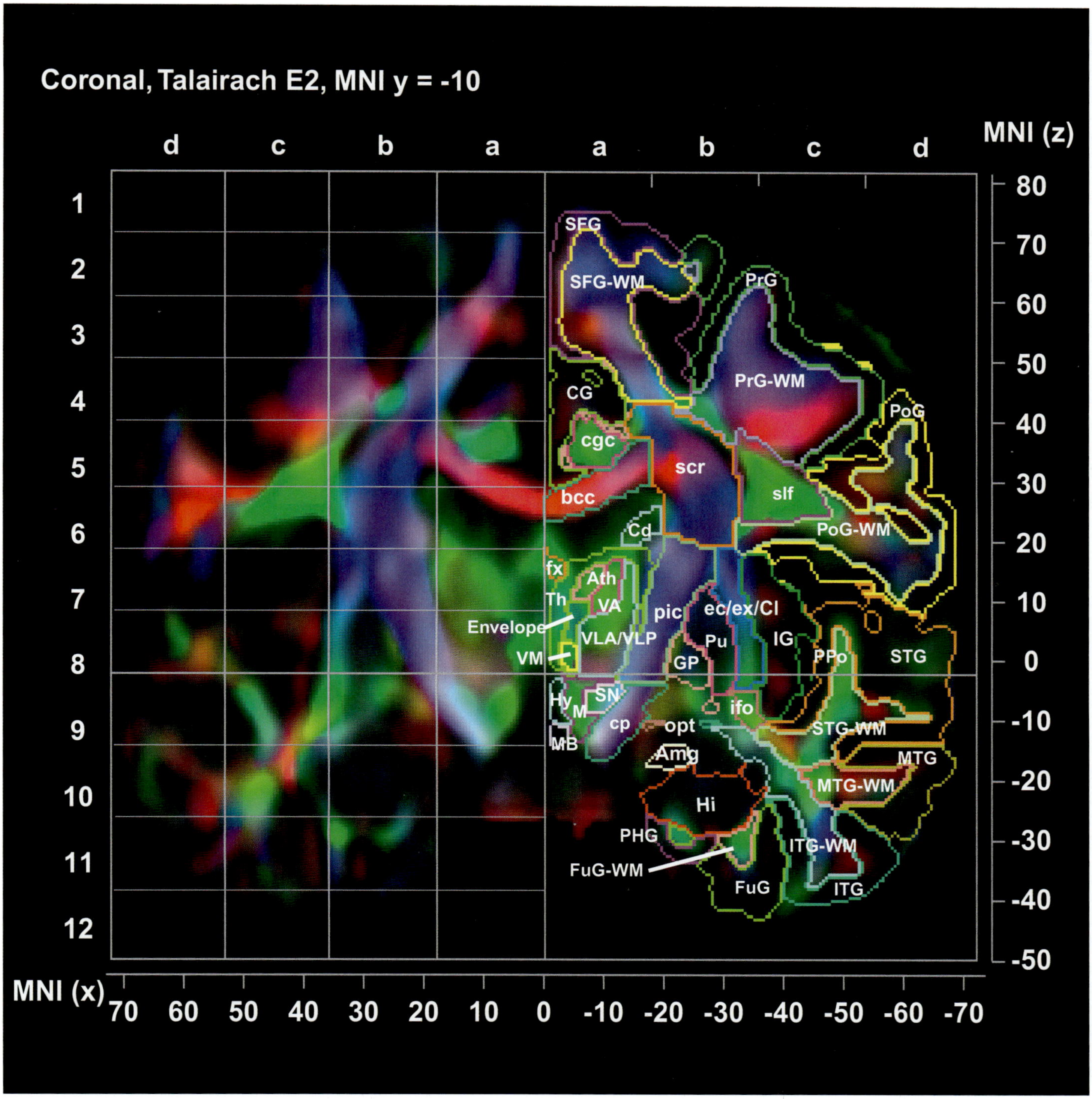

Amg - amygdala
Ath - anterior nuclear group (thalamus)
bcc - body of the corpus callosum
Cd - caudate nucleus
CG - cingulate gyrus
cgc - cingulum (cingulate gyrus part)
cp - cerebral peduncle
ec/ex/Cl - external capsule / extreme capsule / claustrum (cannot be resolved with current resolution)
Envelope - nuclei of the internal medullary lamina (thalamus)
FuG - fusiform gyrus
FuG-WM - fusiform gyrus white matter
fx - fornix (column and body)
GP - globus pallidus
Hi - hippocampus
Hy - hypothalamus
ifo - inferior fronto-occipital fasciculus
IG - insular gyrus
ITG - inferior temporal gyrus
ITG-WM - inferior temporal gyrus white matter
M - midbrain
MB - mammillary body
MTG - middle temporal gyrus
MTG-WM - middle temporal gyrus white matter
opt - optic tract
PHG - parahippocampal gyrus
pic - posterior limb of the internal capsule
PoG - postcentral gyrus
PoG-WM - postcentral gyrus white matter
PPo - planum polare
PrG - precentral gyrus
PrG-WM - precentral gyrus white matter
Pu - putamen
scr - superior corona radiata
SFG - superior frontal gyrus
SFG-WM - superior frontal gyrus white matter
slf - superior longitudinal fasciculus
SN - substantia nigra
STG - superior temporal gyrus
STG-WM - superior temporal gyrus white matter
Th - thalamus
VA - ventroanterior thalamic nucleus
VLA/VLP - anterior and posterior ventrolateral thalamic nuclei
VM - ventromedial thalamic nucleus

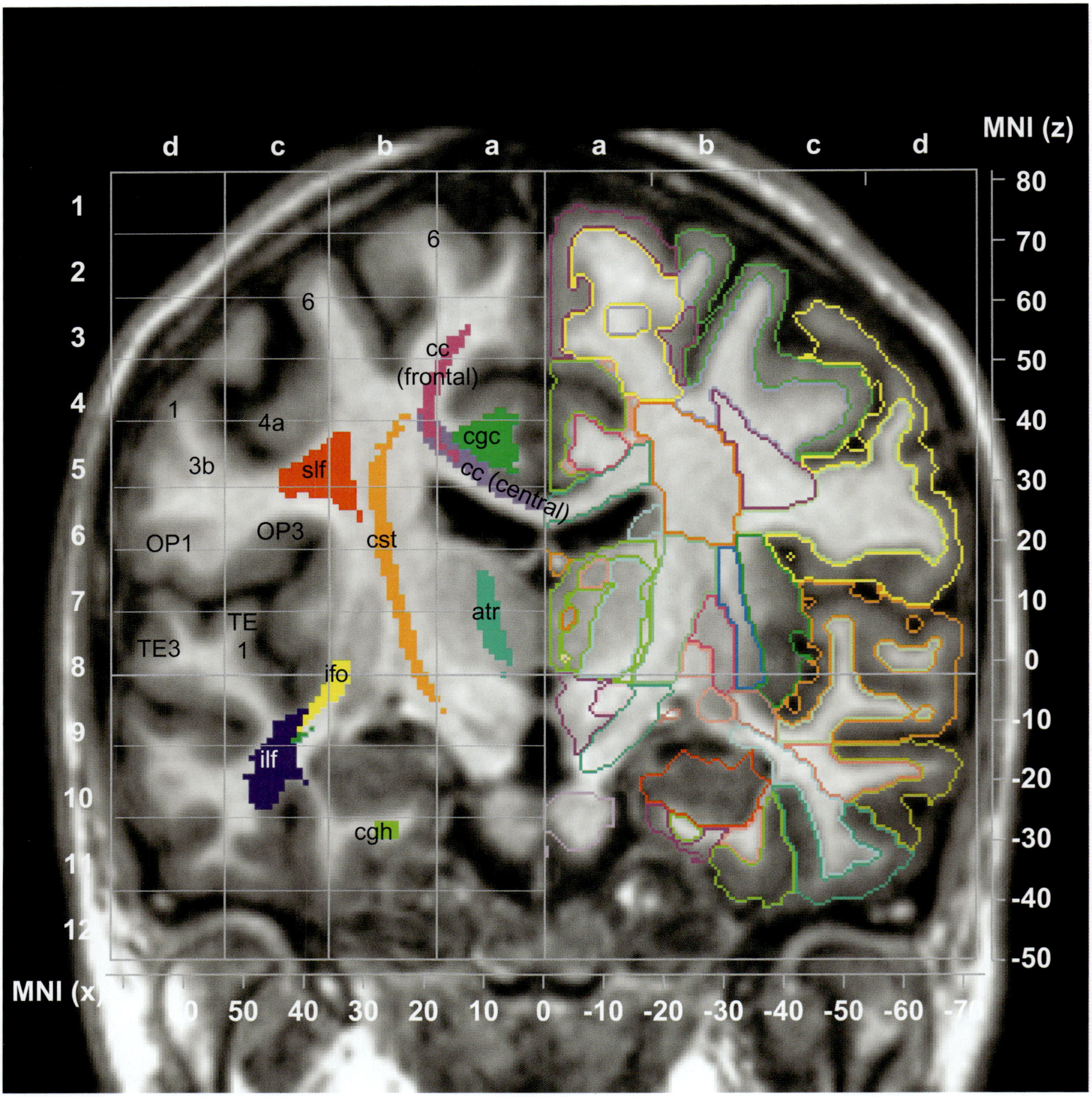

atr - anterior thalamic radiation
cc (central) - corpus callosum (connecting bilateral central areas)
cc (frontal) - corpus callosum (connecting bilateral frontal areas)
cgc - cingulum (cingulate gyrus part)
cgh - cingulum (hippocampal part)
cst - corticospinal tract
ifo - inferior fronto-occipital fasciculus
ilf - inferior longitudinal fasciculus
slf - superior longitudinal fasciculus

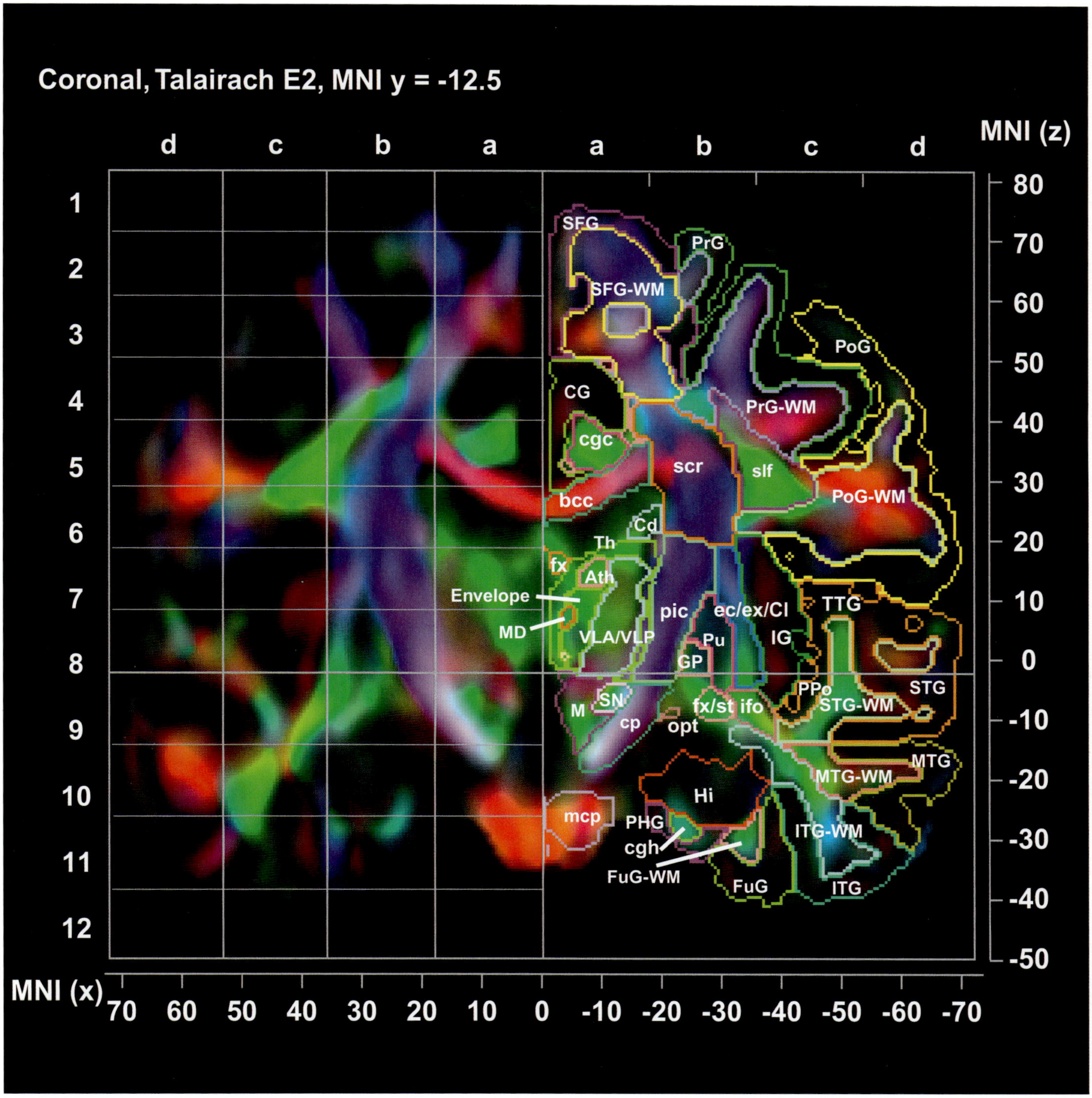

Ath - anterior nuclear group (thalamus)
bcc - body of the corpus callosum
Cd - caudate nucleus
CG - cingulate gyrus
cgc - cingulum (cingulate gyrus part)
cgh - cingulum (hippocampal part)
cp - cerebral peduncle
ec/ex/Cl - external capsule / extreme capsule / claustrum (cannot be resolved with current resolution)
Envelope - nuclei of the internal medullary lamina (thalamus)
FuG - fusiform gyrus
FuG-WM - fusiform gyrus white matter
fx - fornix (column and body)
GP - globus pallidus
Hi - hippocampus
ifo - inferior fronto-occipital fasciculus
IG - insular gyrus
ITG - inferior temporal gyrus
ITG-WM - inferior temporal gyrus white matter
M - midbrain
mcp - middle cerebellar peduncle
MD - mediodorsal thalamic nucleus
MTG - middle temporal gyrus
MTG-WM - middle temporal gyrus white matter
opt - optic tract
PHG - parahippocampal gyrus
pic - posterior limb of the internal capsule
PoG - postcentral gyrus
PoG-WM - postcentral gyrus white matter
PPo - planum polare
PrG - precentral gyrus
PrG-WM - precentral gyrus white matter
Pu - putamen
scr - superior corona radiata
SFG - superior frontal gyrus
SFG-WM - superior frontal gyrus white matter
slf - superior longitudinal fasciculus
SN - substantia nigra
STG - superior temporal gyrus
STG-WM - superior temporal gyrus white matter
Th - thalamus
TTG - transverse temporal gyrus
VLA/VLP - anterior and posterior ventrolateral thalamic nuclei

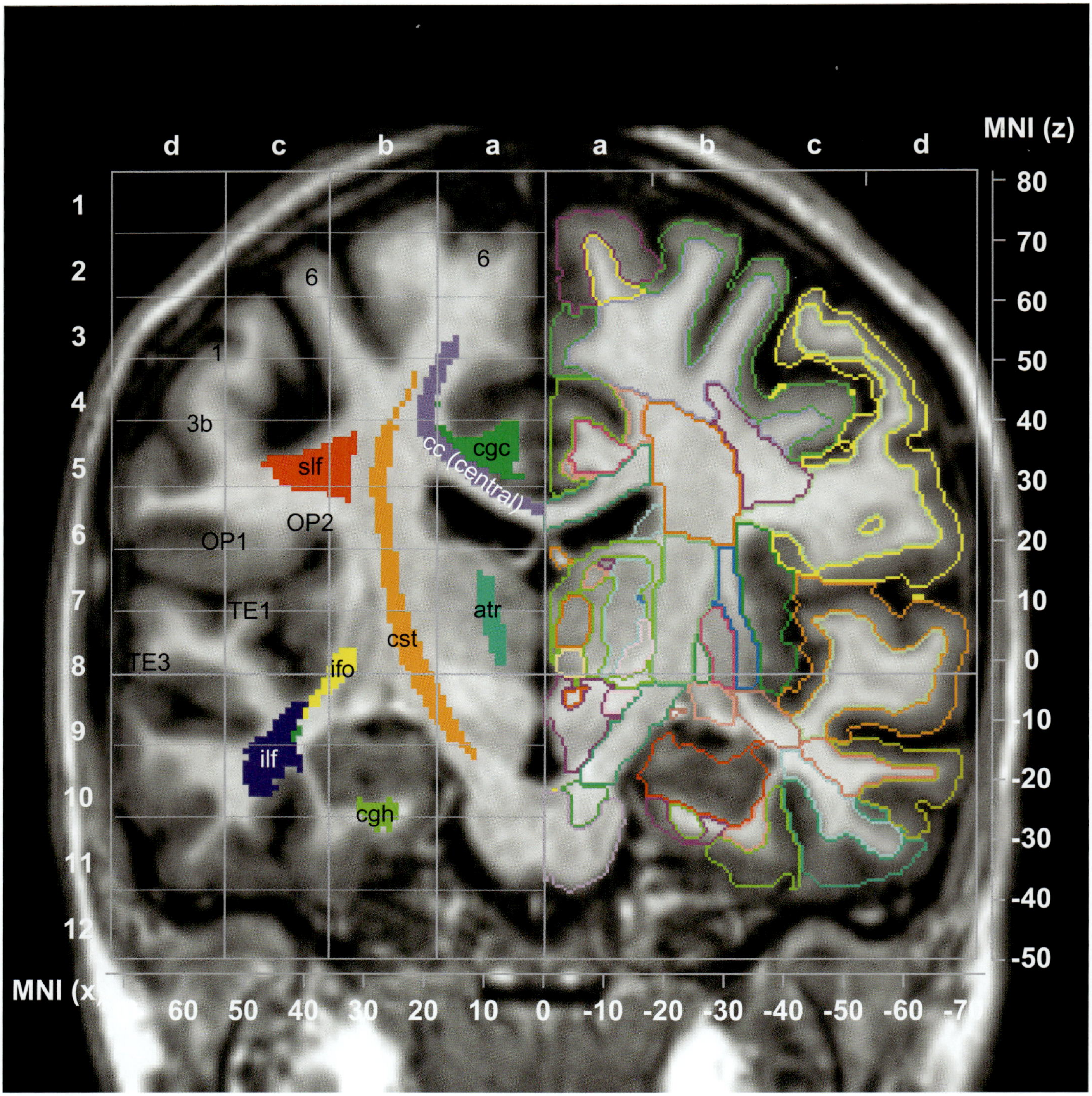

atr - anterior thalamic radiation
cc (central) - corpus callosum (connecting bilateral central areas)
cgc - cingulum (cingulate gyrus part)
cgh - cingulum (hippocampal part)
cst - corticospinal tract
ifo - inferior fronto-occipital fasciculus
ilf - inferior longitudinal fasciculus
slf - superior longitudinal fasciculus

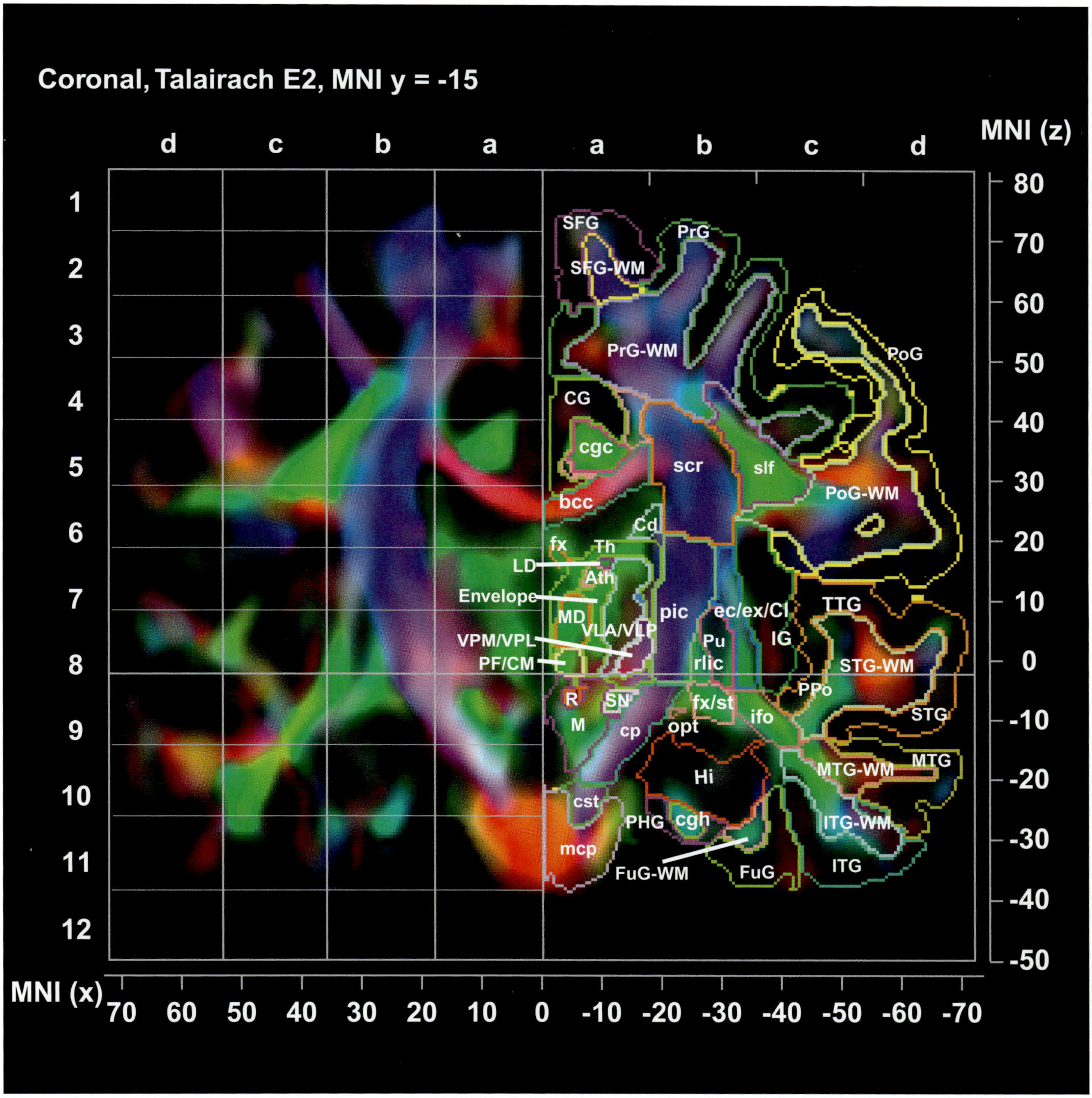

Ath - anterior nuclear group (thalamus)
bcc - body of the corpus callosum
Cd - caudate nucleus
CG - cingulate gyrus
cgc - cingulum (cingulate gyrus part)
cgh - cingulum (hippocampal part)
cp - cerebral peduncle
cst - corticospinal tract
ec/ex/Cl - external capsule / extreme capsule / claustrum (cannot be resolved with current resolution)
Envelope - nuclei of the internal medullary lamina (thalamus)
FuG - fusiform gyrus
FuG-WM - fusiform gyrus white matter
fx - fornix (column and body)
fx/st - fornix (crus) / stria terminalis (cannot be resolved with current resolution)
Hi - hippocampus
ifo - inferior fronto-occipital fasciculus
IG - insular gyrus
ITG - inferior temporal gyrus
ITG-WM - inferior temporal gyrus white matter
LD - laterodorsal thalamic nucleus
M - midbrain
mcp - middle cerebellar peduncle
MD - mediodorsal thalamic nucleus
MTG - middle temporal gyrus
MTG-WM - middle temporal gyrus white matter
opt - optic tract
PF/CM - parafascicular nucleus / centromedian nucleus (thalamus)
PHG - parahippocampal gyrus
pic - posterior limb of the internal capsule
PoG - postcentral gyrus
PoG-WM - postcentral gyrus white matter
PPo - planum polare
PrG - precentral gyrus
PrG-WM - precentral gyrus white matter
Pu - putamen
R - red nucleus
rlic - retrolenticular part of the internal capsule
scr - superior corona radiata
SFG - superior frontal gyrus
SFG-WM - superior frontal gyrus white matter
slf - superior longitudinal fasciculus
SN - substantia nigra
STG - superior temporal gyrus
STG-WM - superior temporal gyrus white matter
Th - thalamus
TTG - transverse temporal gyrus
VLA/VLP - anterior and posterior ventrolateral thalamic nuclei
VPM/VPL - ventroposterior medialis / ventroposterior lateralis (thalamus)

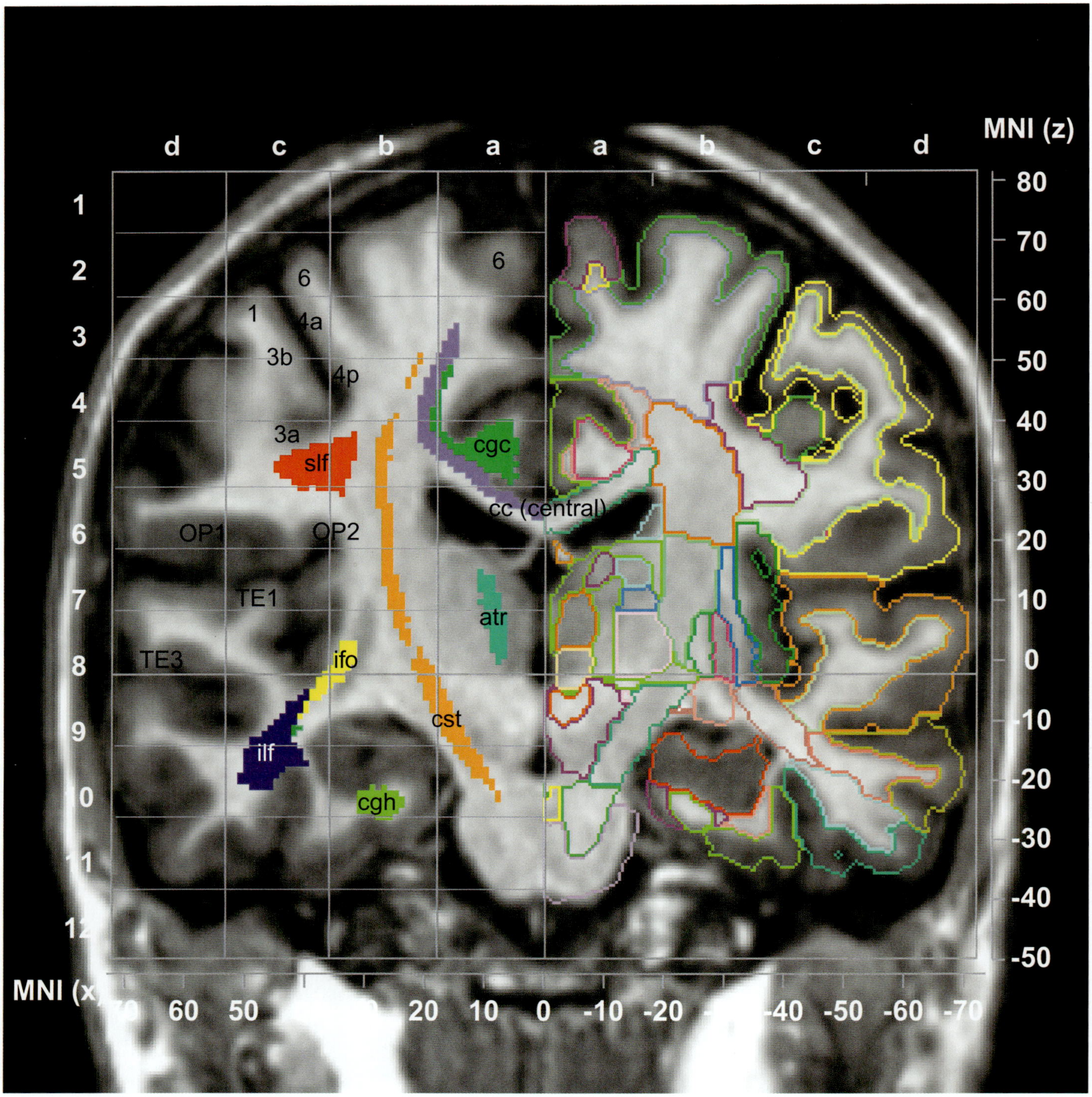

atr	-	anterior thalamic radiation
cc (central)	-	corpus callosum (connecting bilateral central areas)
cgc	-	cingulum (cingulate gyrus part)
cgh	-	cingulum (hippocampal part)
cst	-	corticospinal tract
ifo	-	inferior fronto-occipital fasciculus
ilf	-	inferior longitudinal fasciculus
slf	-	superior longitudinal fasciculus

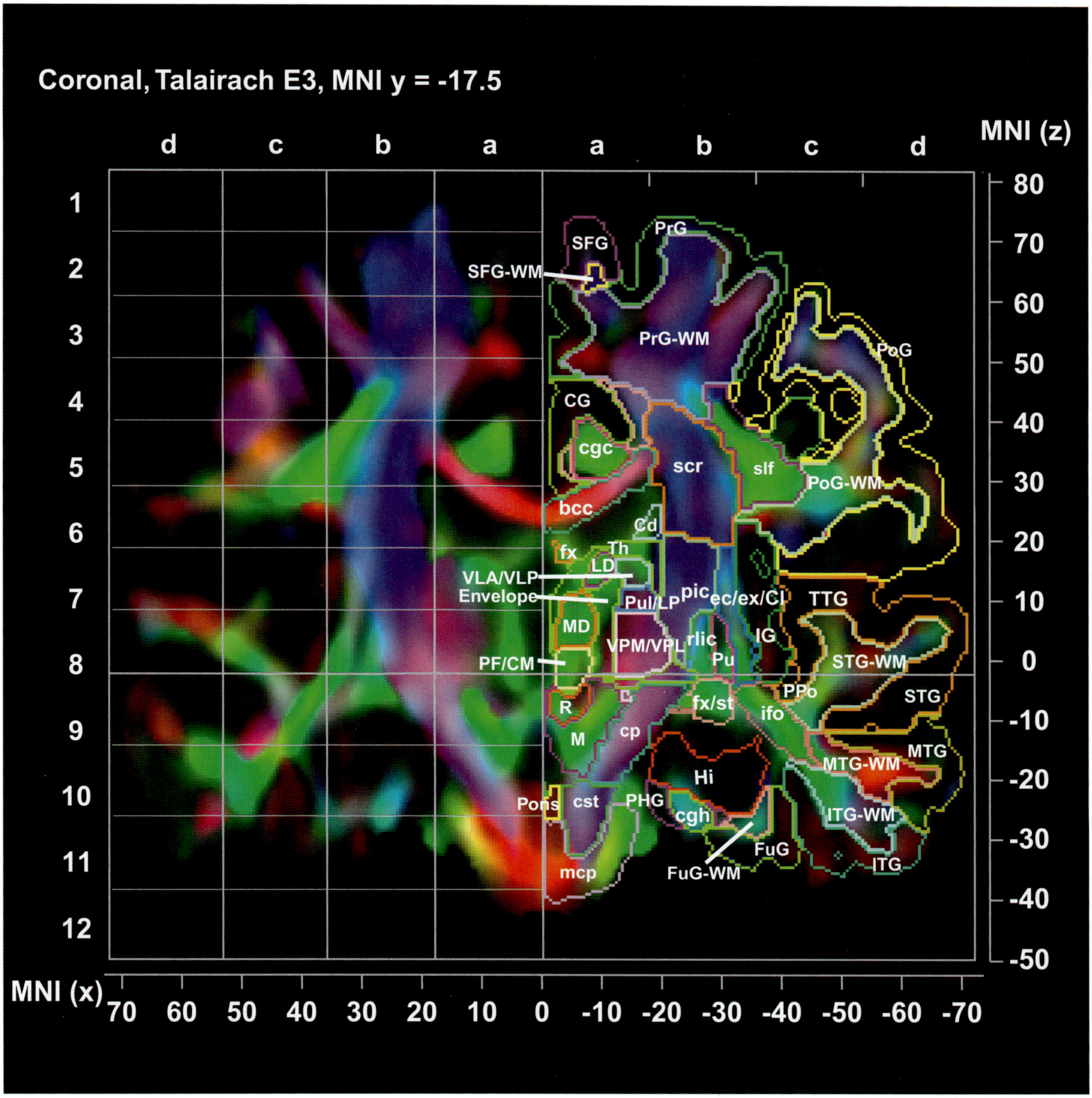

bcc - body of the corpus callosum
Cd - caudate nucleus
CG - cingulate gyrus
cgc - cingulum (cingulate gyrus part)
cgh - cingulum (hippocampal part)
cp - cerebral peduncle
cst - corticospinal tract
ec/ex/Cl - external capsule / extreme capsule / claustrum (cannot be resolved with current resolution)
Envelope - nuclei of the internal medullary lamina (thalamus)
FuG - fusiform gyrus
FuG-WM - fusiform gyrus white matter
fx - fornix (column and body)
fx/st - fornix (crus) / stria terminalis (cannot be resolved with current resolution)
Hi - hippocampus
IG - insular gyrus
ITG - inferior temporal gyrus
ITG-WM - inferior temporal gyrus white matter
LD - laterodorsal thalamic nucleus
M - midbrain
mcp - middle cerebellar peduncle
MD - mediodorsal thalamic nucleus
MTG - middle temporal gyrus
MTG-WM - middle temporal gyrus white matter
PF/CM - parafascicular nucleus / centromedian nucleus (thalamus)
PHG - parahippocampal gyrus
PoG - postcentral gyrus
PoG-WM - postcentral gyrus white matter
Pons - pons
PPo - planum polare
PrG - precentral gyrus
PrG-WM - precentral gyrus white matter
Pu - putamen
Pul/LP - pulvinar / lateral posterior thalamic nucleus
R - red nucleus
scr - superior corona radiata
SFG - superior frontal gyrus
SFG-WM - superior frontal gyrus white matter
slf - superior longitudinal fasciculus
STG - superior temporal gyrus
STG-WM - superior temporal gyrus white matter
Th - thalamus
TTG - transverse temporal gyrus
VLA/VLP - anterior and posterior ventrolateral thalamic nuclei
VPM/VPL - ventroposterior medialis / ventroposterior lateralis (thalamus)

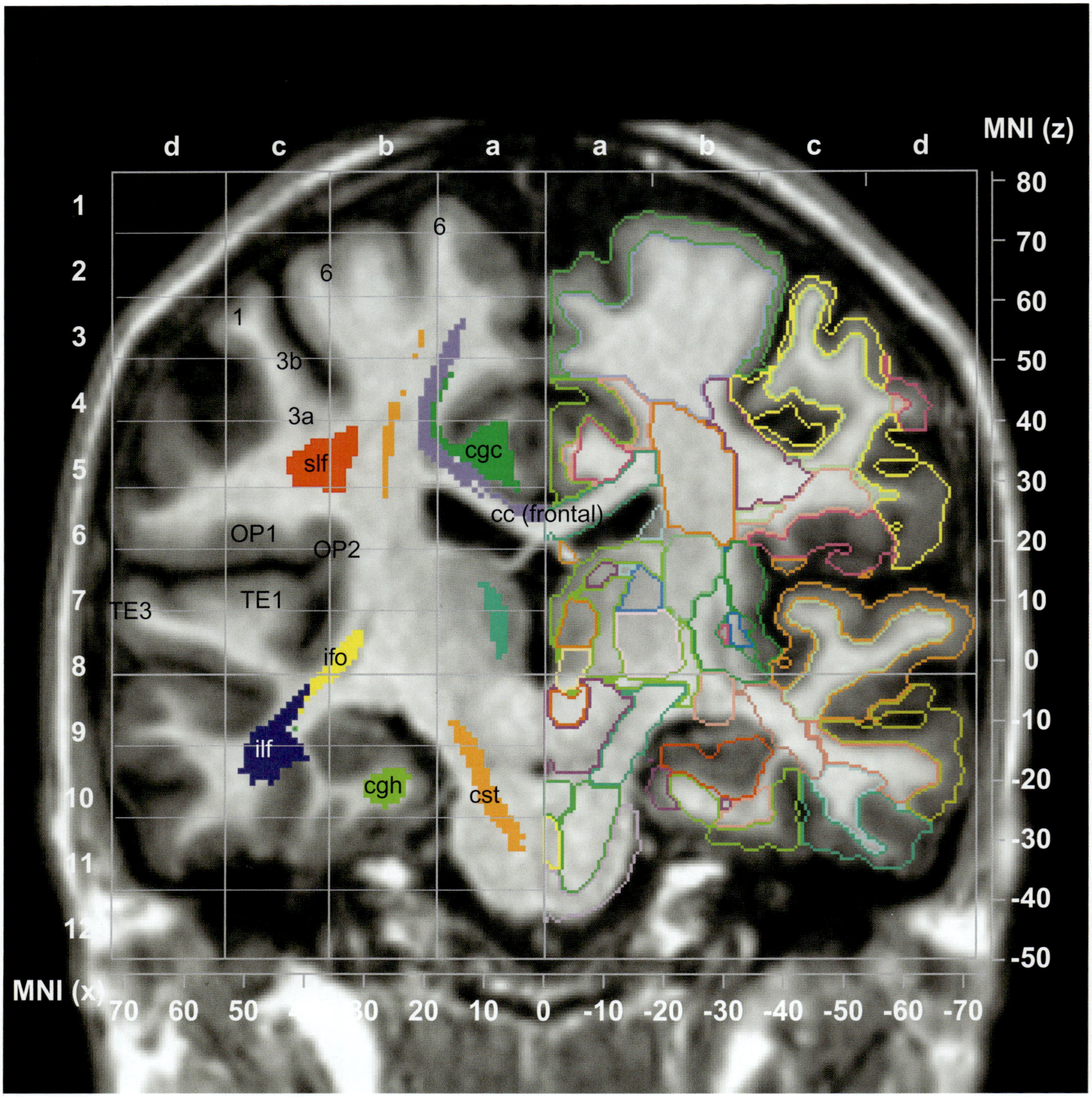

cc (frontal) - corpus callosum (connecting bilateral frontal areas)
cgc - cingulum (cingulate gyrus part)
cgh - cingulum (hippocampal part)
cst - corticospinal tract
ifo - inferior fronto-occipital fasciculus
ilf - inferior longitudinal fasciculus
slf - superior longitudinal fasciculus

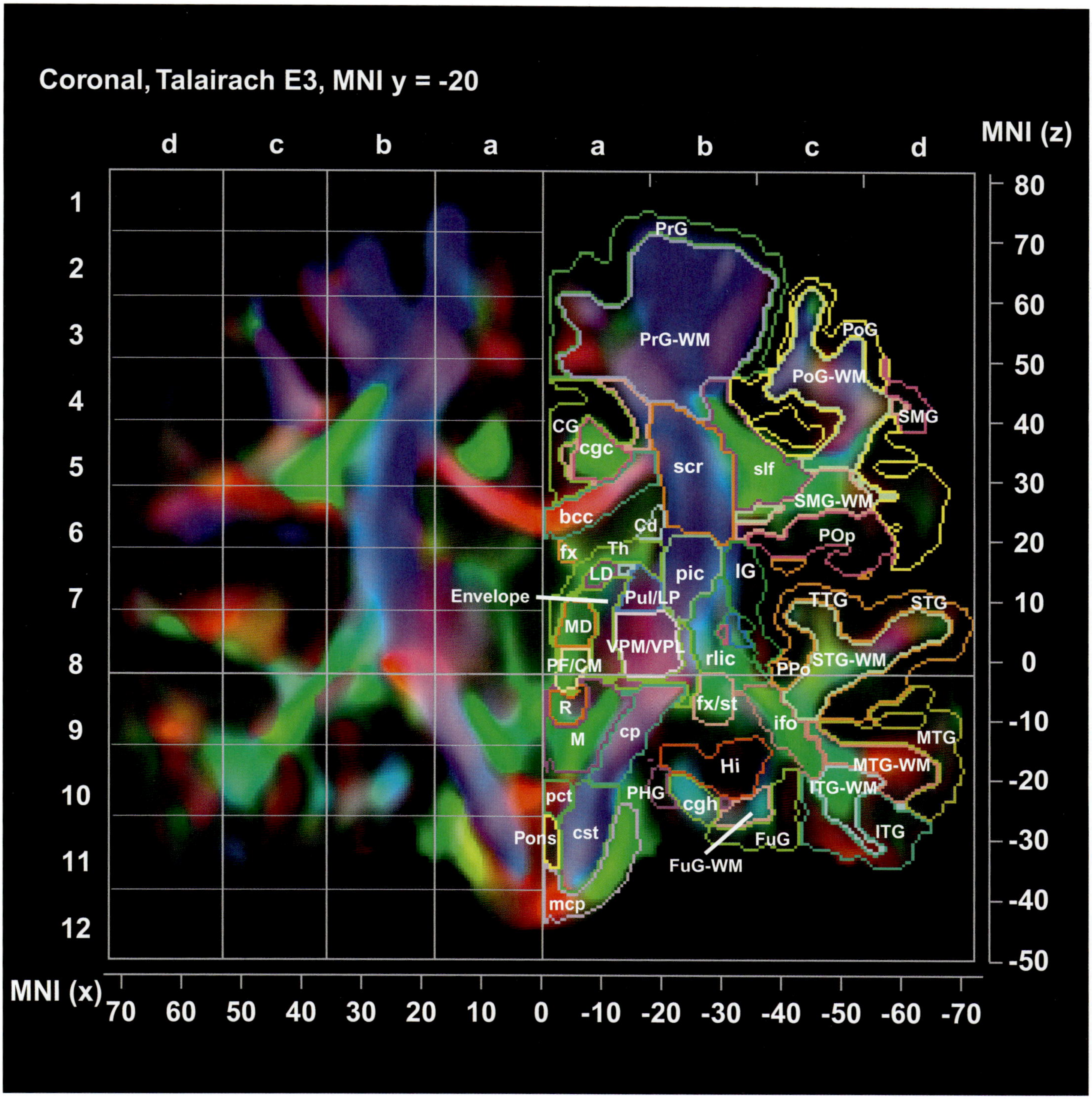

bcc	-	body of the corpus callosum
Cd	-	caudate nucleus
CG	-	cingulate gyrus
cgc	-	cingulum (cingulate gyrus part)
cgh	-	cingulum (hippocampal part)
cp	-	cerebral peduncle
cst	-	corticospinal tract
Envelope	-	nuclei of the internal medullary lamina (thalamus)
FuG	-	fusiform gyrus
FuG-WM	-	fusiform gyrus white matter
fx	-	fornix (column and body)
fx/st	-	fornix (crus) / stria terminalis (cannot be resolved with current resolution)
Hi	-	hippocampus
IG	-	insular gyrus
ifo	-	inferior fronto-occipital fasciculus
ITG	-	inferior temporal gyrus
ITG-WM	-	inferior temporal gyrus white matter
LD	-	laterodorsal thalamic nucleus
M	-	midbrain
mcp	-	middle cerebellar peduncle
MD	-	mediodorsal thalamic nucleus
MTG	-	middle temporal gyrus
MTG-WM	-	middle temporal gyrus white matter
pct	-	pontine crossing tract (a part of mcp)
pic	-	posterior limb of the internal capsule
PF/CM	-	parafascicular nucleus / centromedian nucleus (thalamus)
PHG	-	parahippocampal gyrus
PoG	-	postcentral gyrus
PoG-WM	-	postcentral gyrus white matter
Pons	-	pons
POp	-	parietal operculum
PPo	-	planum polare
PrG	-	precentral gyrus
PrG-WM	-	precentral gyrus white matter
Pul/LP	-	pulvinar / lateral posterior thalamic nucleus
R	-	red nucleus
rlic	-	retrolenticular part of the internal capsule
scr	-	superior corona radiata
slf	-	superior longitudinal fasciculus
SMG	-	supramarginal gyrus
SMG-WM	-	supramarginal gyrus white matter
STG	-	superior temporal gyrus
STG-WM	-	superior temporal gyrus white matter
Th	-	thalamus
TTG	-	transverse temporal gyrus
VPM/VPL	-	ventroposterior medialis / ventroposterior lateralis (thalamus)

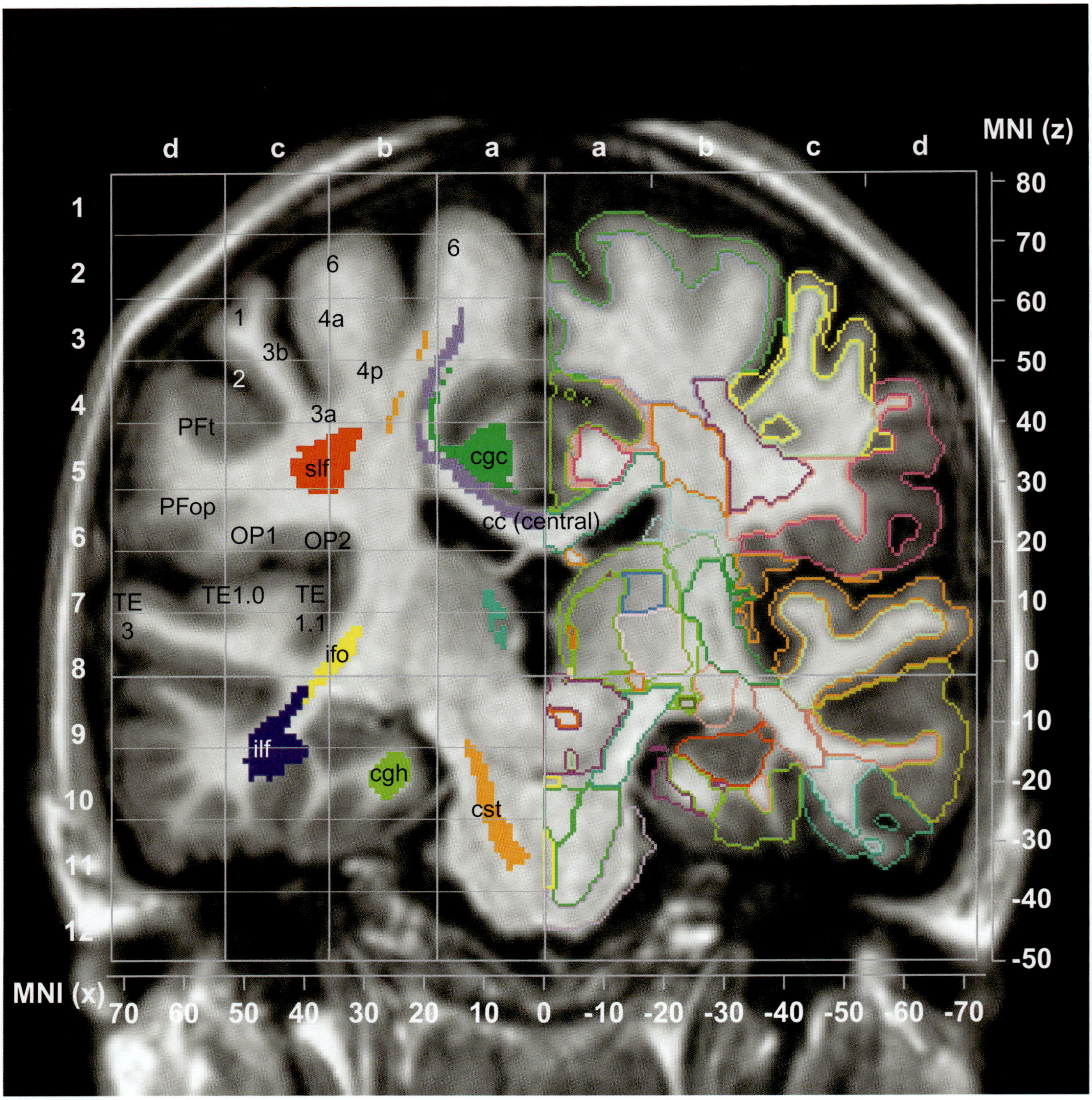

cc (central) - corpus callosum (connecting bilateral central areas)
cgc - cingulum (cingulate gyrus part)
cgh - cingulum (hippocampal part)
cst - corticospinal tract
ifo - inferior fronto-occipital fasciculus
ilf - inferior longitudinal fasciculus
slf - superior longitudinal fasciculus

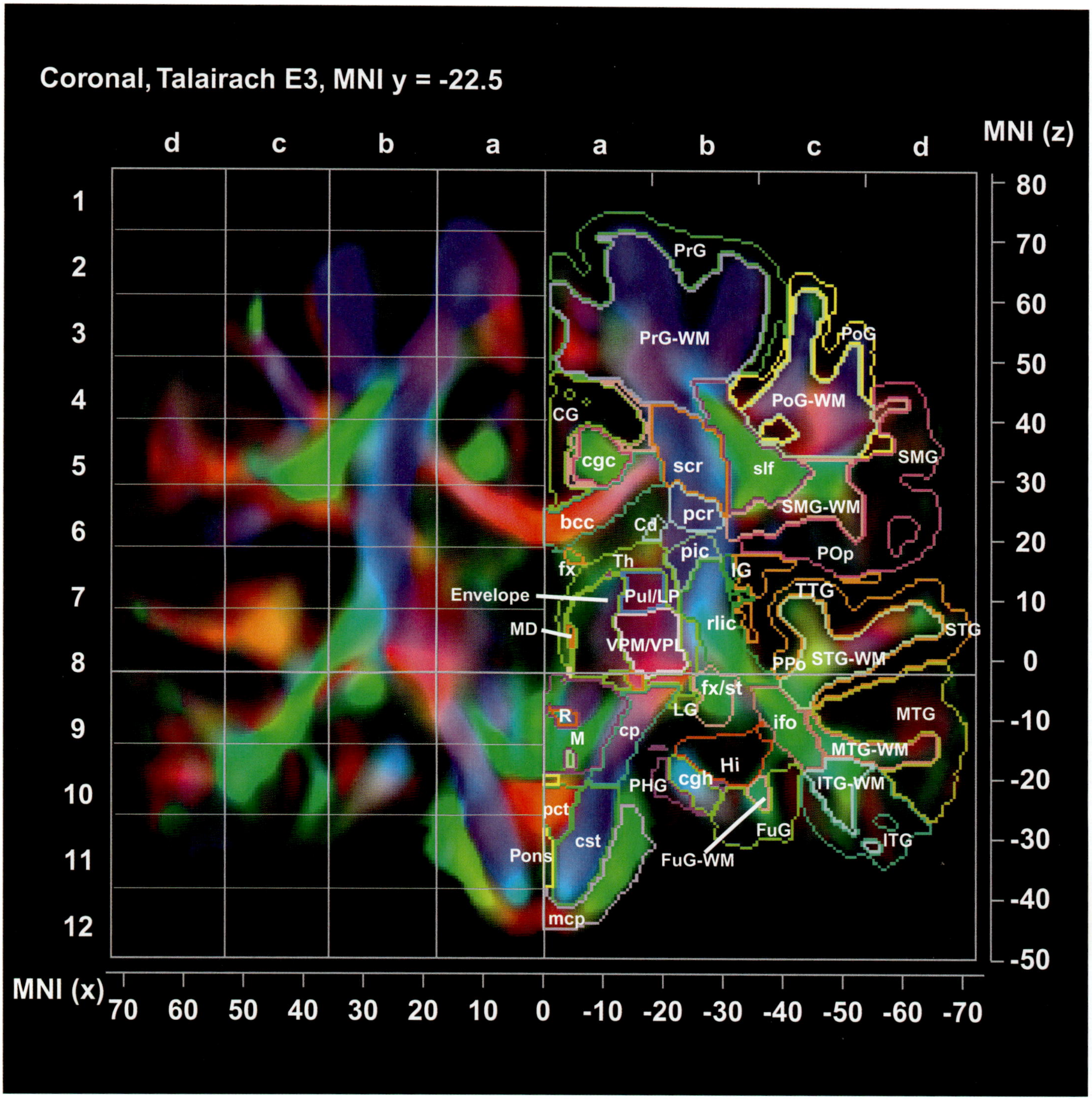

bcc	- body of the corpus callosum
Cd	- caudate nucleus
CG	- cingulate gyrus
cgc	- cingulum (cingulate gyrus part)
cgh	- cingulum (hippocampal part)
cp	- cerebral peduncle
cst	- corticospinal tract
Envelope	- nuclei of the internal medullary lamina (thalamus)
FuG	- fusiform gyrus
FuG-WM	- fusiform gyrus white matter
fx	- fornix (column and body)
fx/st	- fornix (crus) / stria terminalis (cannot be resolved with current resolution)
Hi	- hippocampus
IG	- insular gyrus
ITG	- inferior temporal gyrus
ITG-WM	- inferior temporal gyrus white matter
LG	- lateral geniculate nucleus (thalamus)
M	- midbrain
mcp	- middle cerebellar peduncle
MD	- mediodorsal thalamic nucleus
MTG	- middle temporal gyrus
MTG-WM	- middle temporal gyrus white matter
pcr	- posterior corona radiata
pct	- pontine crossing tract (a part of mcp)
PHG	- parahippocampal gyrus
PoG	- postcentral gyrus
PoG-WM	- postcentral gyrus white matter
Pons	- pons
POp	- parietal operculum
PPo	- planum polare
PrG	- precentral gyrus
PrG-WM	- precentral gyrus white matter
Pul/LP	- pulvinar / lateral posterior thalamic nucleus
R	- red nucleus
scr	- superior corona radiata
slf	- superior longitudinal fasciculus
SMG	- supramarginal gyrus
SMG-WM	- supramarginal gyrus white matter
STG	- superior temporal gyrus
STG-WM	- superior temporal gyrus white matter
Th	- thalamus
TTG	- transverse temporal gyrus
VPM/VPL	- ventroposterior medialis / ventroposterior lateralis (thalamus)

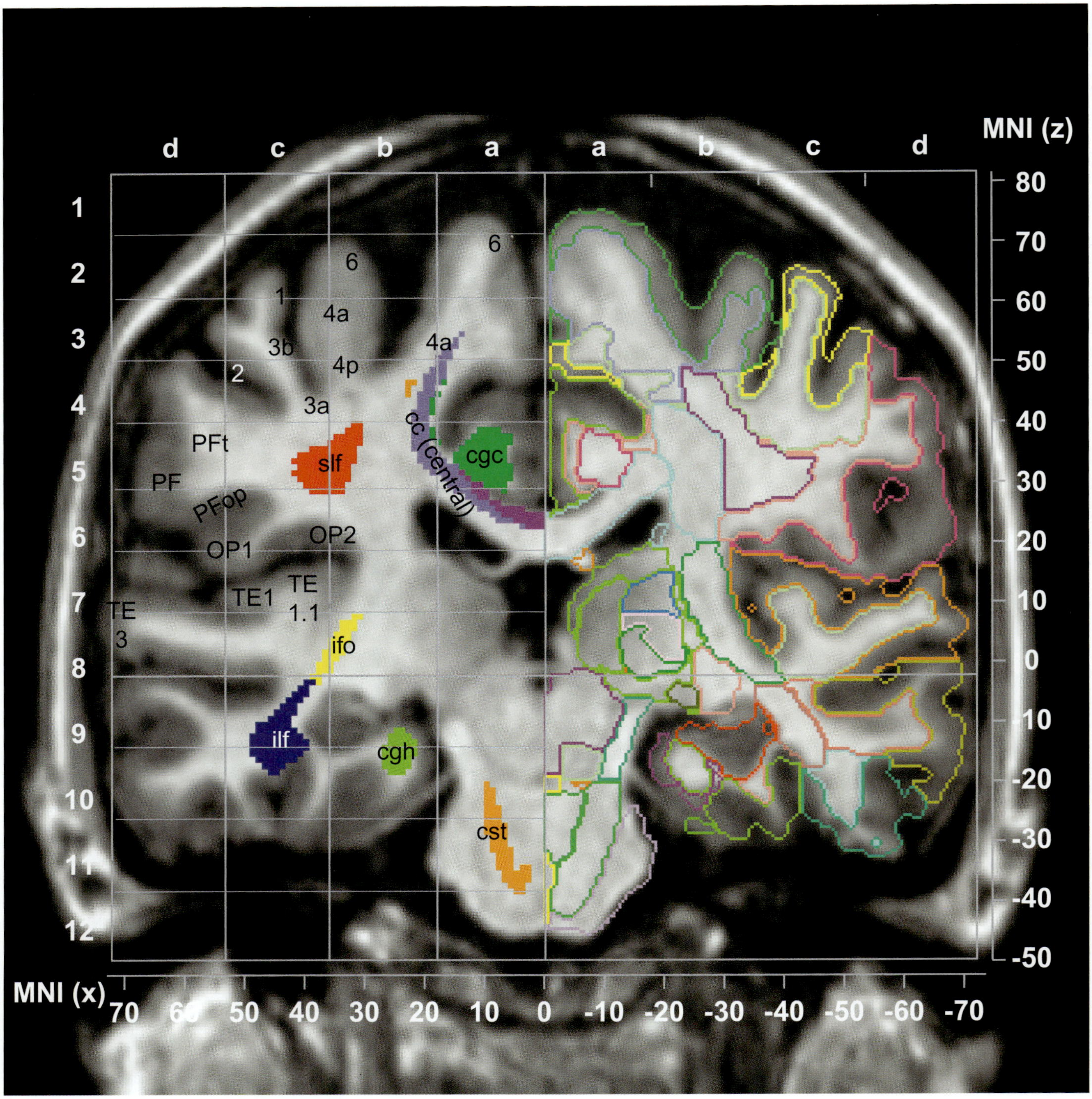

cc (central) - corpus callosum (connecting bilateral central areas)
cgc - cingulum (cingulate gyrus part)
cgh - cingulum (hippocampal part)
cst - corticospinal tract
ifo - inferior fronto-occipital fasciculus
ilf - inferior longitudinal fasciculus
slf - superior longitudinal fasciculus

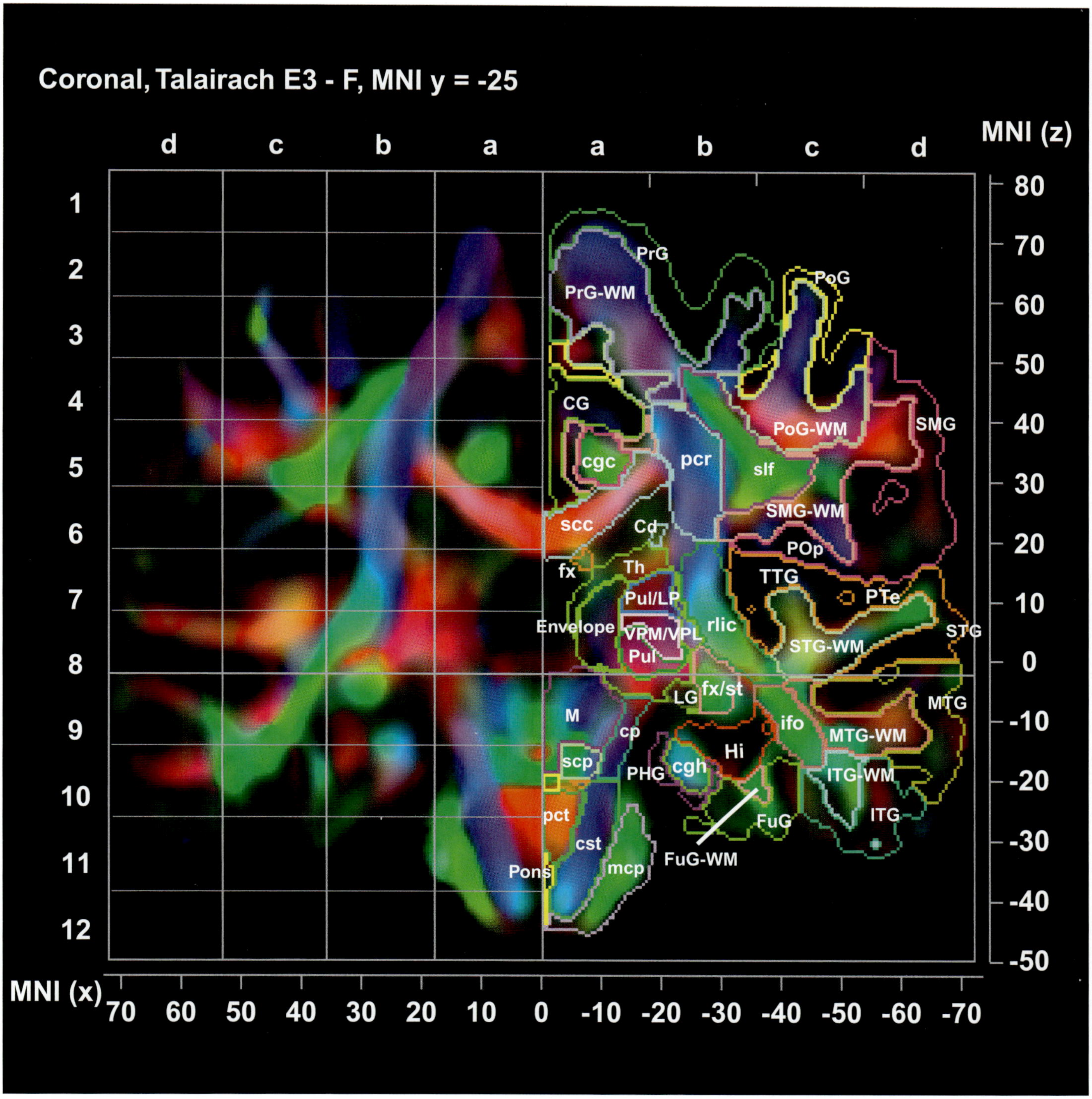

Cd - caudate nucleus
CG - cingulate gyrus
cgc - cingulum (cingulate gyrus part)
cgh - cingulum (hippocampal part)
cp - cerebral peduncle
cst - corticospinal tract
Envelope - nuclei of the internal medullary lamina (thalamus)
FuG - fusiform gyrus
FuG-WM - fusiform gyrus white matter
fx - fornix (column and body)
fx/st - fornix (crus) / stria terminalis (cannot be resolved with current resolution)
Hi - hippocampus
ifo - inferior fronto-occipital fasciculus
ITG - inferior temporal gyrus
ITG-WM - inferior temporal gyrus white matter
LG - lateral geniculate nucleus (thalamus)
M - midbrain
mcp - middle cerebellar peduncle
MTG - middle temporal gyrus
MTG-WM - middle temporal gyrus white matter
pcr - posterior corona radiata
pct - pontine crossing tract (a part of mcp)
PHG - parahippocampal gyrus
PoG - postcentral gyrus
PoG-WM - postcentral gyrus white matter
Pons - pons
POp - parietal operculum
PrG - precentral gyrus
PrG-WM - precentral gyrus white matter
PTe - planum temporale
Pul - pulvinar (thalamus)
Pul/LP - pulvinar / lateral posterior thalamic nucleus
rlic - retrolenticular part of the internal capsule
scc - splenium of the corpus callosum
scp - superior cerebellar peduncle
slf - superior longitudinal fasciculus
SMG - supramarginal gyrus
SMG-WM - supramarginal gyrus white matter
STG - superior temporal gyrus
STG-WM - superior temporal gyrus white matter
Th - thalamus
TTG - transverse temporal gyrus
VPM/VPL - ventroposterior medialis / ventroposterior lateralis (thalamus)

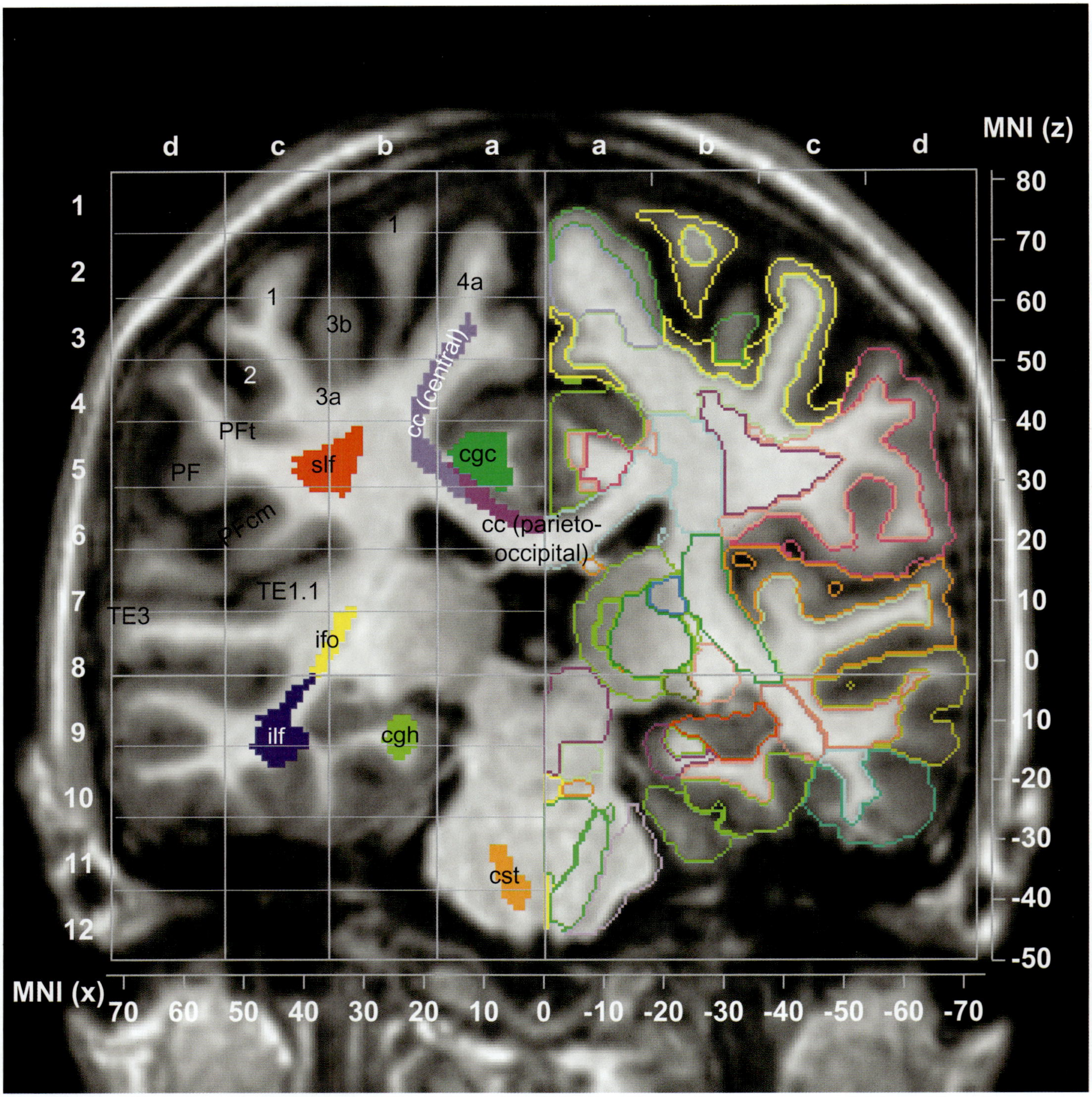

cc (central) - corpus callosum (connecting bilateral central areas)
cc (parieto-occipital) - corpus callosum (connecting bilateral parieto-occipital areas)
cgc - cingulum (cingulate gyrus part)
cgh - cingulum (hippocampal part)
cst - corticospinal tract
ifo - inferior fronto-occipital fasciculus
ilf - inferior longitudinal fasciculus
slf - superior longitudinal fasciculus

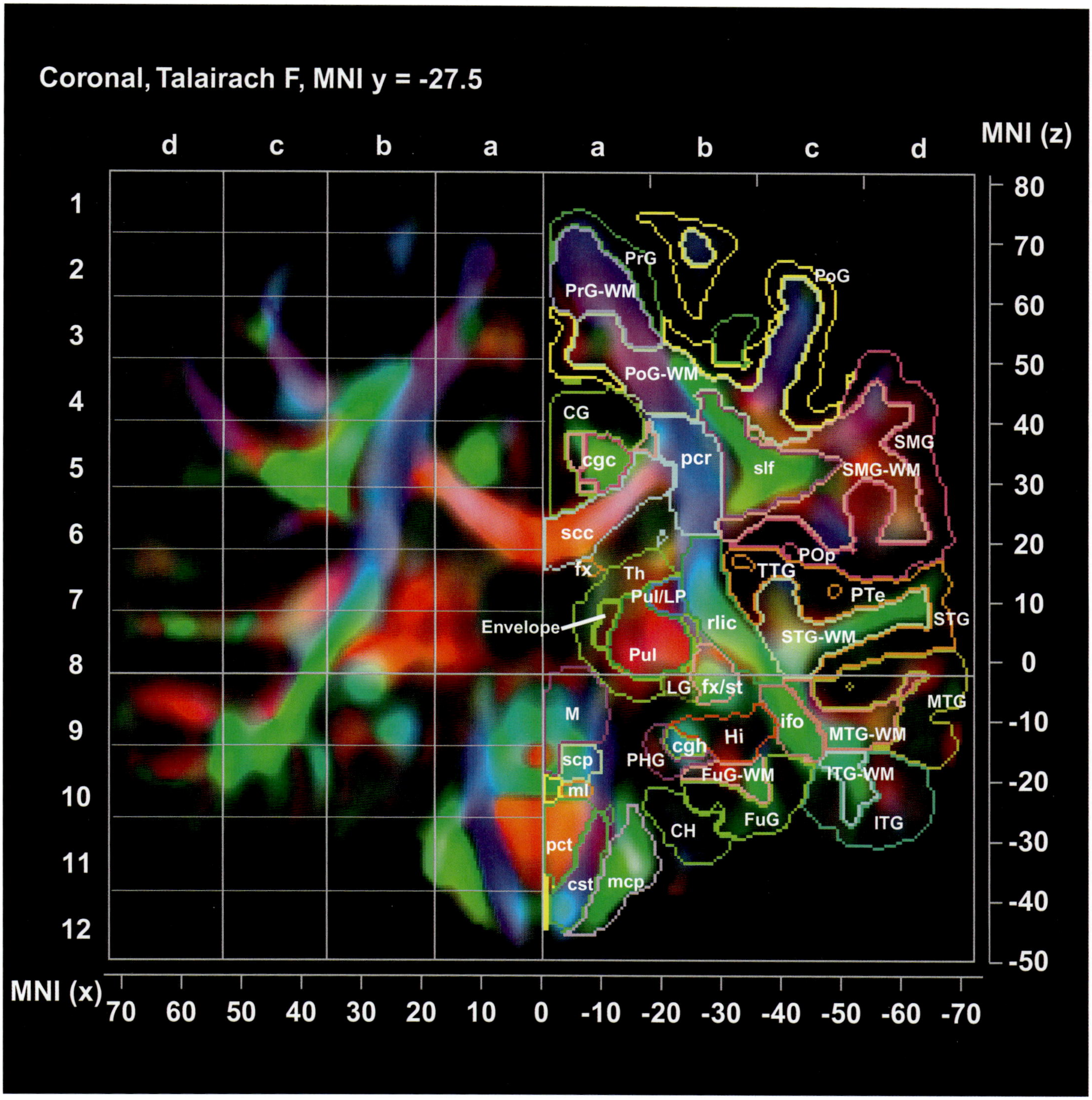

CG - cingulate gyrus
cgc - cingulum (cingulate gyrus part)
cgh - cingulum (hippocampal part)
CH - cerebellar hemisphere
cst - corticospinal tract
Envelope - nuclei of the internal medullary lamina (thalamus)
FuG - fusiform gyrus
FuG-WM - fusiform gyrus white matter
fx - fornix (column and body)
fx/st - fornix (crus) / stria terminalis (cannot be resolved with current resolution)
Hi - hippocampus
ITG - inferior temporal gyrus
ITG-WM - inferior temporal gyrus white matter
LG - lateral geniculate nucleus (thalamus)
M - midbrain
mcp - middle cerebellar peduncle
ml - medial lemniscus
MTG - middle temporal gyrus
MTG-WM - middle temporal gyrus white matter
pcr - posterior corona radiata
pct - pontine crossing tract (a part of mcp)
PHG - parahippocampal gyrus
PoG - postcentral gyrus
PoG-WM - postcentral gyrus white matter
POp - parietal operculum
PrG - precentral gyrus
PrG-WM - precentral gyrus white matter
PTe - planum temporale
Pul - pulvinar (thalamus)
Pul/LP - pulvinar / lateral posterior thalamic nucleus
scc - splenium of the corpus callosum
scp - superior cerebellar peduncle
slf - superior longitudinal fasciculus
SMG - supramarginal gyrus
SMG-WM - supramarginal gyrus white matter
STG - superior temporal gyrus
STG-WM - superior temporal gyrus white matter
Th - thalamus
TTG - transverse temporal gyrus

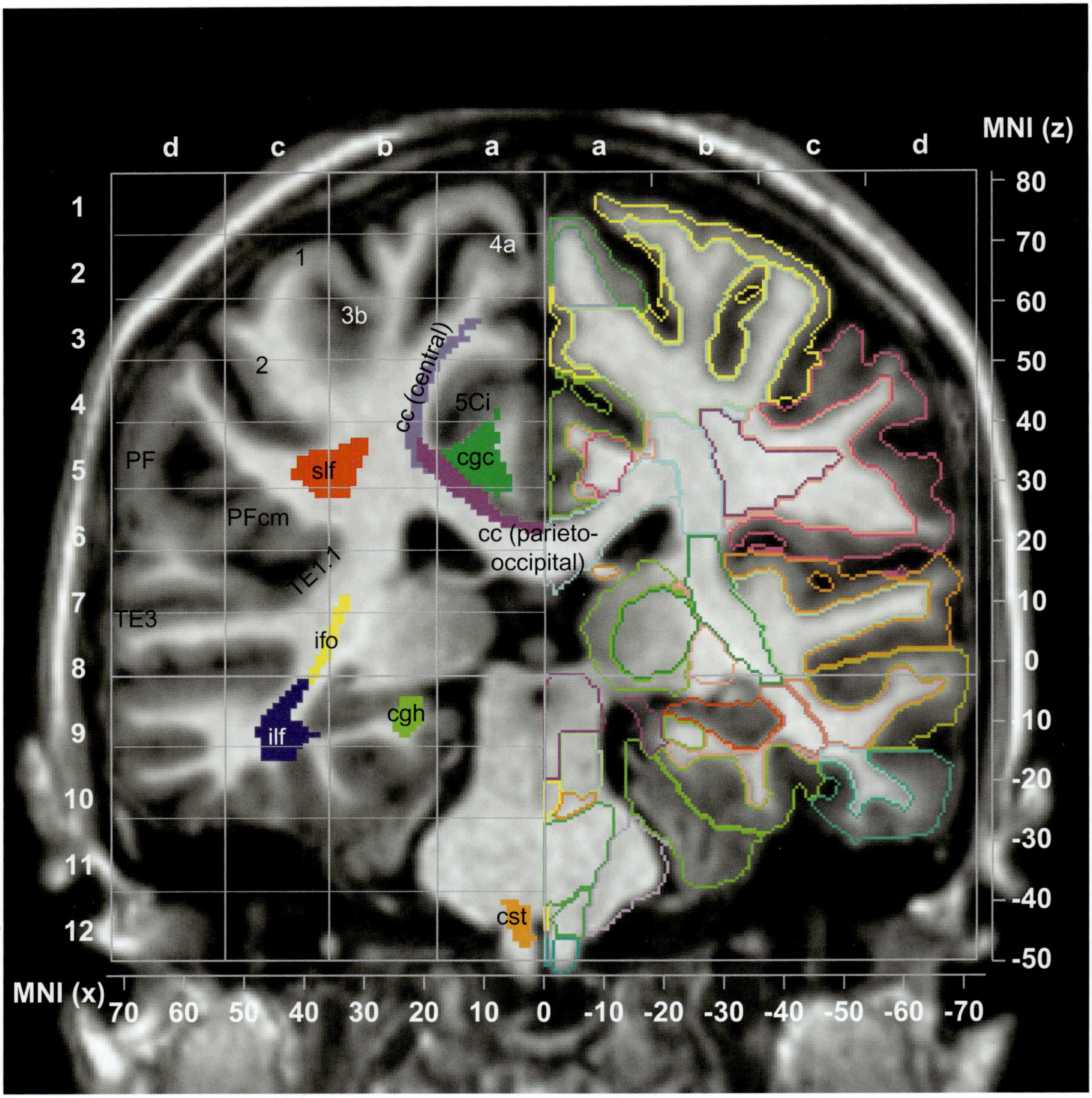

cc (central) - corpus callosum (connecting bilateral central areas)	cgc - cingulum (cingulate gyrus part)	ifo - inferior fronto-occipital fasciculus
cc (parieto-occipital) - corpus callosum (connecting bilateral parieto-occipital areas)	cgh - cingulum (hippocampal part)	ilf - inferior longitudinal fasciculus
	cst - corticospinal tract	slf - superior longitudinal fasciculus

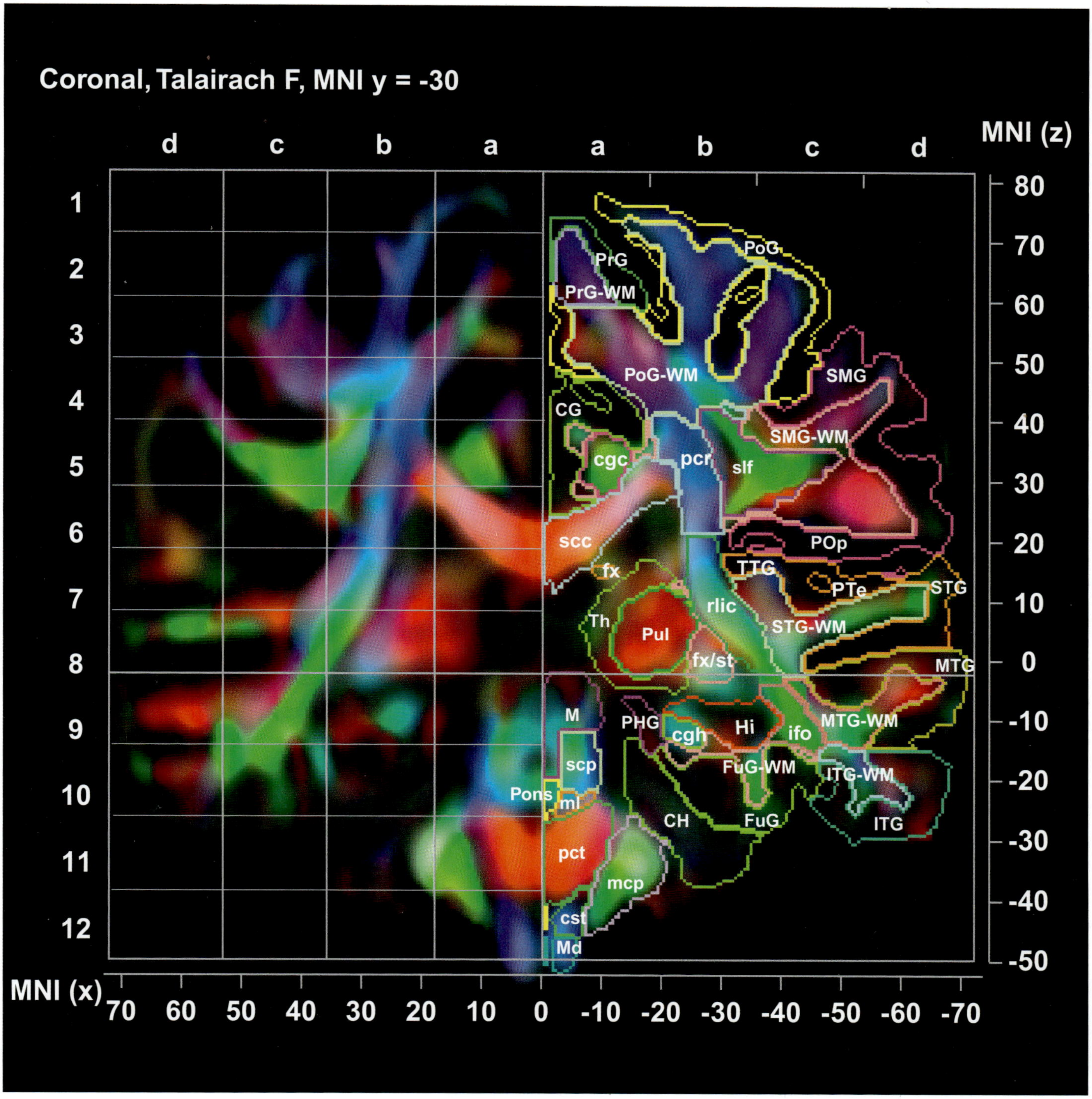

CG - cingulate gyrus
cgc - cingulum (cingulate gyrus part)
cgh - cingulum (hippocampal part)
CH - cerebellar hemisphere
cst - corticospinal tract
FuG - fusiform gyrus
FuG-WM - fusiform gyrus white matter
fx - fornix (column and body)
fx/st - fornix (crus) / stria terminalis (cannot be resolved with current resolution)
Hi - hippocampus
ifo - Inferior fronto-occipital fasciculus
ITG - inferior temporal gyrus
ITG-WM - inferior temporal gyrus white matter
M - midbrain
mcp - middle cerebellar peduncle
Md - medulla oblongata
ml - medial lemniscus
MTG - middle temporal gyrus
MTG-WM - middle temporal gyrus white matter
pcr - posterior corona radiata
pct - pontine crossing tract (a part of mcp)
PHG - parahippocampal gyrus
PoG - postcentral gyrus
PoG-WM - postcentral gyrus white matter
Pons - pons
POp - parietal operculum
PrG - precentral gyrus
PrG-WM - precentral gyrus white matter
PTe - planum temporale
Pul - pulvinar (thalamus)
rlic - retrolenticular part of the internal capsule
scc - splenium of the corpus callosum
scp - superior cerebellar peduncle
slf - superior longitudinal fasciculus
SMG - supramarginal gyrus
SMG-WM - supramarginal gyrus white matter
STG - superior temporal gyrus
STG-WM - superior temporal gyrus white matter
Th - thalamus
TTG - transverse temporal gyrus

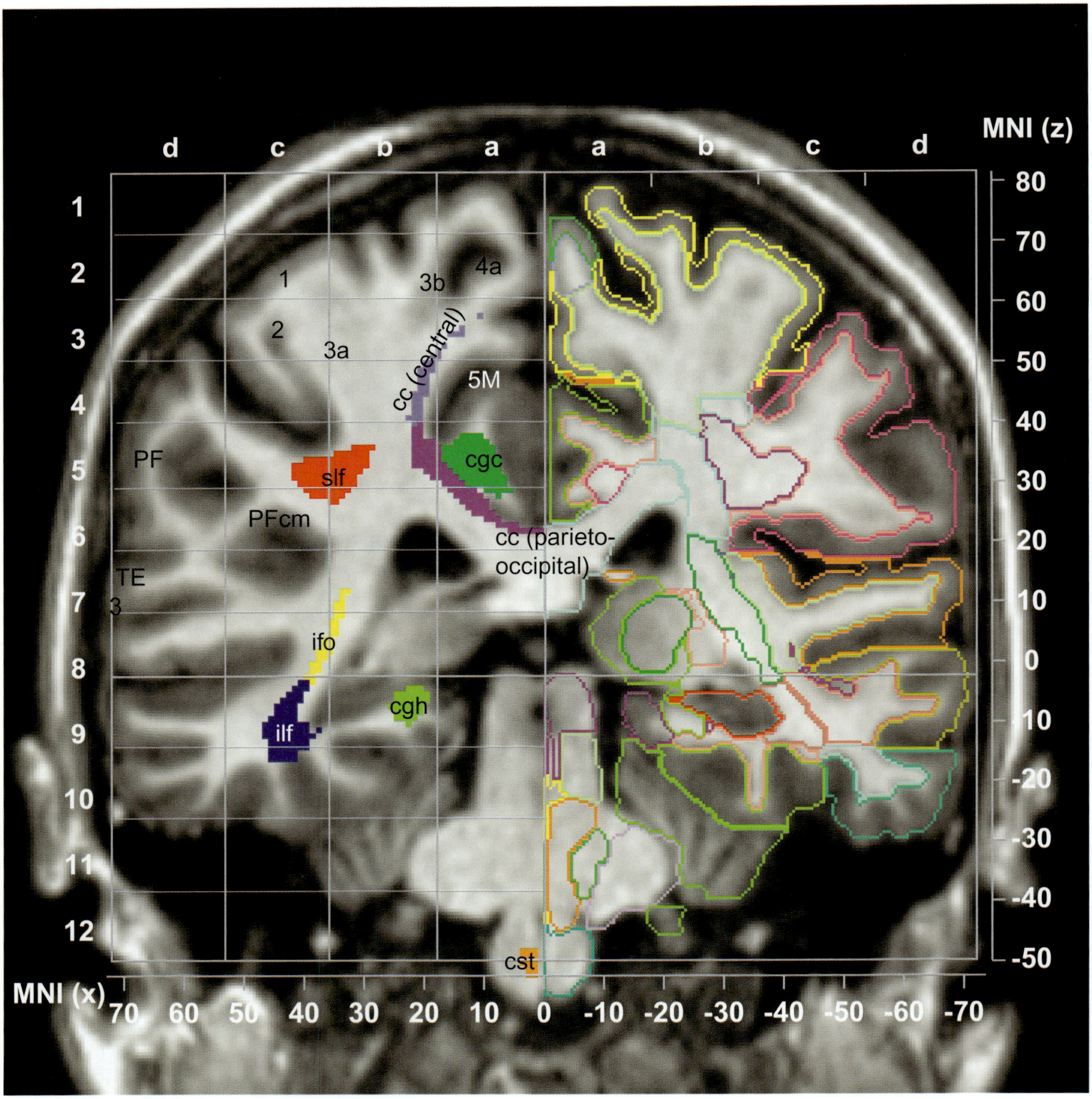

cc (central) - corpus callosum (connecting bilateral central areas)
cc (parieto-occipital) - corpus callosum (connecting bilateral parieto-occipital areas)
cgc - cingulum (cingulate gyrus part)
cgh - cingulum (hippocampal part)
cst - corticospinal tract
ifo - inferior fronto-occipital fasciculus
ilf - inferior longitudinal fasciculus
slf - superior longitudinal fasciculus

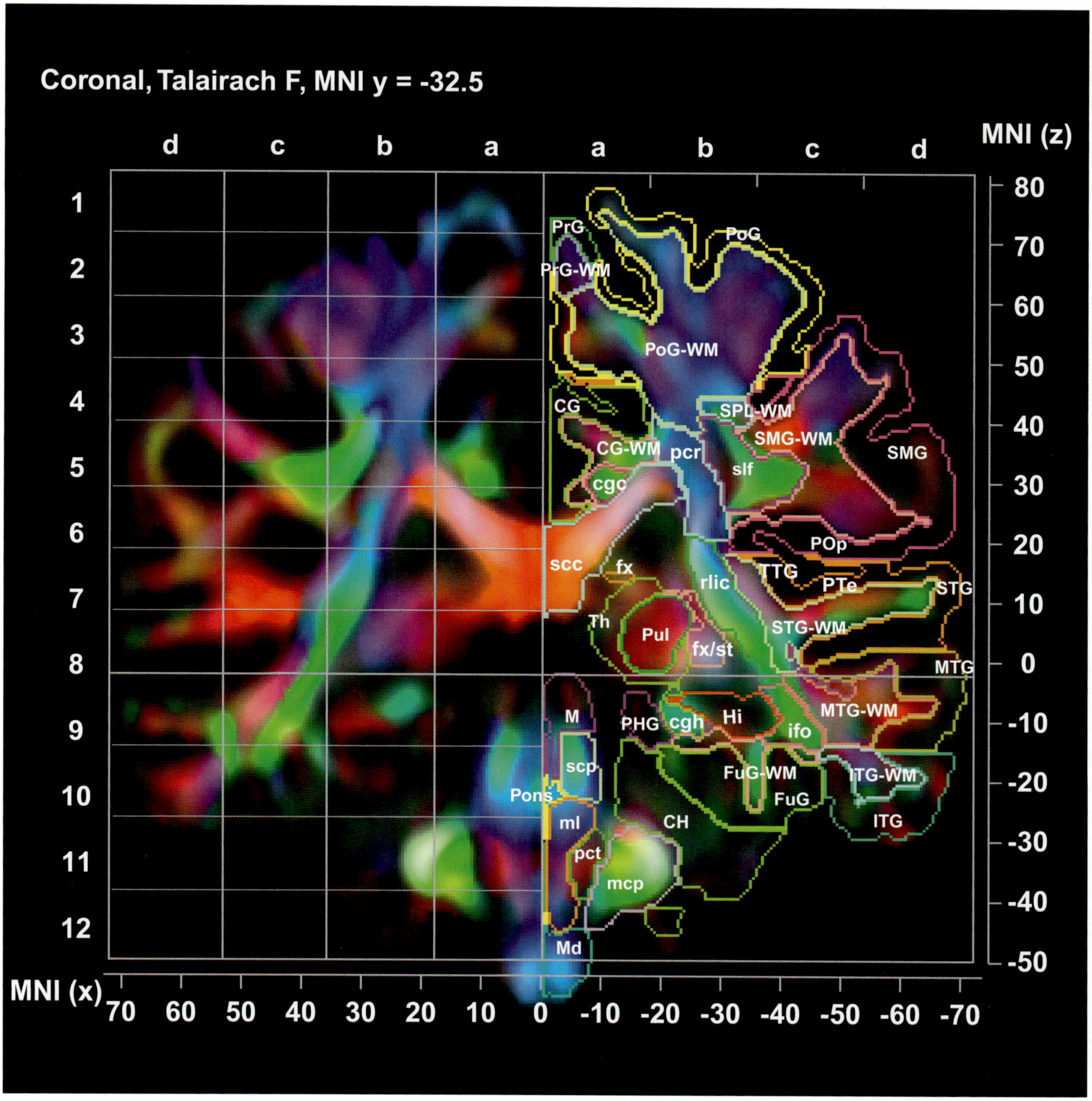

CG - cingulate gyrus
cgc - cingulum (cingulate gyrus part)
cgh - cingulum (hippocampal part)
CG-WM - cingulate gyrus white matter
CH - cerebellar hemisphere
FuG - fusiform gyrus
FuG-WM - fusiform gyrus white matter
fx - fornix (column and body)
fx/st - fornix (crus) / stria terminalis (cannot be resolved with current resolution)
Hi - hippocampus
ITG - inferior temporal gyrus
ITG-WM - inferior temporal gyrus white matter
M - midbrain
mcp - middle cerebellar peduncle
Md - medulla oblongata
ml - medial lemniscus
MTG - middle temporal gyrus
MTG-WM - middle temporal gyrus white matter
pcr - posterior corona radiata
pct - pontine crossing tract (a part of mcp)
PHG - parahippocampal gyrus
PoG - postcentral gyrus
PoG-WM - postcentral gyrus white matter
Pons - pons
POp - parietal operculum
PrG - precentral gyrus
PrG-WM - precentral gyrus white matter
PTe - planum temporale
Pul - pulvinar (thalamus)
scc - splenium of the corpus callosum
scp - superior cerebellar peduncle
slf - superior longitudinal fasciculus
SMG - supramarginal gyrus
SMG-WM - supramarginal gyrus white matter
SPL-WM - superior parietal lobule white matter
STG - superior temporal gyrus
STG-WM - superior temporal gyrus white matter
Th - thalamus
TTG - transverse temporal gyrus

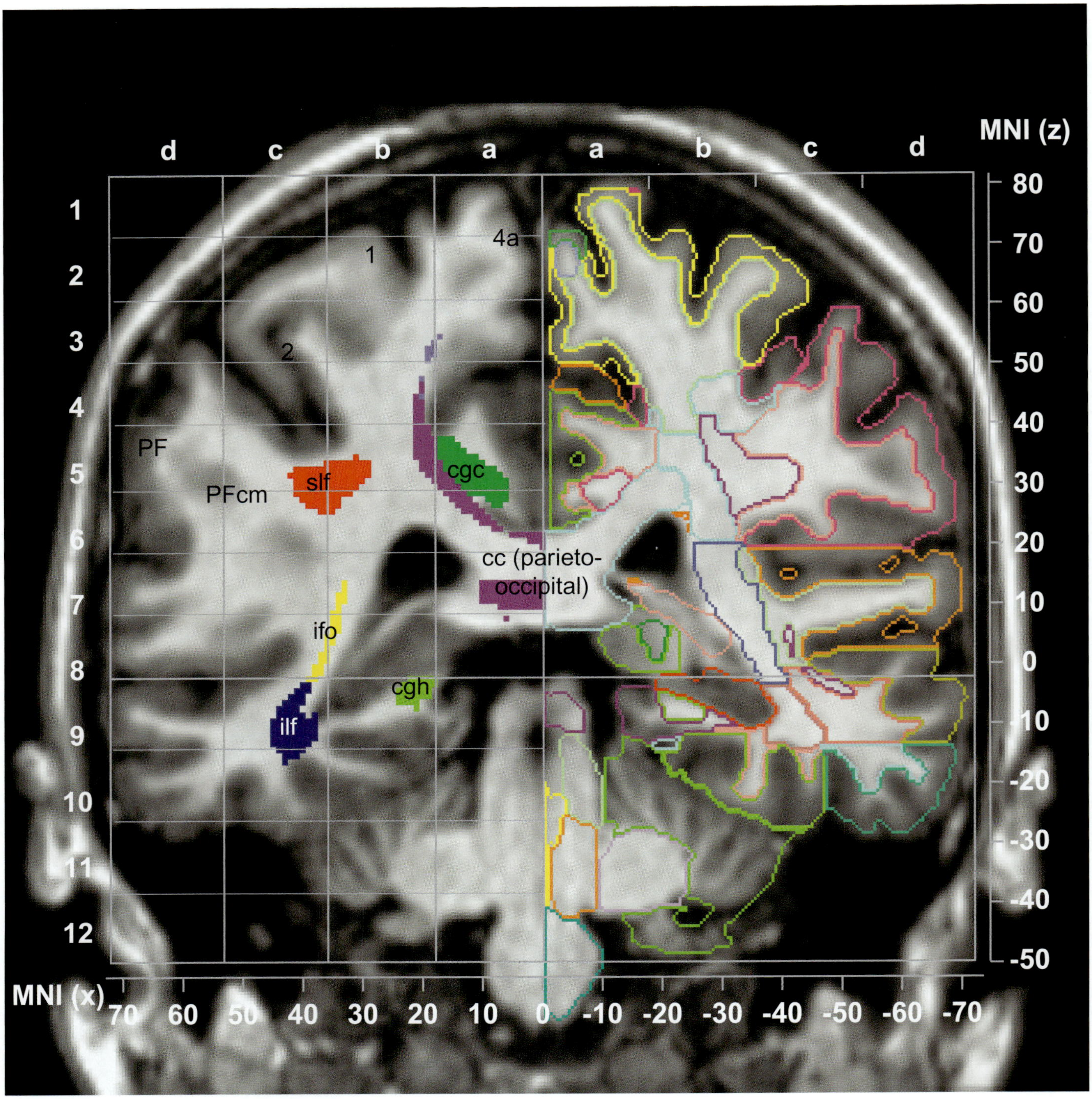

cc (parieto-occipital) - corpus callosum (connecting bilateral parieto-occipital areas)
cgc - cingulum (cingulate gyrus part)
cgh - cingulum (hippocampal part)
ifo - inferior fronto-occipital fasciculus
ilf - inferior longitudinal fasciculus
slf - superior longitudinal fasciculus

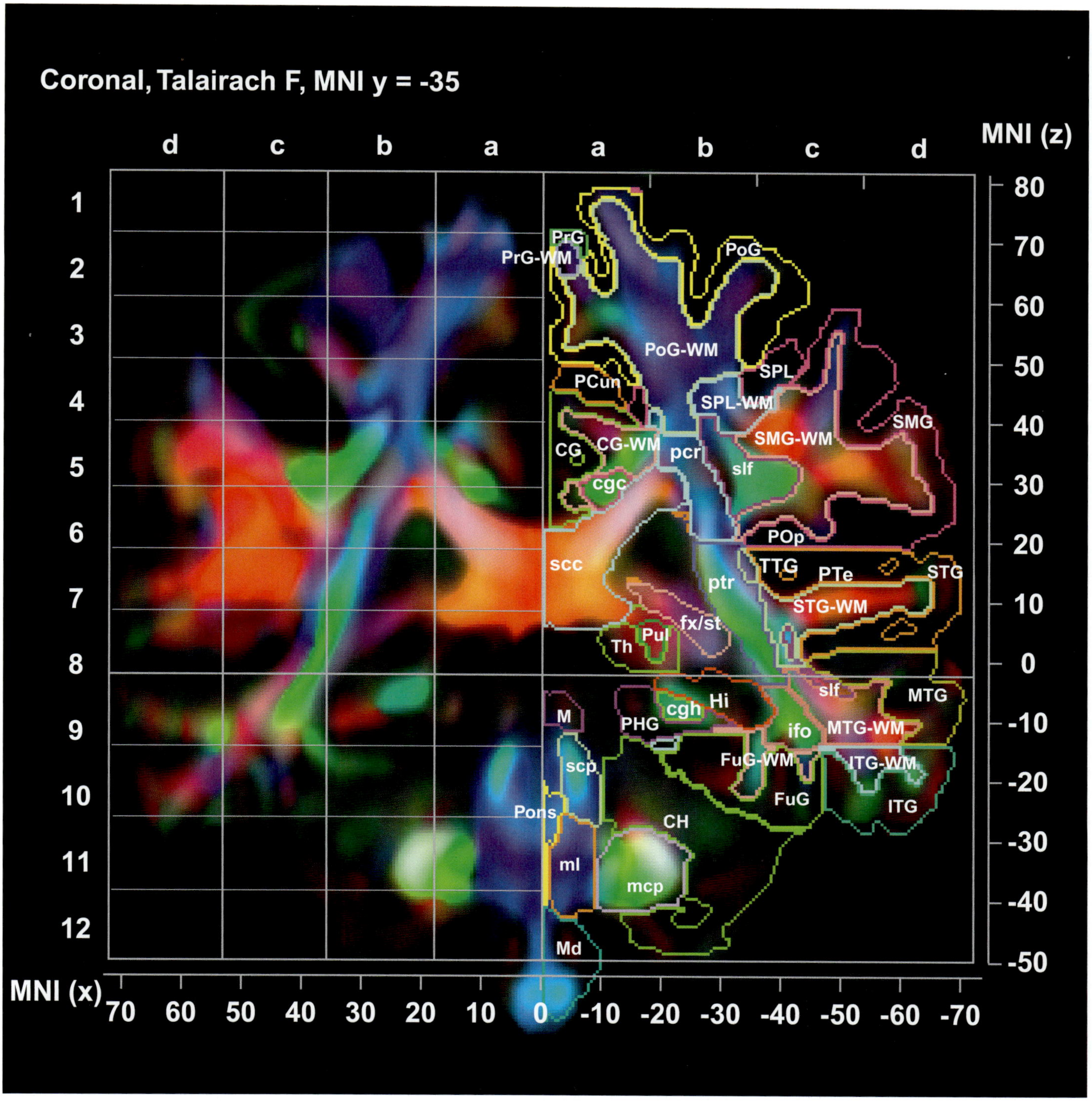

CG - cingulate gyrus
cgc - cingulum (cingulate gyrus part)
cgh - cingulum (hippocampal part)
CG-WM - cingulate gyrus white matter
CH - cerebellar hemisphere
FuG - fusiform gyrus
FuG-WM - fusiform gyrus white matter
fx/st - fornix (crus) / stria terminalis (cannot be resolved with current resolution)
Hi - hippocampus
ifo - inferior fronto-occipital fasciculus
ITG - inferior temporal gyrus
ITG-WM - inferior temporal gyrus white matter
M - midbrain
mcp - middle cerebellar peduncle
Md - medulla oblongata
ml - medial lemniscus
MTG - middle temporal gyrus
MTG-WM - middle temporal gyrus white matter
pcr - posterior corona radiata
PCun - precuneus
PHG - parahippocampal gyrus
PoG - postcentral gyrus
PoG-WM - postcentral gyrus white matter
Pons - pons
POp - parietal operculum
PrG - precentral gyrus
PrG-WM - precentral gyrus white matter
PTe - planum temporale
ptr - posterior thalamic radiation (include the optic radiation)
Pul - pulvinar (thalamus)
scc - splenium of the corpus callosum
scp - superior cerebellar peduncle
slf - superior longitudinal fasciculus
SMG - supramarginal gyrus
SMG-WM - supramarginal gyrus white matter
SPL - superior parietal lobule
SPL-WM - superior parietal lobule white matter
STG - superior temporal gyrus
STG-WM - superior temporal gyrus white matter
Th - thalamus
TTG - transverse temporal gyrus

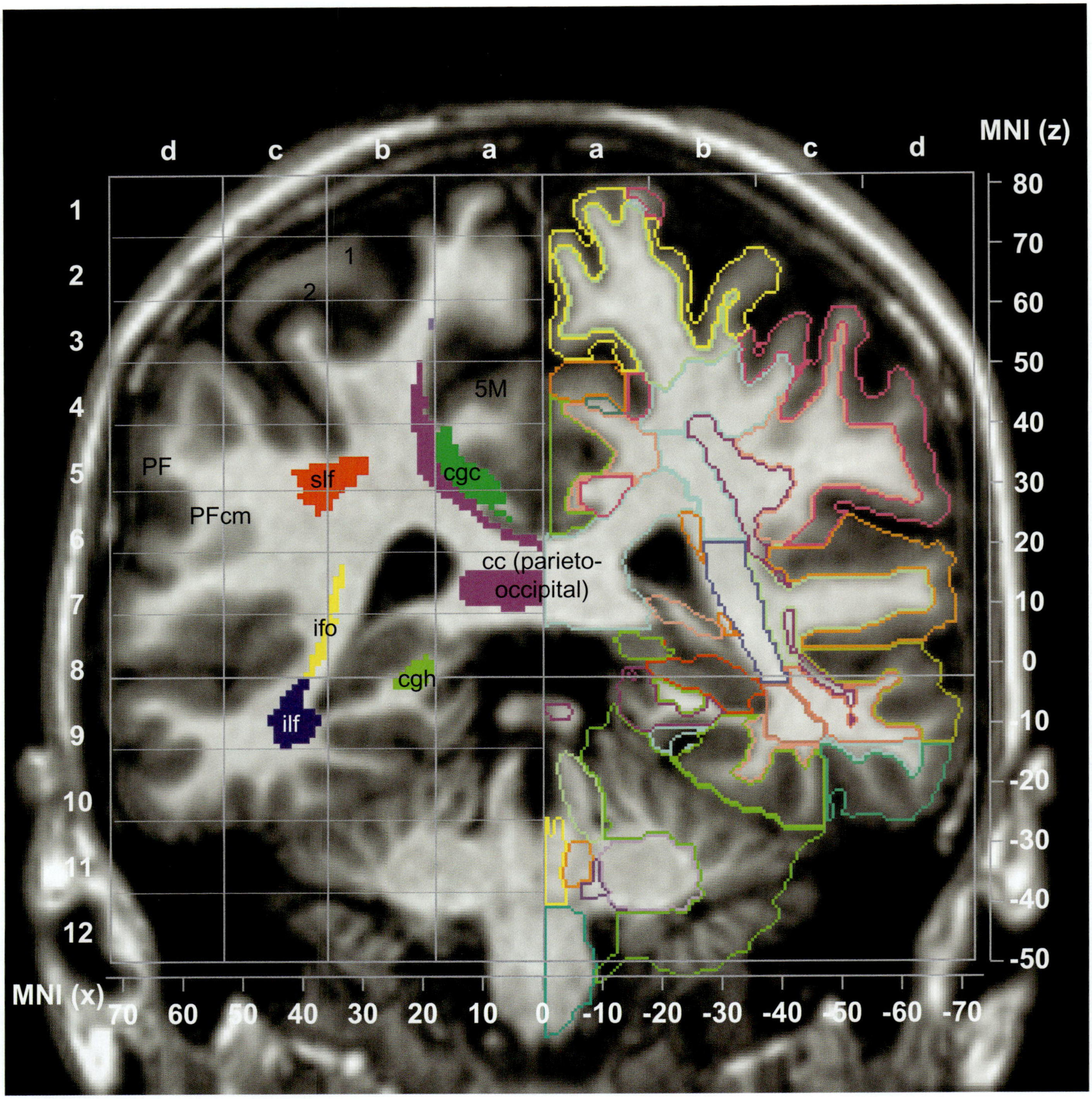

cc (parieto-occipital) - corpus callosum (connecting bilateral parieto-occipital areas)
cgc - cingulum (cingulate gyrus part)
cgh - cingulum (hippocampal part)
ifo - inferior fronto-occipital fasciculus
ilf - inferior longitudinal fasciculus
slf - superior longitudinal fasciculus

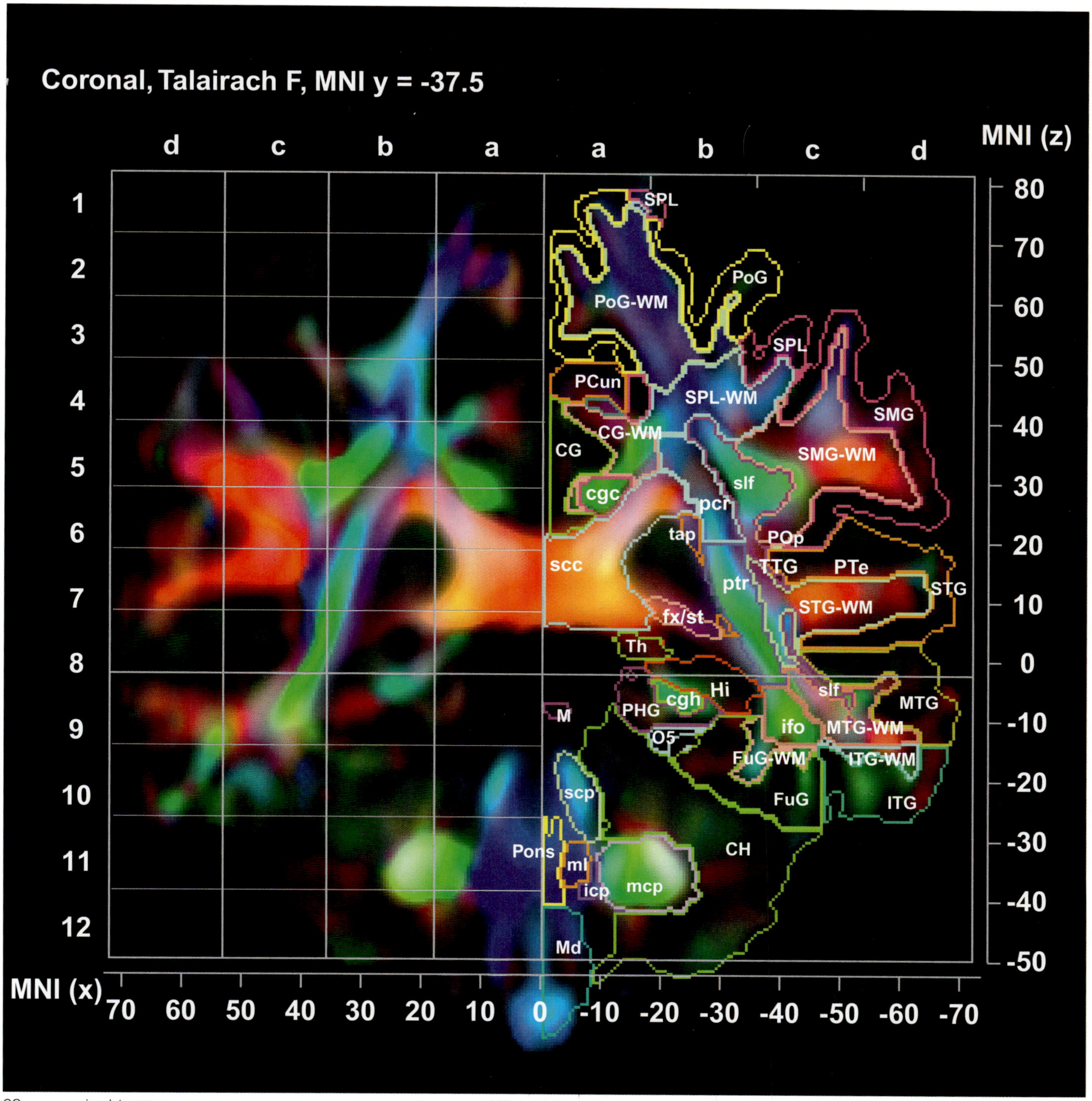

CG - cingulate gyrus
cgc - cingulum (cingulate gyrus part)
cgh - cingulum (hippocampal part)
CG-WM - cingulate gyrus white matter
CH - cerebellar hemisphere
FuG - fusiform gyrus
FuG-WM - fusiform gyrus white matter
fx/st - fornix (crus) / stria terminalis (cannot be resolved with current resolution)
Hi - hippocampus
ITG - inferior temporal gyrus
ITG-WM - inferior temporal gyrus white matter
M - midbrain
mcp - middle cerebellar peduncle
Md - medulla oblongata
ml - medial lemniscus
MTG - middle temporal gyrus
MTG-WM - middle temporal gyrus white matter
O5 - lyngual gyrus
pcr - posterior corona radiata
PCun - precuneus
PHG - parahippocampal gyrus
PoG - postcentral gyrus
PoG-WM - postcentral gyrus white matter
Pons - pons
POp - parietal operculum
PTe - planum temporale
ptr - posterior thalamic radiation (include the optic radiation)
scc - splenium of the corpus callosum
scp - superior cerebellar peduncle
slf - superior longitudinal fasciculus
SMG - supramarginal gyrus
SMG-WM - supramarginal gyrus white matter
SPL - superior parietal lobule
SPL-WM - superior parietal lobule white matter
STG - superior temporal gyrus
STG-WM - superior temporal gyrus white matter
tap - tapetum of the corpus callosum
Th - thalamus
TTG - transverse temporal gyrus

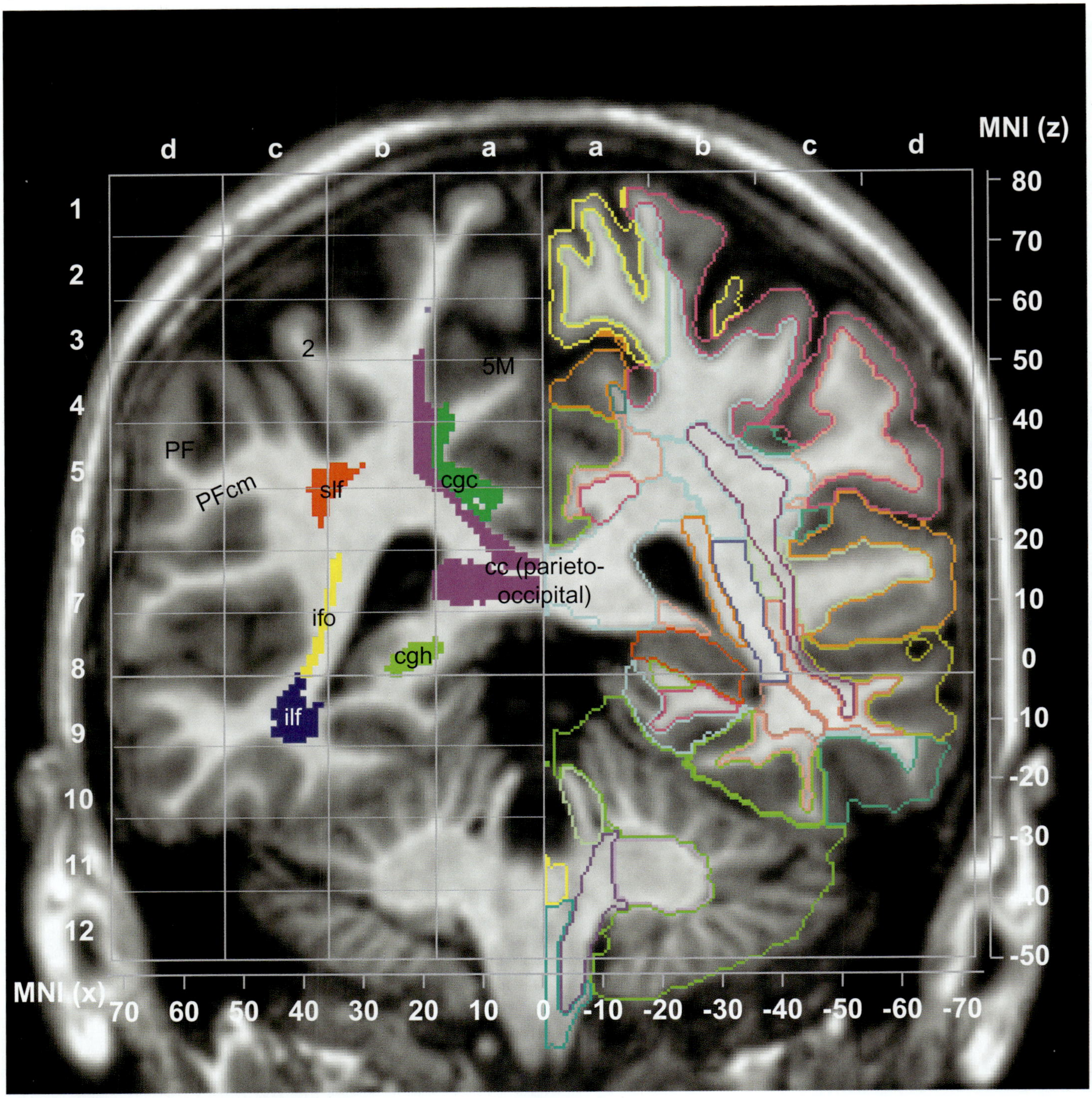

cc (parieto-occipital) - corpus callosum (connecting bilateral parieto-occipital areas)
cgc - cingulum (cingulate gyrus part)
cgh - cingulum (hippocampal part)
ifo - inferior fronto-occipital fasciculus
ilf - inferior longitudinal fasciculus
slf - superior longitudinal fasciculus

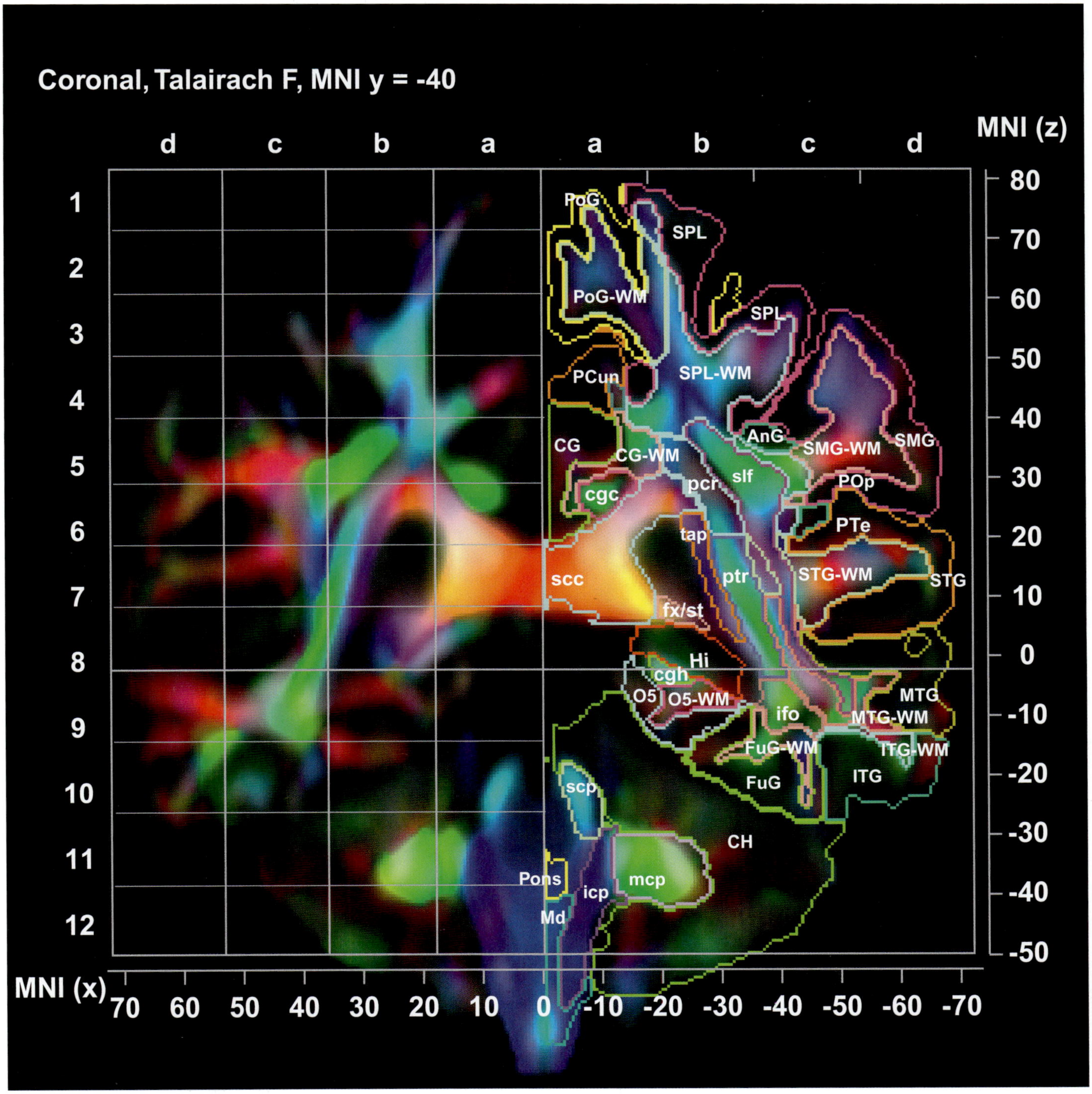

AnG - angular gyrus
CG - cingulate gyrus
cgc - cingulum (cingulate gyrus part)
cgh - cingulum (hippocampal part)
CG-WM - cingulate gyrus white matter
CH - cerebellar hemisphere
FuG - fusiform gyrus
FuG-WM - fusiform gyrus white matter
fx/st - fornix (crus) / stria terminalis (cannot be resolved with current resolution)
Hi - hippocampus
ITG - inferior temporal gyrus
ITG-WM - inferior temporal gyrus white matter

mcp - middle cerebellar peduncle
Md - medulla oblongata
MTG - middle temporal gyrus
MTG-WM - middle temporal gyrus white matter
O5 - lyngual gyrus
O5-WM - lyngual gyrus white matter
pcr - posterior corona radiata
PCun - precuneus
PoG - postcentral gyrus
PoG-WM - postcentral gyrus white matter
Pons - pons
POp - parietal operculum
PTe - planum temporale

ptr - posterior thalamic radiation (include the optic radiation)
scc - splenium of the corpus callosum
scp - superior cerebellar peduncle
slf - superior longitudinal fasciculus
SMG - supramarginal gyrus
SMG-WM - supramarginal gyrus white matter
SPL - superior parietal lobule
SPL-WM - superior parietal lobule white matter
STG - superior temporal gyrus
STG-WM - superior temporal gyrus white matter
tap - tapetum of the corpus callosum

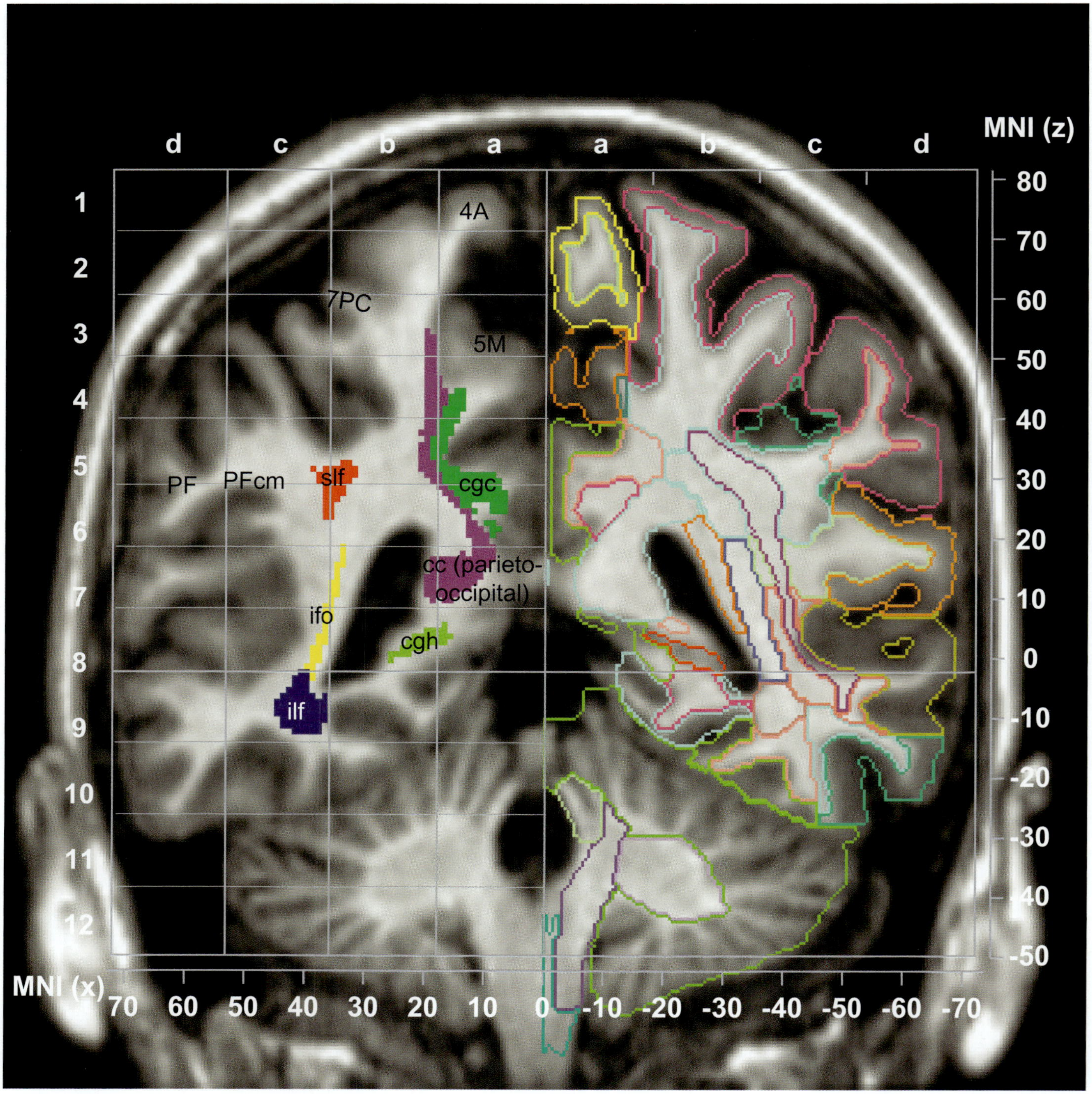

cc (parieto-occipital) - corpus callosum (connecting bilateral parieto-occipital areas)
cgc - cingulum (cingulate gyrus part)
cgh - cingulum (hippocampal part)
ifo - inferior fronto-occipital fasciculus
ilf - inferior longitudinal fasciculus
slf - superior longitudinal fasciculus

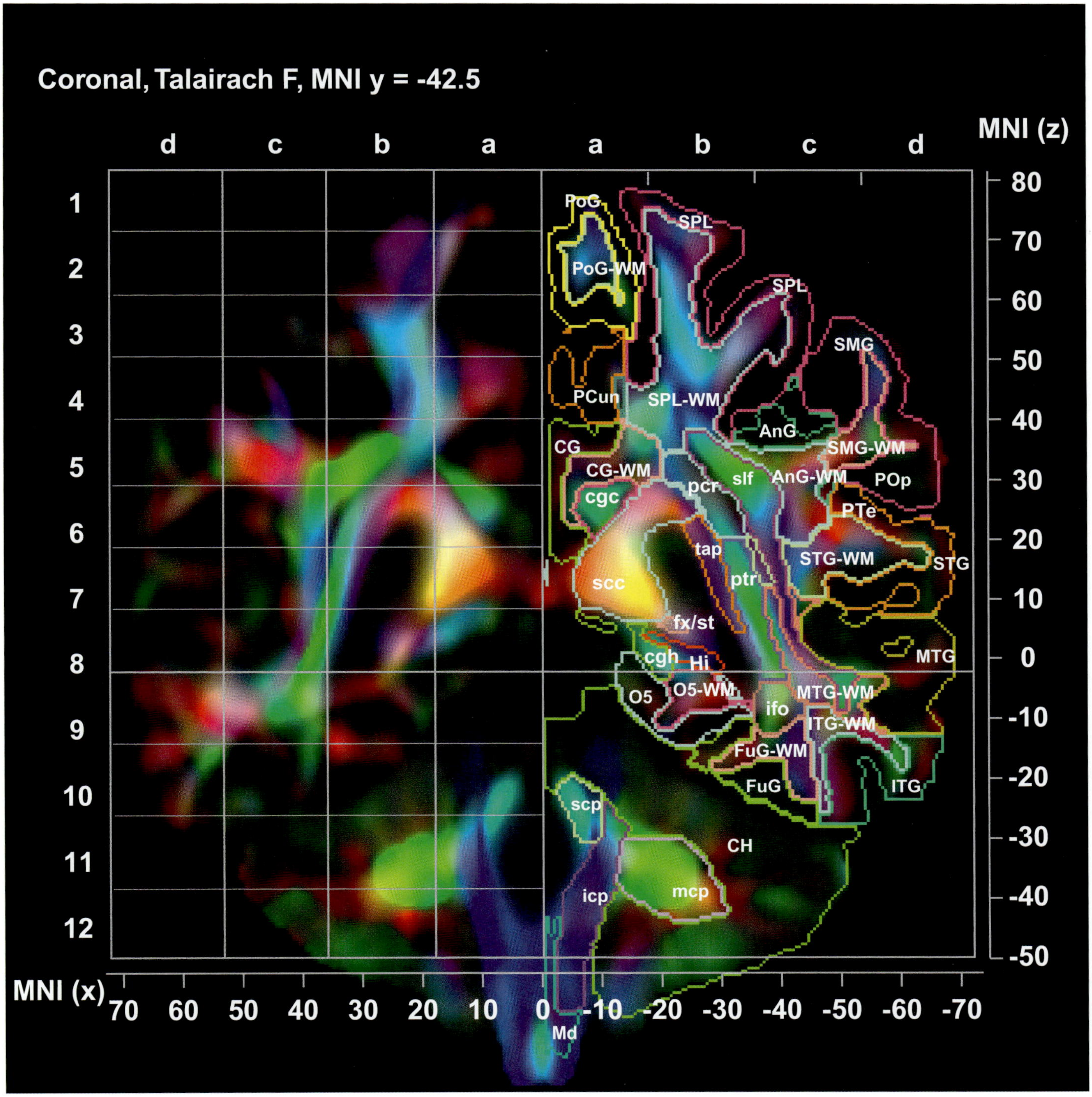

AnG - angular gyrus
AnG-WM - angular gyrus white matter
CG - cingulate gyrus
cgc - cingulum (cingulate gyrus part)
cgh - cingulum (hippocampal part)
CG-WM - cingulate gyrus white matter
CH - cerebellar hemisphere
FuG - fusiform gyrus
FuG-WM - fusiform gyrus white matter
fx/st - fornix (crus) / stria terminalis (cannot be resolved with current resolution)
Hi - hippocampus
icp - inferior cerebellar peduncle
ifo - inferior fronto-occipital fasciculus
ITG - inferior temporal gyrus
ITG-WM - inferior temporal gyrus white matter
mcp - middle cerebellar peduncle
Md - medulla oblongata
MTG - middle temporal gyrus
MTG-WM - middle temporal gyrus white matter
O5 - lyngual gyrus
O5-WM - lyngual gyrus white matter
pcr - posterior corona radiata
PCun - precuneus
PoG - postcentral gyrus
PoG-WM - postcentral gyrus white matter
POp - parietal operculum
PTe - planum temporale
ptr - posterior thalamic radiation (include the optic radiation)
scc - splenium of the corpus callosum
scp - superior cerebellar peduncle
slf - superior longitudinal fasciculus
SMG - supramarginal gyrus
SMG-WM - supramarginal gyrus white matter
SPL - superior parietal lobule
SPL-WM - superior parietal lobule white matter
STG - superior temporal gyrus
STG-WM - superior temporal gyrus white matter
tap - tapetum of the corpus callosum

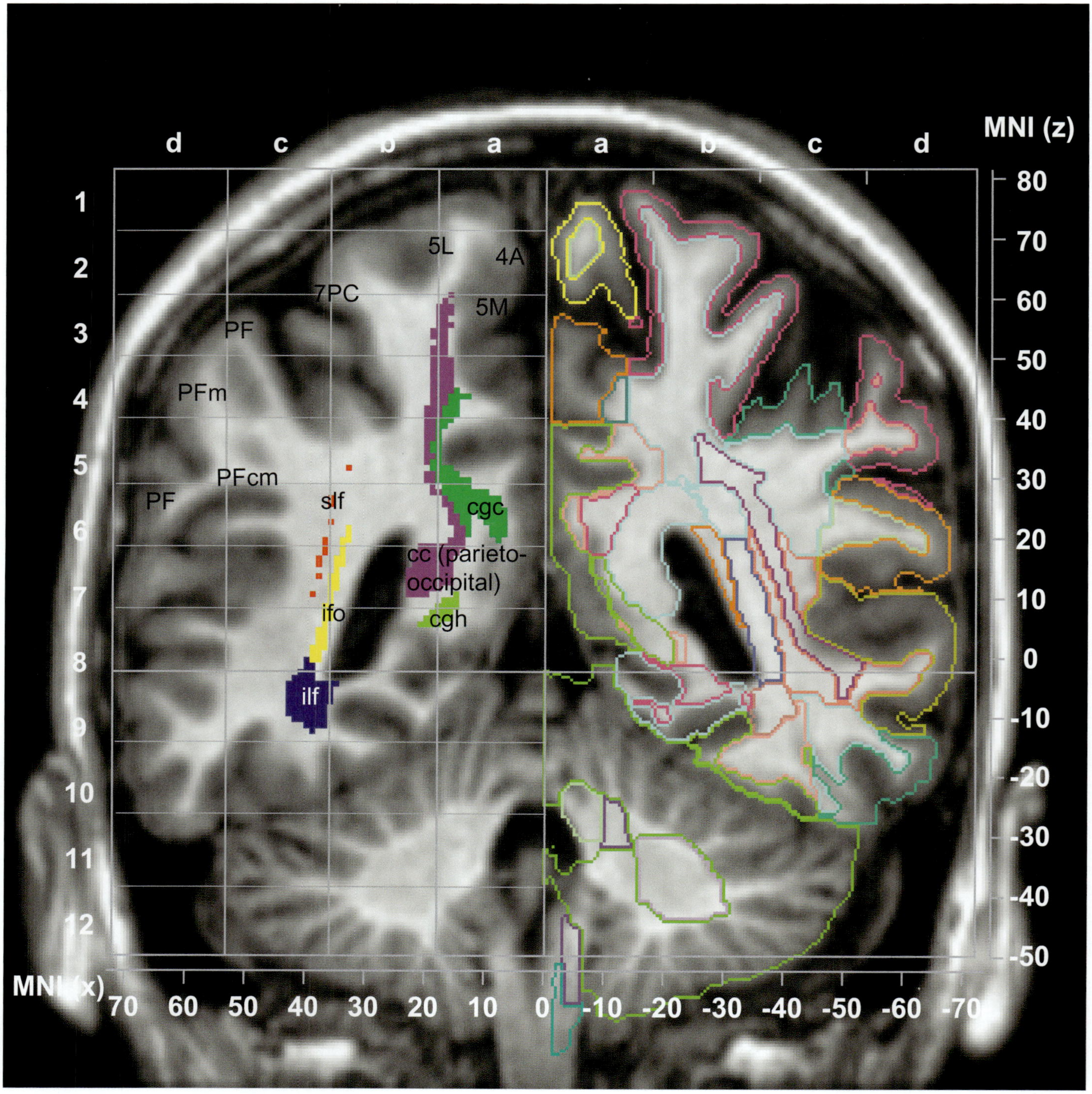

cc (parieto-occipital) - corpus callosum (connecting bilateral parieto-occipital areas)
cgc - cingulum (cingulate gyrus part)
cgh - cingulum (hippocampal part)
ifo - inferior fronto-occipital fasciculus
ilf - inferior longitudinal fasciculus
slf - superior longitudinal fasciculus

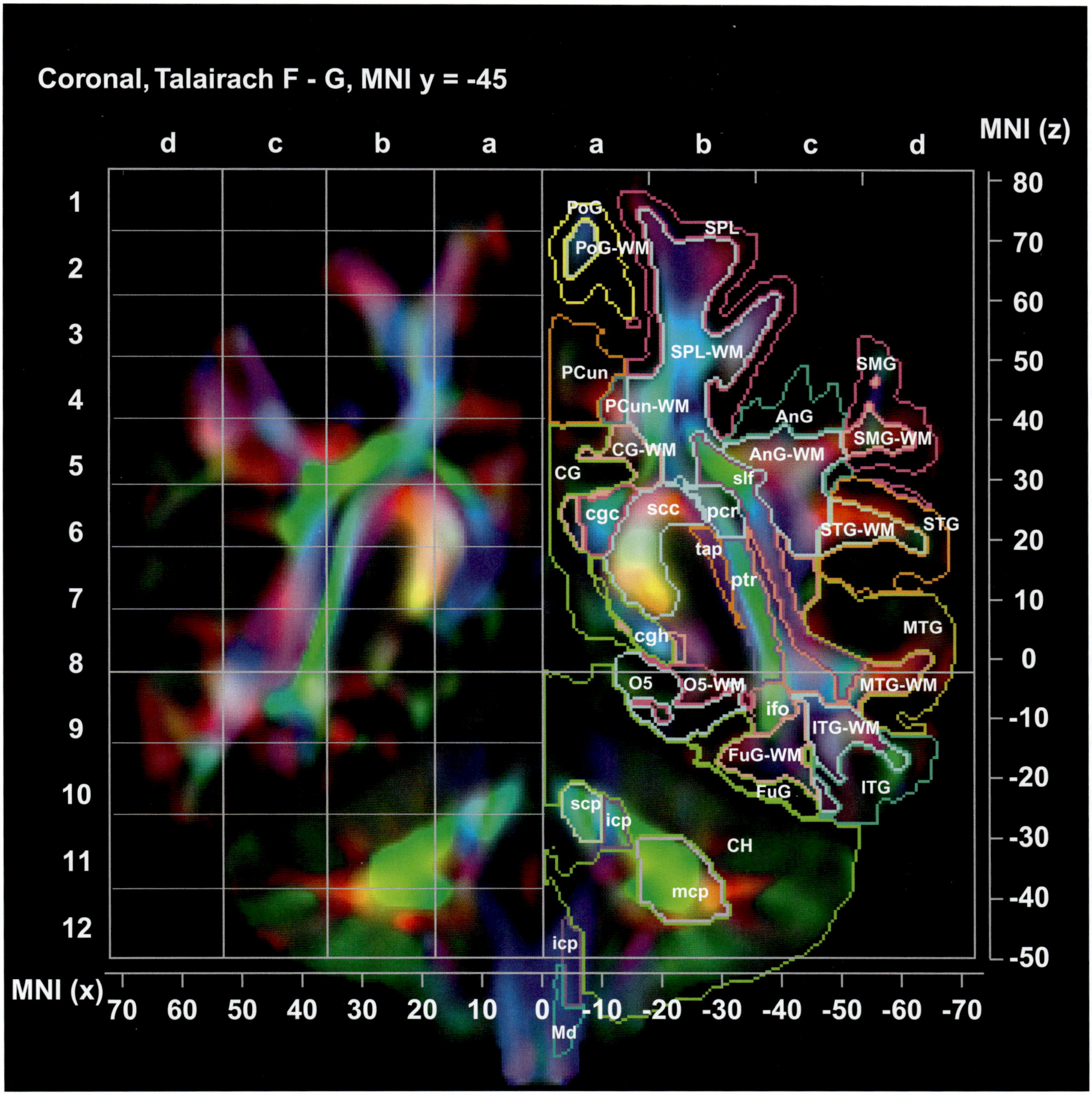

AnG - angular gyrus
AnG-WM - angular gyrus white matter
CG - cingulate gyrus
cgc - cingulum (cingulate gyrus part)
cgh - cingulum (hippocampal part)
CG-WM - cingulate gyrus white matter
CH - cerebellar hemisphere
FuG - fusiform gyrus
FuG-WM - fusiform gyrus white matter
icp - inferior cerebellar peduncle
ifo - inferior fronto-occipital fasciculus
ITG - inferior temporal gyrus
ITG-WM - inferior temporal gyrus white matter
mcp - middle cerebellar peduncle
Md - medulla oblongata
MTG - middle temporal gyrus
MTG-WM - middle temporal gyrus white matter
O5 - lyngual gyrus
O5-WM - lyngual gyrus white matter
pcr - posterior corona radiata
PCun - precuneus
PCun-WM - precuneus white matter
PoG - postcentral gyrus
PoG-WM - postcentral gyrus white matter
ptr - posterior thalamic radiation (include the optic radiation)
scc - splenium of the corpus callosum
scp - superior cerebellar peduncle
slf - superior longitudinal fasciculus
SMG - supramarginal gyrus
SMG-WM - supramarginal gyrus white matter
SPL - superior parietal lobule
SPL-WM - superior parietal lobule white matter
STG - superior temporal gyrus
STG-WM - superior temporal gyrus white matter
tap - tapetum of the corpus callosum

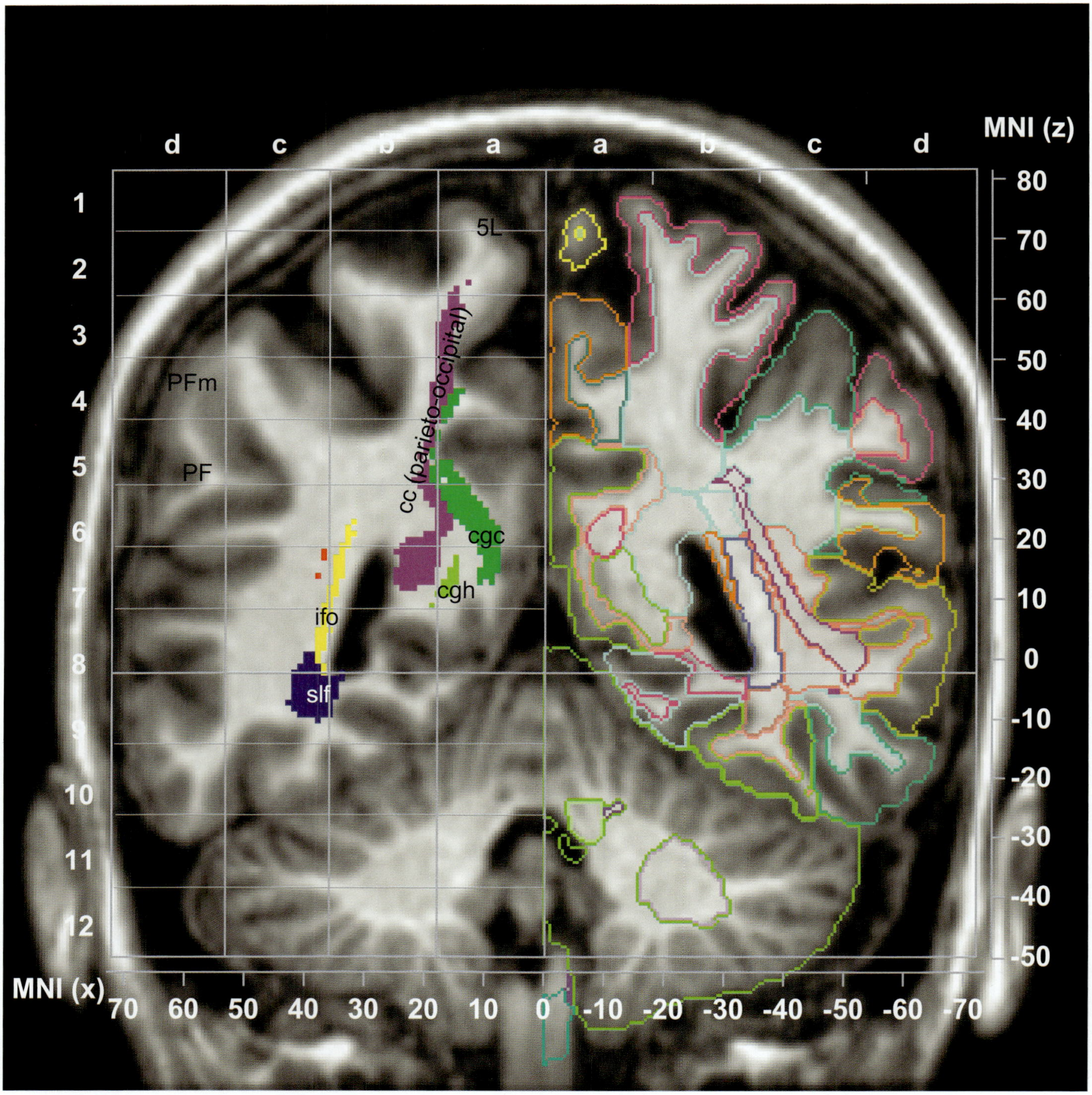

cc (parieto-occipital) - corpus callosum (connecting bilateral parieto-occipital areas)
cgc - cingulum (cingulate gyrus part)
cgh - cingulum (hippocampal part)
ifo - inferior fronto-occipital fasciculus
slf - superior longitudinal fasciculus

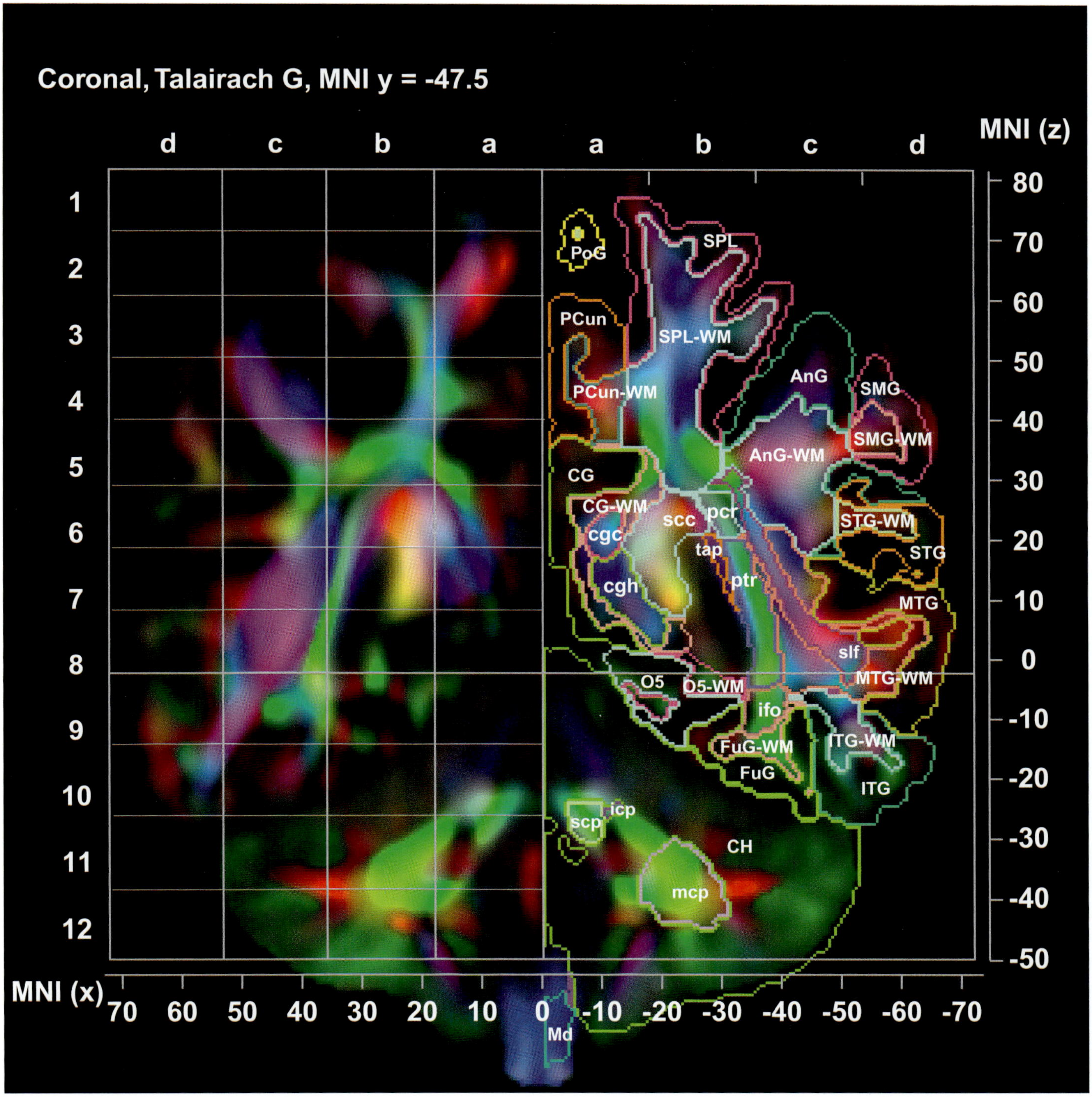

AnG	- angular gyrus	ITG-WM	- inferior temporal gyrus white matter	ptr	- posterior thalamic radiation (include the optic radiation)
AnG-WM	- angular gyrus white matter	mcp	- middle cerebellar peduncle	scc	- splenium of the corpus callosum
CG	- cingulate gyrus	Md	- medulla oblongata	scp	- superior cerebellar peduncle
cgc	- cingulum (cingulate gyrus part)	MTG	- middle temporal gyrus	slf	- superior longitudinal fasciculus
cgh	- cingulum (hippocampal part)	MTG-WM	- middle temporal gyrus white matter	SMG	- supramarginal gyrus
CG-WM	- cingulate gyrus white matter	O5	- lyngual gyrus	SMG-WM	- supramarginal gyrus white matter
CH	- cerebellar hemisphere	O5-WM	- lyngual gyrus white matter	SPL	- superior parietal lobule
FuG	- fusiform gyrus	pcr	- posterior corona radiata	SPL-WM	- superior parietal lobule white matter
FuG-WM	- fusiform gyrus white matter	PCun	- precuneus	STG	- superior temporal gyrus
icp	- inferior cerebellar peduncle	PCun-WM	- precuneus white matter	STG-WM	- superior temporal gyrus white matter
ifo	- inferior fronto-occipital fasciculus	PoG	- postcentral gyrus	tap	- tapetum of the corpus callosum
ITG	- inferior temporal gyrus				

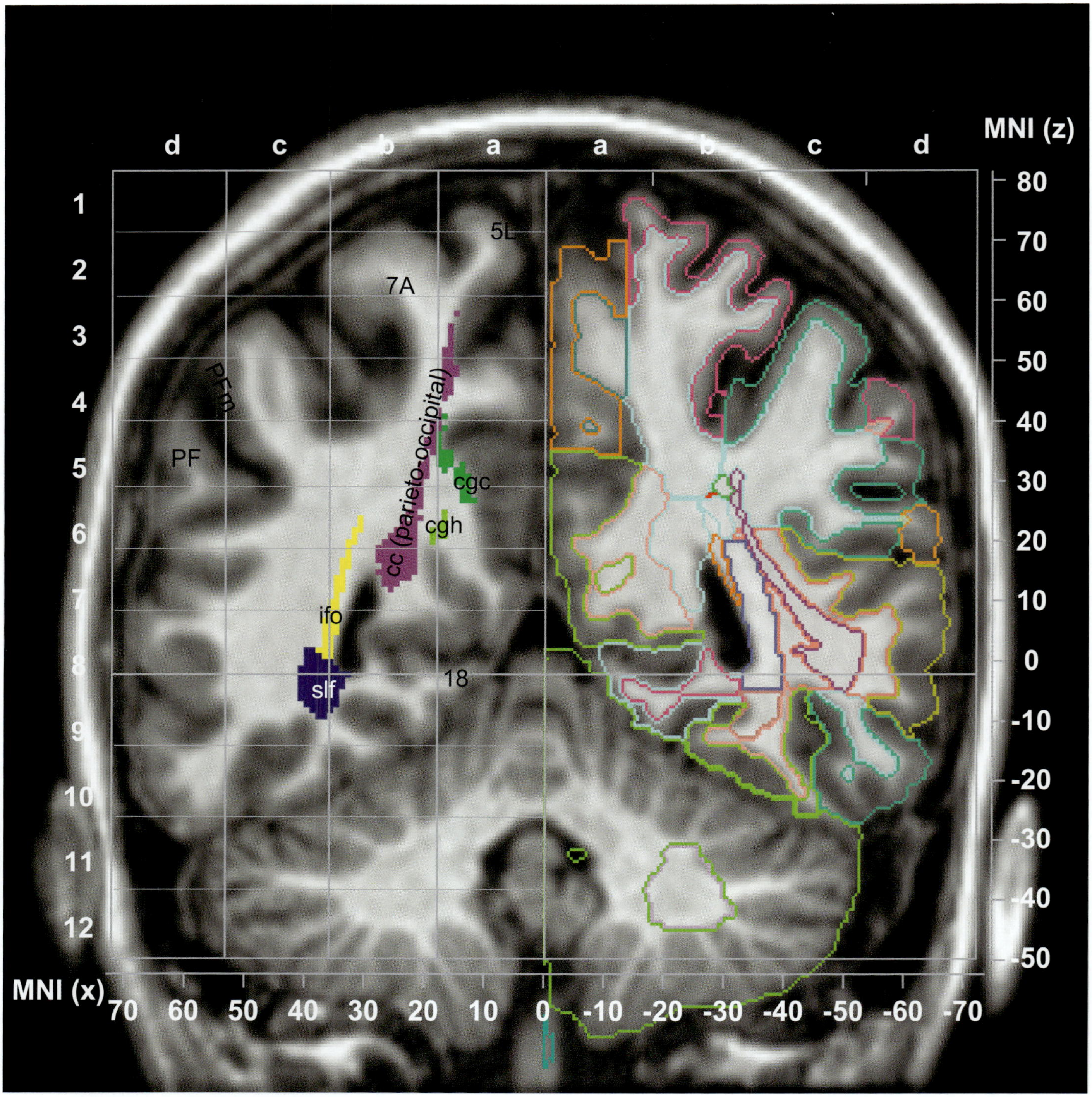

cc (parieto-occipital) - corpus callosum (connecting bilateral parieto-occipital areas)
cgc - cingulum (cingulate gyrus part)
cgh - cingulum (hippocampal part)
ifo - Inferior fronto-occipital fasciculus
slf - superior longitudinal fasciculus

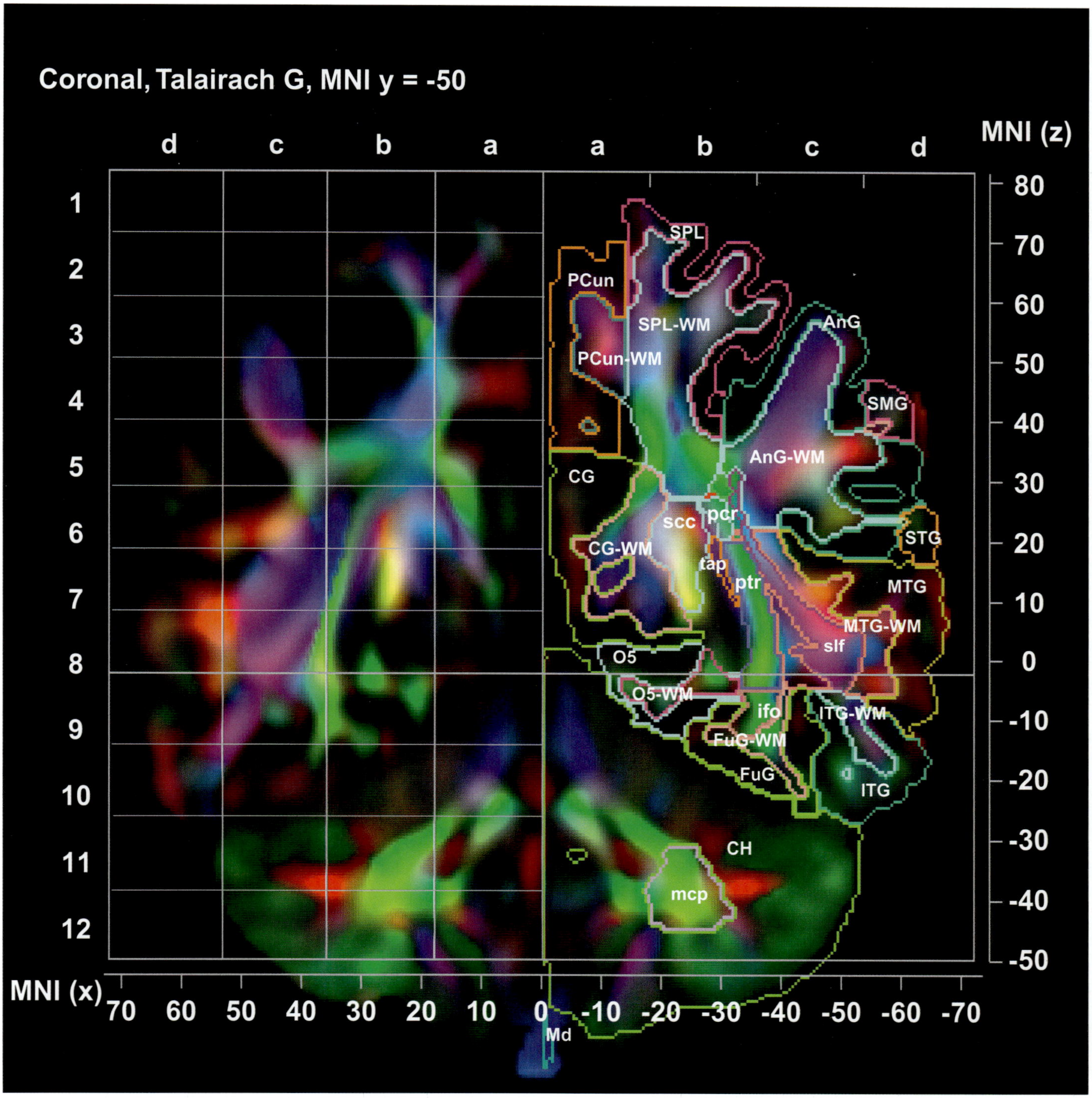

AnG - angular gyrus
AnG-WM - angular gyrus white matter
CG - cingulate gyrus
CG-WM - cingulate gyrus white matter
CH - cerebellar hemisphere
FuG - fusiform gyrus
FuG-WM - fusiform gyrus white matter
ifo - inferior fronto-occipital fasciculus
ITG - inferior temporal gyrus
ITG-WM - inferior temporal gyrus white matter
mcp - middle cerebellar peduncle
Md - medulla oblongata
MTG - middle temporal gyrus
MTG-WM - middle temporal gyrus white matter
O5 - lyngual gyrus
O5-WM - lyngual gyrus white matter
pcr - posterior corona radiata
PCun - precuneus
PCun-WM - precuneus white matter
ptr - posterior thalamic radiation (include the optic radiation)
scc - splenium of the corpus callosum
slf - superior longitudinal fasciculus
SMG - supramarginal gyrus
SPL - superior parietal lobule
SPL-WM - superior parietal lobule white matter
STG - superior temporal gyrus
tap - tapetum of the corpus callosum

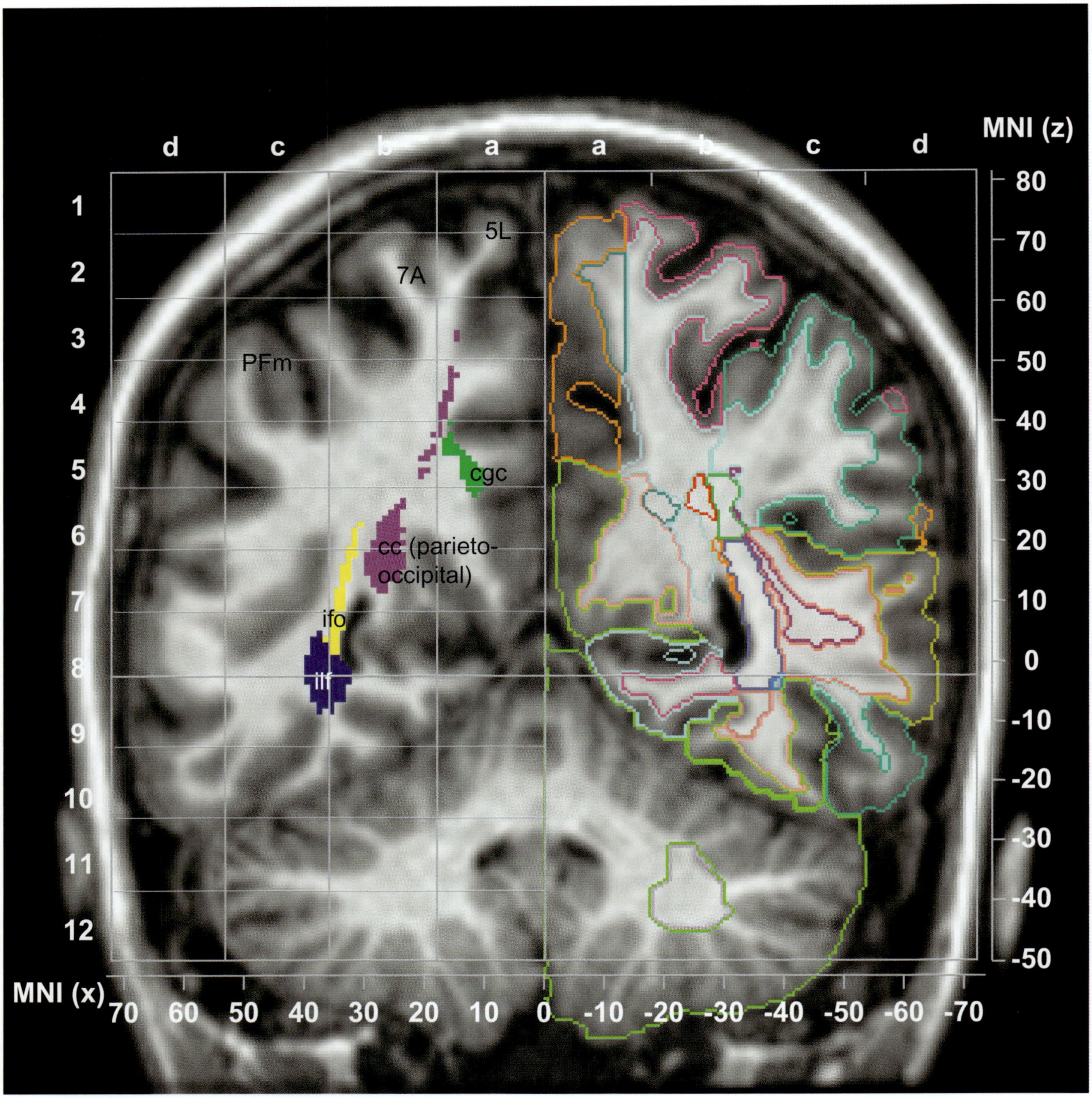

cc (parieto-occipital) - corpus callosum (connecting bilateral parieto-occipital areas)
cgc - cingulum (cingulate gyrus part)
ifo - inferior fronto-occipital fasciculus
ilf - inferior longitudinal fasciculus

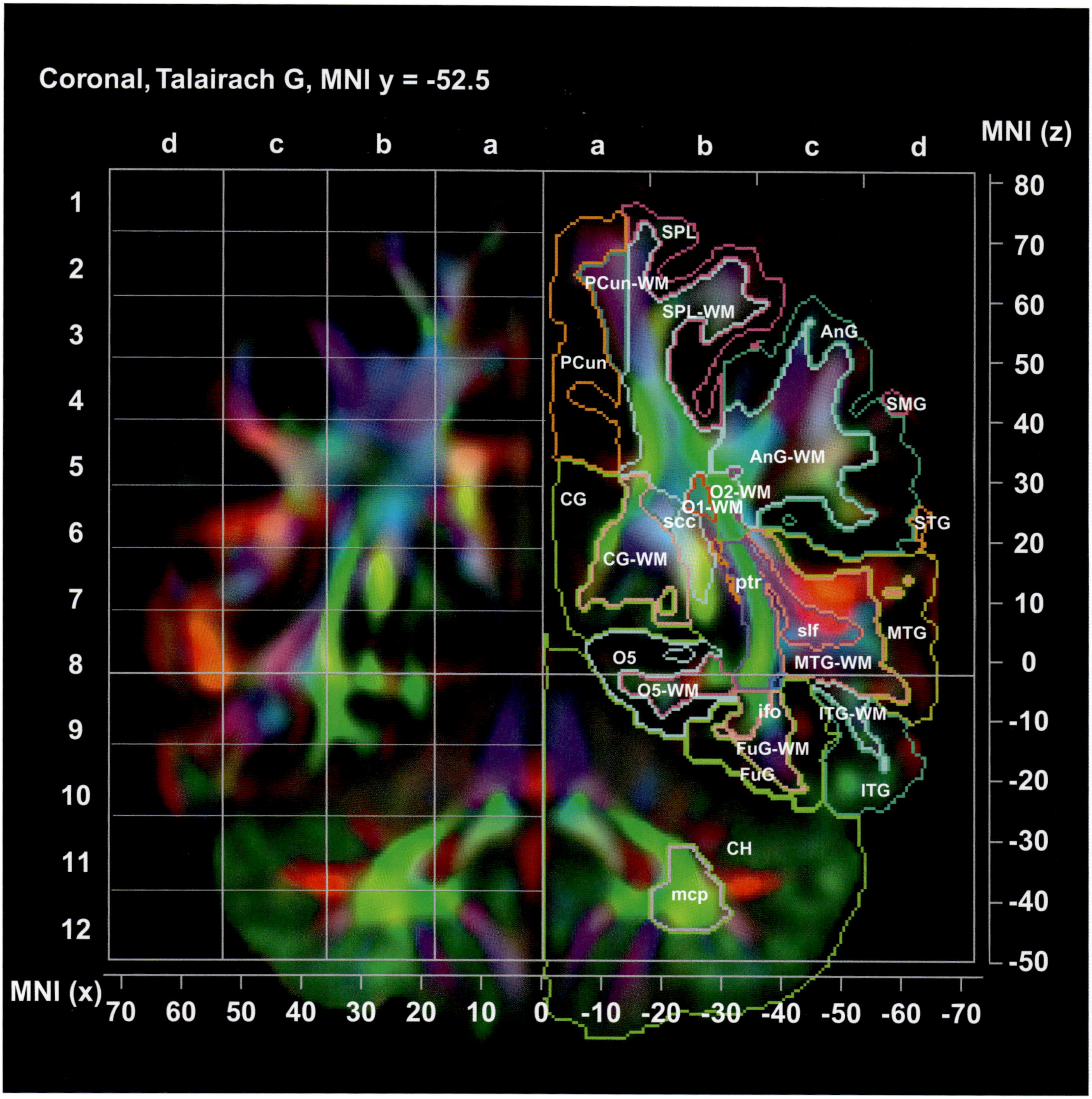

AnG - angular gyrus
AnG-WM - angular gyrus white matter
CG - cingulate gyrus
CG-WM - cingulate gyrus white matter
CH - cerebellar hemisphere
FuG - fusiform gyrus
FuG-WM - fusiform gyrus white matter
ifo - inferior fronto-occipital fasciculus
ITG - inferior temporal gyrus
ITG-WM - inferior temporal gyrus white matter
mcp - middle cerebellar peduncle
MTG - middle temporal gyrus
MTG-WM - middle temporal gyrus white matter
O1-WM - superior occipital gyrus white matter
O2-WM - middle or lateral occipital gyrus white matter
O5 - lyngual gyrus
O5-WM - lyngual gyrus white matter
PCun - precuneus
PCun-WM - precuneus white matter
ptr - posterior thalamic radiation (include the optic radiation)
scc - splenium of the corpus callosum
slf - superior longitudinal fasciculus
SMG - supramarginal gyrus
SPL - superior parietal lobule
SPL-WM - superior parietal lobule white matter
STG - superior temporal gyrus

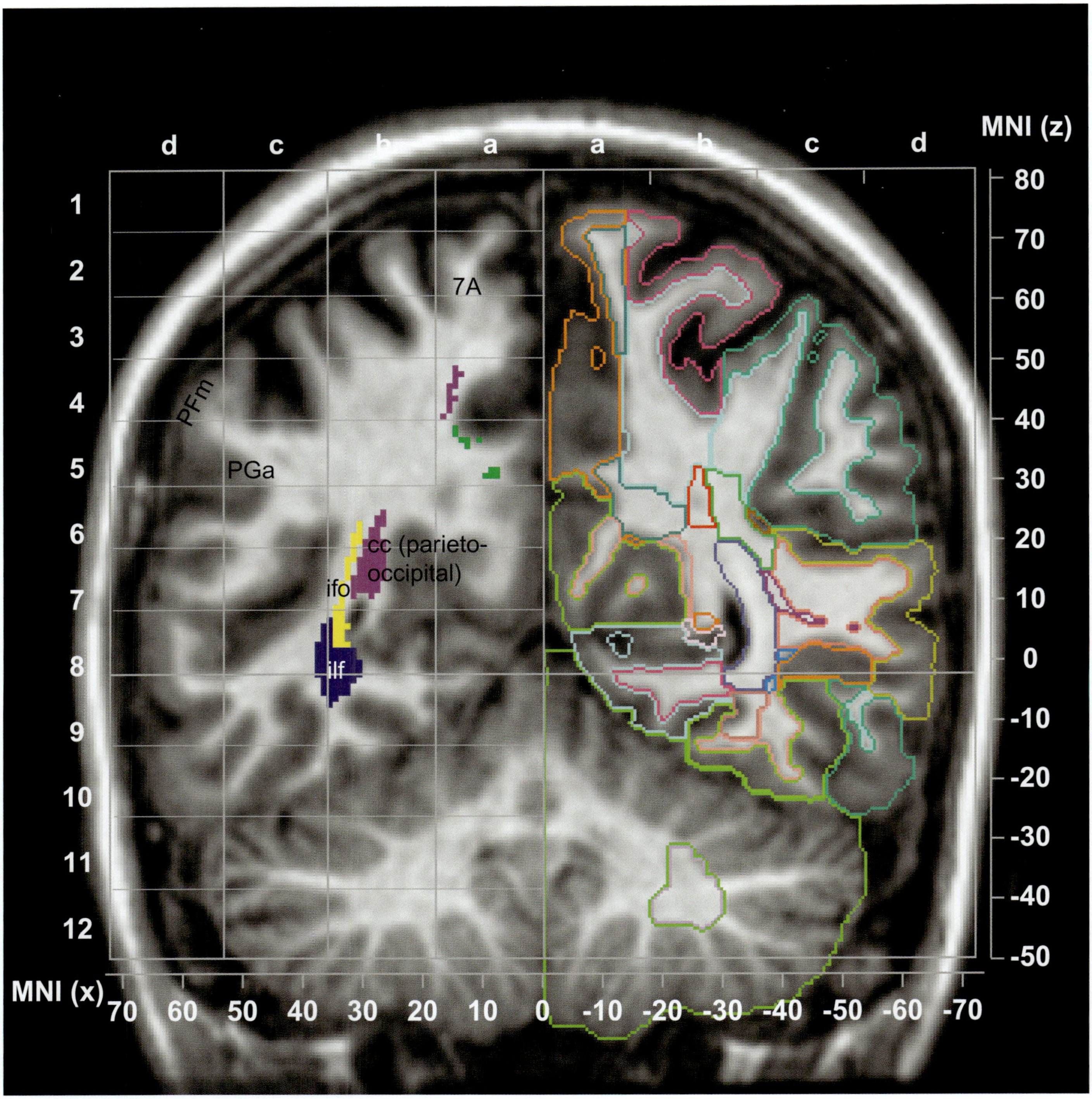

cc (parieto-occipital) - corpus callosum (connecting bilateral parieto-occipital areas)
ifo - inferior fronto-occipital fasciculus
ilf - inferior longitudinal fasciculus

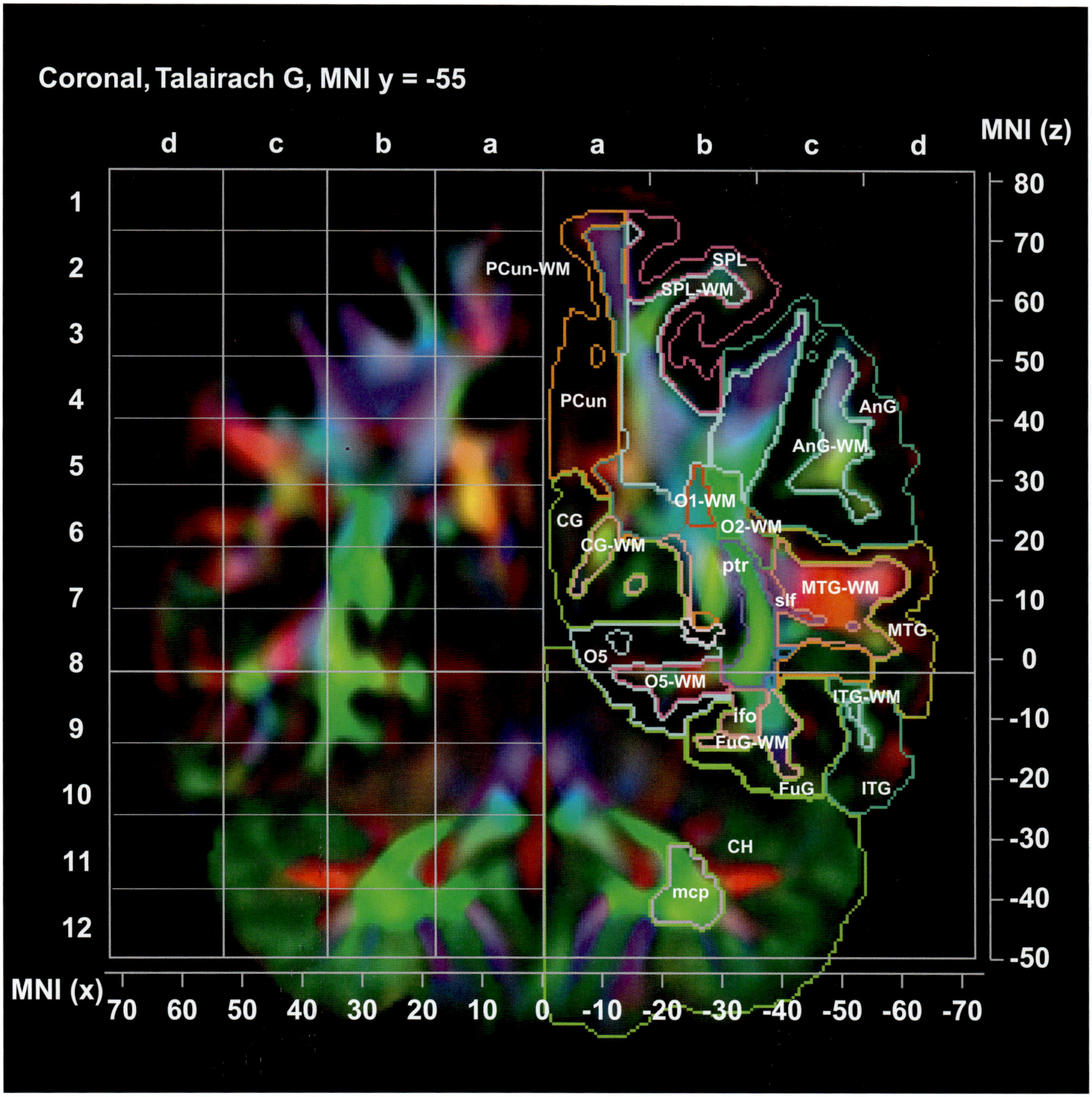

AnG	-	angular gyrus
AnG-WM	-	angular gyrus white matter
CG	-	cingulate gyrus
CG-WM	-	cingulate gyrus white matter
CH	-	cerebellar hemisphere
FuG	-	fusiform gyrus
FuG-WM	-	fusiform gyrus white matter
ifo	-	inferior fronto-occipital fasciculus
ITG	-	inferior temporal gyrus
ITG-WM	-	inferior temporal gyrus white matter
mcp	-	middle cerebellar peduncle
MTG	-	middle temporal gyrus
MTG-WM	-	middle temporal gyrus white matter
O1-WM	-	superior occipital gyrus white matter
O2-WM	-	middle or lateral occipital gyrus white matter
O5	-	lyngual gyrus
O5-WM	-	lyngual gyrus white matter
PCun	-	precuneus
PCun-WM	-	precuneus white matter
ptr	-	posterior thalamic radiation (include the optic radiation)
slf	-	superior longitudinal fasciculus
SPL	-	superior parietal lobule
SPL-WM	-	superior parietal lobule white matter

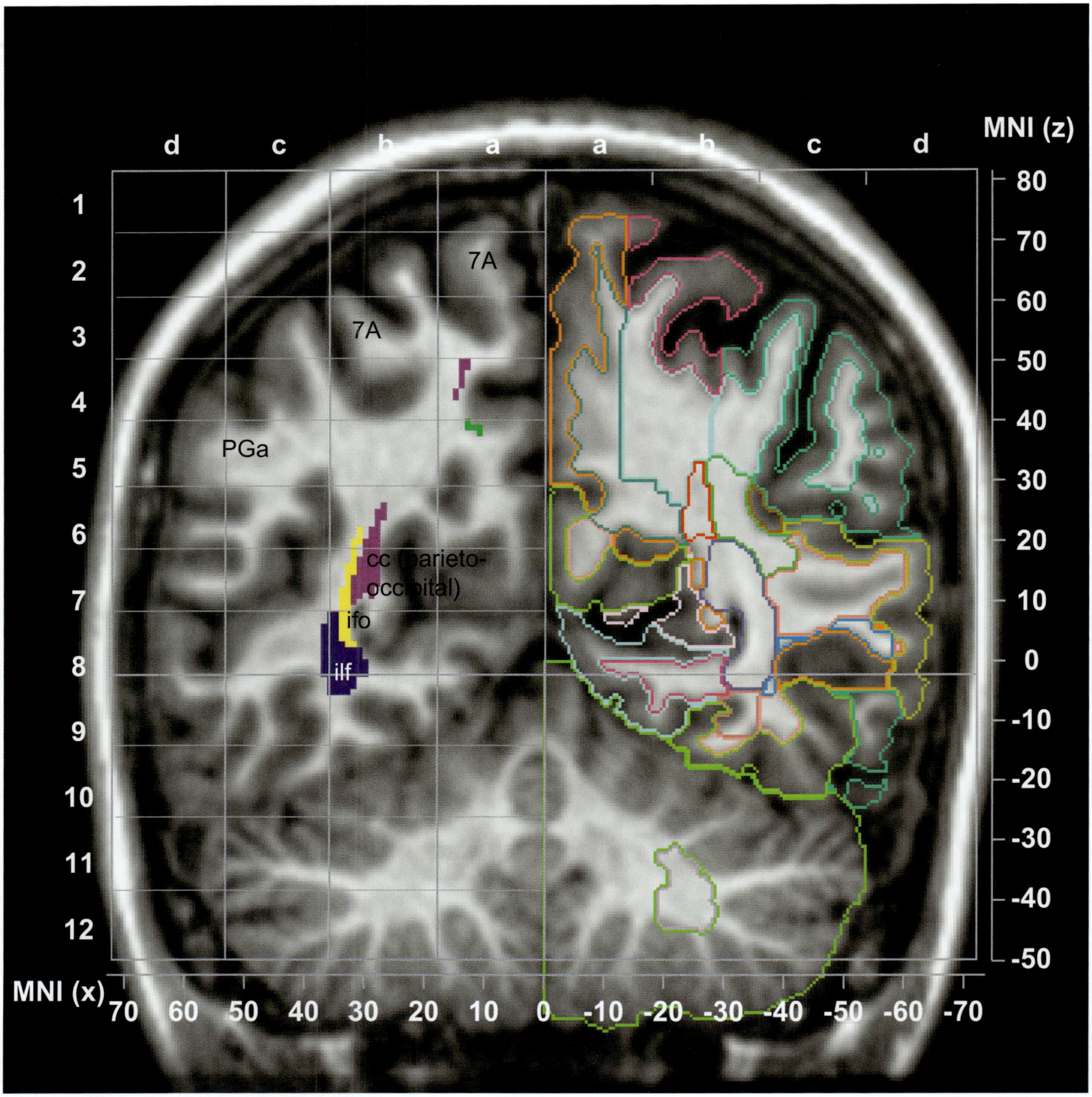

cc (parieto-occipital) - corpus callosum (connecting bilateral parieto-occipital areas)
ifo - inferior fronto-occipital fasciculus
ilf - inferior longitudinal fasciculus

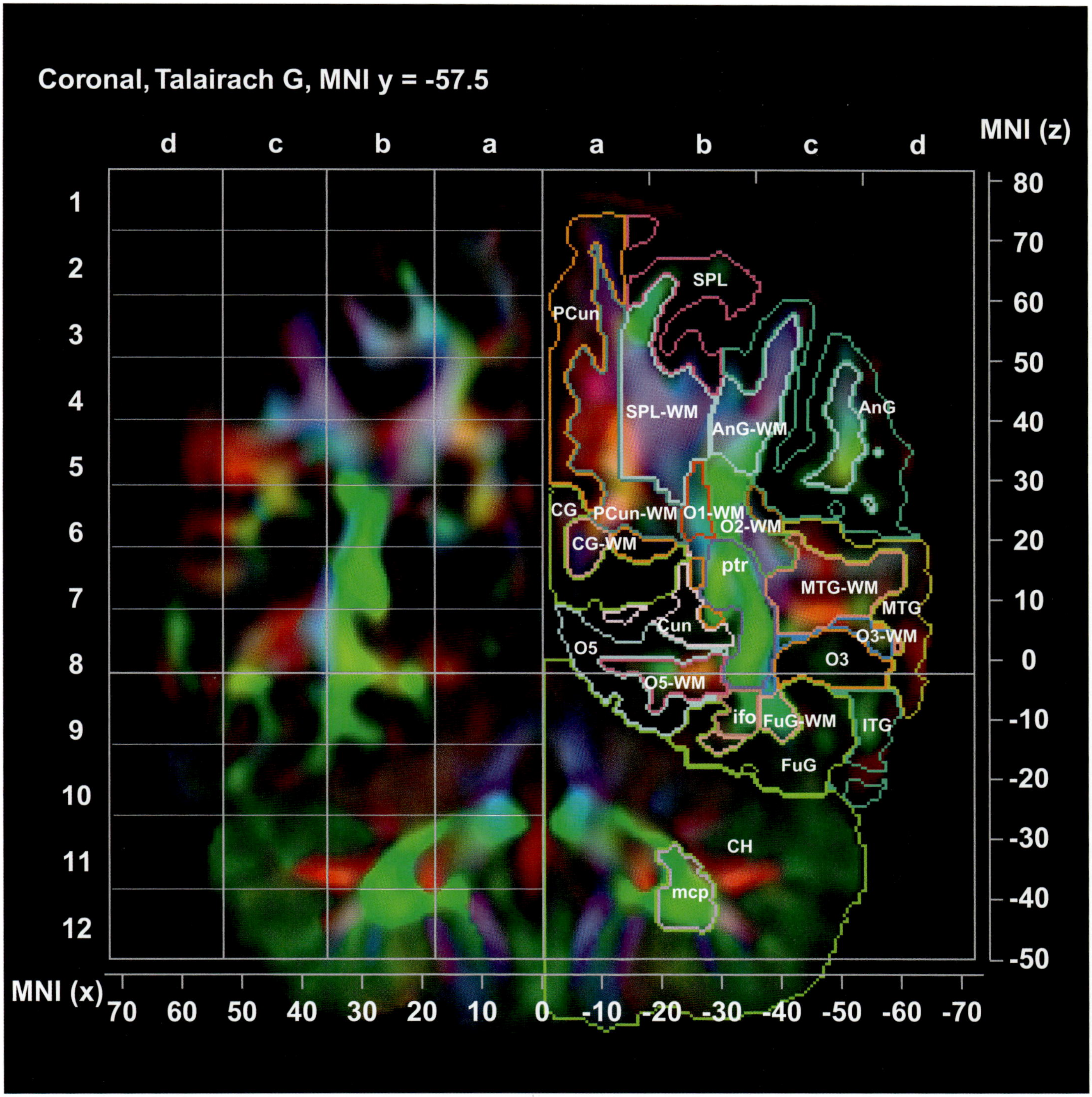

- AnG - angular gyrus
- AnG-WM - angular gyrus white matter
- CG - cingulate gyrus
- CG-WM - cingulate gyrus white matter
- CH - cerebellar hemisphere
- Cun - cuneus
- FuG - fusiform gyrus
- FuG-WM - fusiform gyrus white matter
- ifo - inferior fronto-occipital fasciculus
- ITG - inferior temporal gyrus
- mcp - middle cerebellar peduncle
- MTG - middle temporal gyrus
- MTG-WM - middle temporal gyrus white matter
- O1-WM - superior occipital gyrus white matter
- O2-WM - middle or lateral occipital gyrus white matter
- O3 - inferior occipital gyrus
- O3-WM - inferior occipital gyrus white matter
- O5 - lyngual gyrus
- O5-WM - lyngual gyrus white matter
- PCun - precuneus
- PCun-WM - precuneus white matter
- ptr - posterior thalamic radiation (include the optic radiation)
- SPL - superior parietal lobule
- SPL-WM - superior parietal lobule white matter

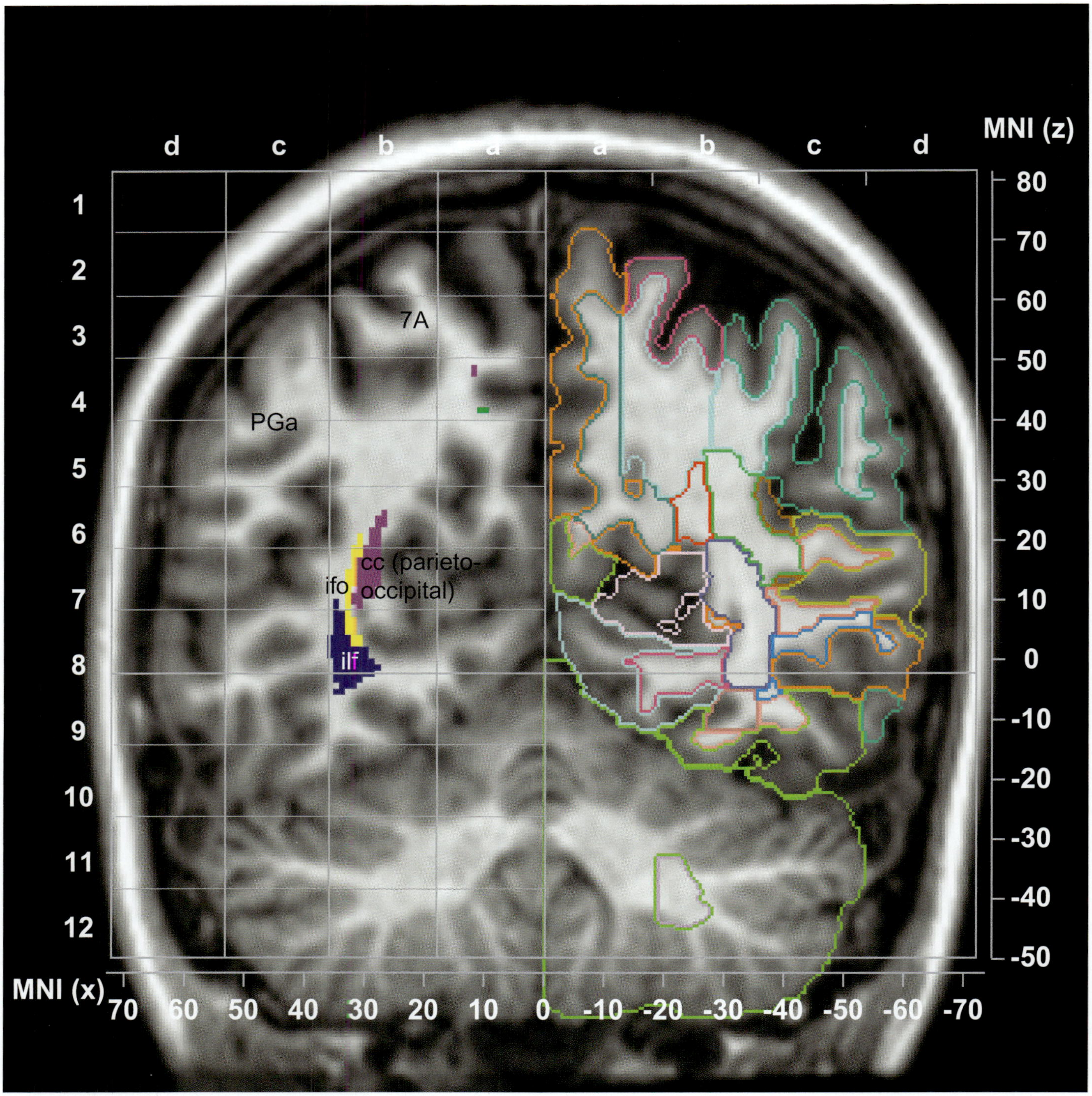

cc (parieto-occipital) - corpus callosum (connecting bilateral parieto-occipital areas)
ifo - inferior fronto-occipital fasciculus
ilf - inferior longitudinal fasciculus

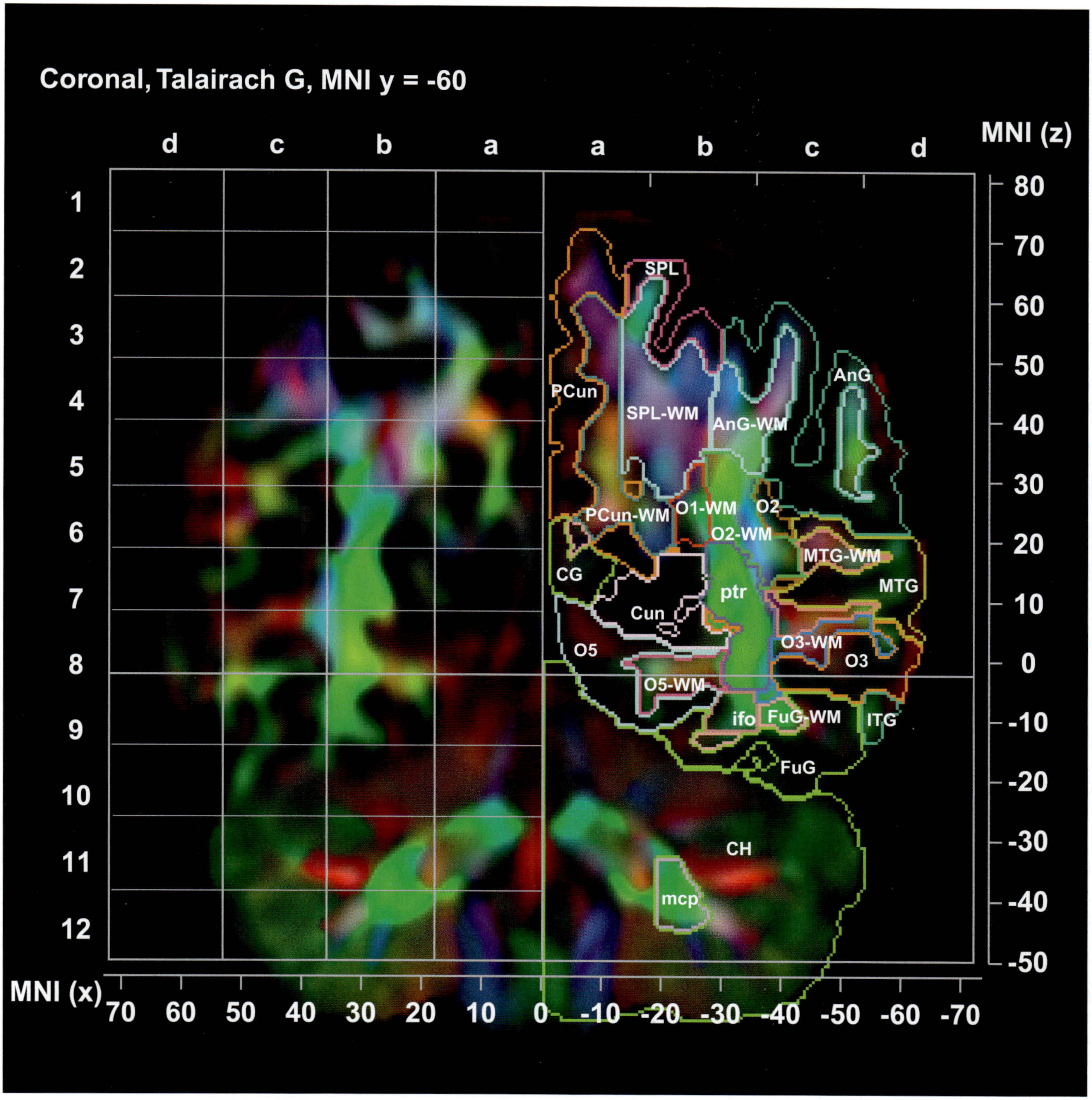

AnG - angular gyrus
AnG-WM - angular gyrus white matter
CG - cingulate gyrus
CH - cerebellar hemisphere
Cun - cuneus
FuG - fusiform gyrus
FuG-WM - fusiform gyrus white matter
ifo - inferior fronto-occipital fasciculus
ITG - inferior temporal gyrus
mcp - middle cerebellar peduncle
MTG - middle temporal gyrus
MTG-WM - middle temporal gyrus white matter
O1-WM - superior occipital gyrus white matter
O2 - middle or lateral occipital gyrus
O2-WM - middle or lateral occipital gyrus white matter
O3 - inferior occipital gyrus
O3-WM - inferior occipital gyrus white matter
O5 - lyngual gyrus
O5-WM - lyngual gyrus white matter
PCun - precuneus
PCun-WM - precuneus white matter
ptr - posterior thalamic radiation (include the optic radiation)
SPL - superior parietal lobule
SPL-WM - superior parietal lobule white matter

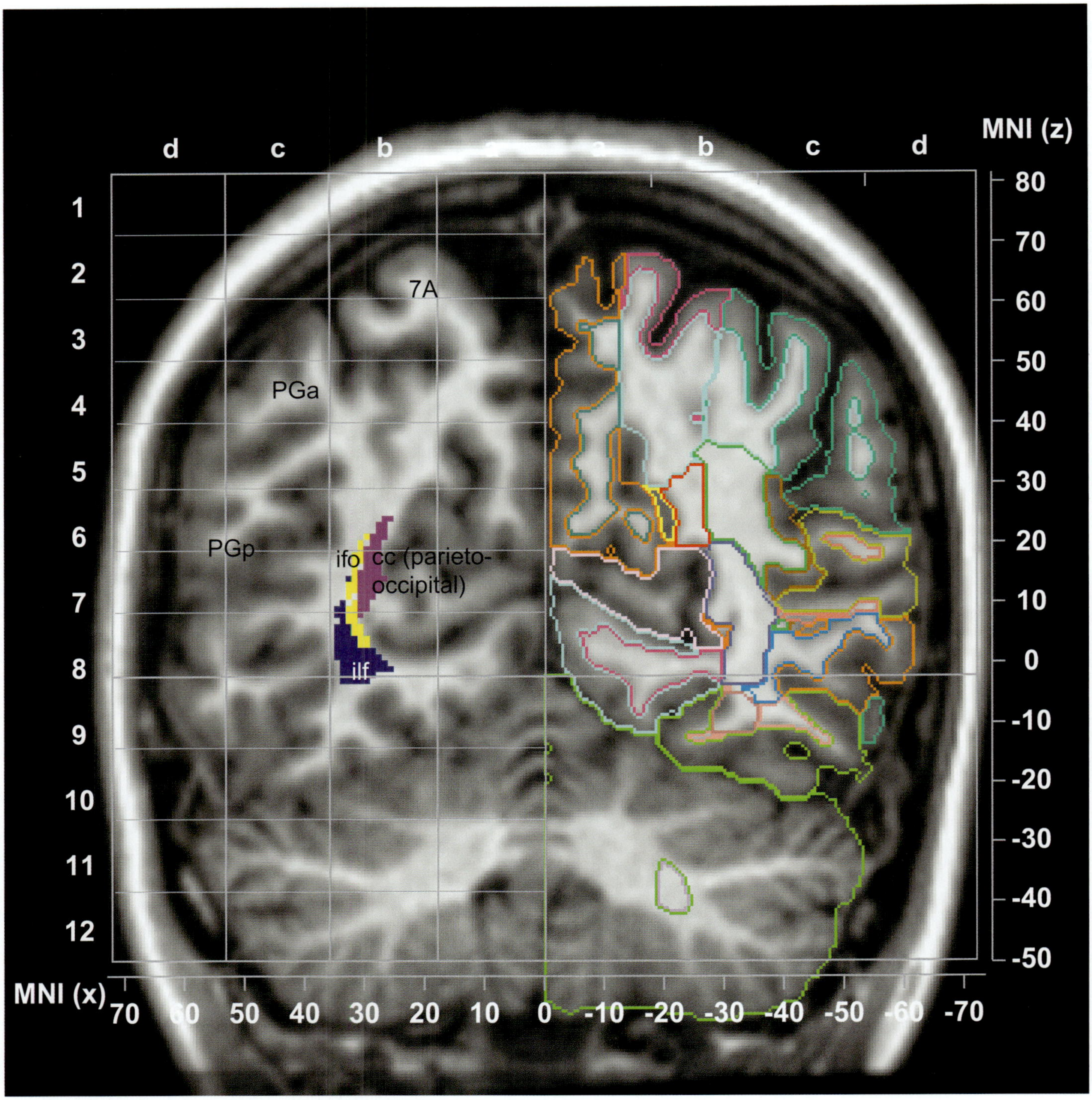

cc (parieto-occipital) - corpus callosum (connecting bilateral parieto-occipital areas)

ifo - inferior fronto-occipital fasciculus

ilf - inferior longitudinal fasciculus

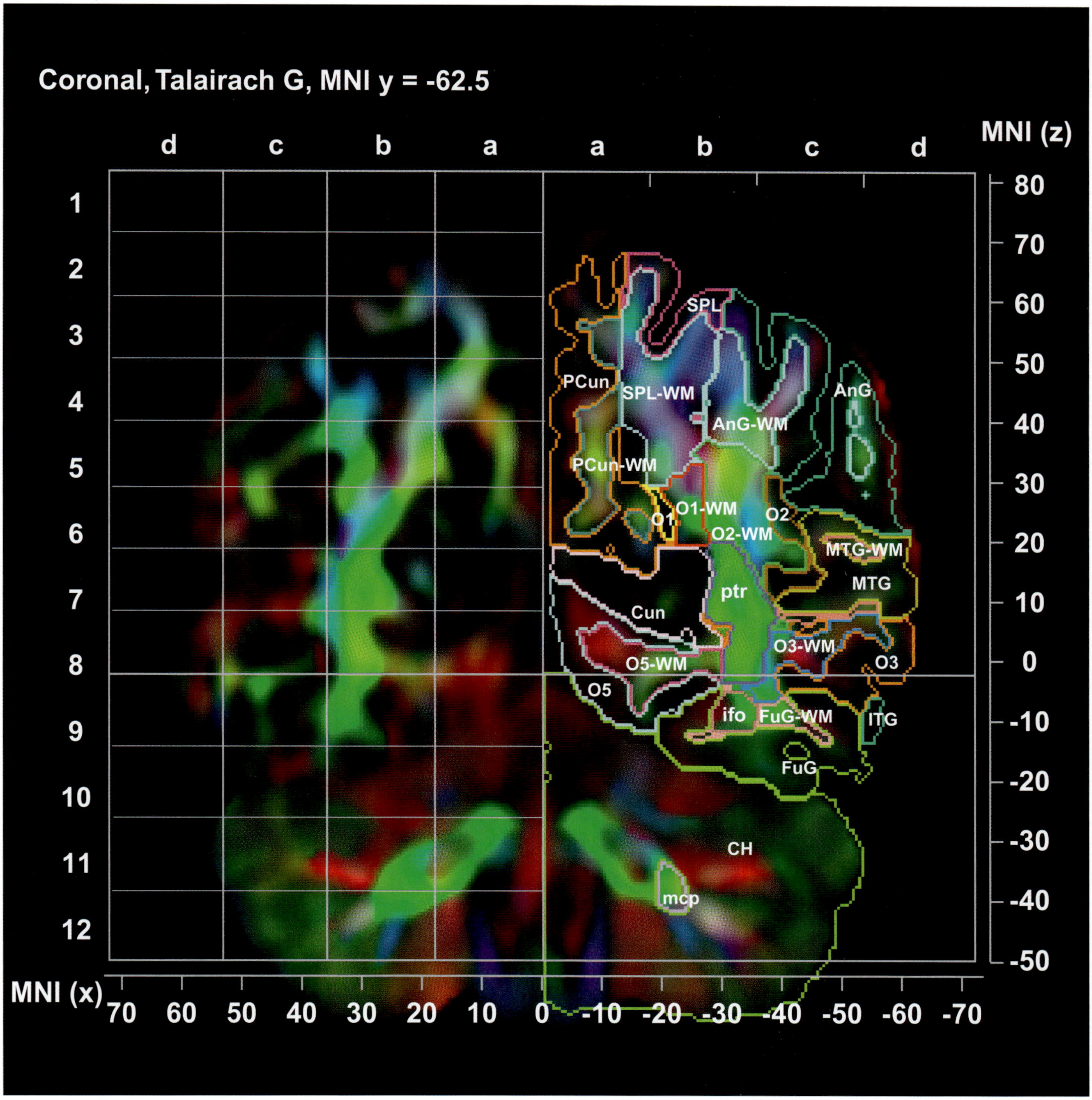

AnG - angular gyrus
AnG-WM - angular gyrus white matter
CH - cerebellar hemisphere
Cun - cuneus
FuG - fusiform gyrus
FuG-WM - fusiform gyrus white matter
ifo - inferior fronto-occipital fasciculus
ITG - inferior temporal gyrus
mcp - middle cerebellar peduncle
MTG - middle temporal gyrus
MTG-WM - middle temporal gyrus white matter
O1 - superior occipital gyrus
O1-WM - superior occipital gyrus white matter
O2 - middle or lateral occipital gyrus
O2-WM - middle or lateral occipital gyrus white matter
O3 - inferior occipital gyrus
O3-WM - inferior occipital gyrus white matter
O5 - lyngual gyrus
O5-WM - lyngual gyrus white matter
PCun - precuneus
PCun-WM - precuneus white matter
ptr - posterior thalamic radiation (include the optic radiation)
SPL - superior parietal lobule
SPL-WM - superior parietal lobule white matter

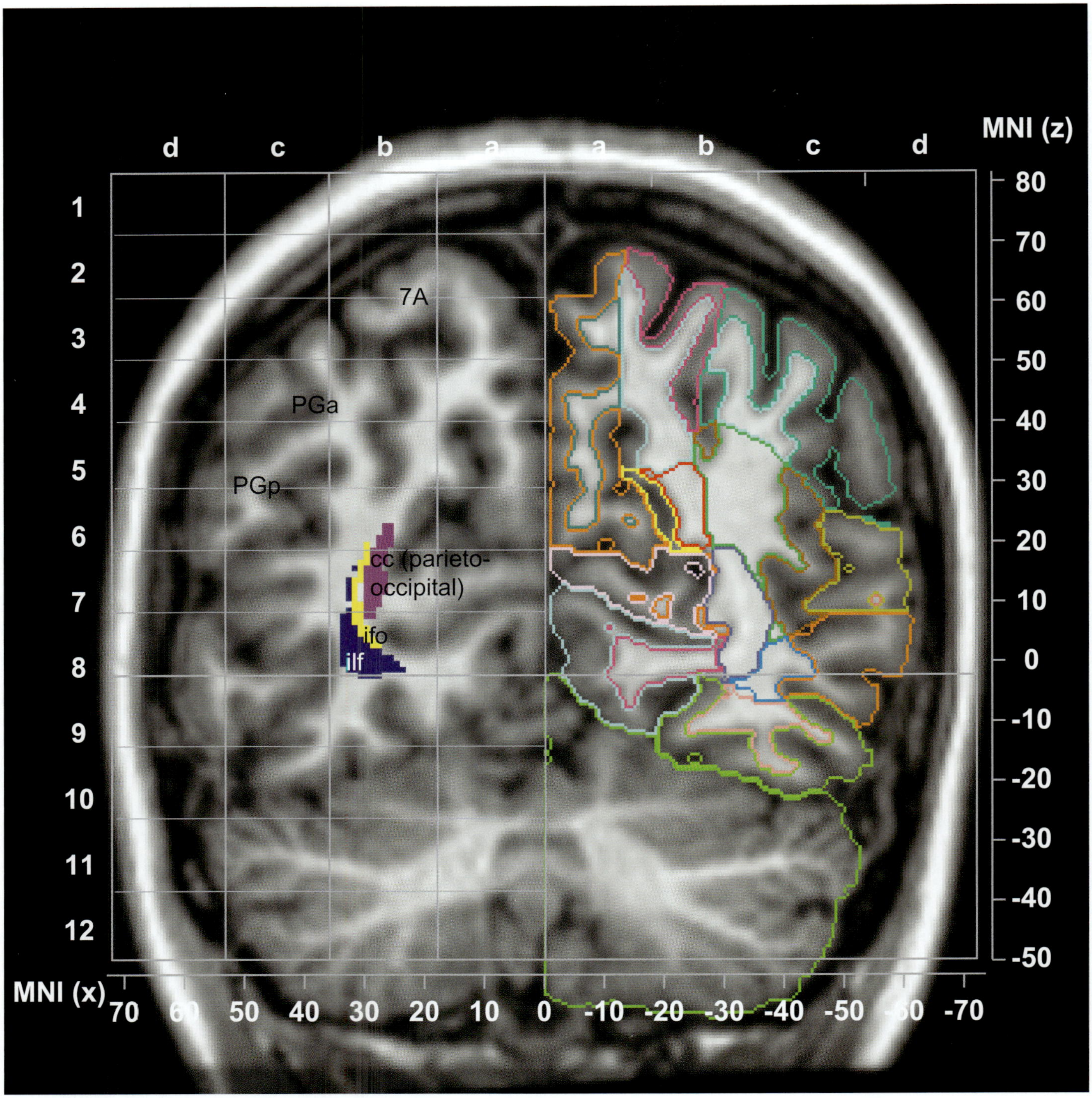

cc (parieto-occipital) - corpus callosum (connecting bilateral parieto-occipital areas)

ifo - inferior fronto-occipital fasciculus

ilf - inferior longitudinal fasciculus

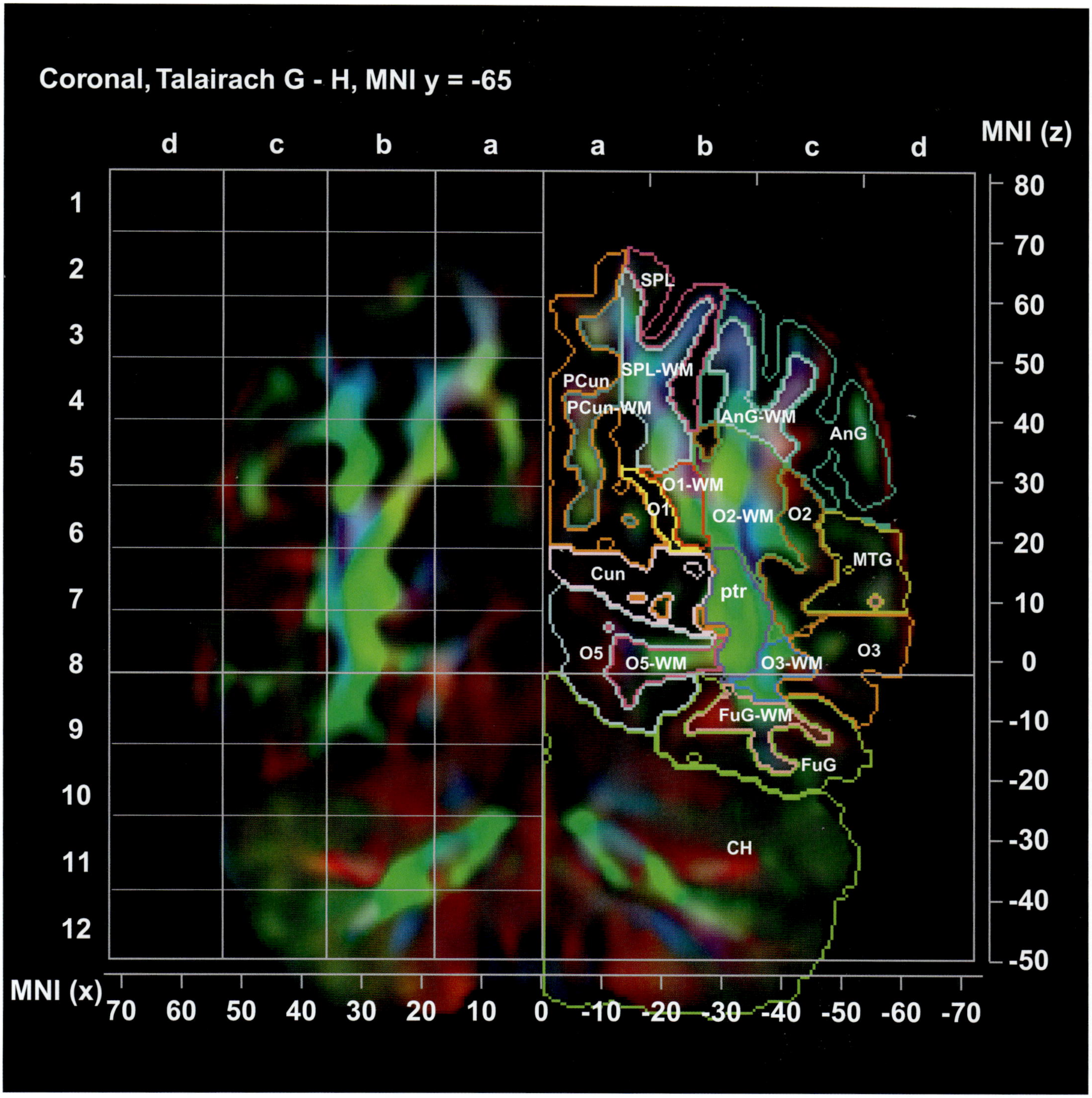

AnG	-	angular gyrus
AnG-WM	-	angular gyrus white matter
CH	-	cerebellar hemisphere
Cun	-	cuneus
FuG	-	fusiform gyrus
FuG-WM	-	fusiform gyrus white matter
MTG	-	middle temporal gyrus
O1	-	superior occipital gyrus
O1-WM	-	superior occipital gyrus white matter
O2	-	middle or lateral occipital gyrus
O2-WM	-	middle or lateral occipital gyrus white matter
O3	-	inferior occipital gyrus
O3-WM	-	inferior occipital gyrus white matter
O5	-	lyngual gyrus
O5-WM	-	lyngual gyrus white matter
PCun	-	precuneus
PCun-WM	-	precuneus white matter
ptr	-	posterior thalamic radiation (include the optic radiation)
SPL	-	superior parietal lobule
SPL-WM	-	superior parietal lobule white matter

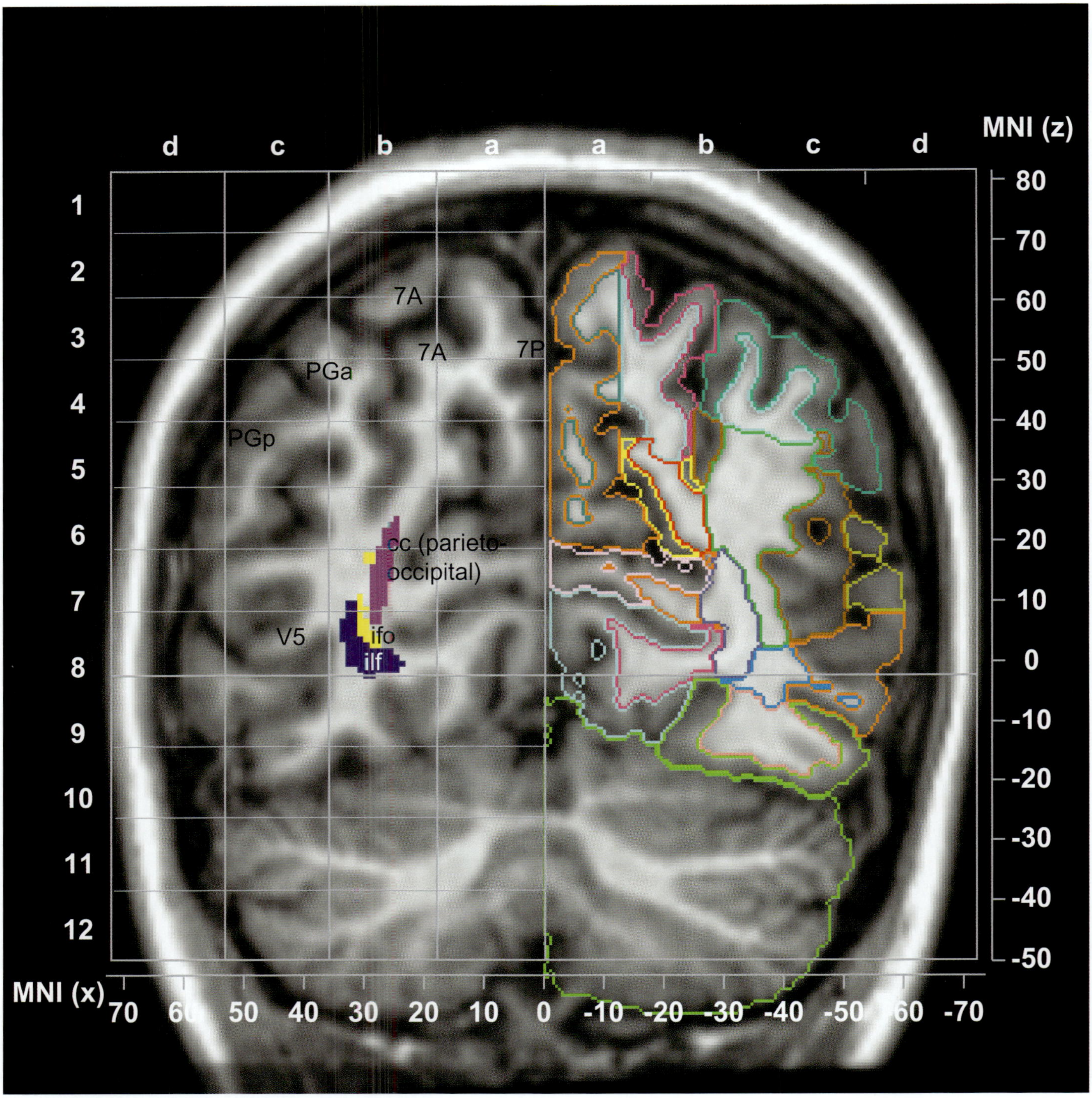

cc (parieto-occipital) - corpus callosum (connecting bilateral parieto-occipital areas)

ifo - inferior fronto-occipital fasciculus

ilf - inferior longitudinal fasciculus

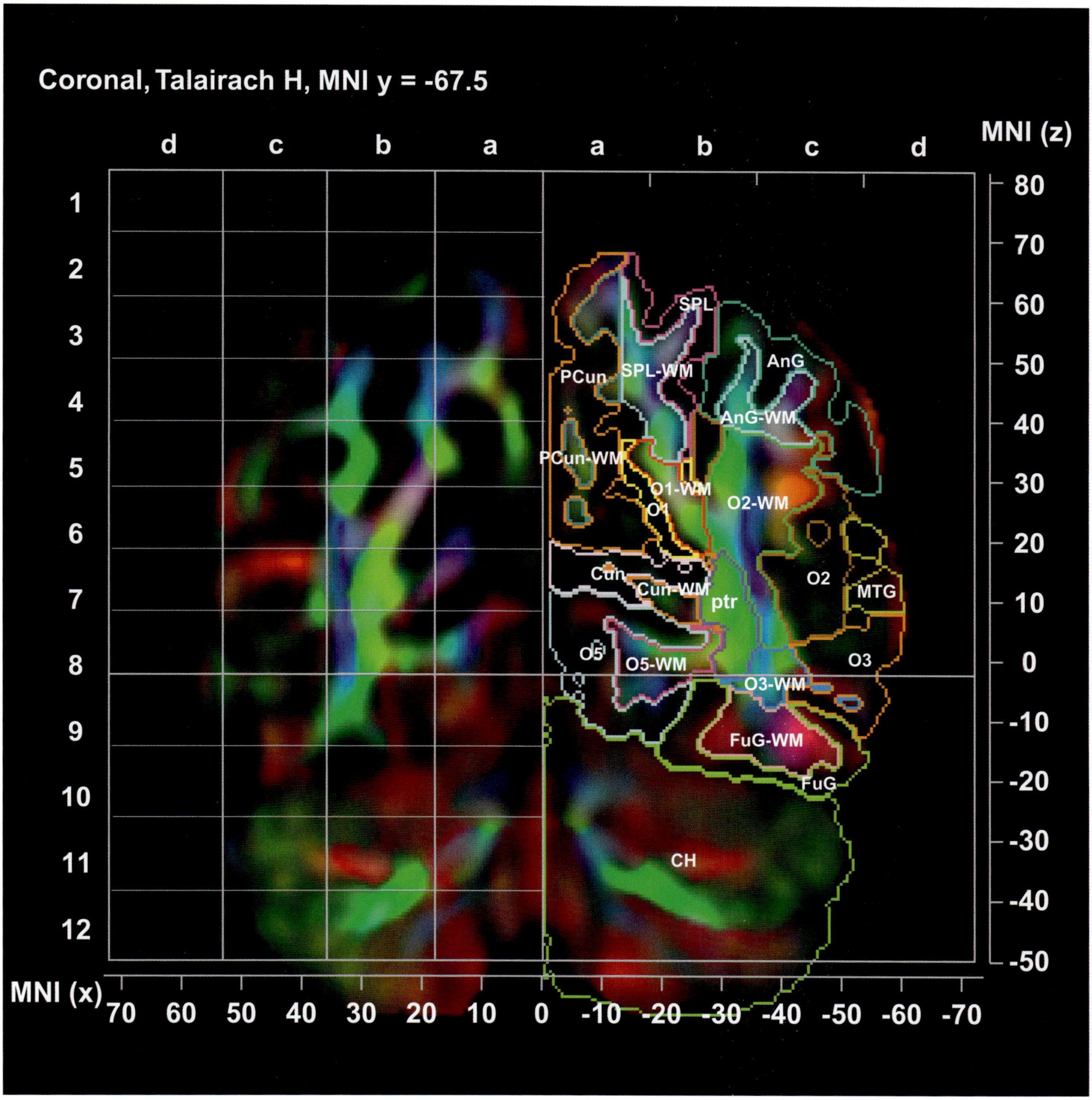

AnG - angular gyrus
AnG-WM - angular gyrus white matter
CH - cerebellar hemisphere
Cun - cuneus
Cun-WM - cuneus white matter
FuG - fusiform gyrus
FuG-WM - fusiform gyrus white matter
MTG - middle temporal gyrus
O1 - superior occipital gyrus
O1-WM - superior occipital gyrus white matter
O2 - middle or lateral occipital gyrus
O2-WM - middle or lateral occipital gyrus white matter
O3 - inferior occipital gyrus
O3-WM - inferior occipital gyrus white matter
O5 - lyngual gyrus
O5-WM - lyngual gyrus white matter
PCun - precuneus
PCun-WM - precuneus white matter
ptr - posterior thalamic radiation (include the optic radiation)
SPL - superior parietal lobule
SPL-WM - superior parietal lobule white matter

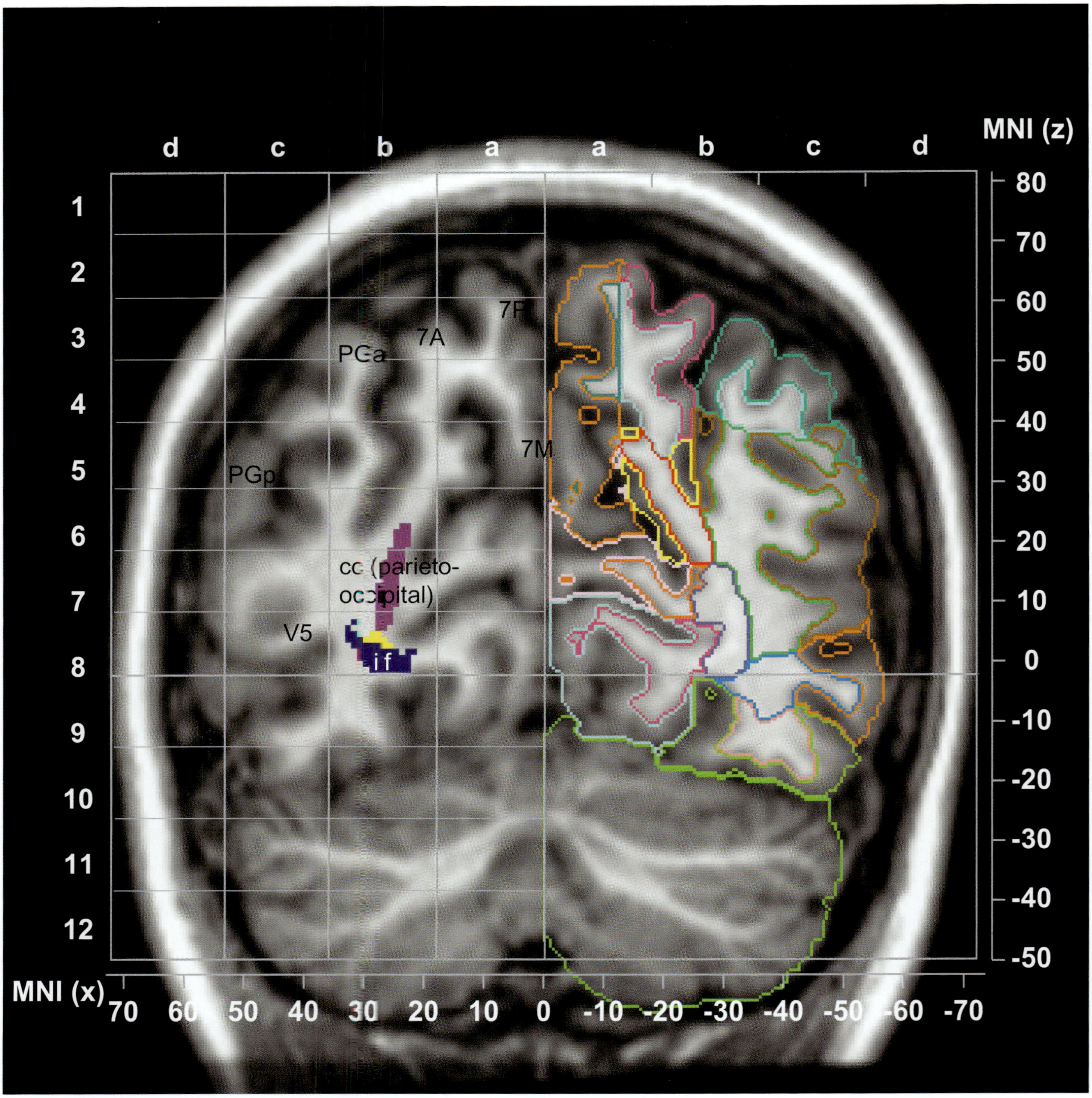

cc (parieto-occipital) - corpus callosum (connecting bilateral parieto-occipital areas)

ilf - inferior longitudinal fasciculus

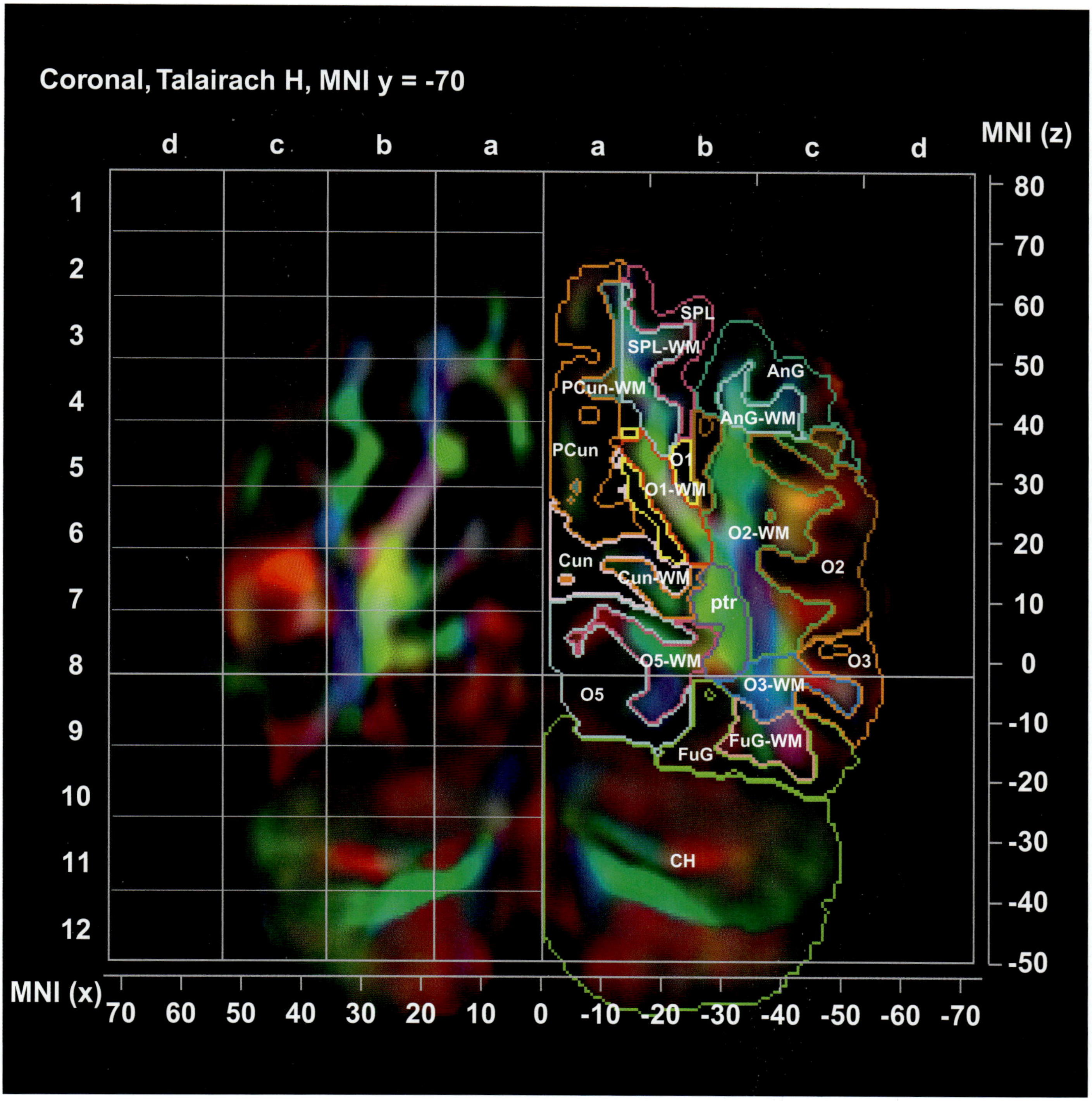

AnG	-	angular gyrus
AnG-WM	-	angular gyrus white matter
CH	-	cerebellar hemisphere
Cun	-	cuneus
Cun-WM	-	cuneus white matter
FuG	-	fusiform gyrus
FuG-WM	-	fusiform gyrus white matter
O1	-	superior occipital gyrus
O1-WM	-	superior occipital gyrus white matter
O2	-	middle or lateral occipital gyrus
O2-WM	-	middle or lateral occipital gyrus white matter
O3	-	inferior occipital gyrus
O3-WM	-	inferior occipital gyrus white matter
O5	-	lyngual gyrus
O5-WM	-	lyngual gyrus white matter
PCun	-	precuneus
PCun-WM	-	precuneus white matter
ptr	-	posterior thalamic radiation (include the optic radiation)
SPL	-	superior parietal lobule
SPL-WM	-	superior parietal lobule white matter

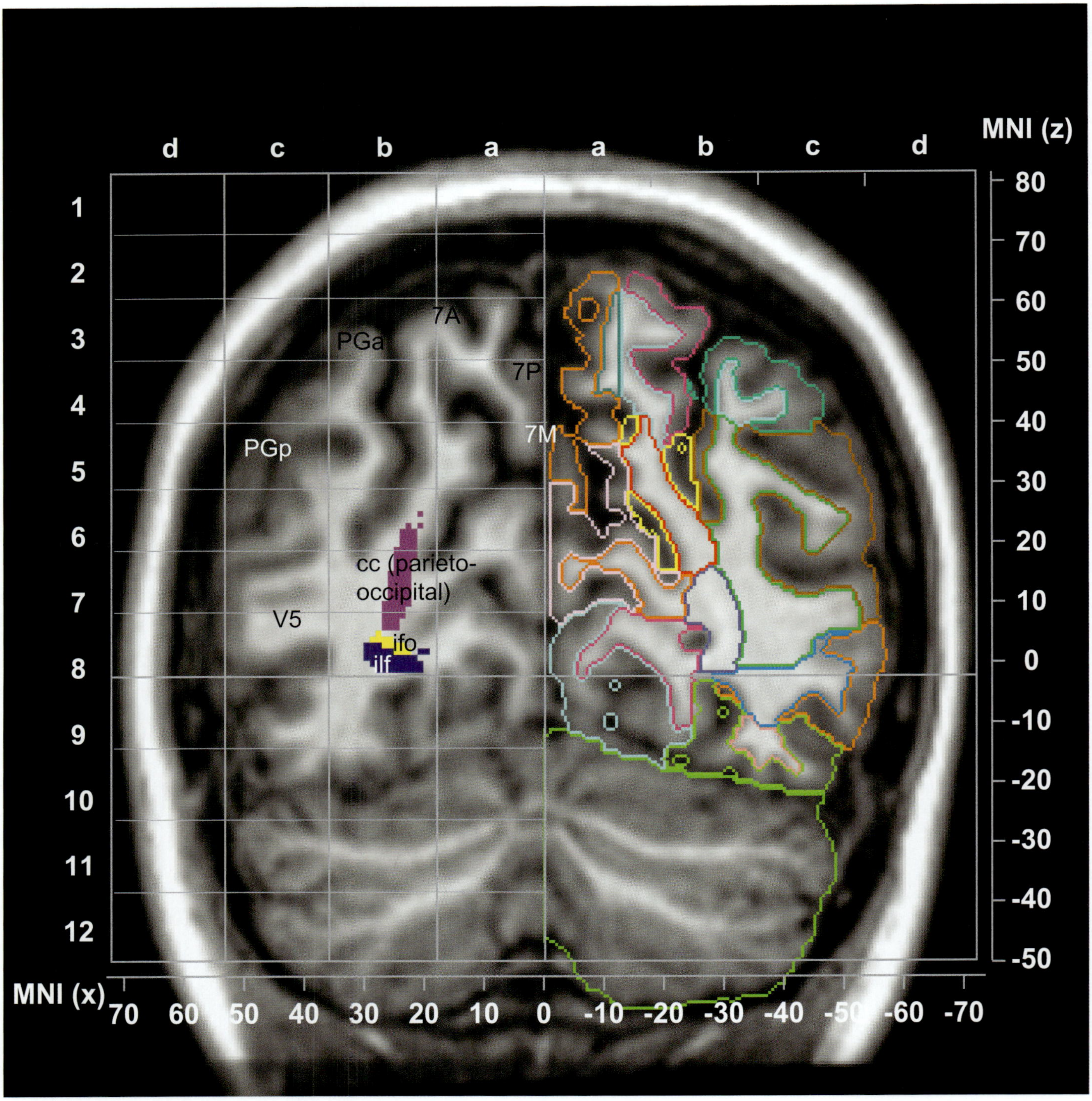

cc (parieto-occipital) - corpus callosum (connecting bilateral parieto-occipital areas)
ifo - inferior fronto-occipital fasciculus
ilf - inferior longitudinal fasciculus

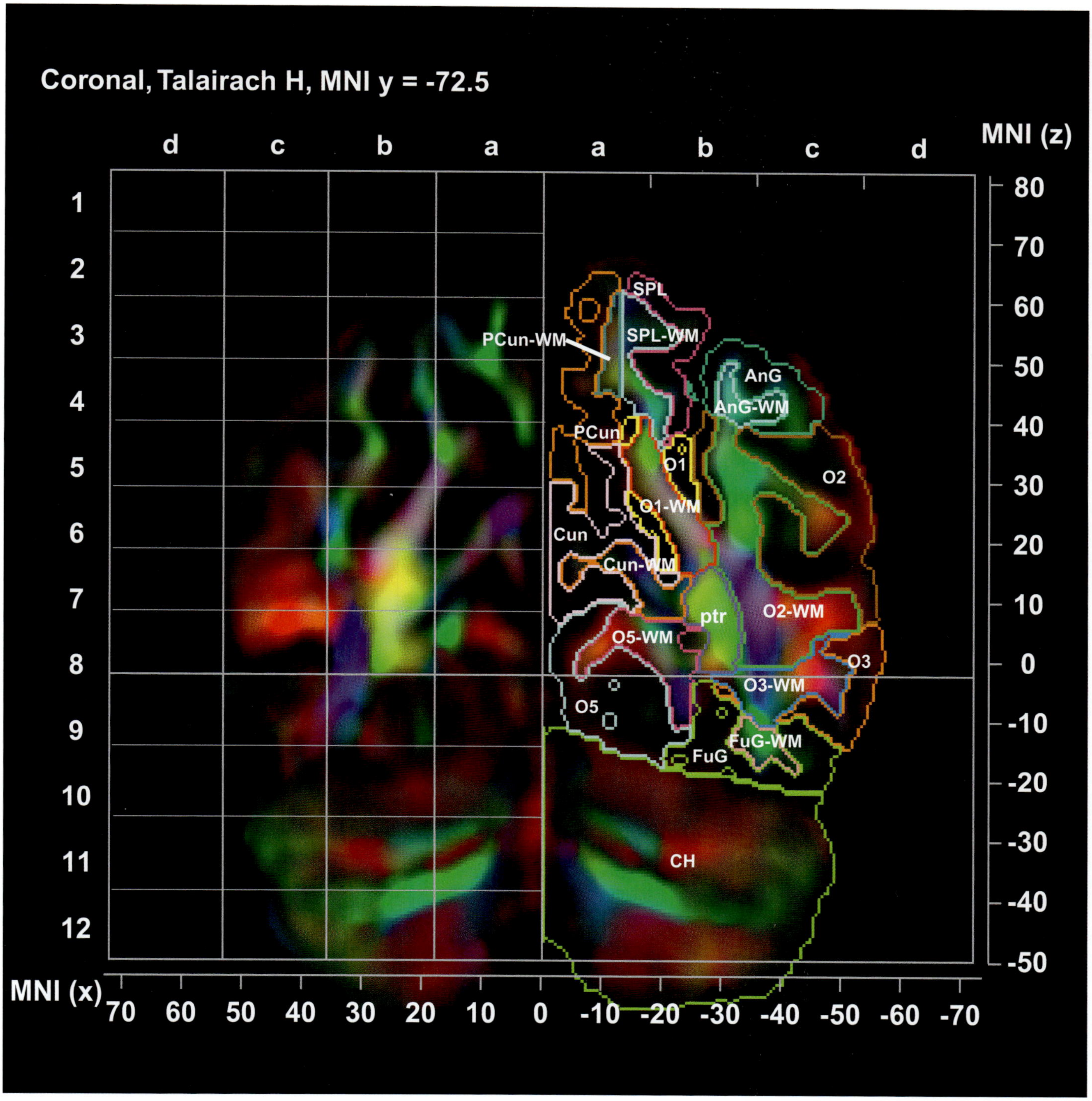

AnG	-	angular gyrus
AnG-WM	-	angular gyrus white matter
CH	-	cerebellar hemisphere
Cun	-	cuneus
Cun-WM	-	cuneus white matter
FuG	-	fusiform gyrus
FuG-WM	-	fusiform gyrus white matter
O1	-	superior occipital gyrus
O1-WM	-	superior occipital gyrus white matter
O2	-	middle or lateral occipital gyrus
O2-WM	-	middle or lateral occipital gyrus white matter
O3	-	inferior occipital gyrus
O3-WM	-	inferior occipital gyrus white matter
O5	-	lyngual gyrus
O5-WM	-	lyngual gyrus white matter
PCun	-	precuneus
PCun-WM	-	precuneus white matter
ptr	-	posterior thalamic radiation (include the optic radiation)
SPL	-	superior parietal lobule
SPL-WM	-	superior parietal lobule white matter

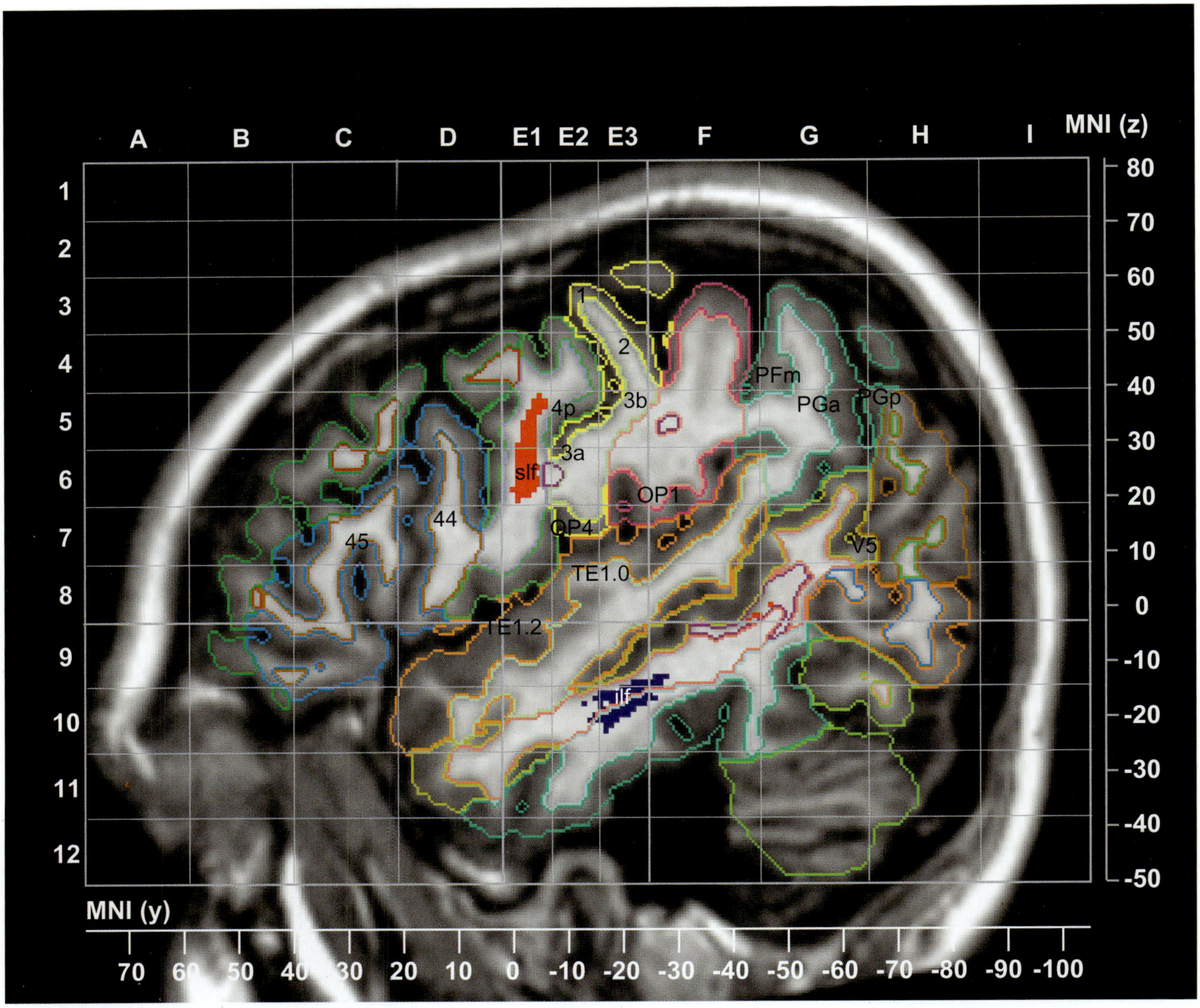

slf - superior longitudinal fasciculus
ilf - inferior longitudinal fasciculus

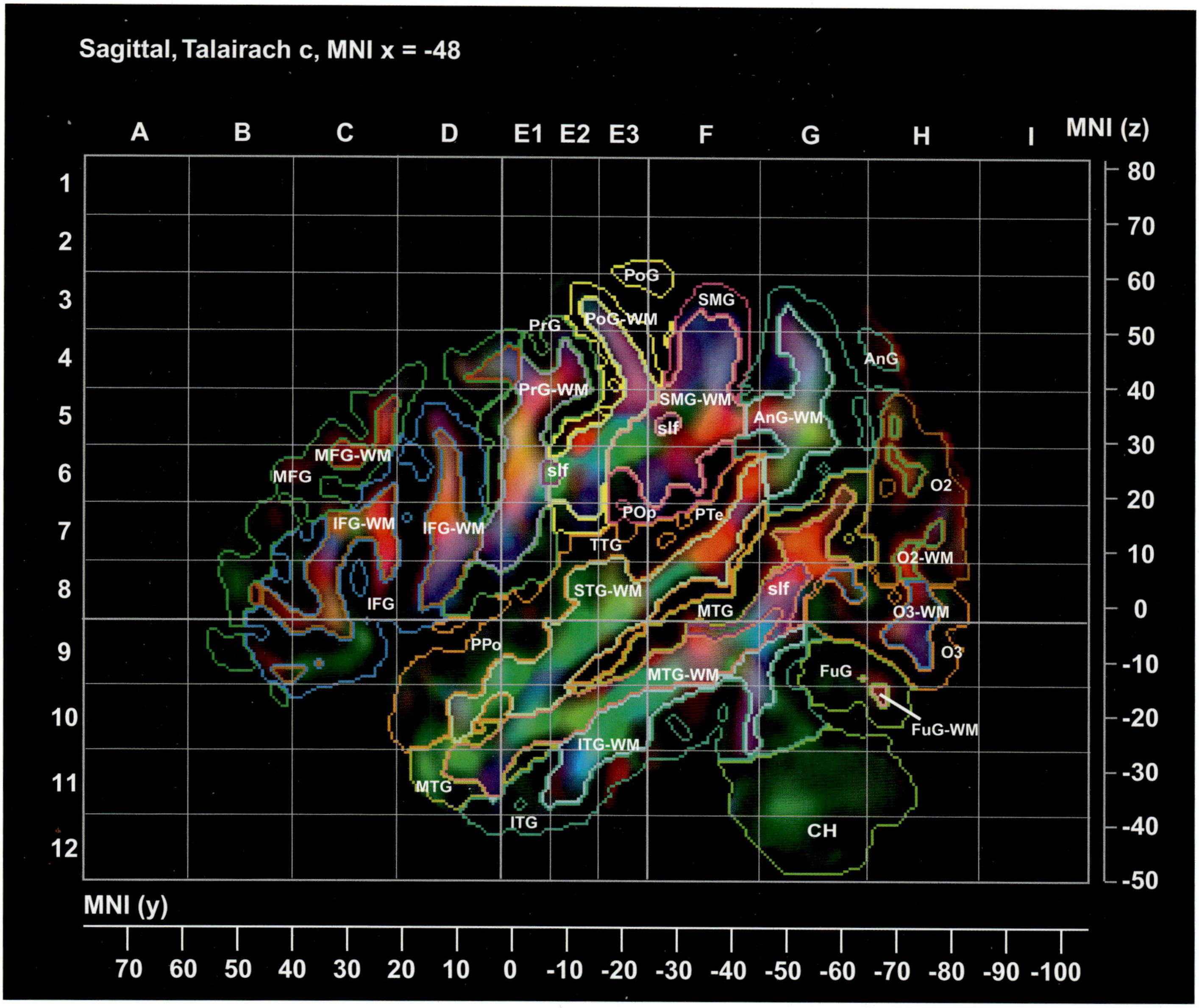

AnG - angular gyrus
AnG-WM - angular gyrus white matter
CH - cerebellar hemisphere
FuG - fusiform gyrus
FuG-WM - fusiform gyrus white matter
IFG - inferior frontal gyrus
IFG-WM - inferior frontal gyrus white matter
ITG - inferior temporal gyrus
ITG-WM - inferior temporal gyrus white matter
MFG - middle frontal gyrus
MFG-WM - middle frontal gyrus white matter
MTG - middle temporal gyrus
MTG-WM - middle temporal gyrus white matter
O2 - middle or lateral occipital gyrus
O2-WM - middle or lateral occipital gyrus white matter
O3 - inferior occipital gyrus
O3-WM - inferior occipital gyrus white matter
PoG - postcentral gyrus
PoG-WM - postcentral gyrus white matter
POp - parietal operculum
PPo - planum polare
PrG - precentral gyrus
PrG-WM - precentral gyrus white matter
PTe - planum temporale
slf - superior longitudinal fasciculus
SMG - supramarginal gyrus
SMG-WM - supramarginal gyrus white matter
STG-WM - superior temporal gyrus white matter
TTG - transverse temporal gyrus

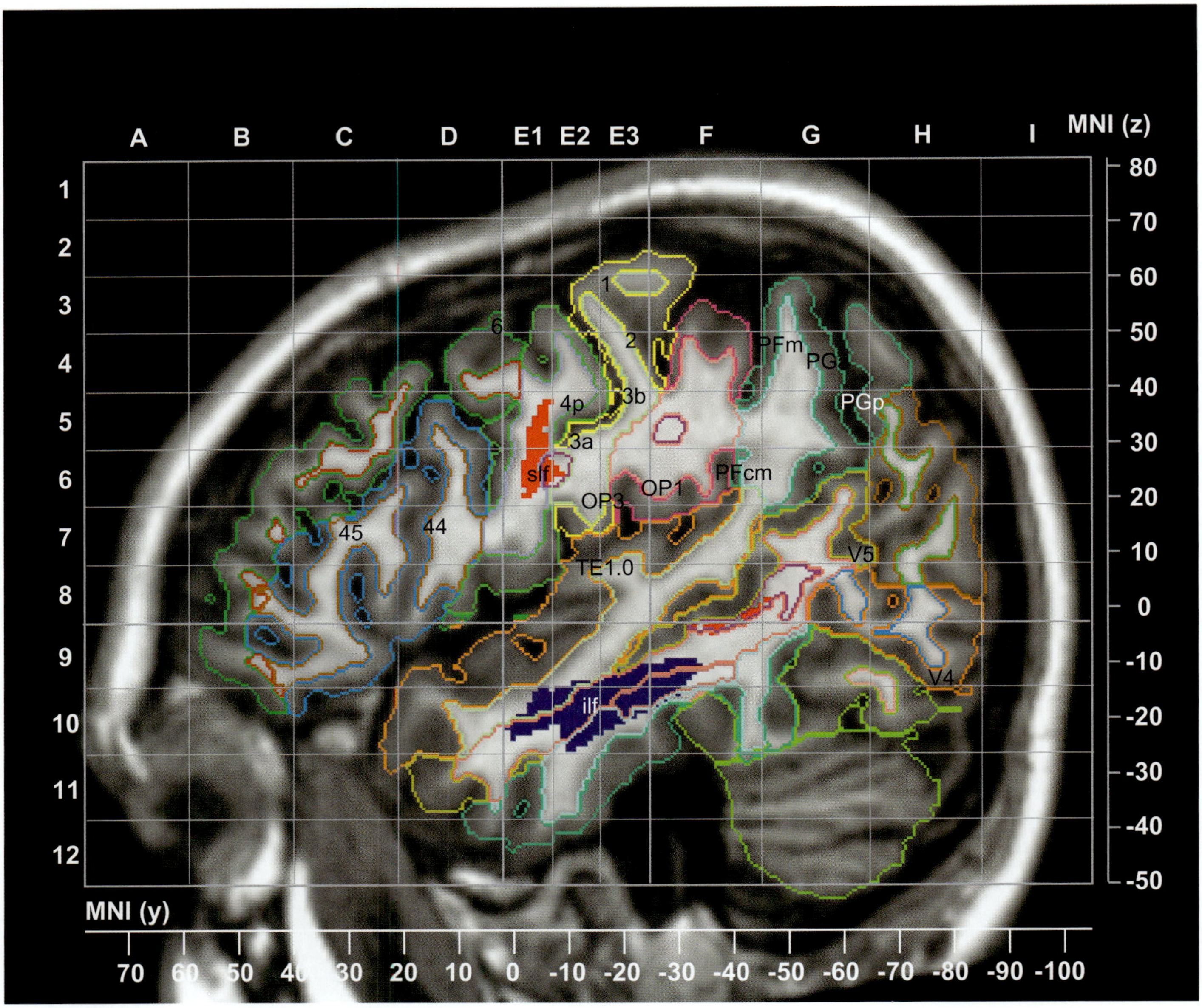

ilf - inferior longitudinal fasciculus
slf - superior longitudinal fasciculus

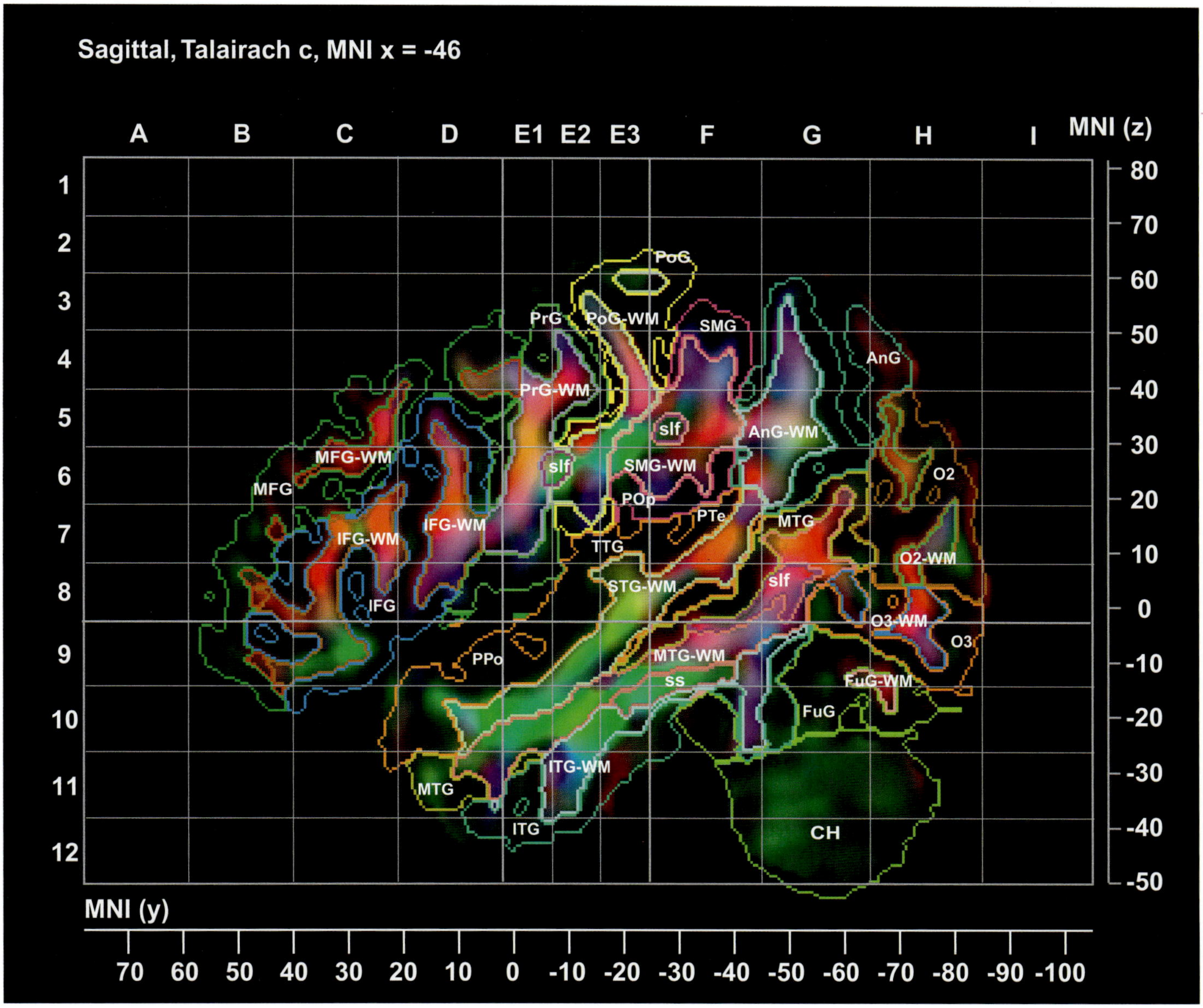

AnG - angular gyrus
AnG-WM - angular gyrus white matter
CH - cerebellar hemisphere
FuG - fusiform gyrus
FuG-WM - fusiform gyrus white matter
IFG - inferior frontal gyrus
IFG-WM - inferior frontal gyrus white matter
ITG - inferior temporal gyrus
ITG-WM - inferior temporal gyrus white matter
MFG - middle frontal gyrus
MFG-WM - middle frontal gyrus white matter
MTG - middle temporal gyrus
MTG-WM - middle temporal gyrus white matter
O2 - middle or lateral occipital gyrus
O2-WM - middle or lateral occipital gyrus white matter
O3 - inferior occipital gyrus
O3-WM - inferior occipital gyrus white matter
PoG - postcentral gyrus
PoG-WM - postcentral gyrus white matter
POp - parietal operculum
PPo - planum polare
PrG - precentral gyrus
PrG-WM - precentral gyrus white matter
PTe - planum temporale
slf - superior longitudinal fasciculus
SMG - supramarginal gyrus
SMG-WM - supramarginal gyrus white matter
ss - sagittal stratum (include inferior longitidinal fasciculus and inferior fronto-occipital fasciculus)
STG-WM - superior temporal gyrus white matter
TTG - transverse temporal gyrus

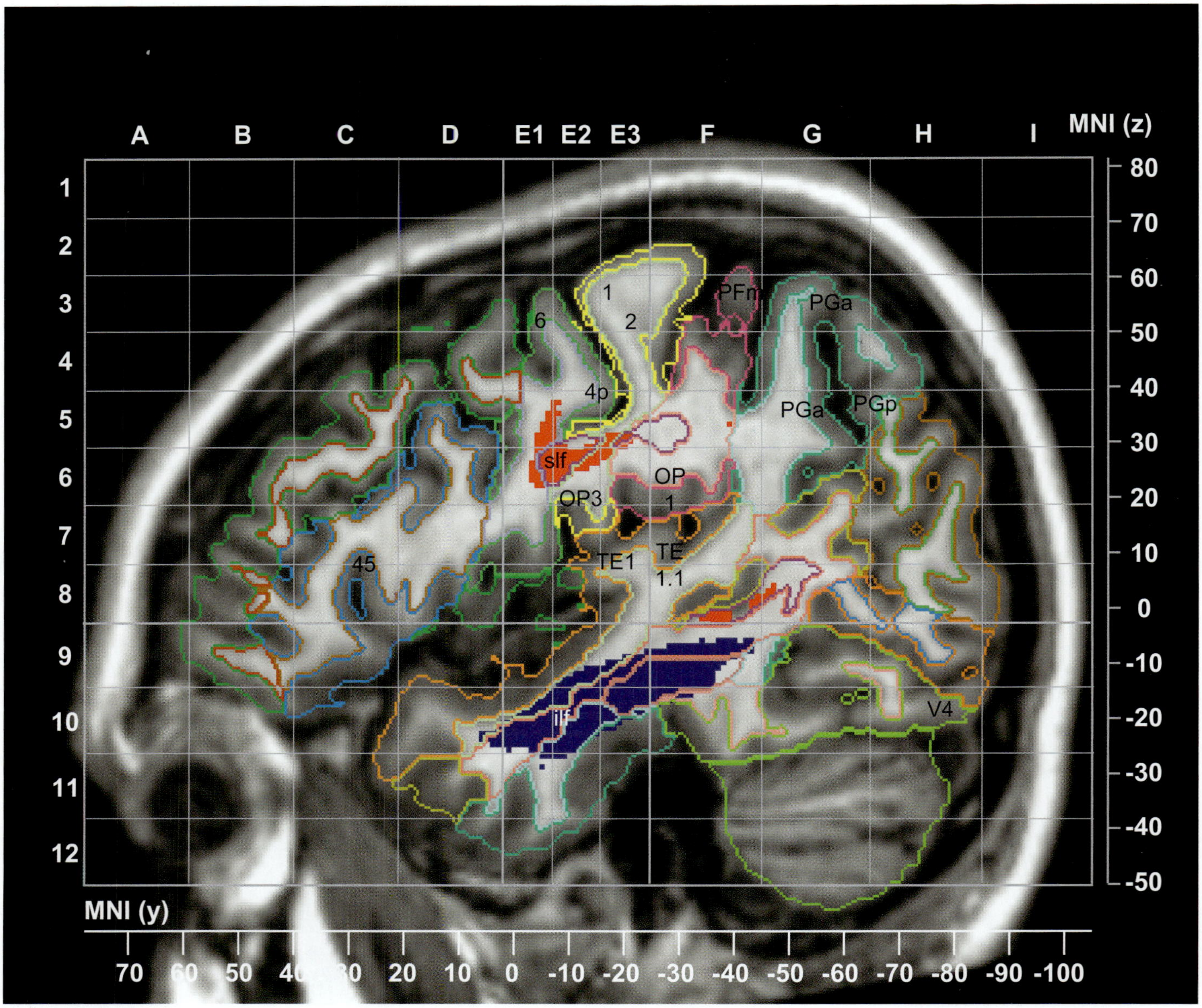

ilf - inferior longitudinal fasciculus
slf - superior longitudinal fasciculus

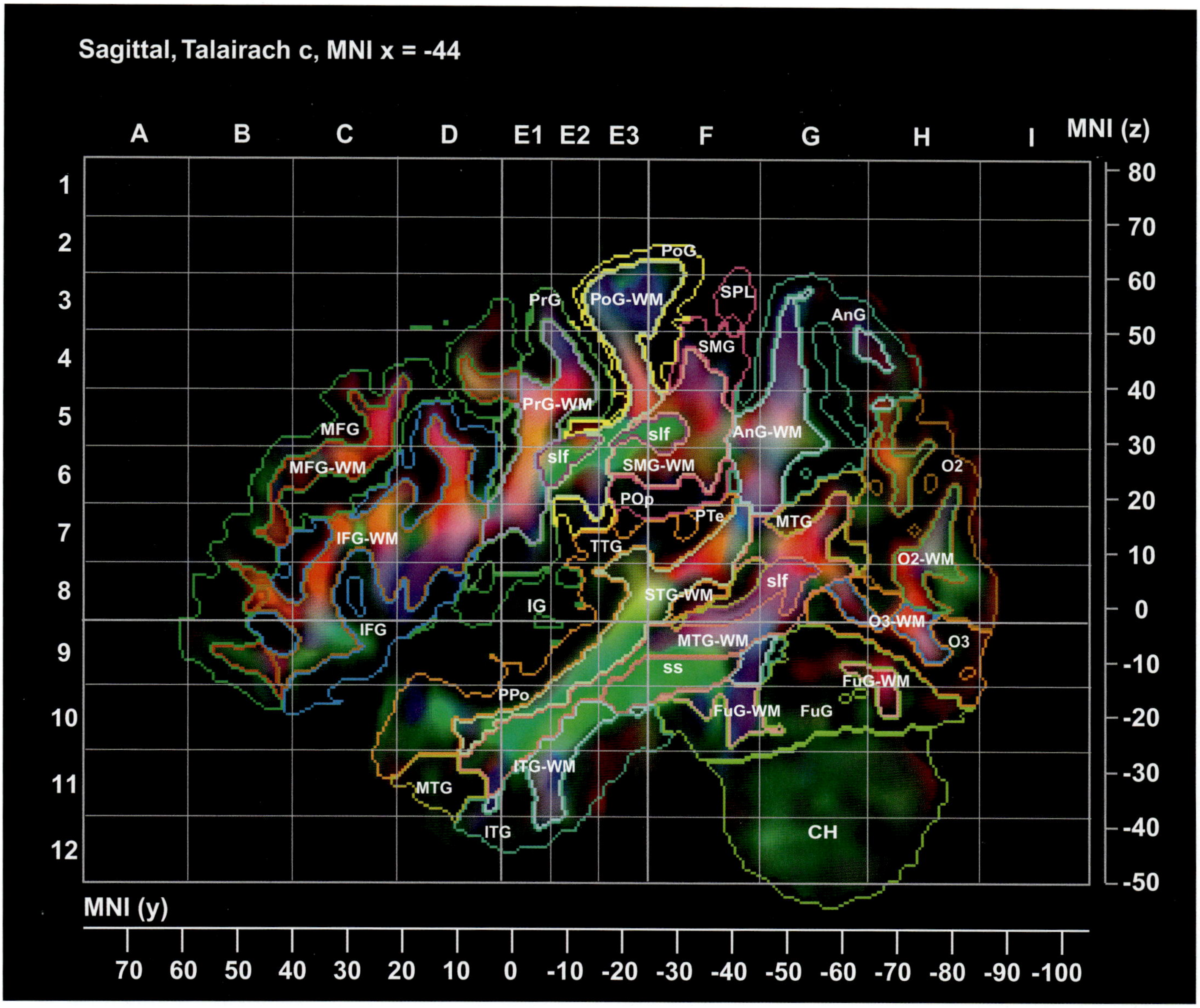

AnG - angular gyrus
AnG-WM - angular gyrus white matter
CH - cerebellar hemisphere
FuG - fusiform gyrus
FuG-WM - fusiform gyrus white matter
IFG - inferior frontal gyrus
IFG-WM - inferior frontal gyrus white matter
IG - insular gyrus
ITG - inferior temporal gyrus
ITG-WM - inferior temporal gyrus white matter
MFG - middle frontal gyrus
MFG-WM - middle frontal gyrus white matter
MTG - middle temporal gyrus
MTG-WM - middle temporal gyrus white matter
O2 - middle or lateral occipital gyrus
O2-WM - middle or lateral occipital gyrus white matter
O3 - inferior occipital gyrus
O3-WM - inferior occipital gyrus white matter
PoG - postcentral gyrus
PoG-WM - postcentral gyrus white matter
POp - parietal operculum
PPo - planum polare
PrG - precentral gyrus
PrG-WM - precentral gyrus white matter
PTe - planum temporale
slf - superior longitudinal fasciculus
SMG - supramarginal gyrus
SMG-WM - supramarginal gyrus white matter
SPL - superior parietal lobule
ss - sagittal stratum (include inferior longitidinal fasciculus and inferior fronto-occipital fasciculus)
STG-WM - superior temporal gyrus white matter
TTG - transverse temporal gyrus

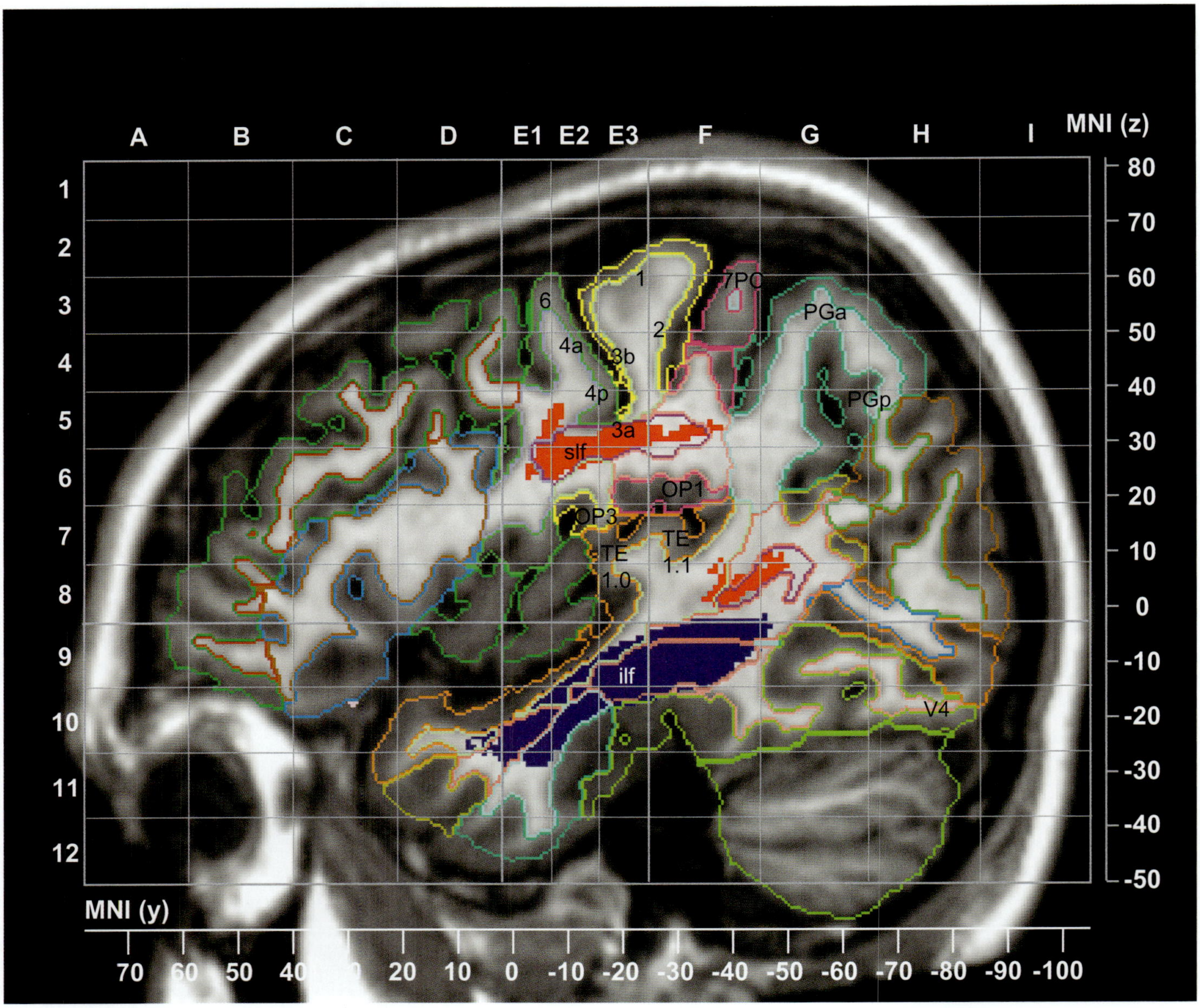

ilf - inferior longitudinal fasciculus
slf - superior longitudinal fasciculus

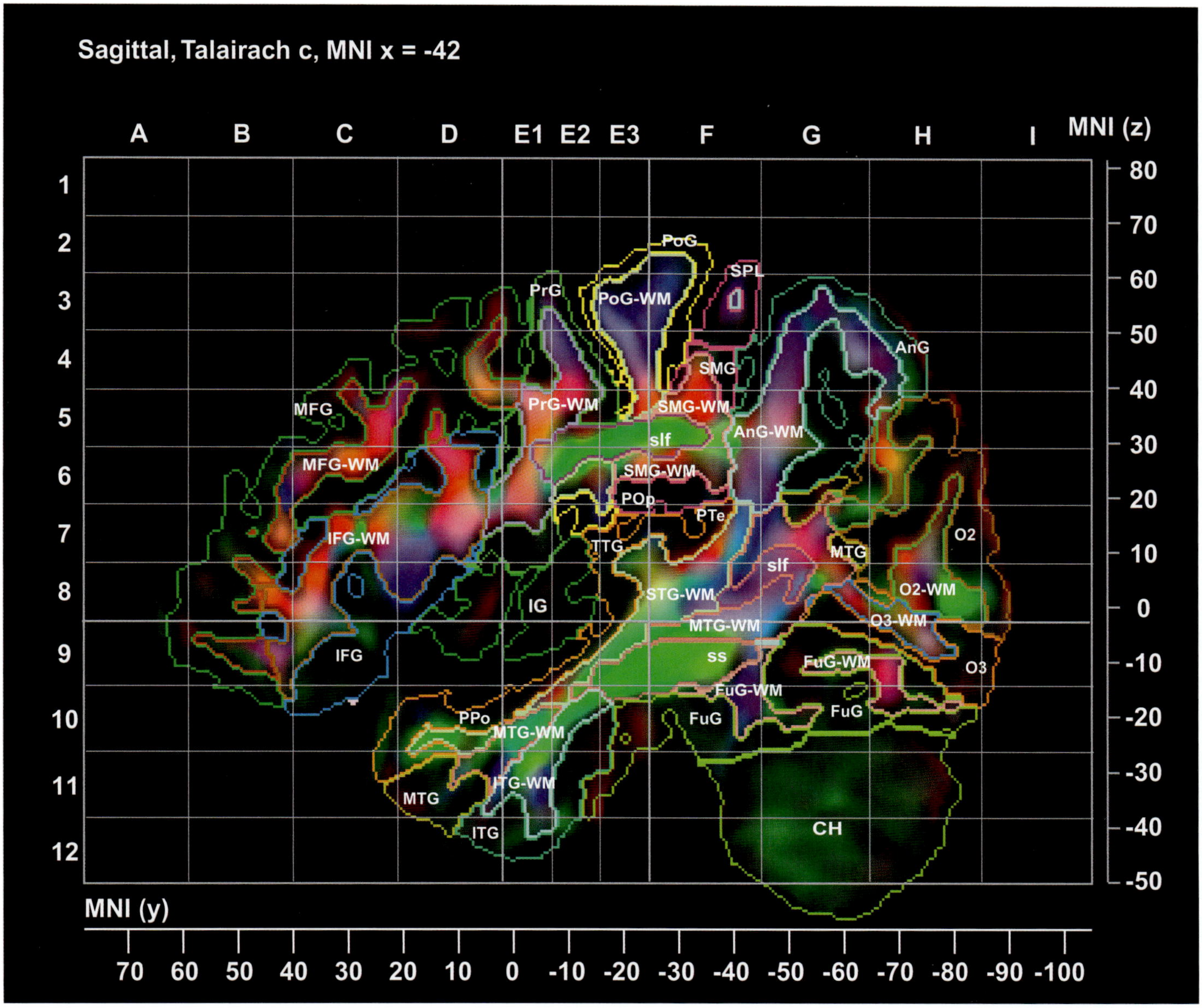

AnG - angular gyrus
AnG-WM - angular gyrus white matter
CH - cerebellar hemisphere
FuG - fusiform gyrus
FuG-WM - fusiform gyrus white matter
IFG - inferior frontal gyrus
IFG-WM - inferior frontal gyrus white matter
IG - insular gyrus
ITG - inferior temporal gyrus
ITG-WM - inferior temporal gyrus white matter
MFG - middle frontal gyrus
MFG-WM - middle frontal gyrus white matter
MTG - middle temporal gyrus
MTG-WM - middle temporal gyrus white matter
O2 - middle or lateral occipital gyrus
O2-WM - middle or lateral occipital gyrus white matter
O3 - inferior occipital gyrus
O3-WM - inferior occipital gyrus white matter
PoG - postcentral gyrus
PoG-WM - postcentral gyrus white matter
POp - parietal operculum
PPo - planum polare
PrG - precentral gyrus
PrG-WM - precentral gyrus white matter
PTe - planum temporale
slf - superior longitudinal fasciculus
SMG - supramarginal gyrus
SMG-WM - supramarginal gyrus white matter
SPL - superior parietal lobule
ss - sagittal stratum (include inferior longitidinal fasciculus and inferior fronto-occipital fasciculus)
STG-WM - superior temporal gyrus white matter
TTG - transverse temporal gyrus

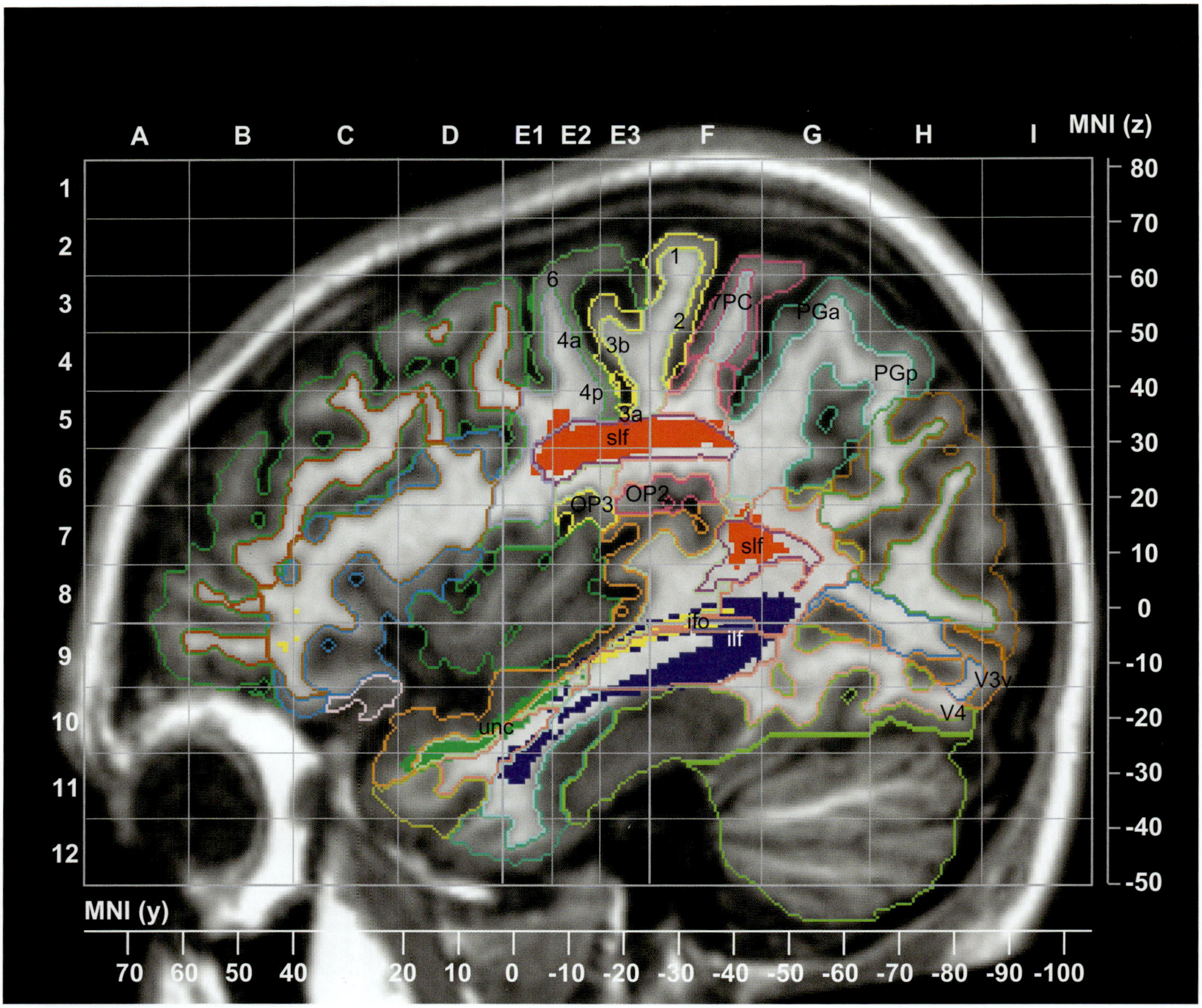

ilf - inferior longitudinal fasciculus
slf - superior longitudinal fasciculus
unc - uncinate fasciculus
ifo - inferior fronto-occipital fasciculus

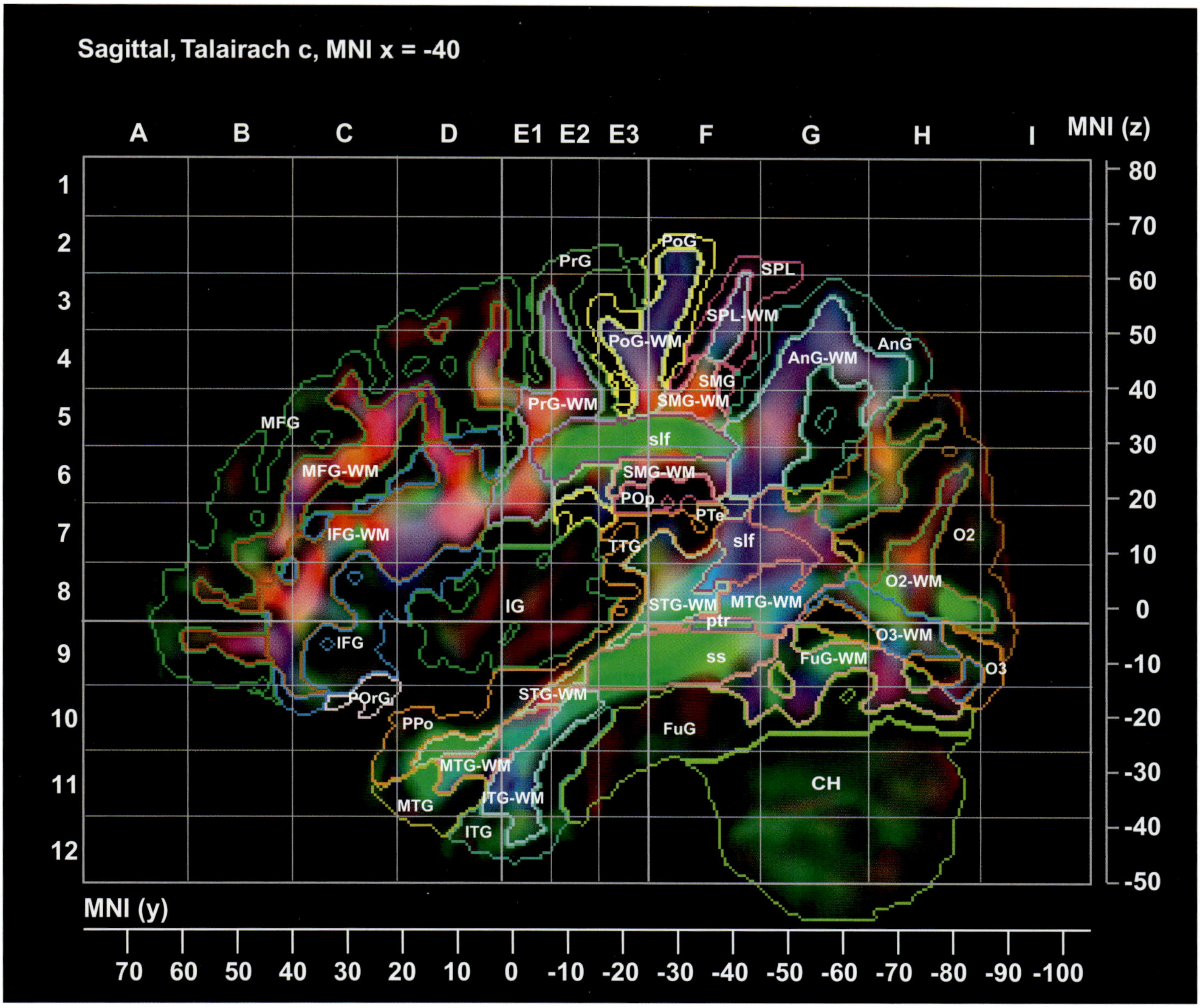

AnG - angular gyrus
AnG-WM - angular gyrus white matter
CH - cerebellar hemisphere
FuG - fusiform gyrus
FuG-WM - fusiform gyrus white matter
IFG - inferior frontal gyrus
IFG-WM - inferior frontal gyrus white matter
IG - insular gyrus
ITG - inferior temporal gyrus
ITG-WM - inferior temporal gyrus white matter
MFG - middle frontal gyrus
MFG-WM - middle frontal gyrus white matter
MTG - middle temporal gyrus
MTG-WM - middle temporal gyrus white matter
O2 - middle or lateral occipital gyrus
O2-WM - middle or lateral occipital gyrus white matter
O3 - inferior occipital gyrus
O3-WM - inferior occipital gyrus white matter
PoG - postcentral gyrus
PoG-WM - postcentral gyrus white matter
POp - parietal operculum
POrG - posterior orbital gyrus
PPo - planum polare
PrG - precentral gyrus
PrG-WM - precentral gyrus white matter
PTe - planum temporale
ptr - posterior thalamic radiation (include the optic radiation)
slf - superior longitudinal fasciculus
SMG - supramarginal gyrus
SMG-WM - supramarginal gyrus white matter
SPL - superior parietal lobule
SPL-WM - superior parietal lobule white matter
ss - sagittal stratum (include inferior longitidinal fasciculus and inferior fronto-occipital fasciculus)
STG-WM - superior temporal gyrus white matter
TTG - transverse temporal gyrus

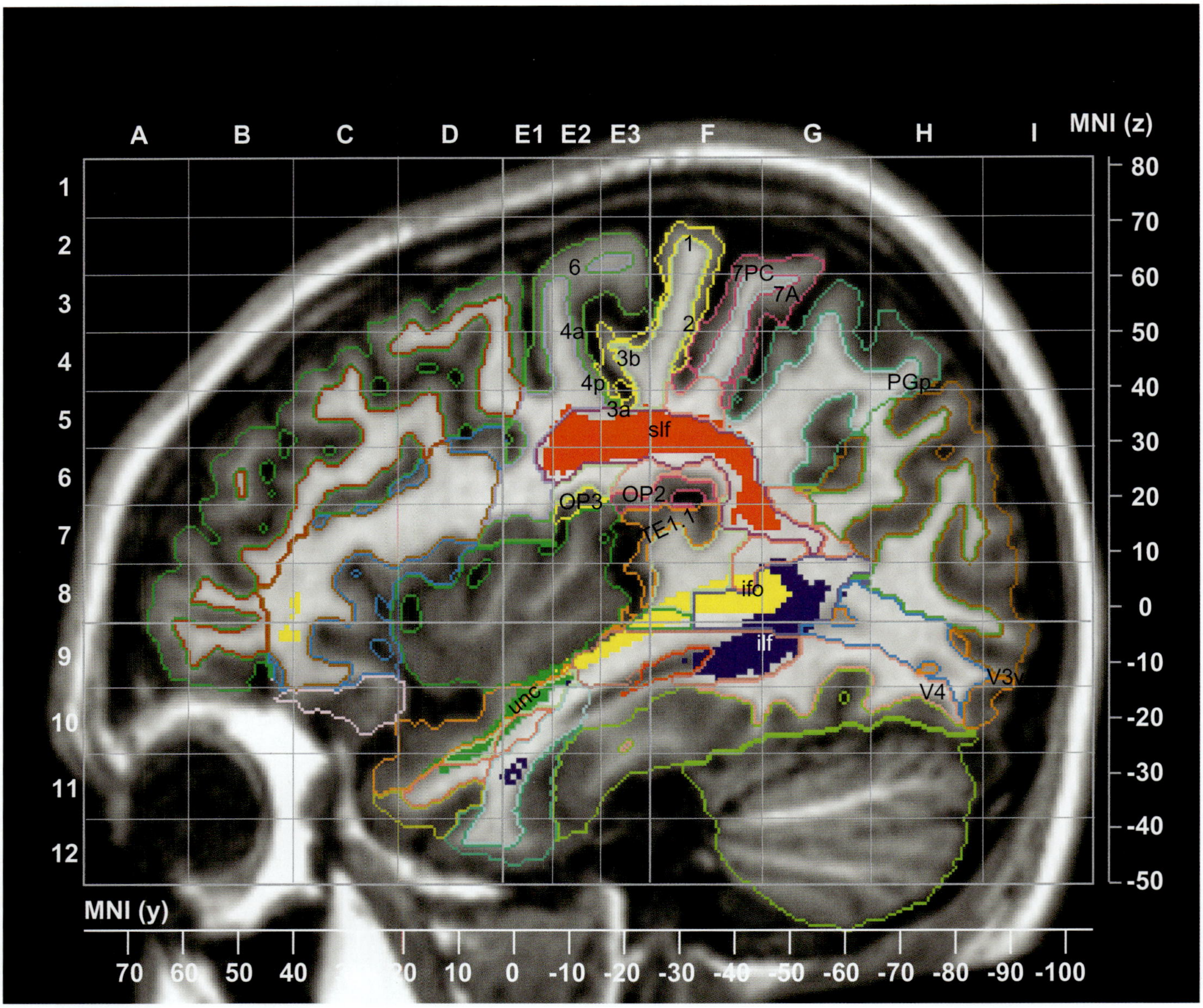

ifo - inferior fronto-occipital fasciculus
ilf - inferior longitudinal fasciculus
slf - superior longitudinal fasciculus
unc - uncinate fasciculus

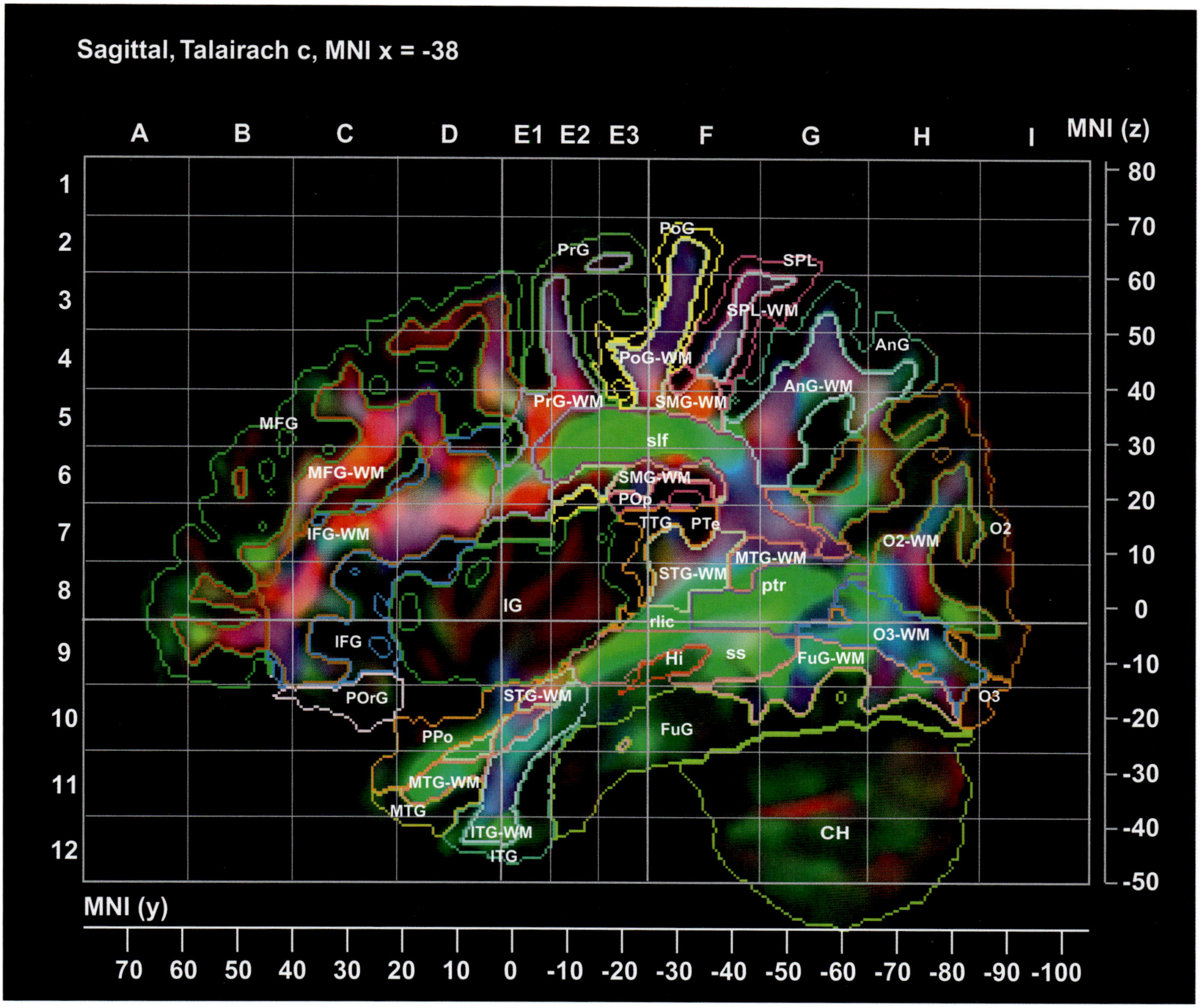

AnG - angular gyrus
AnG-WM - angular gyrus white matter
CH - cerebellar hemisphere
FuG - fusiform gyrus
FuG-WM - fusiform gyrus white matter
Hi - hippocampus
IFG - inferior frontal gyrus
IFG-WM - inferior frontal gyrus white matter
IG - insular gyrus
ITG - inferior temporal gyrus
ITG-WM - inferior temporal gyrus white matter
MFG - middle frontal gyrus
MFG-WM - middle frontal gyrus white matter
MTG - middle temporal gyrus
MTG-WM - middle temporal gyrus white matter
O2 - middle or lateral occipital gyrus
O2-WM - middle or lateral occipital gyrus white matter
O3 - inferior occipital gyrus
O3-WM - inferior occipital gyrus white matter
PoG - postcentral gyrus
PoG-WM - postcentral gyrus white matter
POp - parietal operculum
POrG - posterior orbital gyrus
PPo - planum polare
PrG - precentral gyrus
PrG-WM - precentral gyrus white matter
PTe - planum temporale
ptr - posterior thalamic radiation (include the optic radiation)
rlic - retrolenticular part of the internal capsule
slf - superior longitudinal fasciculus
SMG-WM - supramarginal gyrus white matter
SPL - superior parietal lobule
SPL-WM - superior parietal lobule white matter
ss - sagittal stratum (include inferior longitidinal fasciculus and inferior fronto-occipital fasciculus)
STG-WM - superior temporal gyrus white matter
TTG - transverse temporal gyrus

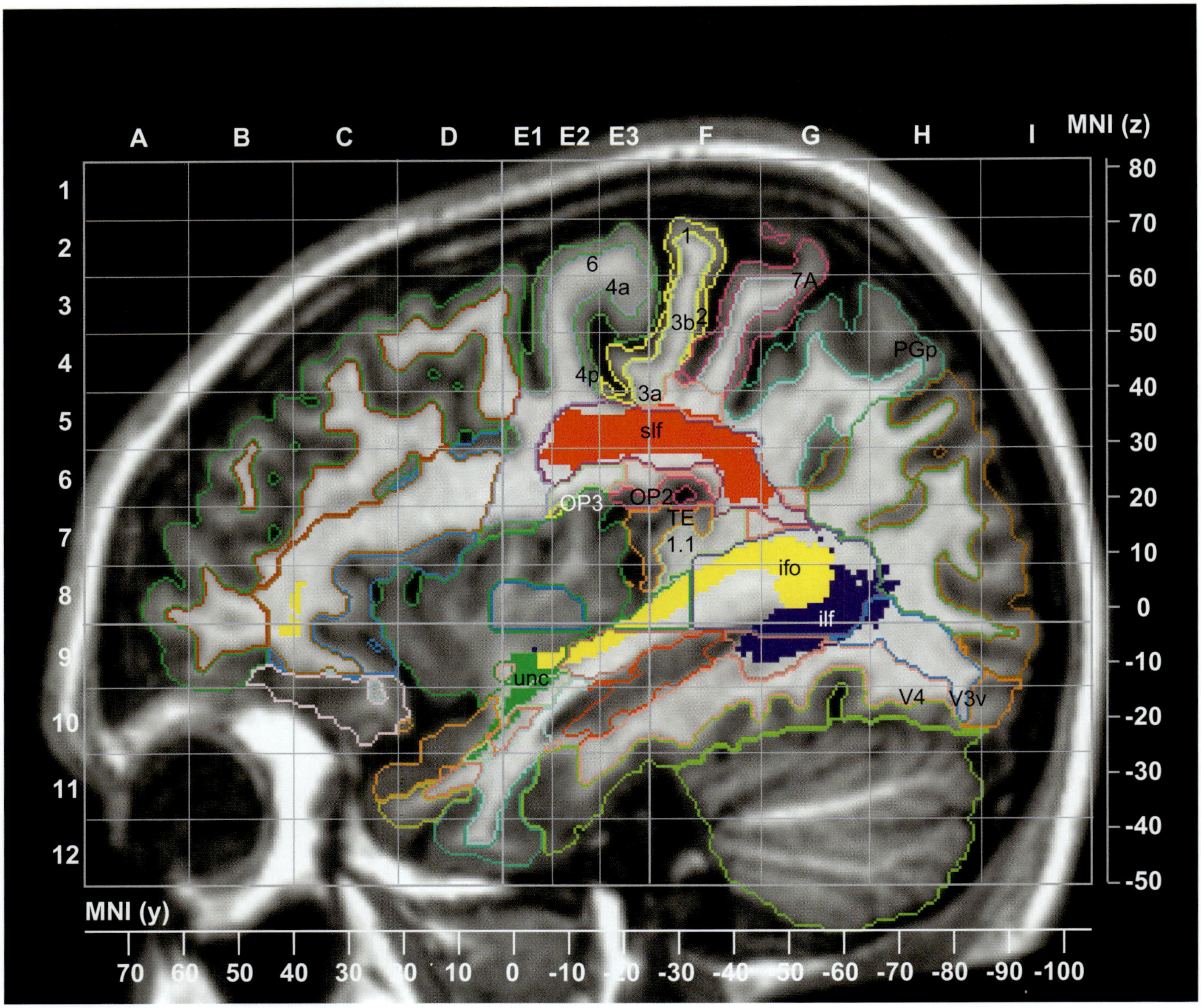

ifo - inferior fronto-occipital fasciculus
ilf - inferior longitudinal fasciculus
slf - superior longitudinal fasciculus
unc - uncinate fasciculus

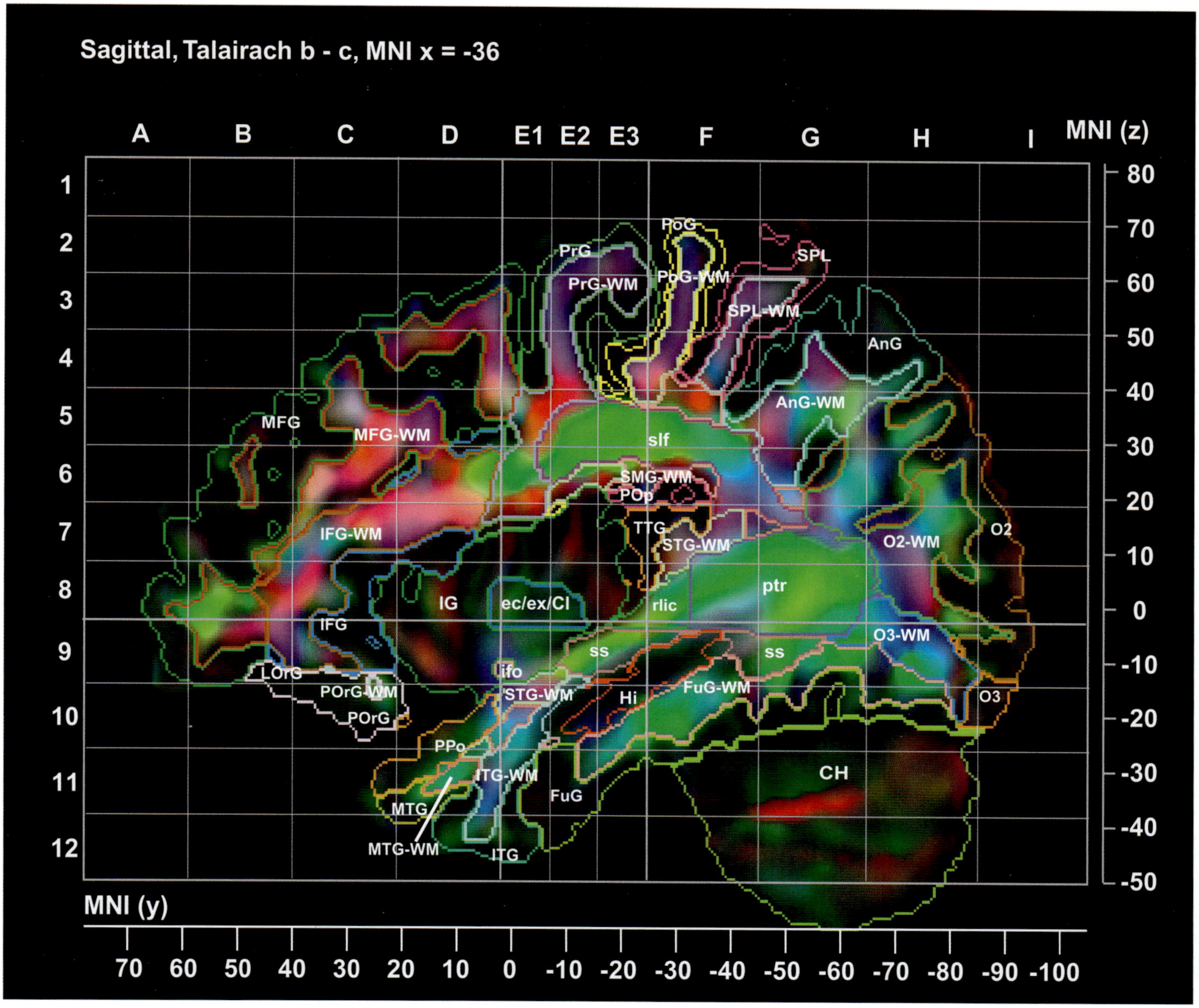

AnG - angular gyrus
AnG-WM - angular gyrus white matter
CH - cerebellar hemisphere
ec/ex/Cl - external capsule / extreme capsule / claustrum (cannot be resolved with current resolution)
FuG - fusiform gyrus
FuG-WM - fusiform gyrus white matter
Hi - hippocampus
IFG - inferior frontal gyrus
IFG-WM - inferior frontal gyrus white matter
ifo - inferior fronto-occipital fasciculus
IG - insular gyrus
ITG - inferior temporal gyrus
ITG-WM - inferior temporal gyrus white matter
LOrG - lateral orbital gyrus
MFG - middle frontal gyrus
MFG-WM - middle frontal gyrus white matter
MTG - middle temporal gyrus
MTG-WM - middle temporal gyrus white matter
O2 - middle or lateral occipital gyrus
O2-WM - middle or lateral occipital gyrus white matter
O3 - inferior occipital gyrus
O3-WM - inferior occipital gyrus white matter
PoG - postcentral gyrus
PoG-WM - postcentral gyrus white matter
POp - parietal operculum
POrG - posterior orbital gyrus
POrG-WM - posterior orbital gyrus white matter
PPo - planum polare
PrG - precentral gyrus
PrG-WM - precentral gyrus white matter
ptr - posterior thalamic radiation (include the optic radiation)
rlic - retrolenticular part of the internal capsule
slf - superior longitudinal fasciculus
SMG-WM - supramarginal gyrus white matter
SPL - superior parietal lobule
SPL-WM - superior parietal lobule white matter
ss - sagittal stratum (include inferior longitidinal fasciculus and inferior fronto-occipital fasciculus)
STG-WM - superior temporal gyrus white matter
TTG - transverse temporal gyrus

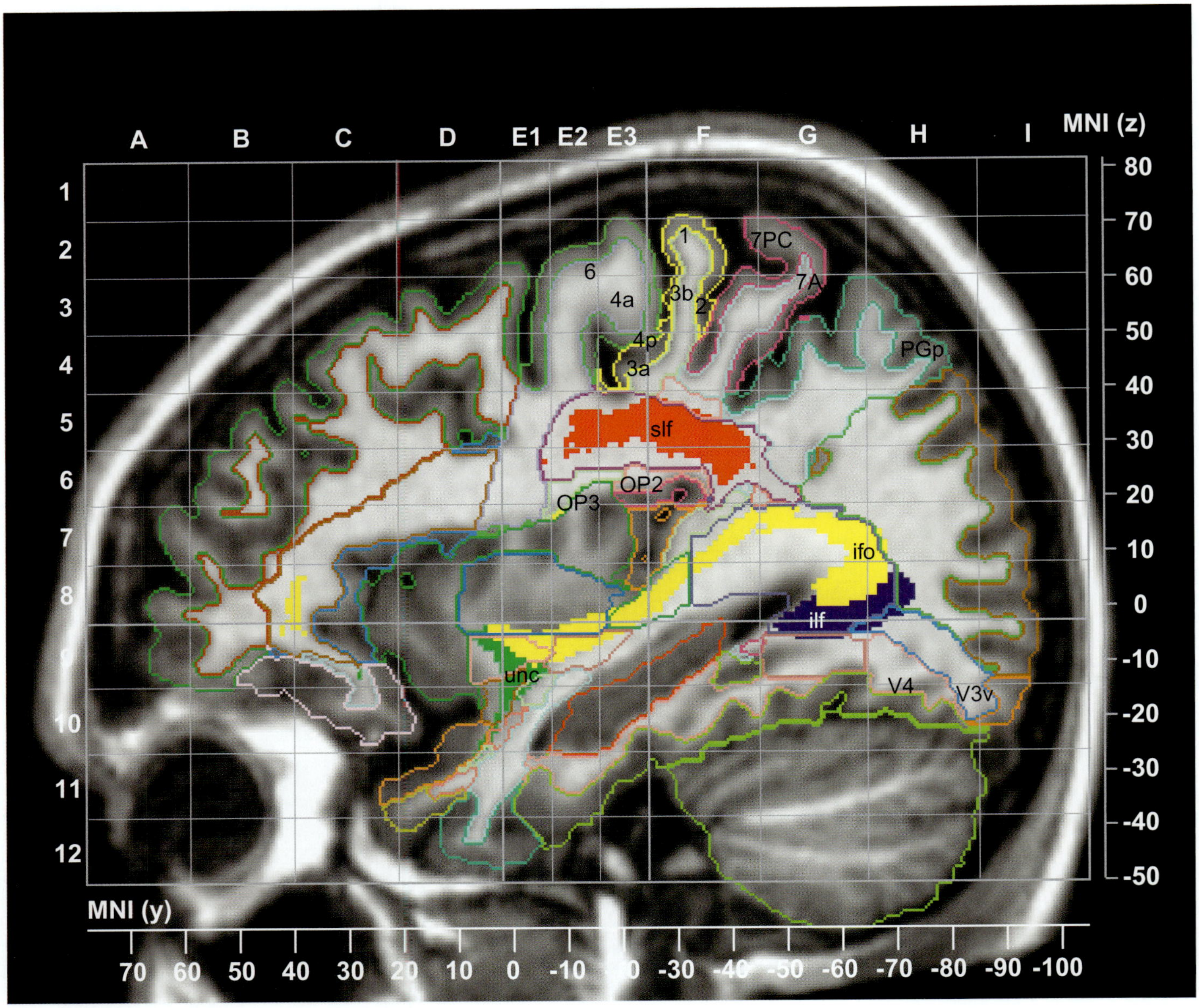

ifo - inferior fronto-occipital fasciculus
ilf - inferior longitudinal fasciculus
slf - superior longitudinal fasciculus
unc - uncinate fasciculus

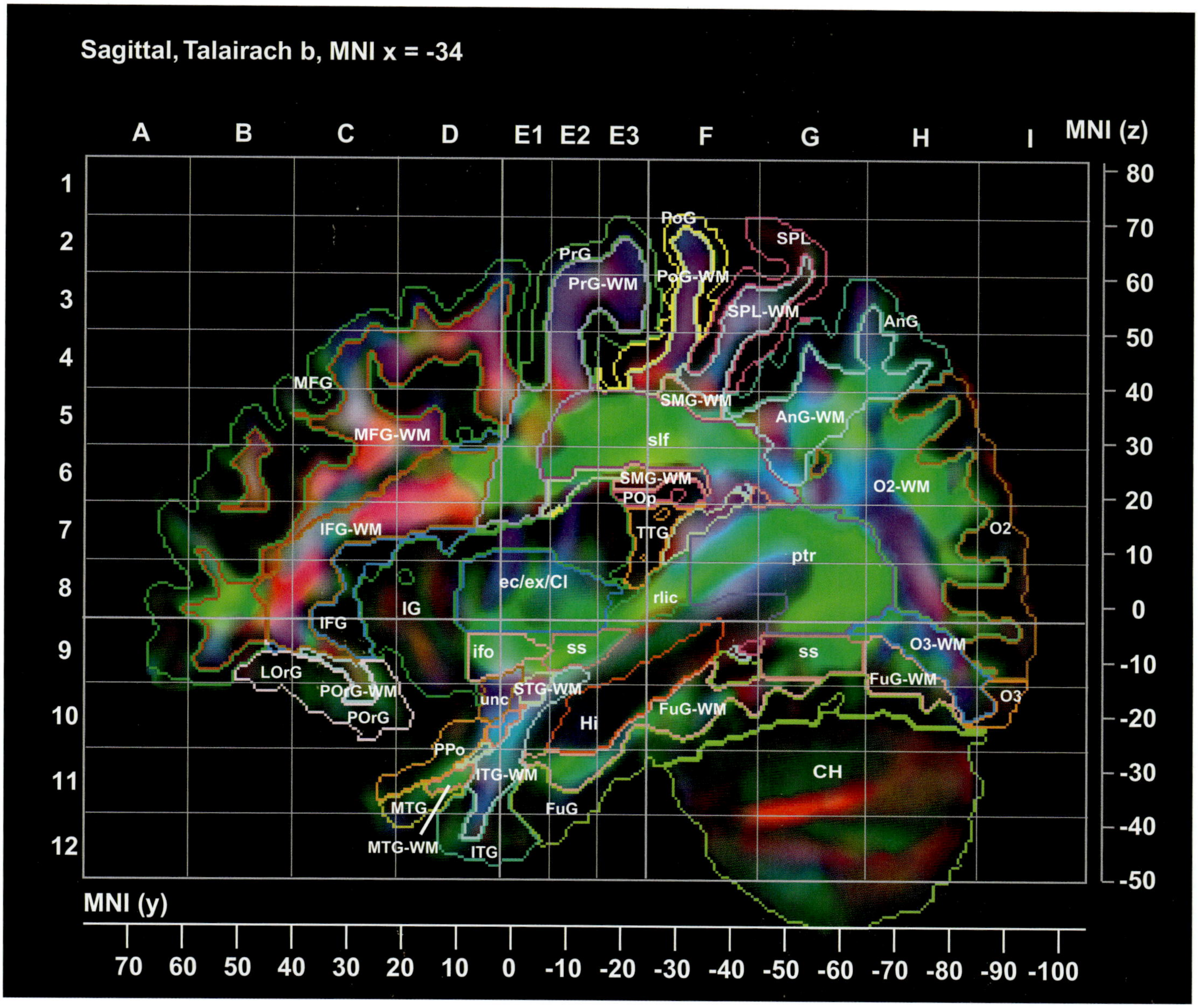

AnG - angular gyrus
AnG-WM - angular gyrus white matter
CH - cerebellar hemisphere
ec/ex/Cl - external capsule / extreme capsule / claustrum (cannot be resolved with current resolution)
FuG - fusiform gyrus
FuG-WM - fusiform gyrus white matter
Hi - hippocampus
IFG - inferior frontal gyrus
IFG-WM - inferior frontal gyrus white matter
ifo - inferior fronto-occipital fasciculus
IG - insular gyrus
ITG - inferior temporal gyrus
ITG-WM - inferior temporal gyrus white matter
LOrG - lateral orbital gyrus
MFG - middle frontal gyrus
MFG-WM - middle frontal gyrus white matter
MTG - middle temporal gyrus
MTG-WM - middle temporal gyrus white matter
O2 - middle or lateral occipital gyrus
O2-WM - middle or lateral occipital gyrus white matter
O3 - inferior occipital gyrus
O3-WM - inferior occipital gyrus white matter
PoG - postcentral gyrus
PoG-WM - postcentral gyrus white matter
POp - parietal operculum
POrG - posterior orbital gyrus
POrG-WM - posterior orbital gyrus white matter
PPo - planum polare
PrG - precentral gyrus
PrG-WM - precentral gyrus white matter
ptr - posterior thalamic radiation (include the optic radiation)
rlic - retrolenticular part of the internal capsule
slf - superior longitudinal fasciculus
SMG-WM - supramarginal gyrus white matter
SPL - superior parietal lobule
SPL-WM - superior parietal lobule white matter
ss - sagittal stratum (include inferior longitidinal fasciculus and inferior fronto-occipital fasciculus)
STG-WM - superior temporal gyrus white matter
TTG - transverse temporal gyrus
unc - uncinate fasciculus

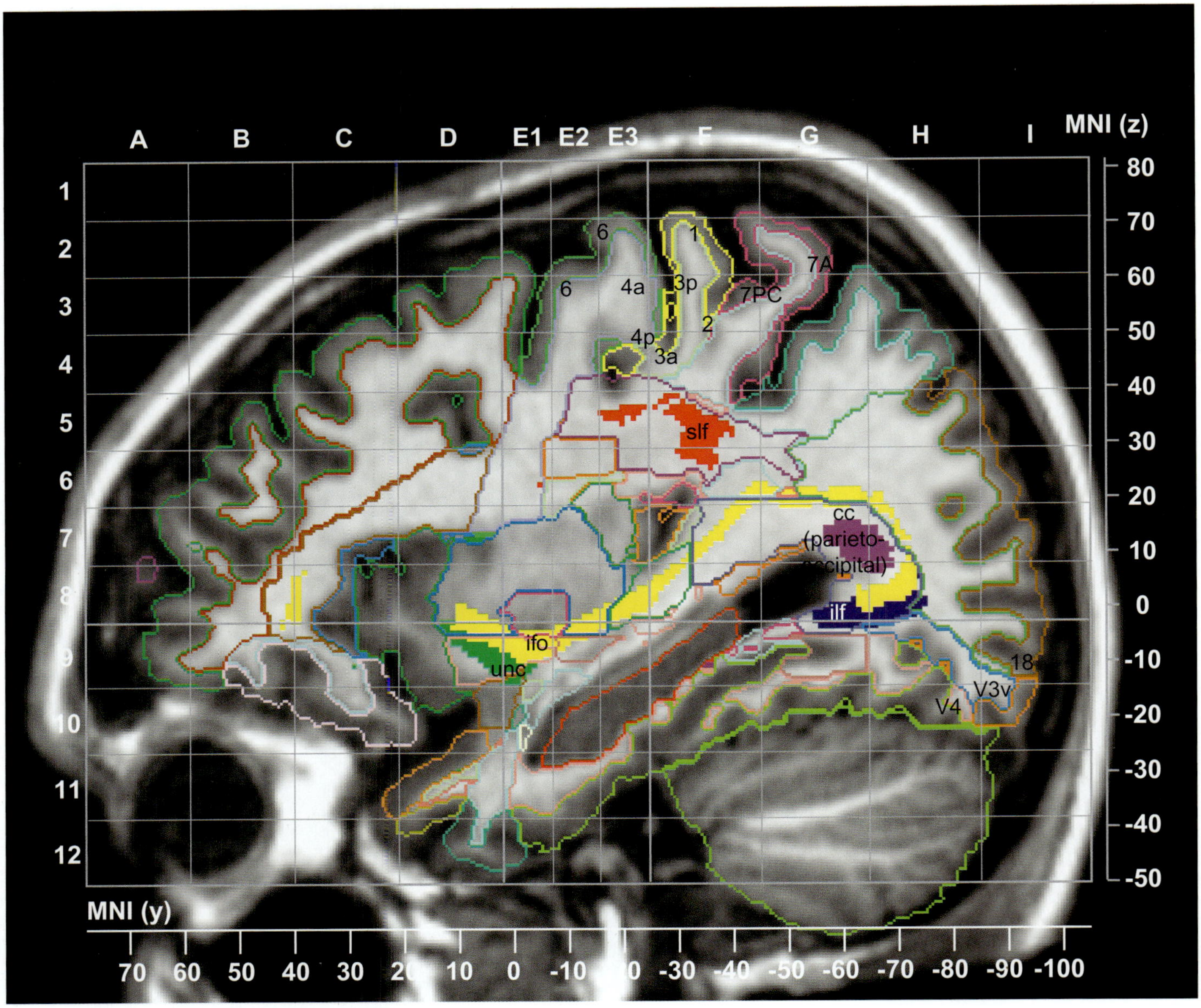

cc (parieto-occipital) - corpus callosum (connecting bilateral parieto-occipital areas)
ifo - inferior fronto-occipital fasciculus
ilf - inferior longitudinal fasciculus
slf - superior longitudinal fasciculus
unc - uncinate fasciculus

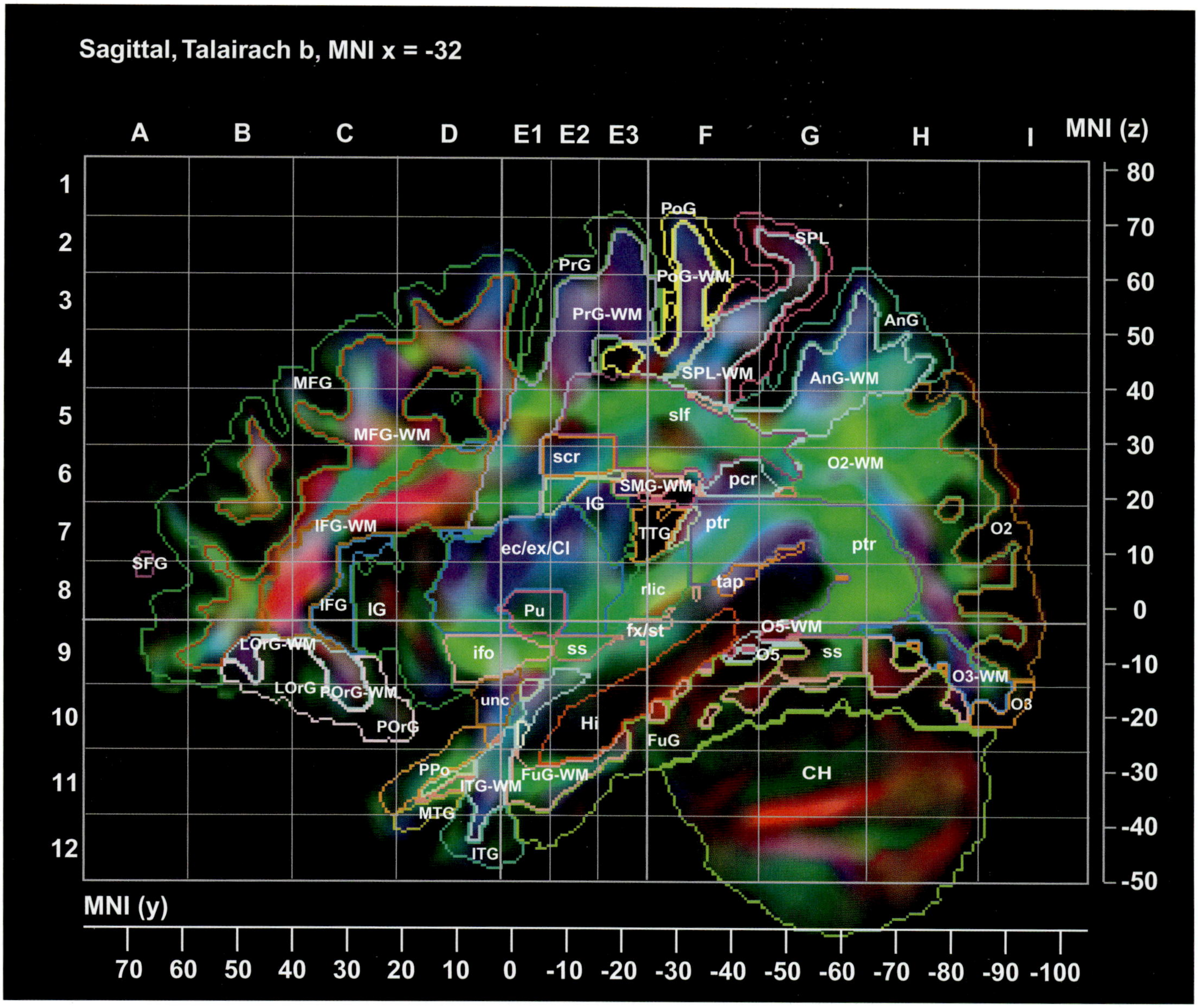

AnG - angular gyrus
AnG-WM - angular gyrus white matter
CH - cerebellar hemisphere
ec/ex/Cl - external capsule / extreme capsule / claustrum (cannot be resolved with current resolution)
FuG - fusiform gyrus
FuG-WM - fusiform gyrus white matter
Hi - hippocampus
IFG - inferior frontal gyrus
IFG-WM - inferior frontal gyrus white matter
ifo - inferior fronto-occipital fasciculus
IG - insular gyrus
ITG - inferior temporal gyrus
ITG-WM - inferior temporal gyrus white matter
LOrG - lateral orbital gyrus
LOrG-WM - lateral orbital gyrus white matter
MFG - middle frontal gyrus
MFG-WM - middle frontal gyrus white matter
MTG - middle temporal gyrus
O2 - middle or lateral occipital gyrus
O2-WM - middle or lateral occipital gyrus white matter
O3 - inferior occipital gyrus
O3-WM - inferior occipital gyrus white matter
PoG - postcentral gyrus
PoG-WM - postcentral gyrus white matter
POrG - posterior orbital gyrus
POrG-WM - posterior orbital gyrus white matter
PPo - planum polare
PrG - precentral gyrus
PrG-WM - precentral gyrus white matter
ptr - posterior thalamic radiation (include the optic radiation)
rlic - retrolenticular part of the internal capsule
slf - superior longitudinal fasciculus
SMG-WM - supramarginal gyrus white matter
SPL - superior parietal lobule
SPL-WM - superior parietal lobule white matter
ss - sagittal stratum (include inferior longitidinal fasciculus and inferior fronto-occipital fasciculus)
TTG - transverse temporal gyrus
unc - uncinate fasciculus

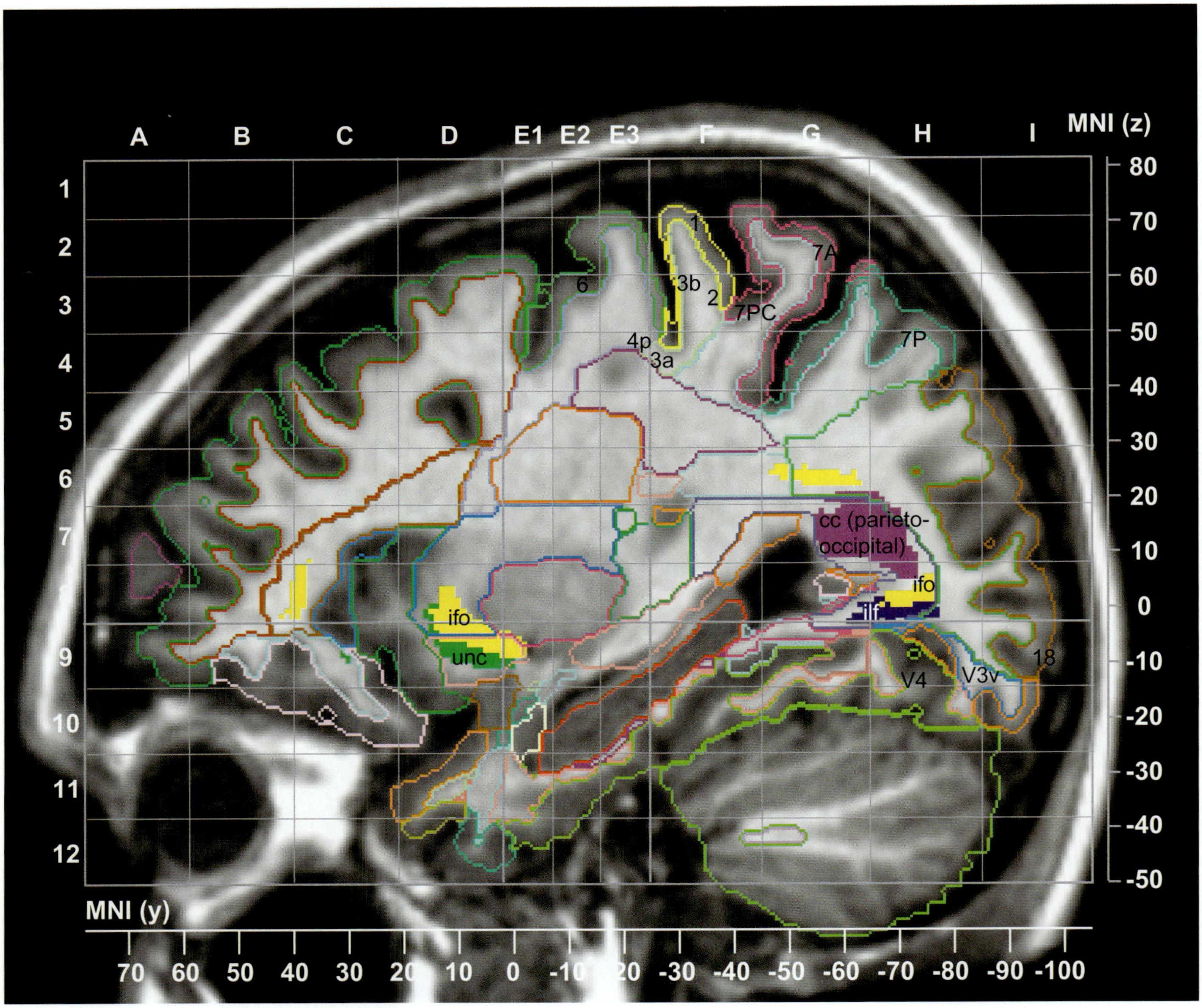

cc (parieto-occipital) - corpus callosum (connecting bilateral parieto-occipital areas)

ifo - inferior fronto-occipital fasciculus

ilf - inferior longitudinal fasciculus

unc - uncinate fasciculus

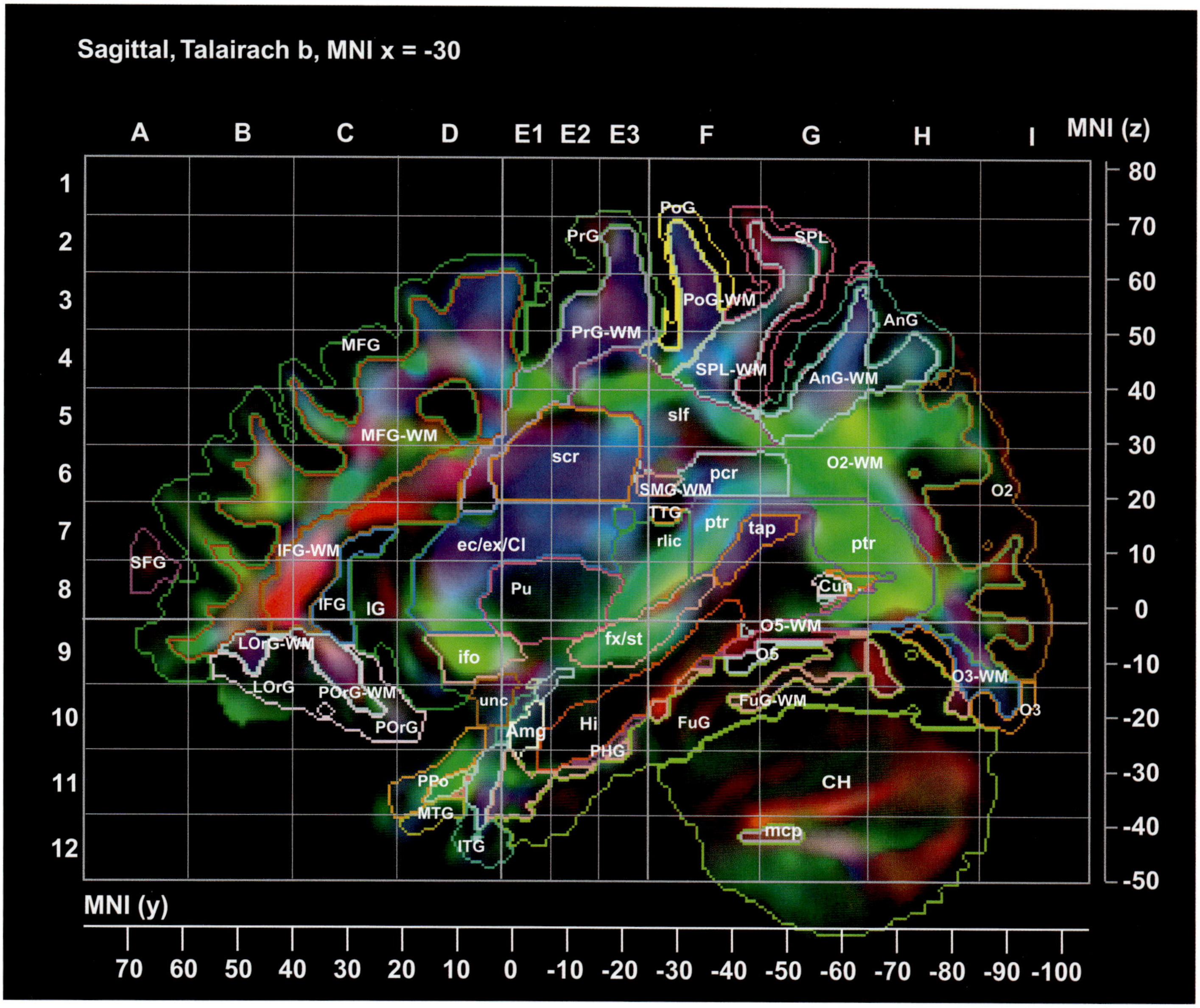

AnG - angular gyrus
AnG-WM - angular gyrus white matter
CH - cerebellar hemisphere
Cun - cuneus
ec/ex/Cl - external capsule / extreme capsule / claustrum (cannot be resolved with current resolution)
FuG - fusiform gyrus
FuG-WM - fusiform gyrus white matter
fx/st - fornix (crus) / stria terminalis (cannot be resolved with current resolution)
Hi - hippocampus
IFG - inferior frontal gyrus
IFG-WM - inferior frontal gyrus white matter
ifo - inferior fronto-occipital fasciculus
IG - insular gyrus
ITG - inferior temporal gyrus
LOrG - lateral orbital gyrus
LOrG-WM - lateral orbital gyrus white matter
mcp - middle cerebellar peduncle
MFG - middle frontal gyrus
MFG-WM - middle frontal gyrus white matter
MTG - middle temporal gyrus
O2 - middle or lateral occipital gyrus
O2-WM - middle or lateral occipital gyrus white matter
O3 - inferior occipital gyrus
O3-WM - inferior occipital gyrus white matter
O5 - lyngual gyrus
O5-WM - lyngual gyrus white matter
pcr - posterior corona radiata
PHG - parahippocampal gyrus
PoG - postcentral gyrus
PoG-WM - postcentral gyrus white matter
POrG - posterior orbital gyrus
POrG-WM - posterior orbital gyrus white matter
PPo - planum polare
PrG - precentral gyrus
PrG-WM - precentral gyrus white matter
ptr - posterior thalamic radiation (include the optic radiation)
Pu - putamen
rlic - retrolenticular part of the internal capsule
scr - superior corona radiata
SFG - superior frontal gyrus
slf - superior longitudinal fasciculus
SMG-WM - supramarginal gyrus white matter
SPL - superior parietal lobule
SPL-WM - superior parietal lobule white matter
tap - tapetum of the corpus callosum
TTG - transverse temporal gyrus
unc - uncinate fasciculus

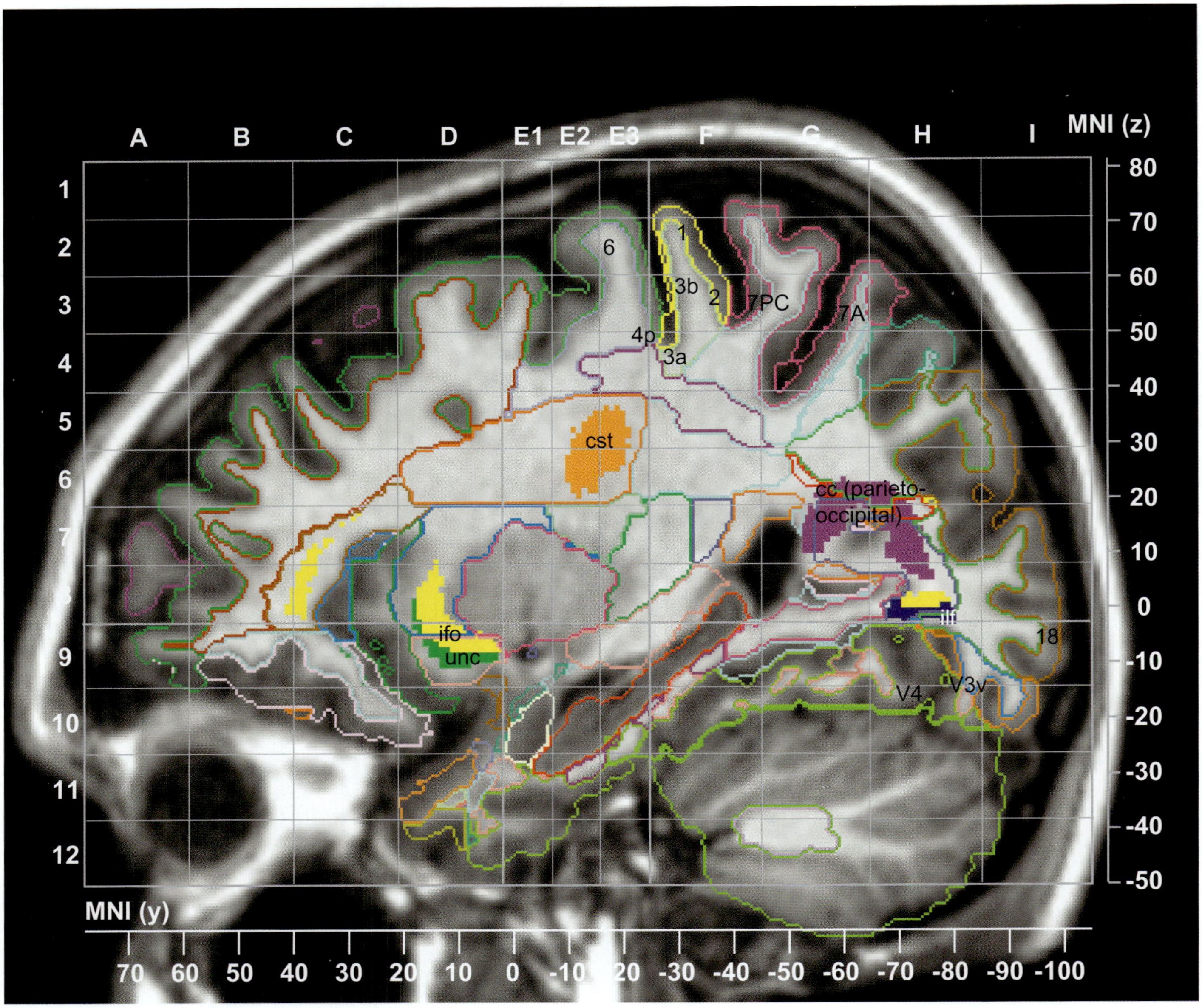

cc (parieto-occipital) - corpus callosum (connecting bilateral parieto-occipital areas)
cst - corticospinal tract
ifo - inferior fronto-occipital fasciculus
ilf - inferior longitudinal fasciculus
unc - uncinate fasciculus

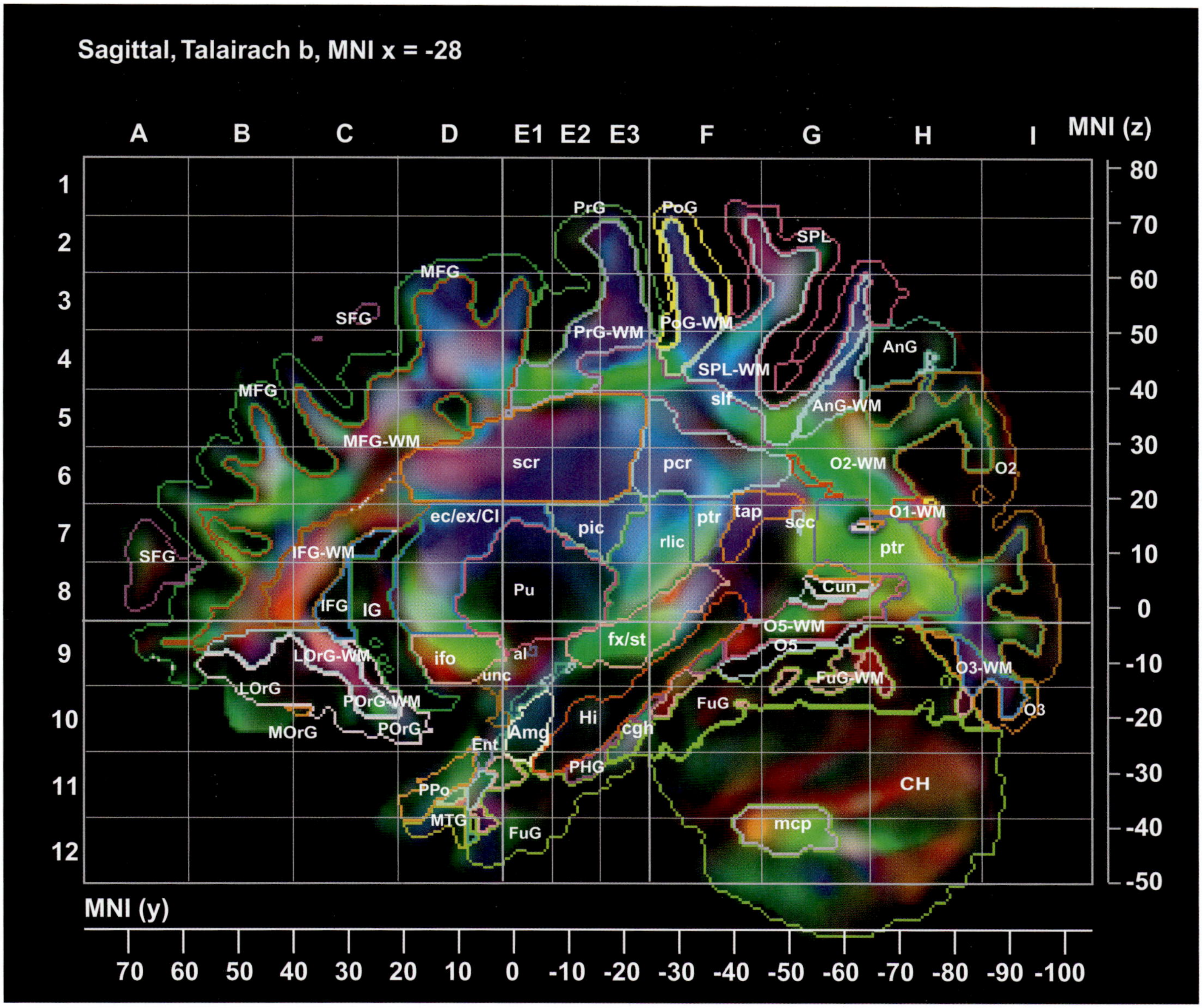

al	-	ansa lenticularis
Amg	-	amygdala
AnG	-	angular gyrus
AnG-WM	-	angular gyrus white matter
cgh	-	cingulum (hippocampal part)
CH	-	cerebellar hemisphere
Cun	-	cuneus
ec/ex/Cl	-	external capsule / extreme capsule / claustrum (cannot be resolved with current resolution)
Ent	-	entorhinal cortex
FuG	-	fusiform gyrus
FuG-WM	-	fusiform gyrus white matter
fx/st	-	fornix (crus) / stria terminalis (cannot be resolved with current resolution)
Hi	-	hippocampus
IFG	-	inferior frontal gyrus
IFG-WM	-	inferior frontal gyrus white matter
ifo	-	inferior fronto-occipital fasciculus
IG	-	insular gyrus
LOrG	-	lateral orbital gyrus
LOrG-WM	-	lateral orbital gyrus white matter
mcp	-	middle cerebellar peduncle
MFG	-	middle frontal gyrus
MFG-WM	-	middle frontal gyrus white matter
MOrG	-	medial orbital gyrus
MTG	-	middle temporal gyrus
O1	-	superior occipital gyrus
O1-WM	-	superior occipital gyrus white matter
O2	-	middle or lateral occipital gyrus
O2-WM	-	middle or lateral occipital gyrus white matter
O3	-	inferior occipital gyrus
O3-WM	-	inferior occipital gyrus white matter
O5	-	lyngual gyrus
O5-WM	-	lyngual gyrus white matter
pcr	-	posterior corona radiata
PHG	-	parahippocampal gyrus
pic	-	posterior limb of the internal capsule
PoG	-	postcentral gyrus
PoG-WM	-	postcentral gyrus white matter
POrG	-	posterior orbital gyrus
POrG-WM	-	posterior orbital gyrus white matter
PPo	-	planum polare
PrG	-	precentral gyrus
PrG-WM	-	precentral gyrus white matter
ptr	-	posterior thalamic radiation (include the optic radiation)
Pu	-	putamen
rlic	-	retrolenticular part of the internal capsule
scc	-	splenium of the corpus callosum
scr	-	superior corona radiata
SFG	-	superior frontal gyrus
slf	-	superior longitudinal fasciculus
SPL	-	superior parietal lobule
SPL-WM	-	superior parietal lobule white matter
tap	-	tapetum of the corpus callosum
unc	-	uncinate fasciculus

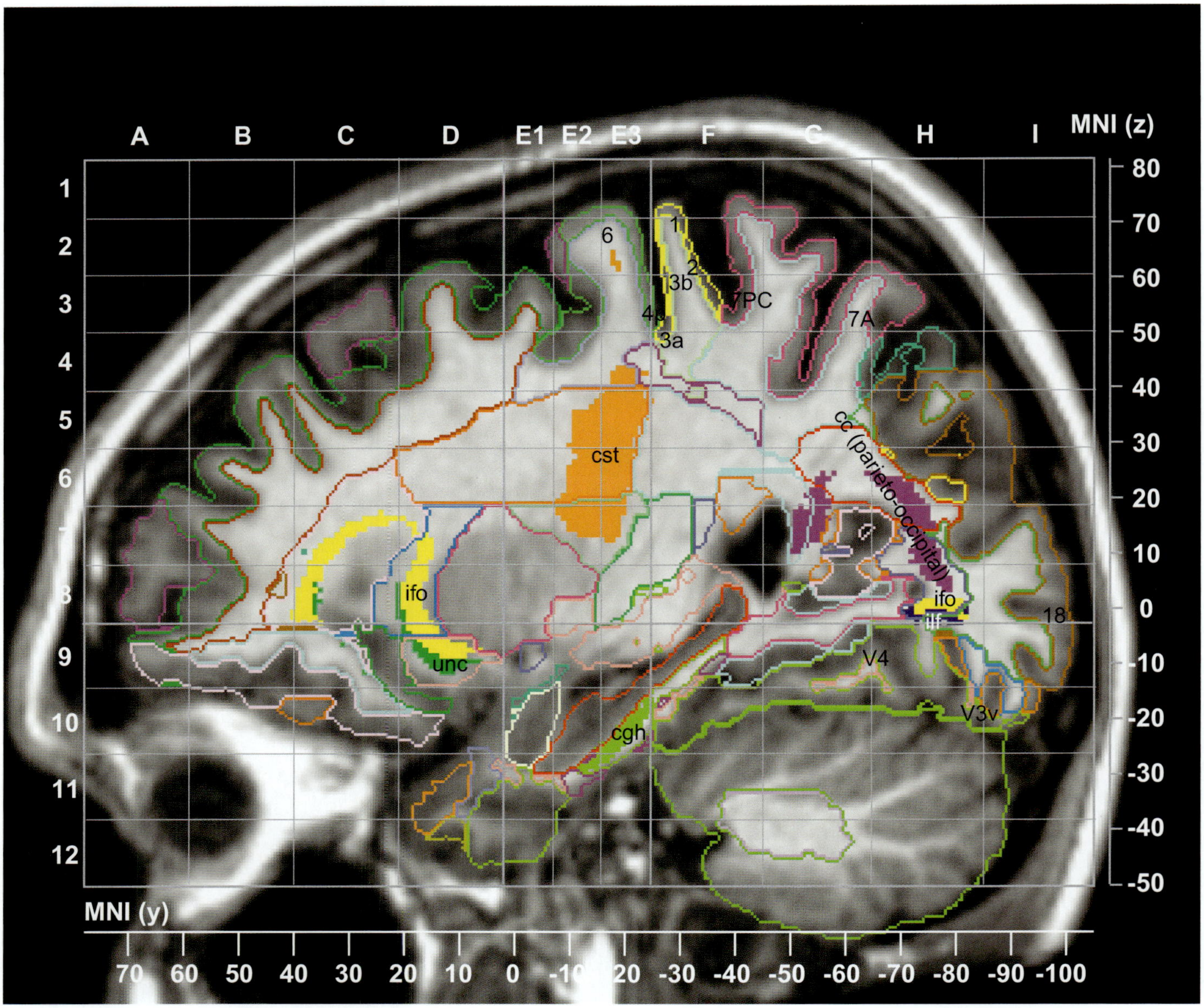

cc (parieto-occipital) - corpus callosum (connecting bilateral parieto-occipital areas)
cgh - cingulum (hippocampal part)
cst - corticospinal tract
ifo - inferior fronto-occipital fasciculus
ilf - inferior longitudinal fasciculus
unc - uncinate fasciculus

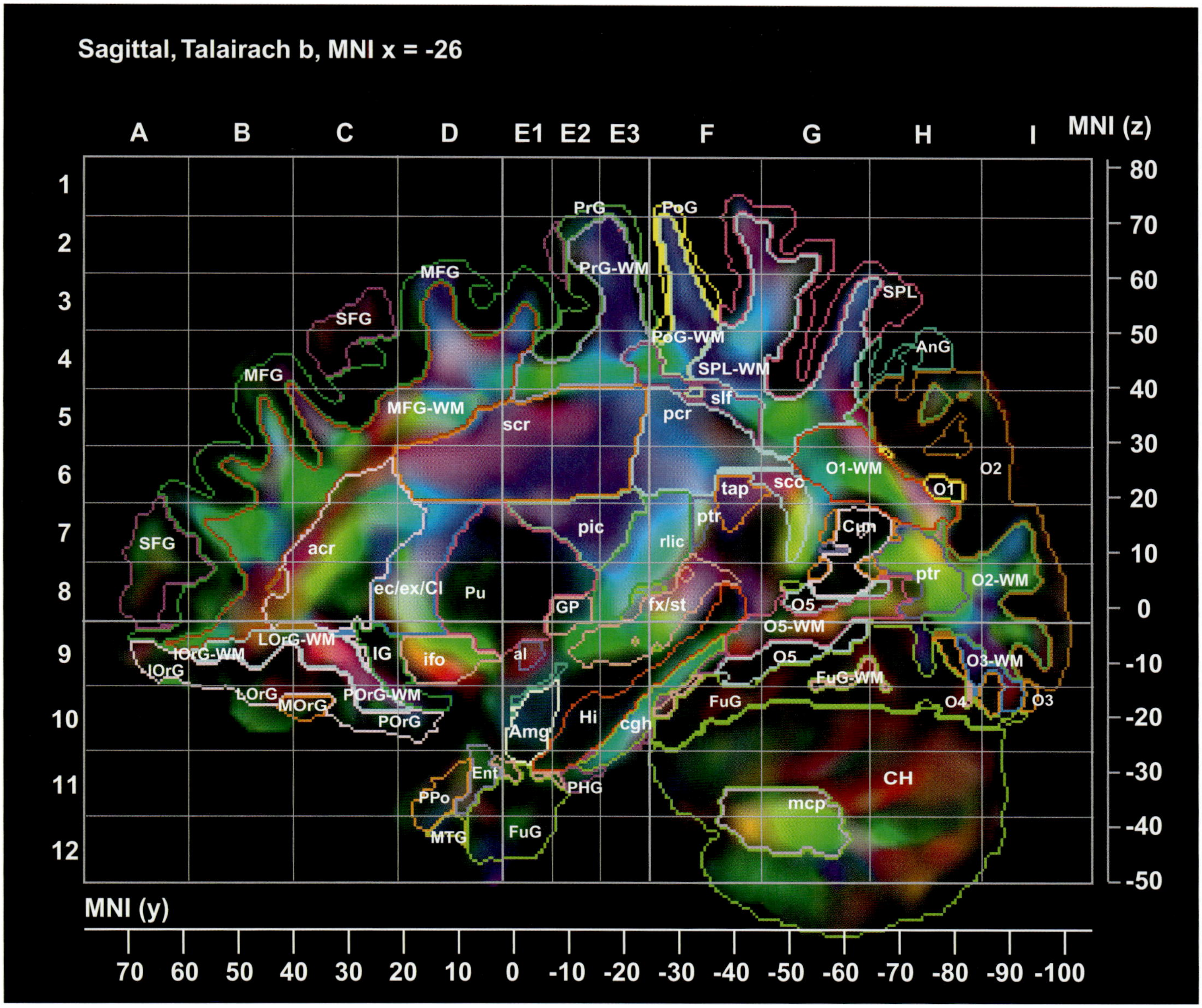

acr - anterior corona radiata
al - ansa lenticularis
Amg - amygdala
AnG - angular gyrus
cgh - cingulum (hippocampal part)
CH - cerebellar hemisphere
Cun - cuneus
ec/ex/Cl - external capsule / extreme capsule / claustrum (cannot be resolved with current resolution)
Ent - entorhinal cortex
FuG - fusiform gyrus
FuG-WM - fusiform gyrus white matter
fx/st - fornix (crus) / stria terminalis (cannot be resolved with current resolution)
GP - globus pallidus
Hi - hippocampus
ifo - inferior fronto-occipital fasciculus
IG - insular gyrus
IOrG - intermediate orbital gyrus
IOrG-WM - intermediate orbital gyrus white matter
LOrG - lateral orbital gyrus
LOrG-WM - lateral orbital gyrus white matter
mcp - middle cerebellar peduncle
MFG - middle frontal gyrus
MFG-WM - middle frontal gyrus white matter
MOrG - medial orbital gyrus
MTG - middle temporal gyrus
O1 - superior occipital gyrus
O1-WM - superior occipital gyrus white matter
O2 - middle or lateral occipital gyrus
O2-WM - middle or lateral occipital gyrus white matter
O3 - inferior occipital gyrus
O3-WM - inferior occipital gyrus white matter
O4 - fourth occipital gyrus (part of the fusiform gyrus)
O5 - lyngual gyrus
O5-WM - lyngual gyrus white matter
pcr - posterior corona radiata
PHG - parahippocampal gyrus
pic - posterior limb of the internal capsule
PoG - postcentral gyrus
PoG-WM - postcentral gyrus white matter
POrG - posterior orbital gyrus
POrG-WM - posterior orbital gyrus white matter
PPo - planum polare
PrG - precentral gyrus
PrG-WM - precentral gyrus white matter
ptr - posterior thalamic radiation (include the optic radiation)
Pu - putamen
rlic - retrolenticular part of the internal capsule
scc - splenium of the corpus callosum
scr - superior corona radiata
SFG - superior frontal gyrus
slf - superior longitudinal fasciculus
SPL - superior parietal lobule
SPL-WM - superior parietal lobule white matter
tap - tapetum of the corpus callosum

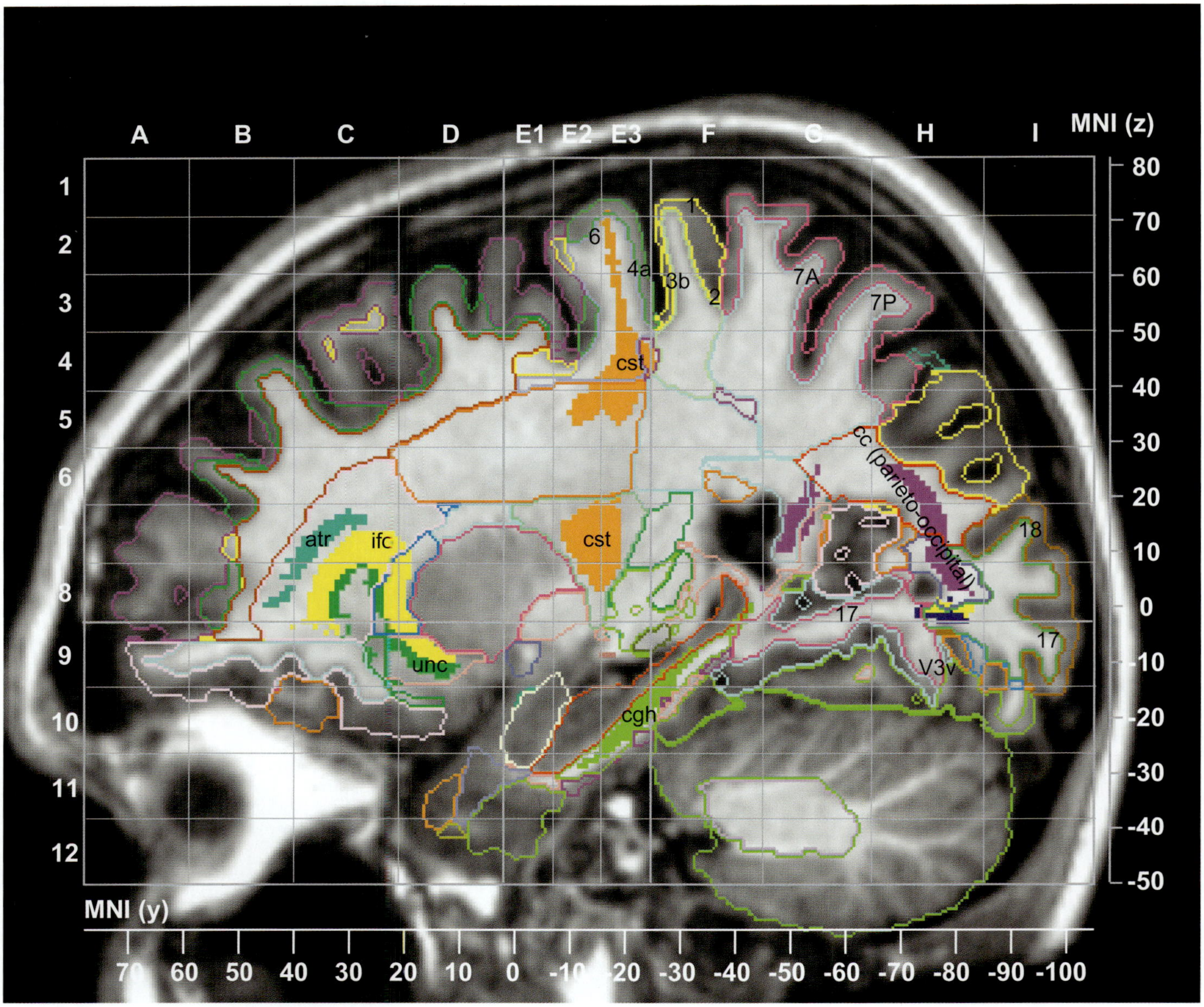

atr	- anterior thalamic radiation	cst	-	corticospinal tract
cc (parieto-occipital)	- corpus callosum (connecting bilateral parieto-occipital areas)	ifo	-	inferior fronto-occipital fasciculus
cgh	- cingulum (hippocampal part)	unc	-	uncinate fasciculus

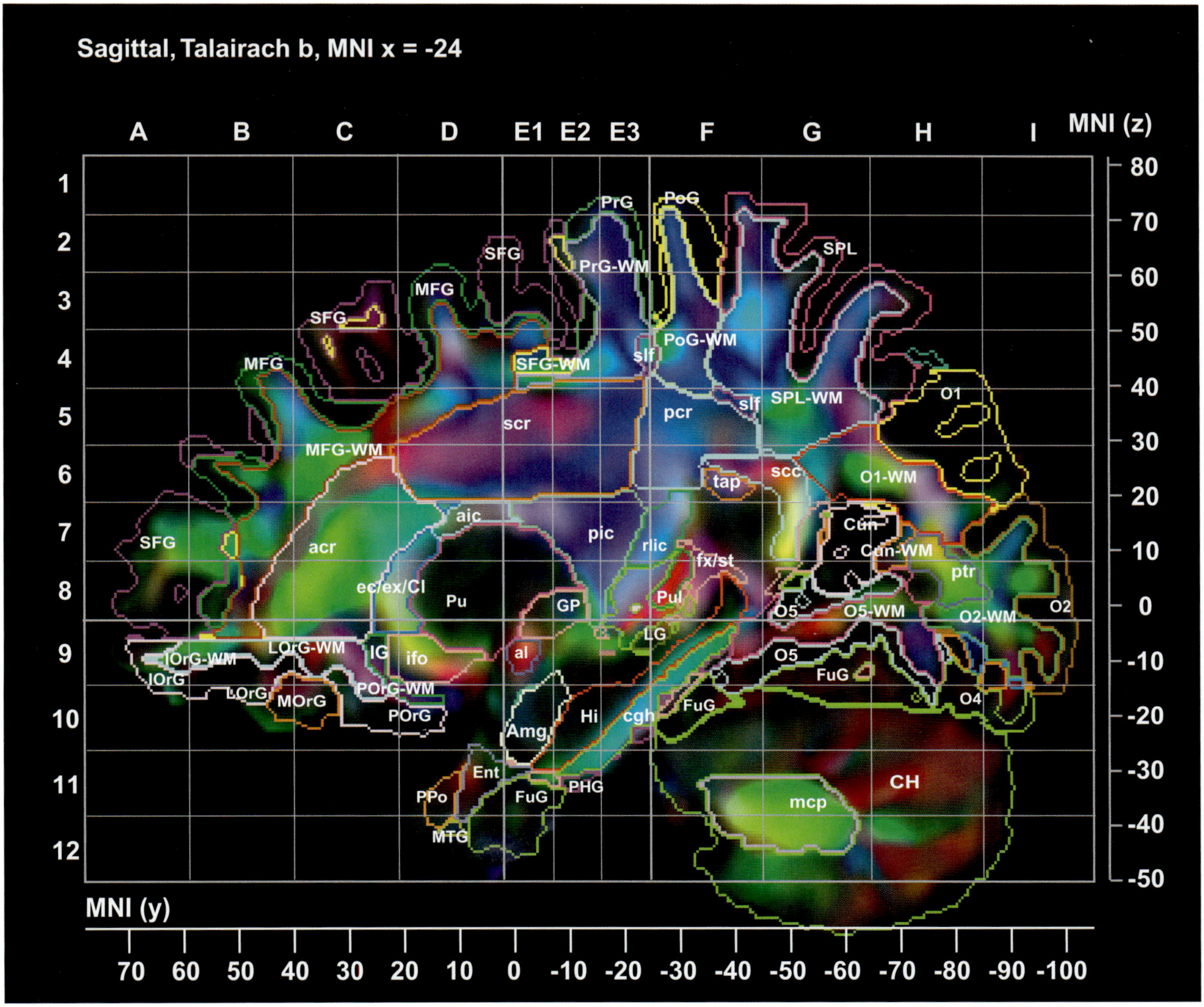

acr - anterior corona radiata
aic - anterior limb of the internal capsule
al - ansa lenticularis
Amg - amygdala
cgh - cingulum (hippocampal part)
CH - cerebellar hemisphere
Cun - cuneus
Cun-WM - cuneus white matter
ec/ex/Cl - external capsule / extreme capsule / claustrum (cannot be resolved with current resolution)
Ent - entorhinal cortex
FuG - fusiform gyrus
fx/st - fornix (crus) / stria terminalis (cannot be resolved with current resolution)
GP - globus pallidus
Hi - hippocampus
ifo - inferior fronto-occipital fasciculus
IG - insular gyrus
IOrG - intermediate orbital gyrus
IOrG-WM - intermediate orbital gyrus white matter
LG - lateral geniculate nucleus (thalamus)
LOrG - lateral orbital gyrus
LOrG-WM - lateral orbital gyrus white matter
mcp - middle cerebellar peduncle
MFG - middle frontal gyrus
MFG-WM - middle frontal gyrus white matter
MOrG - medial orbital gyrus
MTG - middle temporal gyrus
O1 - superior occipital gyrus
O1-WM - superior occipital gyrus white matter
O2 - middle or lateral occipital gyrus
O2-WM - middle or lateral occipital gyrus white matter
O4 - fourth occipital gyrus (part of the fusiform gyrus)
O5 - lyngual gyrus
O5-WM - lyngual gyrus white matter
pcr - posterior corona radiata
PHG - parahippocampal gyrus
pic - posterior limb of the internal capsule
PoG - postcentral gyrus
PoG-WM - postcentral gyrus white matter
POrG - posterior orbital gyrus
POrG-WM - posterior orbital gyrus white matter
PPo - planum polare
PrG - precentral gyrus
PrG-WM - precentral gyrus white matter
ptr - posterior thalamic radiation (include the optic radiation)
Pu - putamen
Pul - pulvinar (thalamus)
rlic - retrolenticular part of the internal capsule
scc - splenium of the corpus callosum
scr - superior corona radiata
SFG - superior frontal gyrus
SFG-WM - superior frontal gyrus white matter
slf - superior longitudinal fasciculus
SPL - superior parietal lobule
SPL-WM - superior parietal lobule white matter
tap - tapetum of the corpus callosum

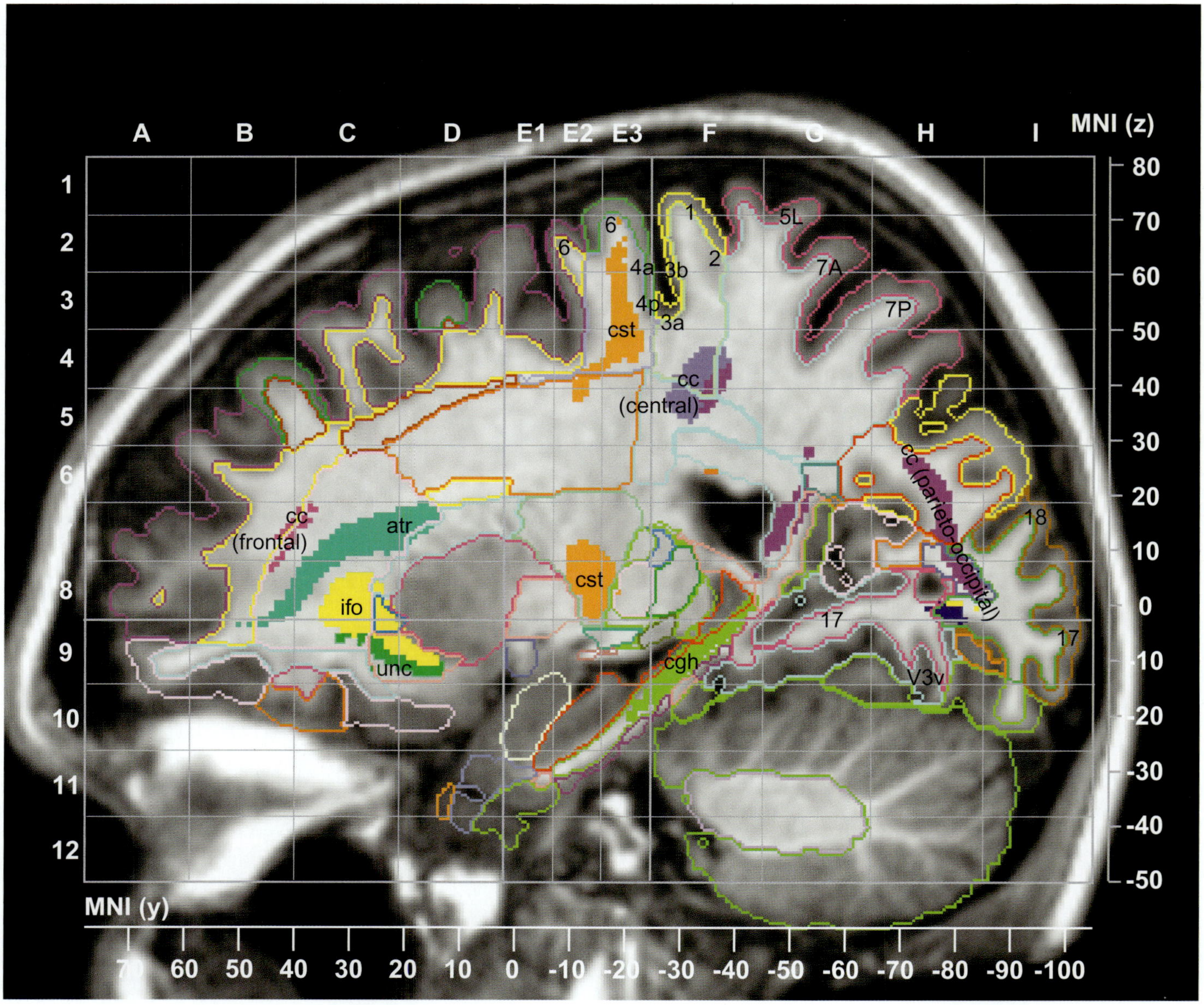

atr - anterior thalamic radiation
cc (central) - corpus callosum (connecting bilateral central areas)
cc (frontal) - corpus callosum (connecting bilateral frontal areas)
cc (parieto-occipital) - corpus callosum (connecting bilateral parieto-occipital areas)
cgh - cingulum (hippocampal part)
cst - corticospinal tract
ifo - inferior fronto-occipital fasciculus
unc - uncinate fasciculus

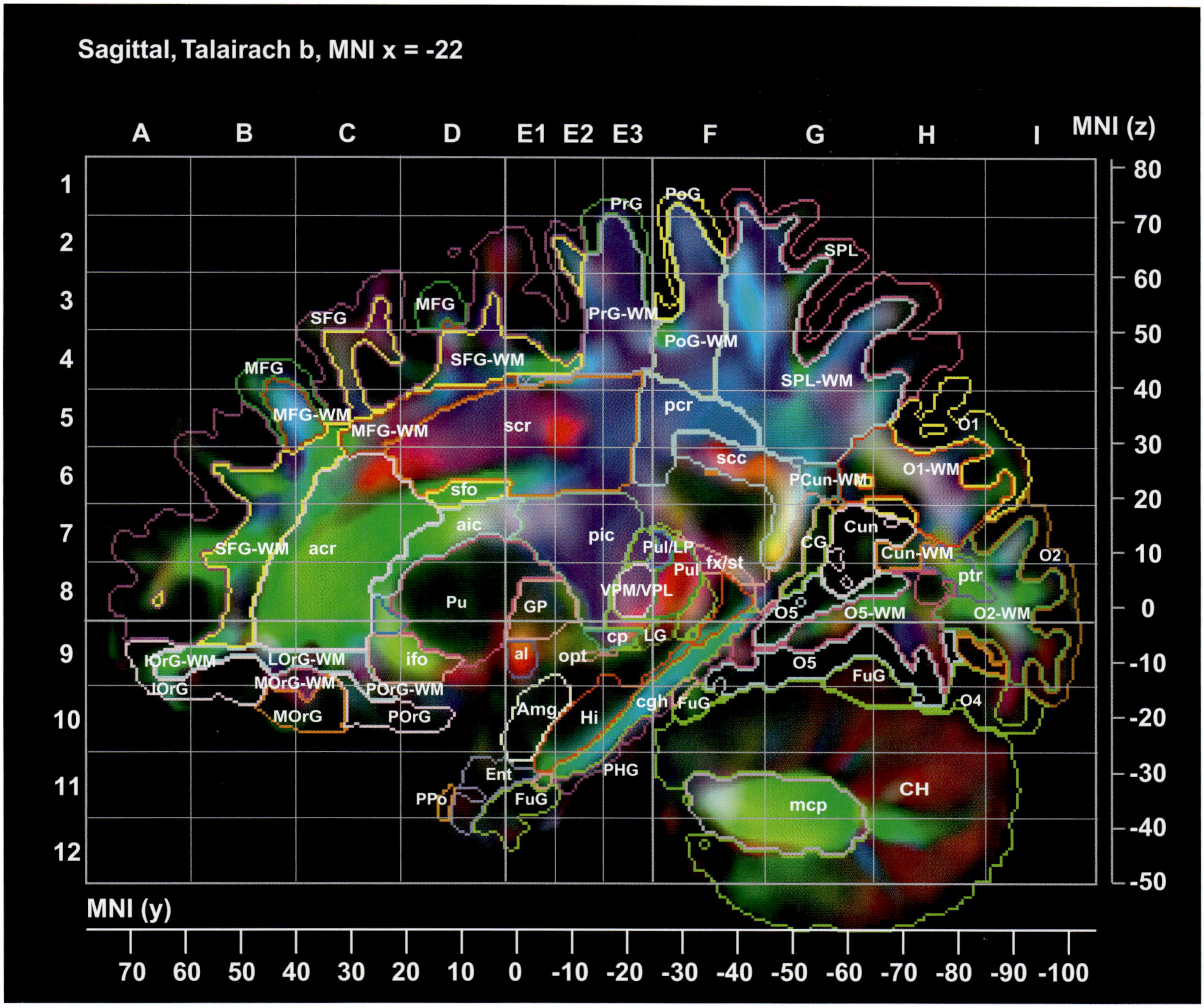

acr - anterior corona radiata
aic - anterior limb of the internal capsule
al - ansa lenticularis
Amg - amygdala
CG - cingulate gyrus
cgh - cingulum (hippocampal part)
CH - cerebellar hemisphere
cp - cerebral peduncle
Cun - cuneus
Cun-WM - cuneus white matter
Ent - entorhinal cortex
FuG - fusiform gyrus
fx/st - fornix (crus) / stria terminalis (can not be resolved with current resolution)
GP - globus pallidus
Hi - hippocampus
ifo - inferior fronto-occipital fasciculus
IOrG - intermediate orbital gyrus
IOrG-WM - intermediate orbital gyrus white matter
LG - lateral geniculate nucleus (thalamus)
LOrG-WM - lateral orbital gyrus white matter
mcp - middle cerebellar peduncle
MFG - middle frontal gyrus
MFG-WM - middle frontal gyrus white matter
MOrG - medial orbital gyrus
MOrG-WM - medial orbital gyrus white matter
O1 - superior occipital gyrus
O1-WM - superior occipital gyrus white matter
O2 - middle or lateral occipital gyrus
O2-WM - middle or lateral occipital gyrus white matter
O4 - fourth occipital gyrus (part of the fusiform gyrus)
O5 - lyngual gyrus
O5-WM - lyngual gyrus white matter
opt - optic tract
pcr - posterior corona radiata
PCun-WM - precuneus white matter
PHG - parahippocampal gyrus
pic - posterior limb of the internal capsule
PoG - postcentral gyrus
PoG-WM - postcentral gyrus white matter
POrG - posterior orbital gyrus
POrG-WM - posterior orbital gyrus white matter
PPo - planum polare
PrG - precentral gyrus
PrG-WM - precentral gyrus white matter
ptr - posterior thalamic radiation (include the optic radiation)
Pu - putamen
Pul - pulvinar (thalamus)
Pul/LP - pulvinar / lateral posterior thalamic nucleus
scc - splenium of the corpus callosum
scr - superior corona radiata
SFG - superior frontal gyrus
SFG-WM - superior frontal gyrus white matter
sfo - superior fronto-occipital fasciculus
SPL - superior parietal lobule
SPL-WM - superior parietal lobule white matter
VPM/VPL - ventroposterior medialis / ventroposterior lateralis (thalamus)

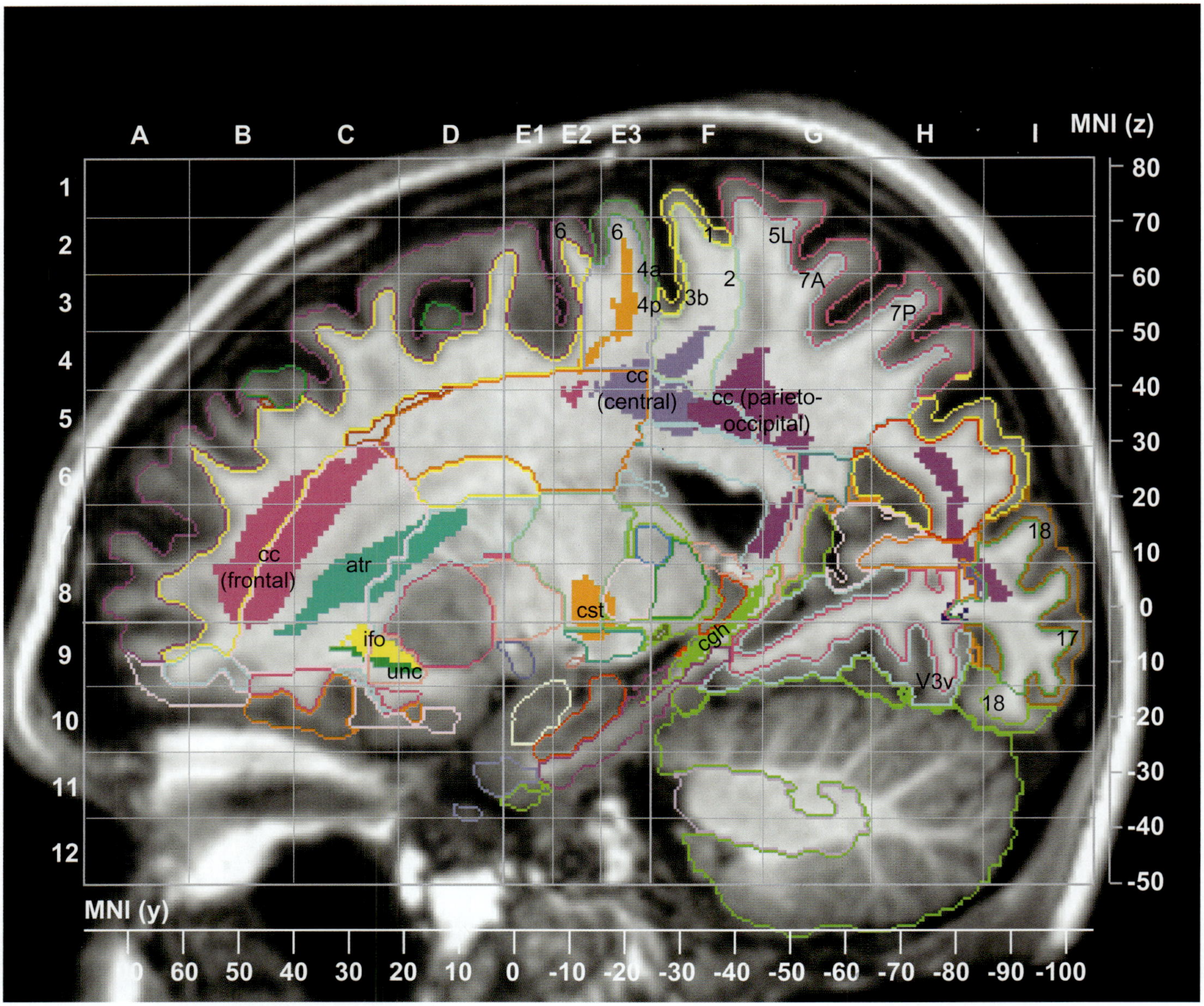

atr	-	anterior thalamic radiation
cc (central)	-	corpus callosum (connecting bilateral central areas)
cc (frontal)	-	corpus callosum (connecting bilateral frontal areas)
cc (parieto-occipital)	-	corpus callosum (connecting bilateral parieto-occipital areas)
cgh	-	cingulum (hippocampal part)
cst	-	corticospinal tract
ifo	-	inferior fronto-occipital fasciculus
unc	-	uncinate fasciculus

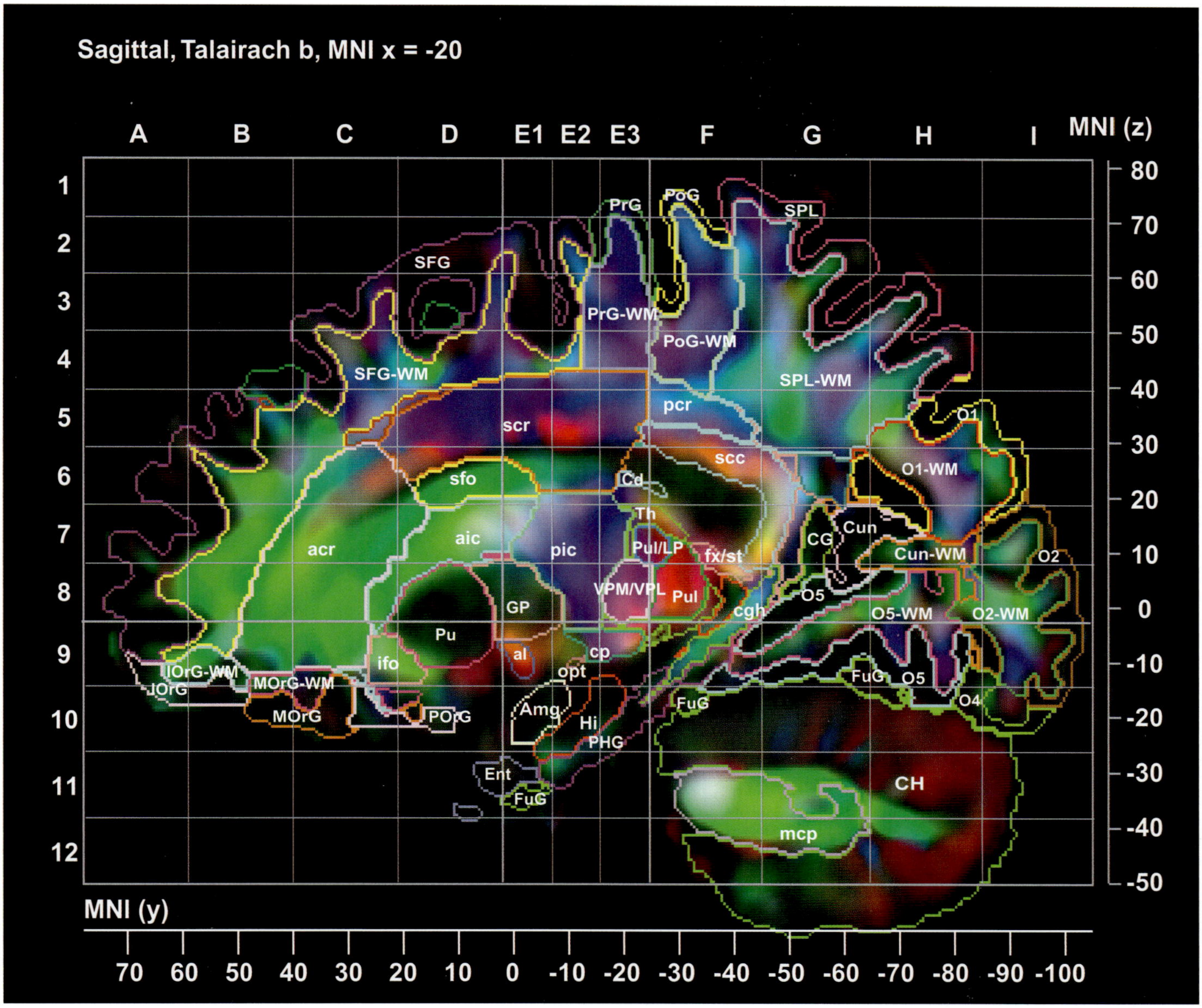

acr - anterior corona radiata
aic - anterior limb of the internal capsule
al - ansa lenticularis
Amg - amygdala
Cd - caudate nucleus
CG - cingulate gyrus
cgh - cingulum (hippocampal part)
CH - cerebellar hemisphere
cp - cerebral peduncle
Cun - cuneus
Cun-WM - cuneus white matter
Ent - entorhinal cortex
FuG - fusiform gyrus
fx/st - fornix (crus) / stria terminalis (cannot be resolved with current resolution)
GP - globus pallidus
Hi - hippocampus
ifo - inferior fronto-occipital fasciculus
IOrG - intermediate orbital gyrus
IOrG-WM - intermediate orbital gyrus white matter
mcp - middle cerebellar peduncle
MOrG - medial orbital gyrus
MOrG-WM - medial orbital gyrus white matter
O1 - superior occipital gyrus
O1-WM - superior occipital gyrus white matter
O2 - middle or lateral occipital gyrus
O2-WM - middle or lateral occipital gyrus white matter
O4 - fourth occipital gyrus (part of the fusiform gyrus)
O5 - lyngual gyrus
O5-WM - lyngual gyrus white matter
opt - optic tract
pcr - posterior corona radiata
PHG - parahippocampal gyrus
pic - posterior limb of the internal capsule
PoG - postcentral gyrus
PoG-WM - postcentral gyrus white matter
POrG - posterior orbital gyrus
PrG - precentral gyrus
PrG-WM - precentral gyrus white matter
Pu - putamen
Pul - pulvinar (thalamus)
Pul/LP - pulvinar / lateral posterior thalamic nucleus
scc - splenium of the corpus callosum
scr - superior corona radiata
SFG - superior frontal gyrus
SFG-WM - superior frontal gyrus white matter
sfo - superior fronto-occipital fasciculus
SPL - superior parietal lobule
SPL-WM - superior parietal lobule white matter
Th - thalamus
VPM/VPL - ventroposterior medialis / ventroposterior lateralis (thalamus)

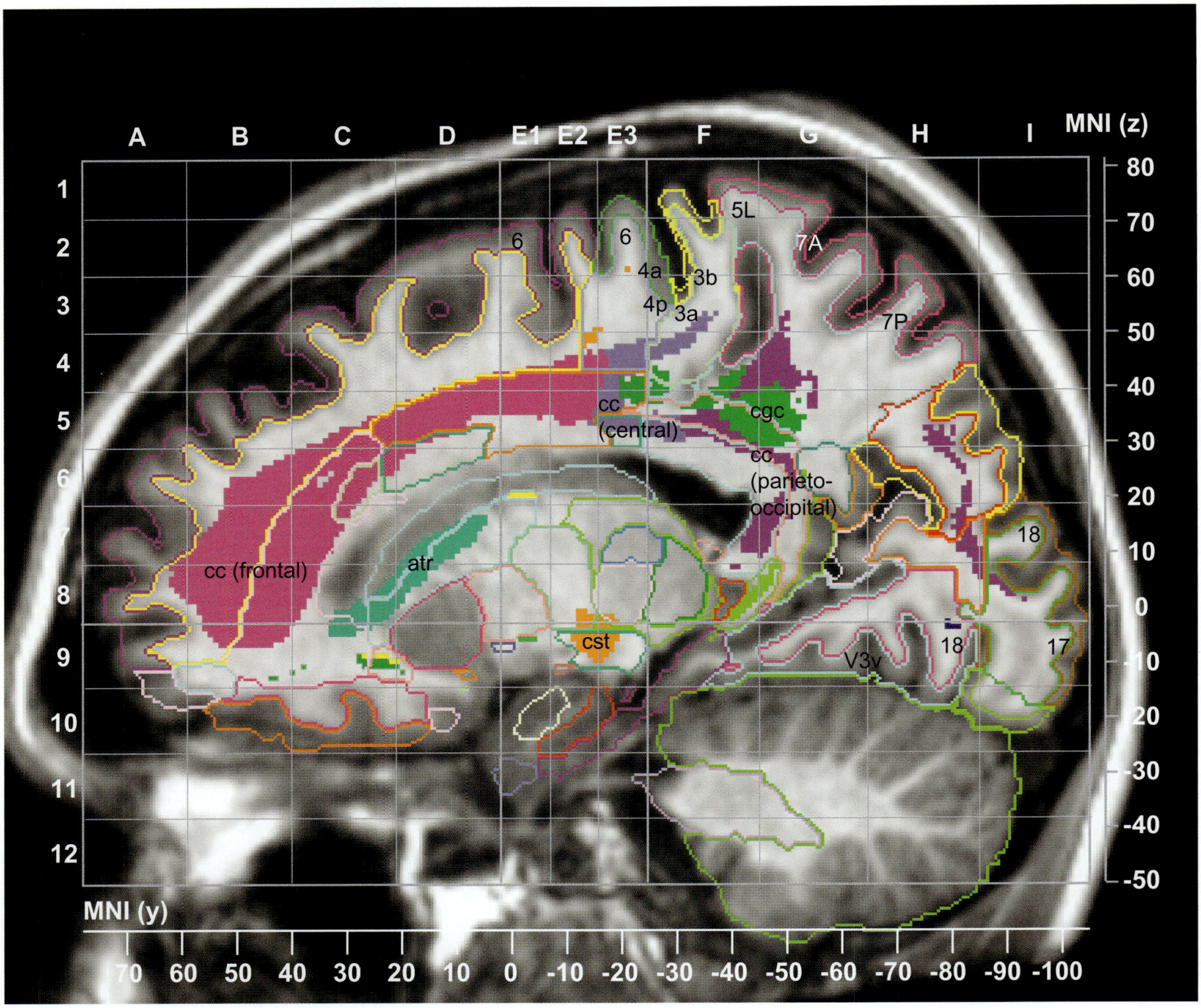

atr - anterior thalamic radiation
cc (central) - corpus callosum (connecting bilateral central areas)
cc (frontal) - corpus callosum (connecting bilateral frontal areas)
cc (parieto-occipital) - corpus callosum (connecting bilateral parieto-occipital areas)
cgc - cingulum (cingulate gyrus part)
cst - corticospinal tract

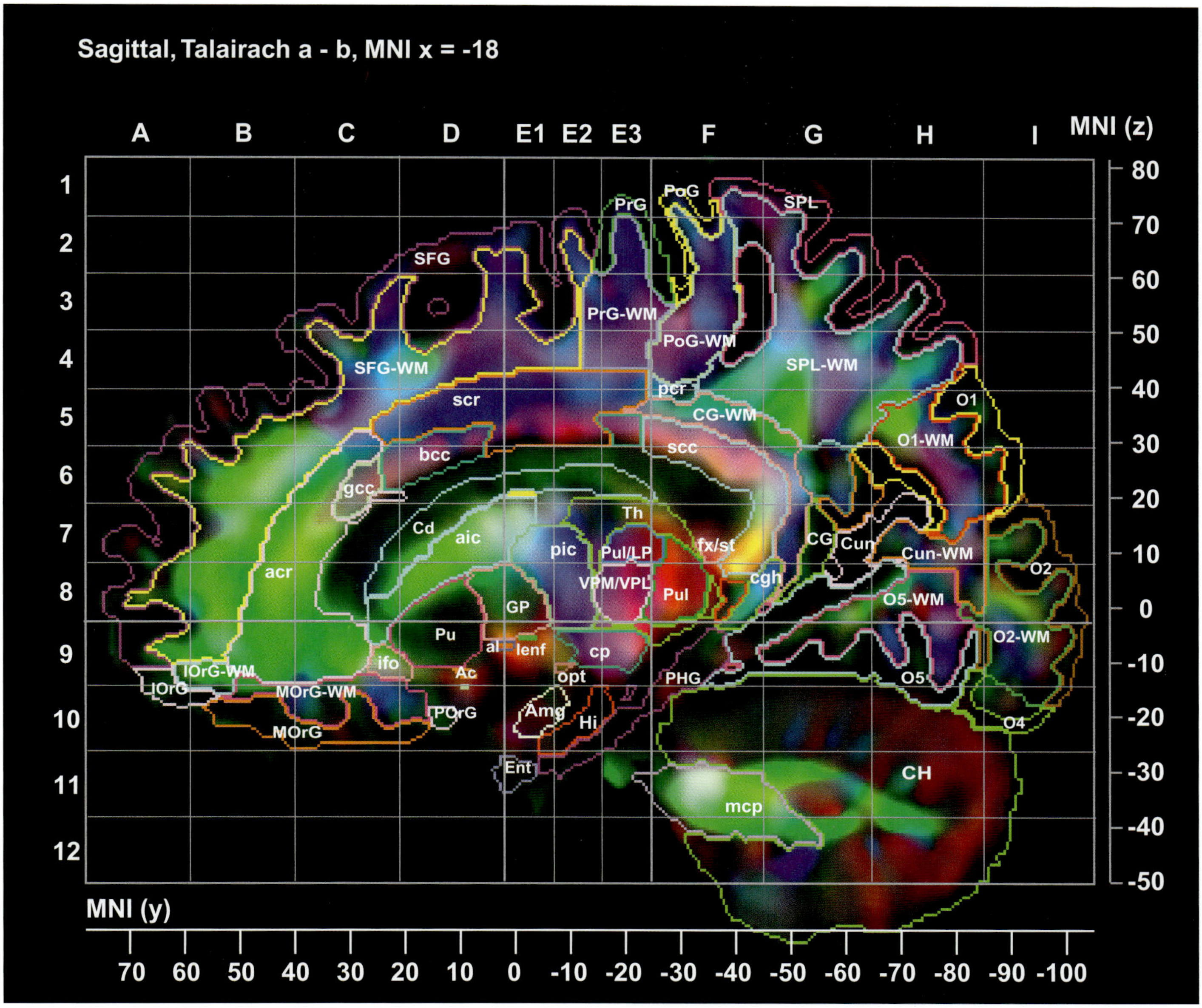

Ac	-	accumbens nucleus
acr	-	anterior corona radiata
aic	-	anterior limb of the internal capsule
al	-	ansa lenticularis
Amg	-	amygdala
bcc	-	body of the corpus callosum
Cd	-	caudate nucleus
CG	-	cingulate gyrus
cgh	-	cingulum (hippocampal part)
CG-WM	-	cingulate gyrus white matter
CH	-	cerebellar hemisphere
cp	-	cerebral peduncle
Cun	-	cuneus
Cun-WM	-	cuneus white matter
Ent	-	entorhinal cortex
fx/st	-	fornix (crus) / stria terminalis (cannot be resolved with current resolution)
gcc	-	genu of the corpus callosum
GP	-	globus pallidus
Hi	-	hippocampus
ifo	-	inferior fronto-occipital fasciculus
IOrG	-	intermediate orbital gyrus
IOrG-WM	-	intermediate orbital gyrus white matter
lenf	-	lenticular fasciculus
mcp	-	middle cerebellar peduncle
MOrG	-	medial orbital gyrus
MOrG-WM	-	medial orbital gyrus white matter
O1	-	superior occipital gyrus
O1-WM	-	superior occipital gyrus white matter
O2	-	middle or lateral occipital gyrus
O2-WM	-	middle or lateral occipital gyrus white matter
O4	-	fourth occipital gyrus (part of the fusiform gyrus)
O5	-	lyngual gyrus
O5-WM	-	lyngual gyrus white matter
opt	-	optic tract
pcr	-	posterior corona radiata
PHG	-	parahippocampal gyrus
pic	-	posterior limb of the internal capsule
PoG	-	postcentral gyrus
PoG-WM	-	postcentral gyrus white matter
POrG	-	posterior orbital gyrus
PrG	-	precentral gyrus
PrG-WM	-	precentral gyrus white matter
Pu	-	putamen
Pul	-	pulvinar (thalamus)
Pul/LP	-	pulvinar / lateral posterior thalamic nucleus
scc	-	splenium of the corpus callosum
scr	-	superior corona radiata
SFG	-	superior frontal gyrus
SFG-WM	-	superior frontal gyrus white matter
SPL	-	superior parietal lobule
SPL-WM	-	superior parietal lobule white matter
Th	-	thalamus
VPM/VPL	-	ventroposterior medialis / ventroposterior lateralis (thalamus)

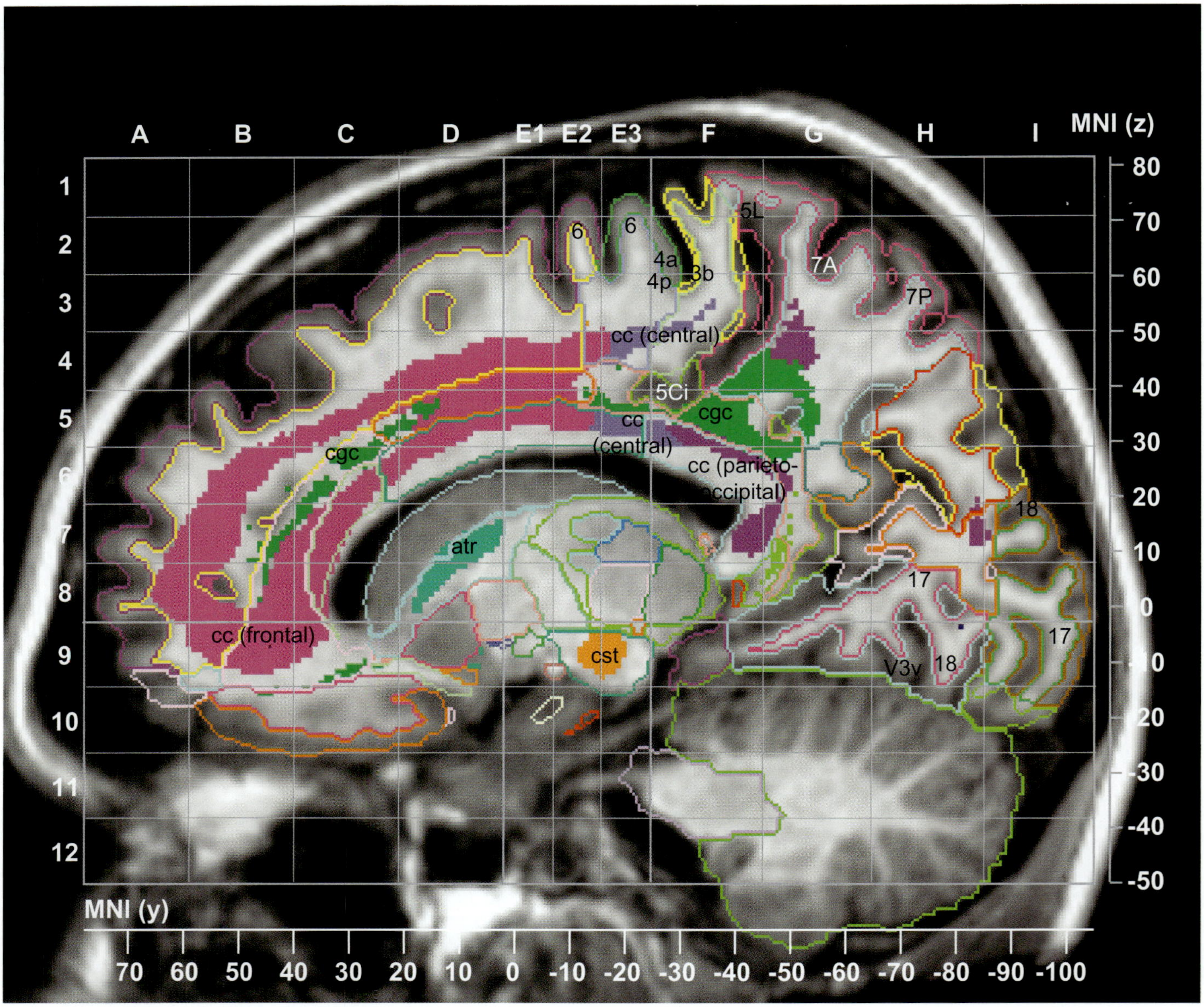

atr - anterior thalamic radiation

cc (central) - corpus callosum (connecting bilateral central areas)

cc (frontal) - corpus callosum (connecting bilateral frontal areas)

cc (parieto-occipital) - corpus callosum (connecting bilateral parieto-occipital areas)

cgc - cingulum (cingulate gyrus part)

cst - corticospinal tract

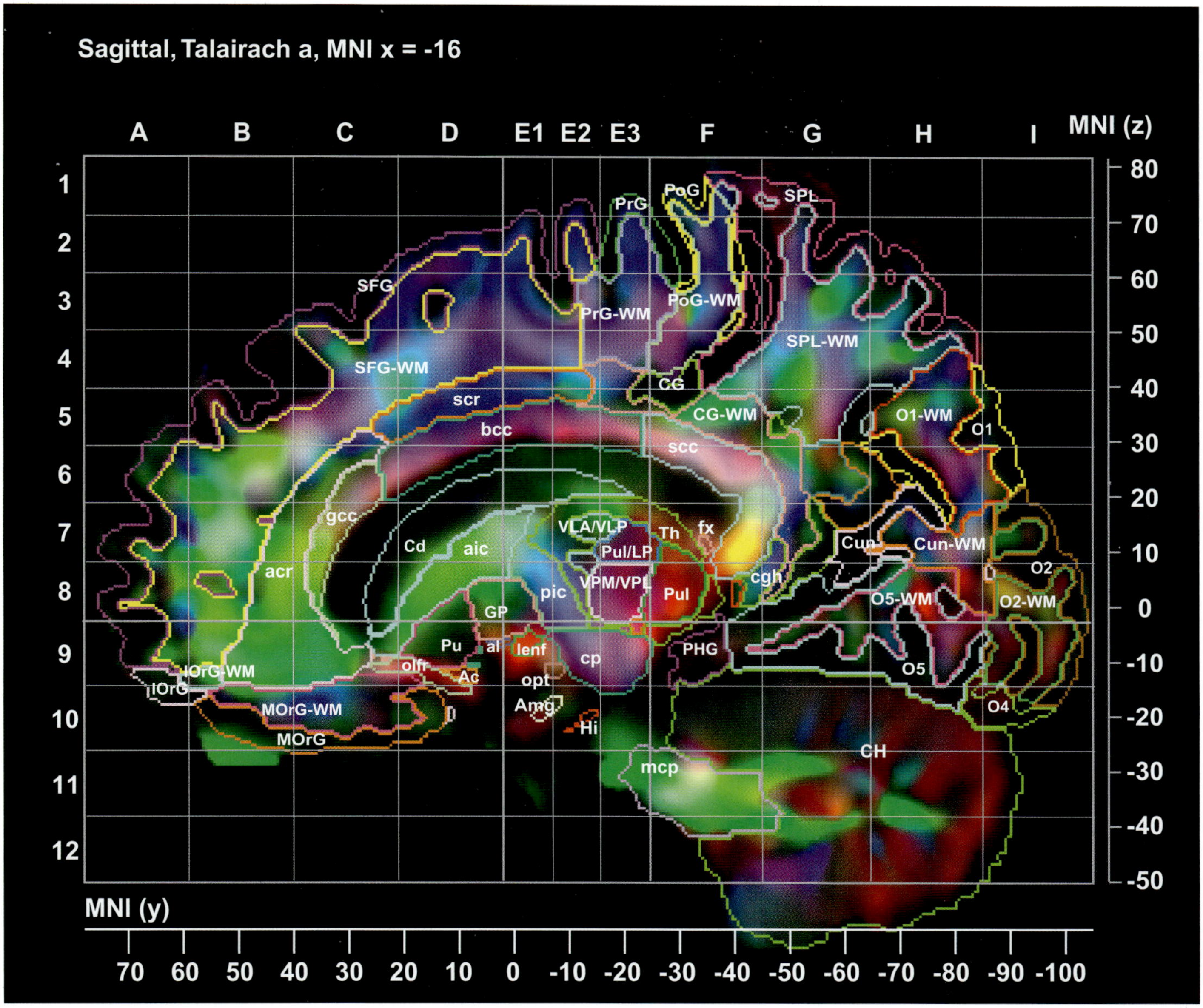

Ac - accumbens nucleus
acr - anterior corona radiata
aic - anterior limb of the internal capsule
al - ansa lenticularis
Amg - amygdala
bcc - body of the corpus callosum
Cd - caudate nucleus
CG - cingulate gyrus
cgh - cingulum (hippocampal part)
CG-WM - cingulate gyrus white matter
CH - cerebellar hemisphere
cp - cerebral peduncle
Cun - cuneus
Cun-WM - cuneus white matter
fx - fornix (column and body)
gcc - genu of the corpus callosum
GP - globus pallidus
Hi - hippocampus
lOrG - intermediate orbital gyrus
lOrG-WM - intermediate orbital gyrus white matter
lenf - lenticular fasciculus
mcp - middle cerebellar peduncle
MOrG - medial orbital gyrus
MOrG-WM - medial orbital gyrus white matter
O1 - superior occipital gyrus
O1-WM - superior occipital gyrus white matter
O2 - middle or lateral occipital gyrus
O2-WM - middle or lateral occipital gyrus white matter
O4 - fourth occipital gyrus (part of the fusiform gyrus)
O5 - lyngual gyrus
O5-WM - lyngual gyrus white matter
olfr - olfactory radiation
opt - optic tract
PHG - parahippocampal gyrus
pic - posterior limb of the internal capsule
PoG - postcentral gyrus
PoG-WM - postcentral gyrus white matter
PrG - precentral gyrus
PrG-WM - precentral gyrus white matter
Pu - putamen
Pul - pulvinar (thalamus)
Pul/LP - pulvinar / lateral posterior thalamic nucleus
scc - splenium of the corpus callosum
scr - superior corona radiata
SFG - superior frontal gyrus
SFG-WM - superior frontal gyrus white matter
SPL - superior parietal lobule
SPL-WM - superior parietal lobule white matter
Th - thalamus
VLA/VLP - anterior and posterior ventrolateral thalamic nuclei
VPM/VPL - ventroposterior medialis / ventroposterior lateralis (thalamus)

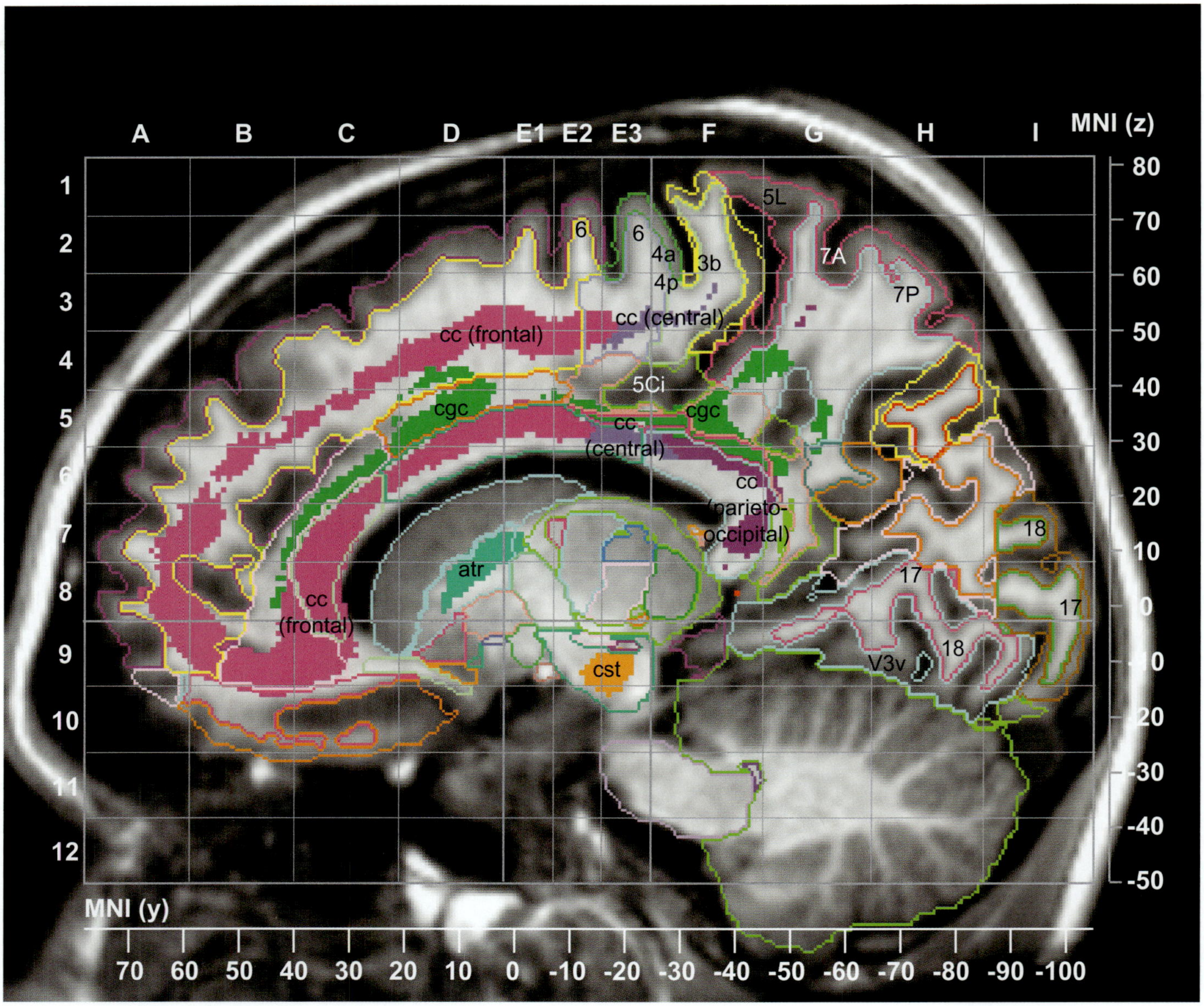

atr	-	anterior thalamic radiation
cc (central)	-	corpus callosum (connecting bilateral central areas)
cc (frontal)	-	corpus callosum (connecting bilateral frontal areas)
cc (parieto-occipital)	-	corpus callosum (connecting bilateral parieto-occipital areas)
cgc	-	cingulum (cingulate gyrus part)
cst	-	corticospinal tract

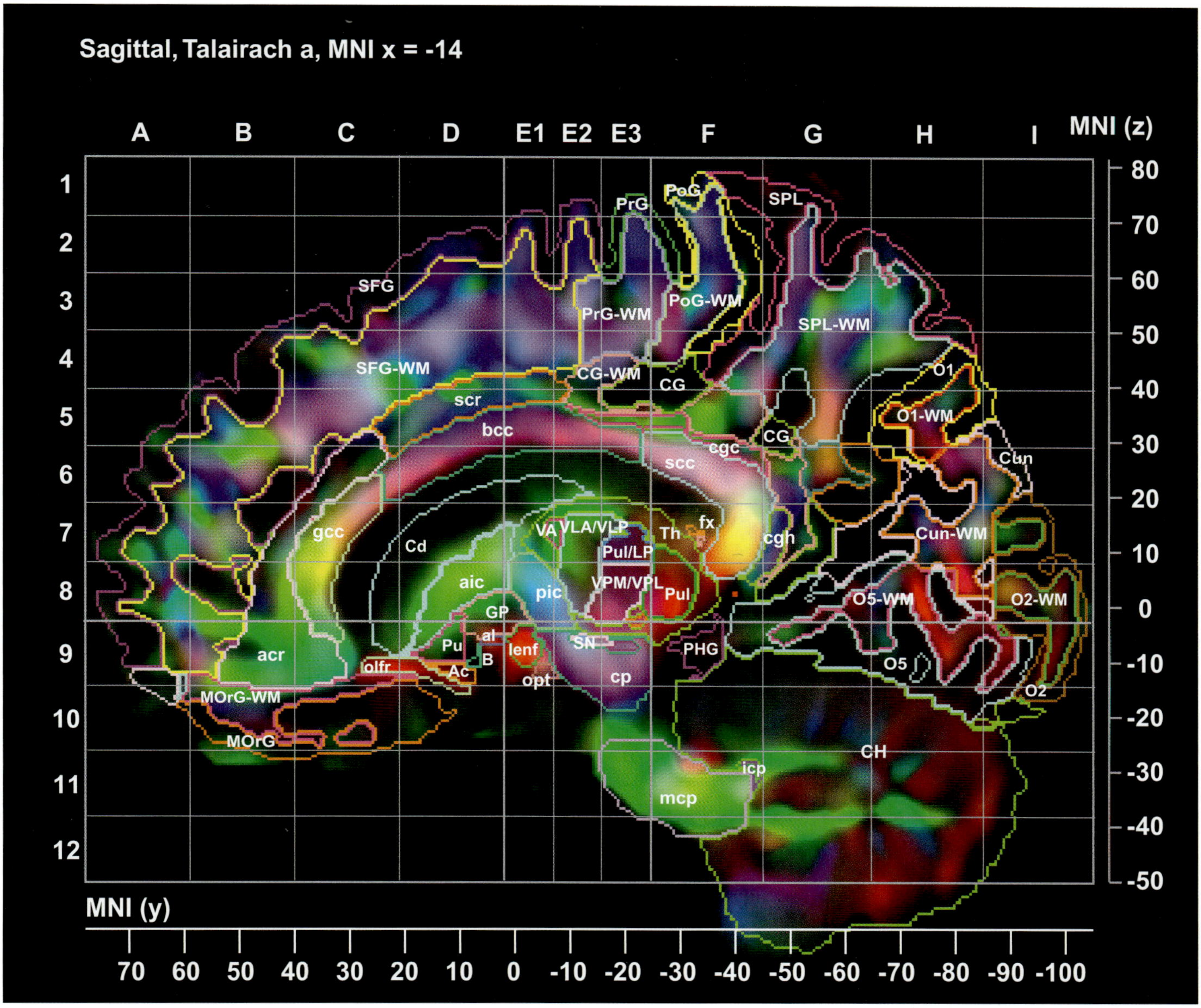

Ac - accumbens nucleus
acr - anterior corona radiata
aic - anterior limb of the internal capsule
al - ansa lenticularis
B - basal nucleus of Meynert
bcc - body of the corpus callosum
Cd - caudate nucleus
CG - cingulate gyrus
cgc - cingulum (cingulate gyrus part)
cgh - cingulum (hippocampal part)
CG-WM - cingulate gyrus white matter
CH - cerebellar hemisphere
cp - cerebral peduncle
Cun - cuneus
Cun-WM - cuneus white matter
fx - fornix (column and body)
gcc - genu of the corpus callosum
GP - globus pallidus
icp - inferior cerebellar peduncle
lenf - lenticular fasciculus
mcp - middle cerebellar peduncle
MOrG - medial orbital gyrus
MOrG-WM - medial orbital gyrus white matter
O1 - superior occipital gyrus
O1-WM - superior occipital gyrus white matter
O2 - middle or lateral occipital gyrus
O2-WM - middle or lateral occipital gyrus white matter
O5 - lyngual gyrus
O5-WM - lyngual gyrus white matter
olfr - olfactory radiation
opt - optic tract
PHG - parahippocampal gyrus
pic - posterior limb of the internal capsule
PoG - postcentral gyrus
PoG-WM - postcentral gyrus white matter
PrG - precentral gyrus
PrG-WM - precentral gyrus white matter
Pu - putamen
Pul - pulvinar (thalamus)
Pul/LP - pulvinar / lateral posterior thalamic nucleus
scc - splenium of the corpus callosum
scr - superior corona radiata
SFG - superior frontal gyrus
SFG-WM - superior frontal gyrus white matter
SN - substantia nigra
SPL - superior parietal lobule
SPL-WM - superior parietal lobule white matter
Th - thalamus
VA - ventroanterior thalamic nucleus
VLA/VLP - anterior and posterior ventrolateral thalamic nuclei
VPM/VPL - ventroposterior medialis / ventroposterior lateralis (thalamus)

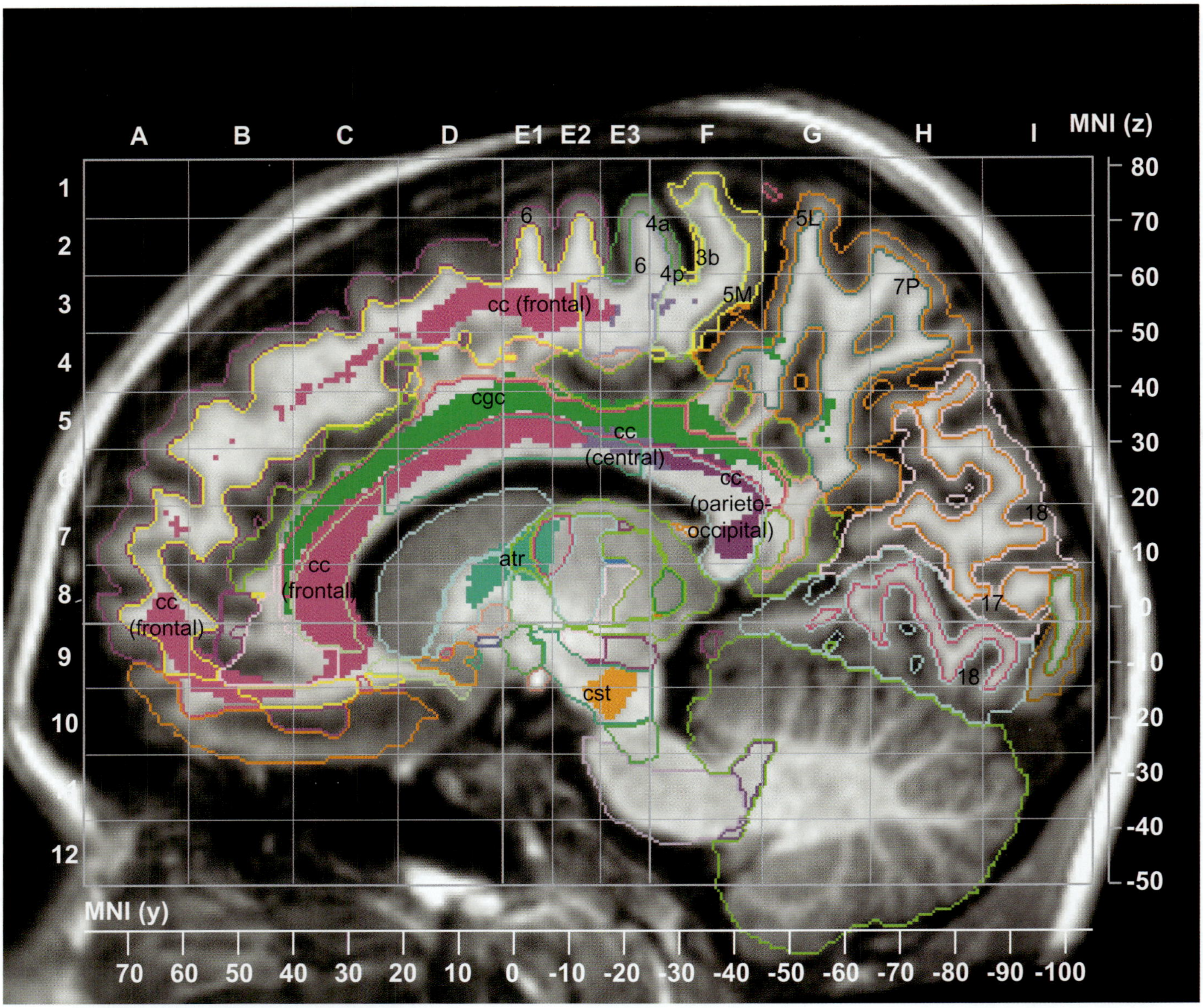

atr	- anterior thalamic radiation	cc (parieto-occipital)	- corpus callosum (connecting bilateral parieto-occipital areas)
cc (central)	- corpus callosum (connecting bilateral central areas)	cgc	- cingulum (cingulate gyrus part)
cc (frontal)	- corpus callosum (connecting bilateral frontal areas)	cst	- corticospinal tract

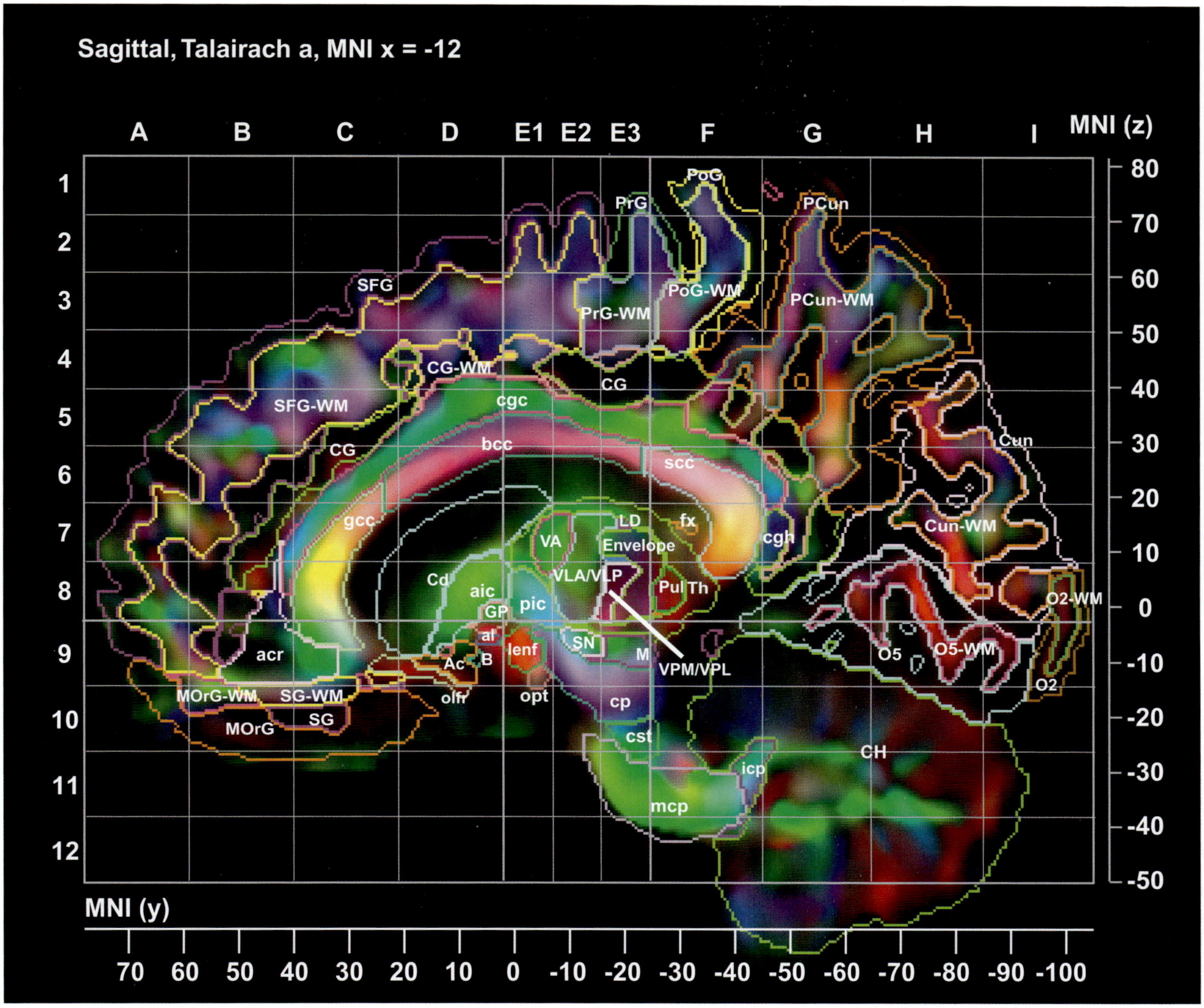

Ac - accumbens nucleus
acr - anterior corona radiata
aic - anterior limb of the internal capsule
al - ansa lenticularis
B - basal nucleus of Meynert
bcc - body of the corpus callosum
Cd - caudate nucleus
CG - cingulate gyrus
cgc - cingulum (cingulate gyrus part)
cgh - cingulum (hippocampal part)
CG-WM - cingulate gyrus white matter
CH - cerebellar hemisphere
cp - cerebral peduncle
cst - corticospinal tract
Cun - cuneus
Cun-WM - cuneus white matter
Envelope - nuclei of the internal medullary lamina (thalamus)
fx - fornix (column and body)
gcc - genu of the corpus callosum
GP - globus pallidus
icp - inferior cerebellar peduncle
LD - laterodorsal thalamic nucleus
lenf - lenticular fasciculus
M - midbrain
mcp - middle cerebellar peduncle
MOrG - medial orbital gyrus
MOrG-WM - medial orbital gyrus white matter
O2 - middle or lateral occipital gyrus
O2-WM - middle or lateral occipital gyrus white matter
O5 - lyngual gyrus
O5-WM - lyngual gyrus white matter
olfr - olfactory radiation
opt - optic tract
PCun - precuneus
PCun-WM - precuneus white matter
pic - posterior limb of the internal capsule
PoG - postcentral gyrus
PoG-WM - postcentral gyrus white matter
PrG - precentral gyrus
PrG-WM - precentral gyrus white matter
Pul - pulvinar (thalamus)
scc - splenium of the corpus callosum
SFG - superior frontal gyrus
SFG-WM - superior frontal gyrus white matter
SG - straight gyrus
SG-WM - straight gyrus white matter
SN - substantia nigra
Th - thalamus
VA - ventroanterior thalamic nucleus
VLA/VLP - anterior and posterior ventrolateral thalamic nuclei
VPM/VPL - ventroposterior medialis / ventroposterior lateralis (thalamus)

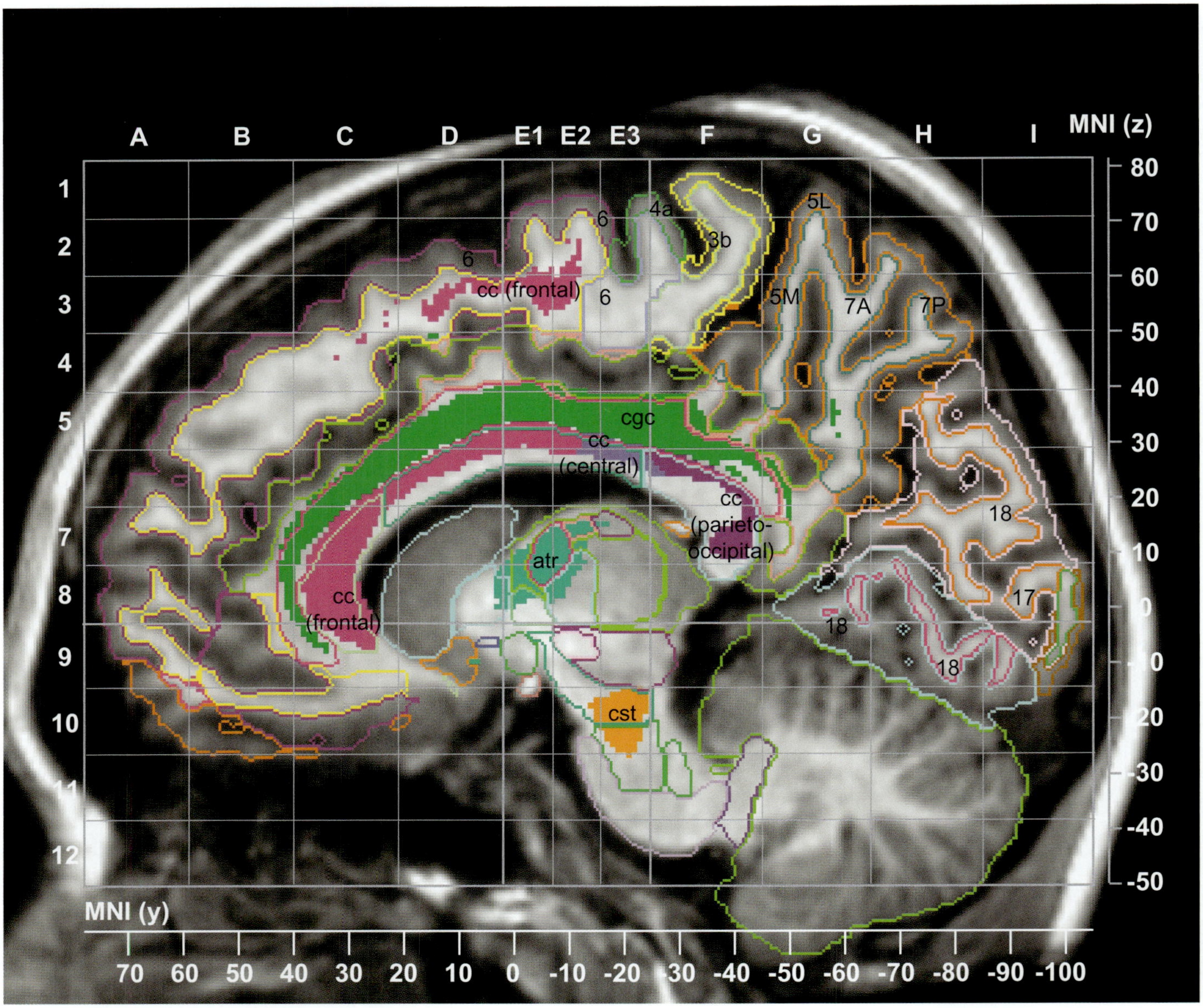

atr	-	anterior thalamic radiation	cc (parieto-occipital)	-	corpus callosum (connecting bilateral parieto-occipital areas)
cc (central)	-	corpus callosum (connecting bilateral central areas)	cgc	-	cingulum (cingulate gyrus part)
cc (frontal)	-	corpus callosum (connecting bilateral frontal areas)	cst	-	corticospinal tract

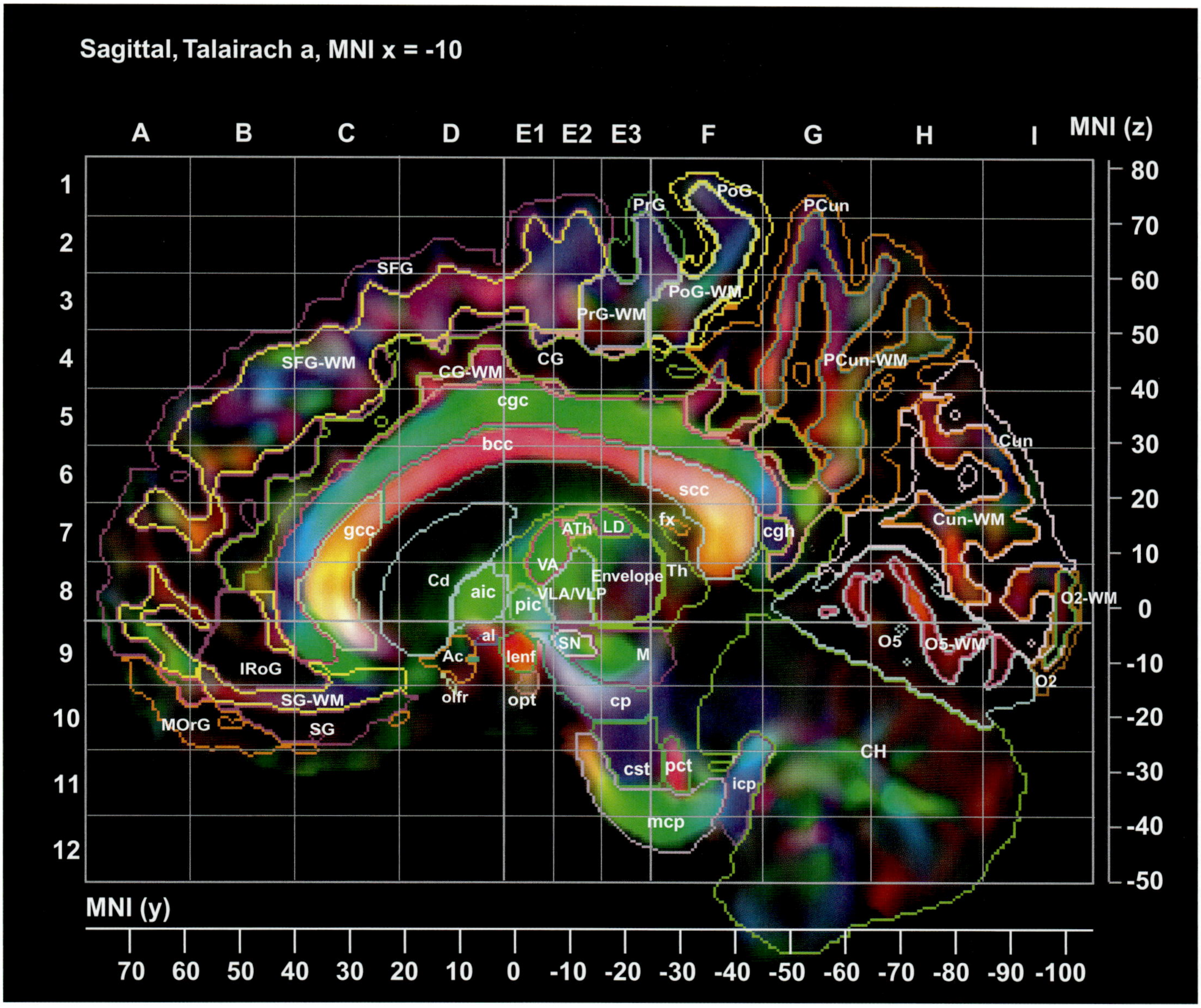

Ac - accumbens nucleus
aic - anterior limb of the internal capsule
al - ansa lenticularis
Ath - anterior nuclear group (thalamus)
bcc - body of the corpus callosum
Cd - caudate nucleus
CG - cingulate gyrus
cgc - cingulum (cingulate gyrus part)
cgh - cingulum (hippocampal part)
CG-WM - cingulate gyrus white matter
CH - cerebellar hemisphere
cp - cerebral peduncle
cst - corticospinal tract
Cun - cuneus
Cun-WM - cuneus white matter
Envelope - nuclei of the internal medullary lamina (thalamus)
fx - fornix (column and body)
gcc - genu of the corpus callosum
icp inferior cerebellar peduncle
IOrG - intermediate orbital gyrus
LD - laterodorsal thalamic nucleus
lenf - lenticular fasciculus
M - midbrain
mcp - middle cerebellar peduncle
MOrG - medial orbital gyrus
O2 - middle or lateral occipital gyrus
O2-WM - middle or lateral occipital gyrus white matter
O5 - lyngual gyrus
O5-WM - lyngual gyrus white matter
olfr - olfactory radiation
opt - optic tract
pct pontine crossing tract (a part of mcp)
PCun - precuneus
PCun-WM - precuneus white matter
pic - posterior limb of the internal capsule
PoG - postcentral gyrus
PoG-WM - postcentral gyrus white matter
PrG - precentral gyrus
PrG-WM - precentral gyrus white matter
scc - splenium of the corpus callosum
SFG - superior frontal gyrus
SFG-WM - superior frontal gyrus white matter
SG - straight gyrus
SG-WM - straight gyrus white matter
SN - substantia nigra
Th - thalamus
VA - ventroanterior thalamic nucleus
VLA/VLP - anterior and posterior ventrolateral thalamic nuclei

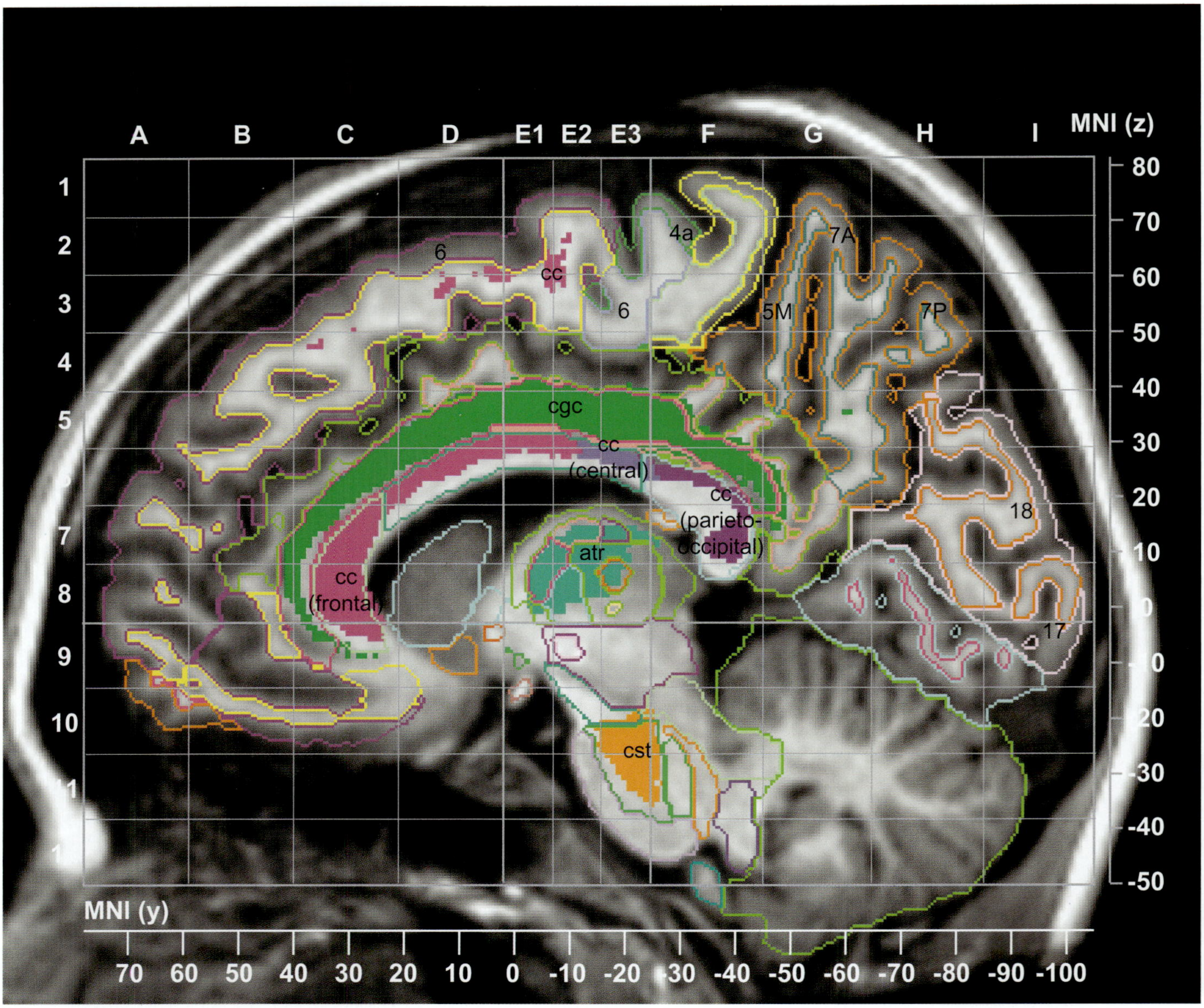

atr - anterior thalamic radiation
cc - corpus callosum
cc (central) - corpus callosum (connecting bilateral central areas)
cc (frontal) - corpus callosum (connecting bilateral frontal areas)
cc (parieto-occipital) - corpus callosum (connecting bilateral parieto-occipital areas)
cgc - cingulum (cingulate gyrus part)
cst - corticospinal tract

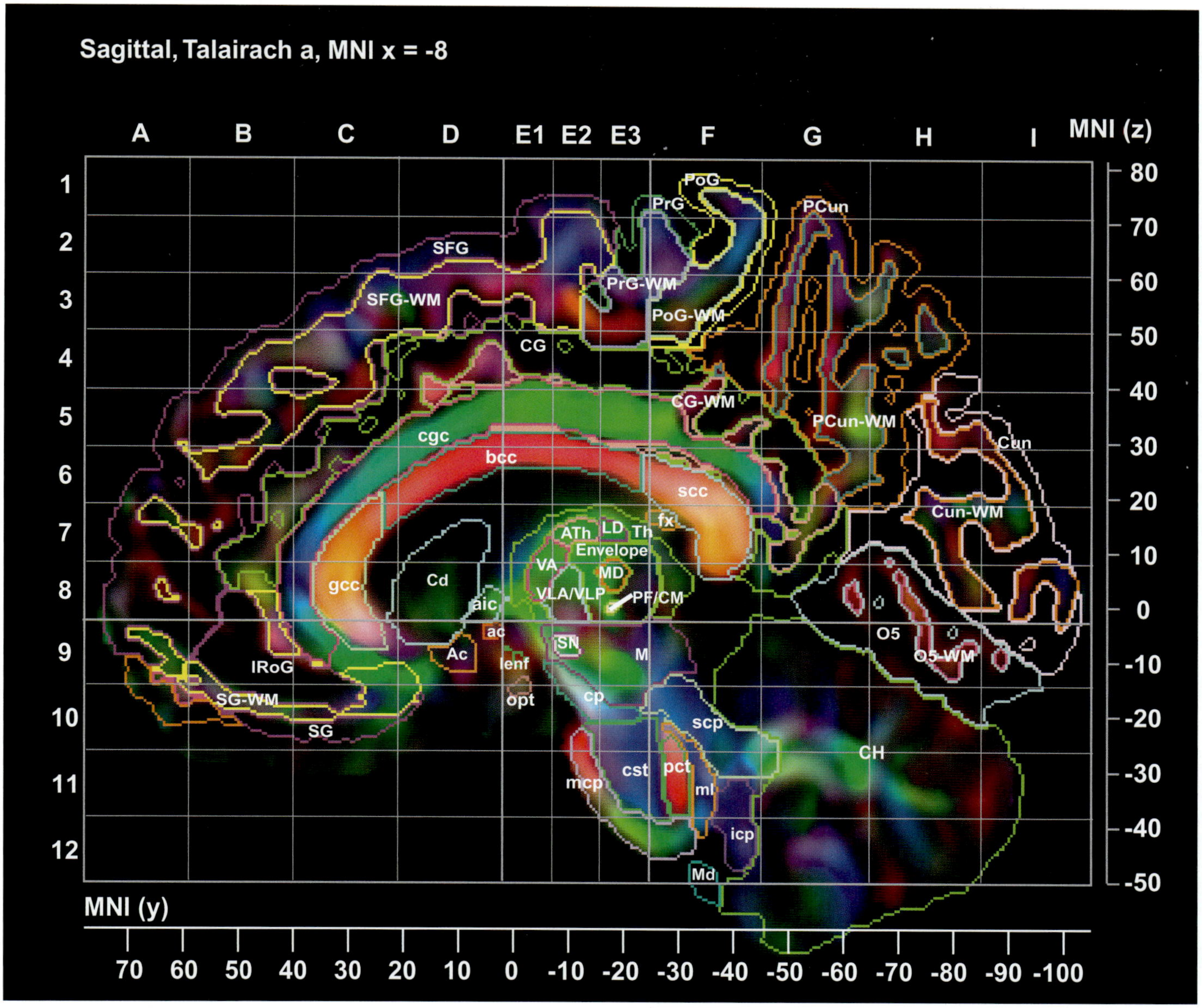

ac - anterior commissure
Ac - accumbens nucleus
aic - anterior limb of the internal capsule
Ath - anterior nuclear group (thalamus)
bcc - body of the corpus callosum
CH - cerebellar hemisphere
Cd - caudate nucleus
CG - cingulate gyrus
cgc - cingulum (cingulate gyrus part)
CG-WM - cingulate gyrus white matter
cp - cerebral peduncle
cst - corticospinal tract
Cun - cuneus
Cun-WM - cuneus white matter
Envelope - nuclei of the internal medullary lamina (thalamus)
fx - fornix (column and body)
gcc - genu of the corpus callosum
icp - inferior cerebellar peduncle
IRoG - inferior rostral gyrus
LD - laterodorsal thalamic nucleus
lenf - lenticular fasciculus
M - midbrain
mcp - middle cerebellar peduncle
MD - mediodorsal thalamic nucleus
Md - medulla oblongata
ml - medial lemniscus
O5 - lyngual gyrus
O5-WM - lyngual gyrus white matter
opt - optic tract
pct - pontine crossing tract (a part of mcp)
PCun - precuneus
PCun-WM - precuneus white matter
PF/CM - parafascicular nucleus / centromedian nucleus (thalamus)
PoG - postcentral gyrus
PoG-WM - postcentral gyrus white matter
PrG - precentral gyrus
PrG-WM - precentral gyrus white matter
scc - splenium of the corpus callosum
scp - superior cerebellar peduncle
SFG - superior frontal gyrus
SFG-WM - superior frontal gyrus white matter
SG - straight gyrus
SG-WM - straight gyrus white matter
SN - substantia nigra
Th - thalamus
VA - ventroanterior thalamic nucleus
VLA/VLP - anterior and posterior ventrolateral thalamic uclei

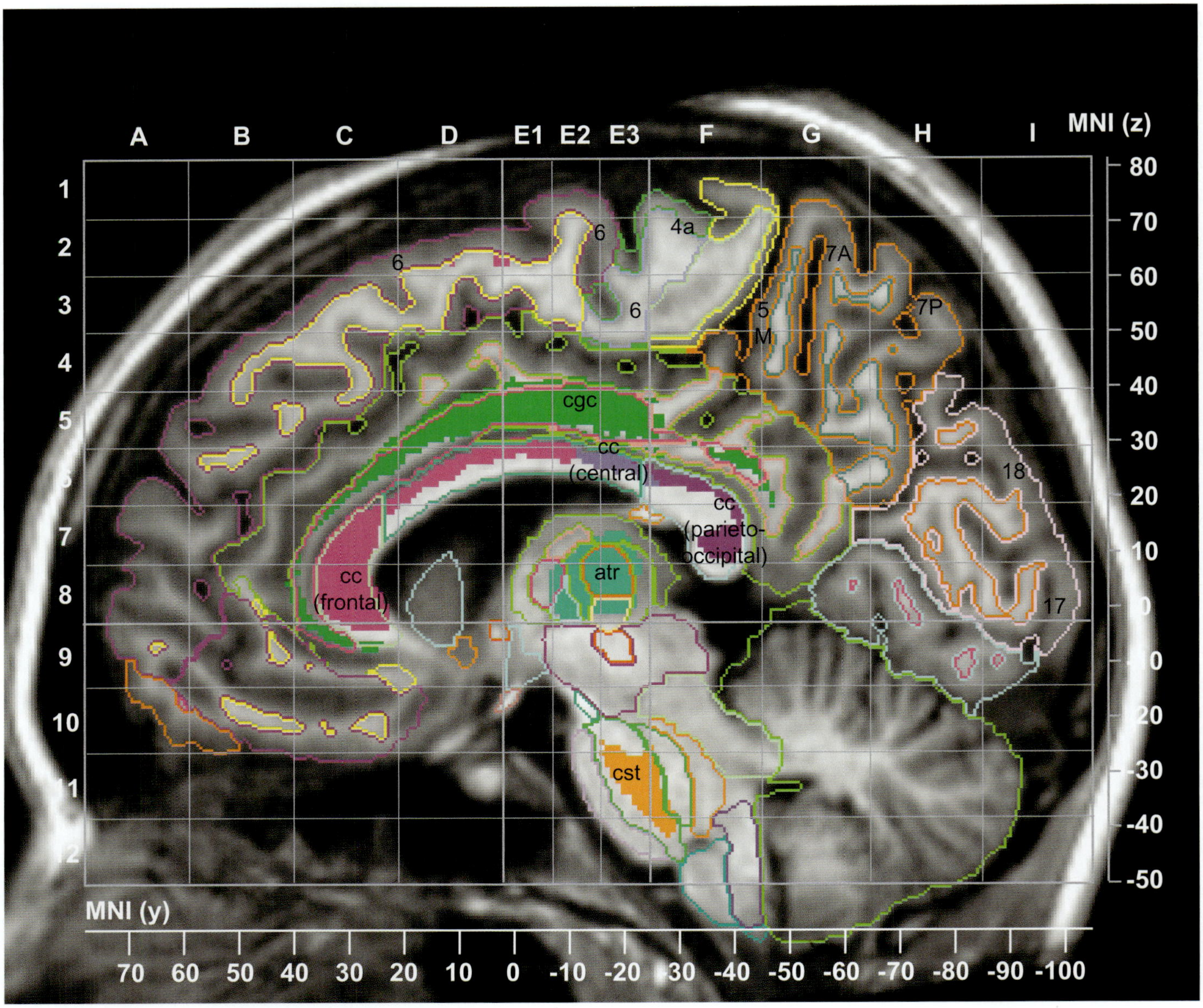

atr - anterior thalamic radiation
cc (central) - corpus callosum (connecting bilateral central areas)
cc (frontal) - corpus callosum (connecting bilateral frontal areas)
cc (parieto-occipital) - corpus callosum (connecting bilateral parieto-occipital areas)
cgc - cingulum (cingulate gyrus part)
cst - corticospinal tract

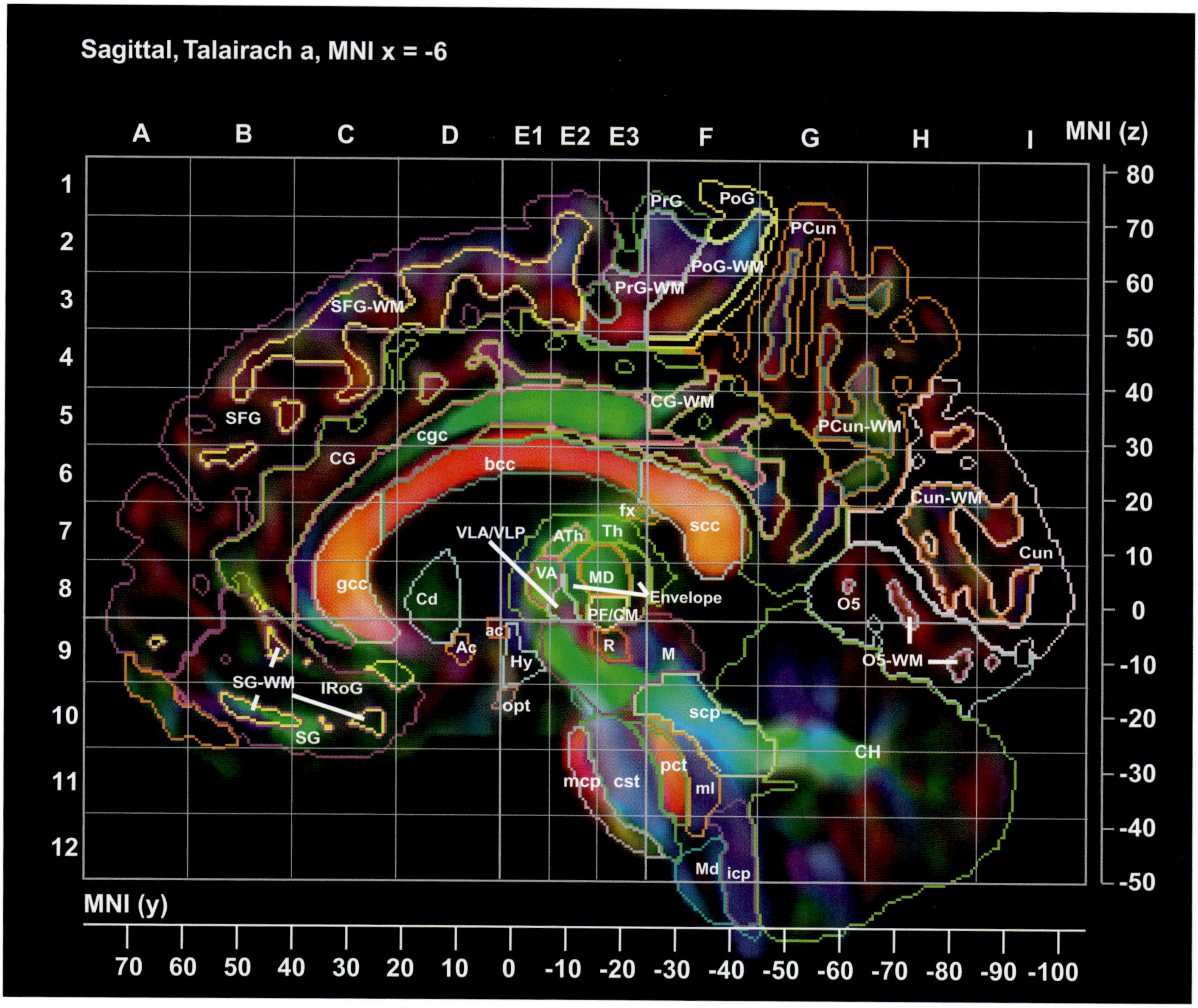

ac - anterior commissure
Ac - accumbens nucleus
Ath - anterior nuclear group (thalamus)
bcc - body of the corpus callosum
Cd - caudate nucleus
CG - cingulate gyrus
cgc - cingulum (cingulate gyrus part)
CG-WM - cingulate gyrus white matter
CH - cerebellar hemisphere
cst - corticospinal tract
Cun - cuneus
Cun-WM - cuneus white matter
Envelope - nuclei of the internal medullary lamina (thalamus)
fx - fornix (column and body)
gcc - genu of the corpus callosum
Hy - hypothalamus
icp - inferior cerebellar peduncle
IRoG - inferior rostral gyrus
M - midbrain
mcp - middle cerebellar peduncle
Md - medulla oblongata
MD - mediodorsal thalamic nucleus
ml - medial lemniscus
O5 - lyngual gyrus
O5-WM - lyngual gyrus white matter
opt - optic tract
PCun - precuneus
PCun-WM - precuneus white matter
PF/CM - parafascicular nucleus / centromedian nucleus (thalamus)
PoG - postcentral gyrus
PoG-WM - postcentral gyrus white matter
PrG - precentral gyrus
PrG-WM - precentral gyrus white matter
R - red nucleus
scc - splenium of the corpus callosum
scp - superior cerebellar peduncle
SFG - superior frontal gyrus
SFG-WM - superior frontal gyrus white matter
SG - straight gyrus
SG-WM - straight gyrus white matter
Th - thalamus
VA - ventroanterior thalamic nucleus
VLA/VLP - anterior and posterior ventrolateral thalamic nuclei

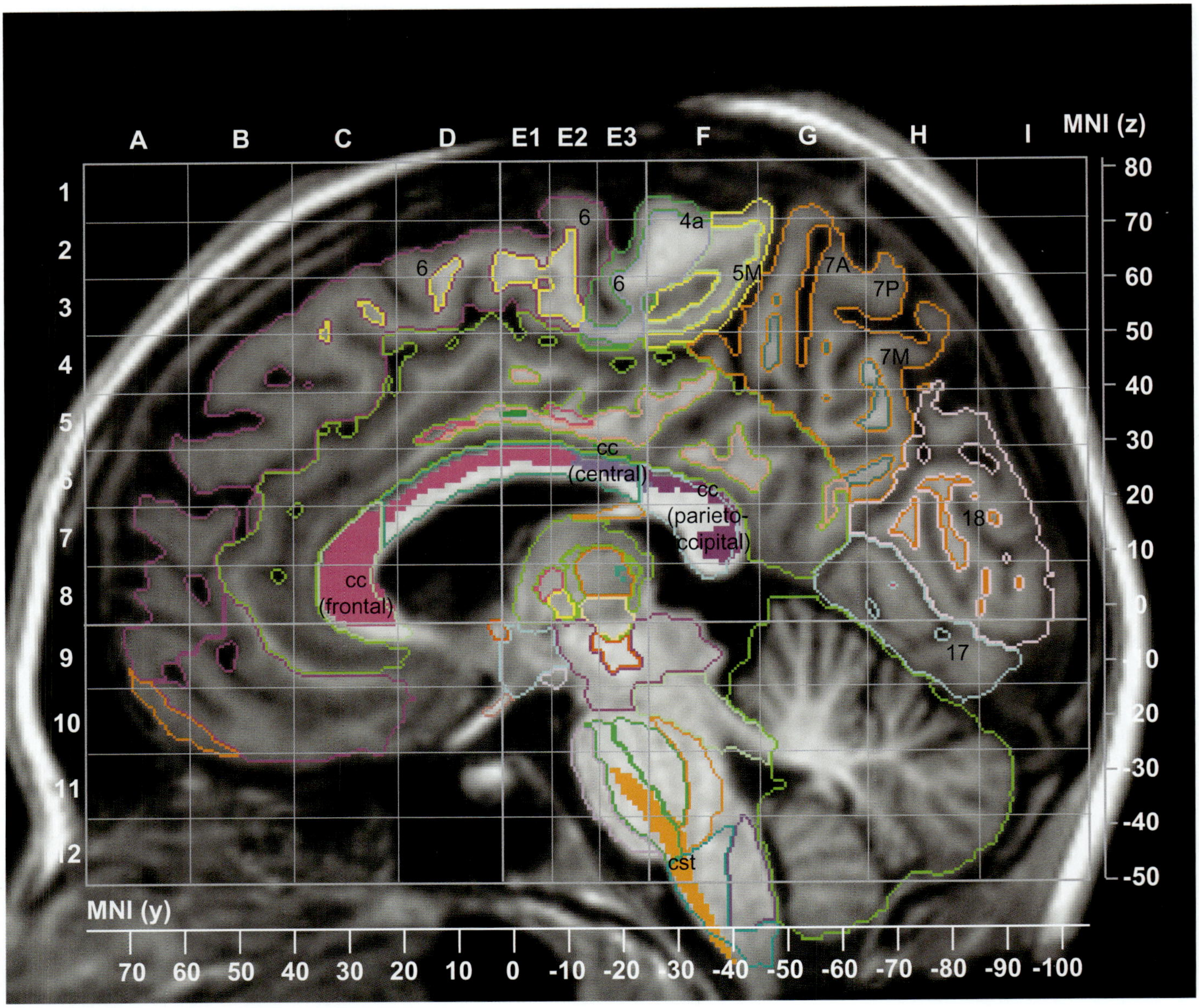

cc (central) - corpus callosum (connecting bilateral central areas)

cc (frontal) - corpus callosum (connecting bilateral frontal areas)

cc (parieto-occipital) - corpus callosum (connecting bilateral parieto-occipital areas)

cst - corticospinal tract

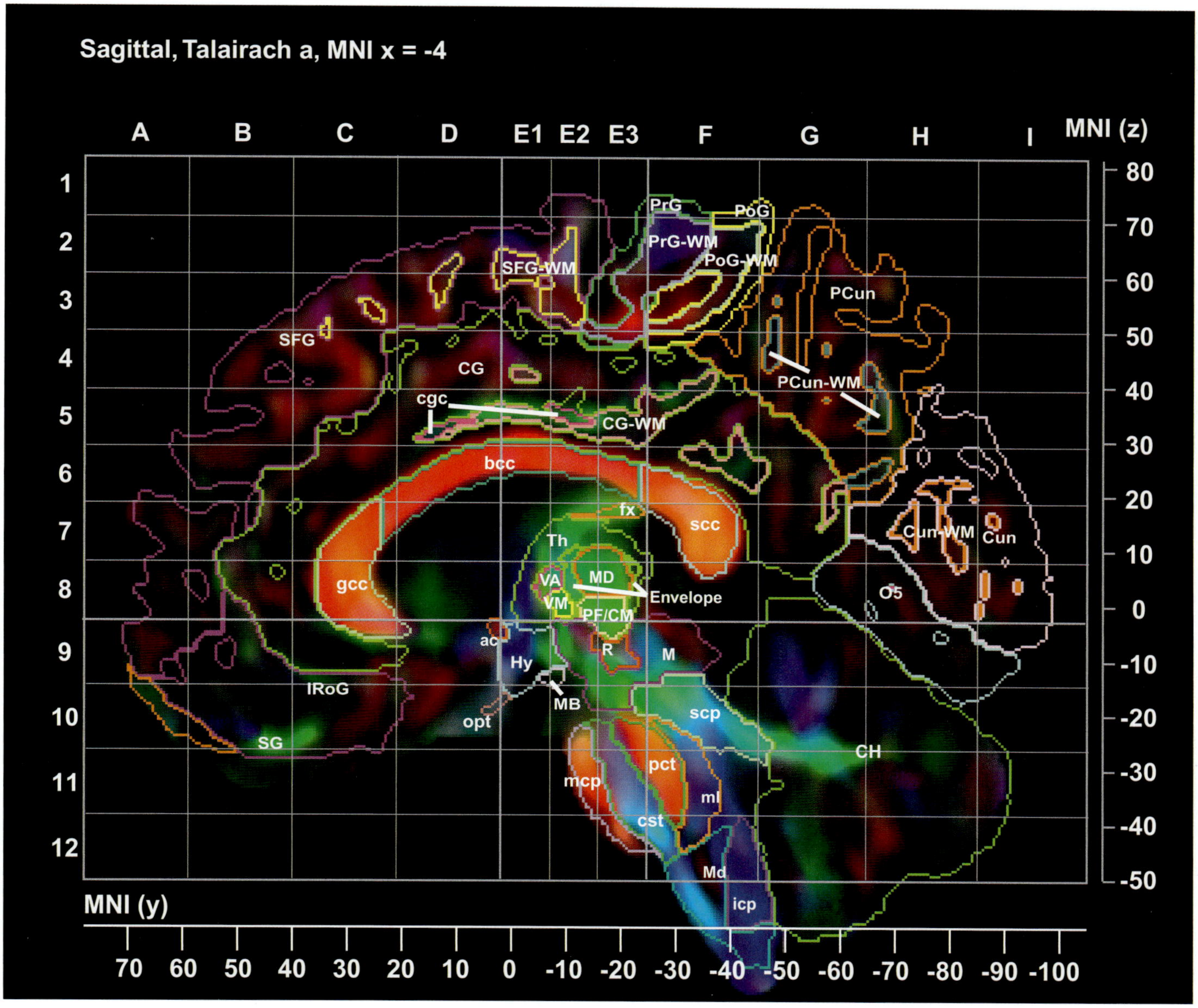

ac - anterior commissure
bcc - body of the corpus callosum
CG - cingulate gyrus
cgc - cingulum (cingulate gyrus part)
CG-WM - cingulate gyrus white matter
CH - cerebellar hemisphere
cst - corticospinal tract
Cun - cuneus
Cun-WM - cuneus white matter
Envelope - nuclei of the internal medullary lamina (thalamus)
fx - fornix (column and body)
gcc - genu of the corpus callosum
Hy - hypothalamus
icp - inferior cerebellar peduncle
IRoG - inferior rostral gyrus
M - midbrain
MB - mammillary body
mcp - middle cerebellar peduncle
Md - medulla oblongata
MD - mediodorsal thalamic nucleus
ml - medial lemniscus
O5 - lyngual gyrus
opt - optic tract
pct - pontine crossing tract (a part of mcp)
PCun - precuneus
PCun-WM - precuneus white matter
PF/CM - parafascicular nucleus / centromedian nucleus (thalamus)
PoG - postcentral gyrus
PoG-WM - postcentral gyrus white matter
PrG - precentral gyrus
PrG-WM - precentral gyrus white matter
R - red nucleus
scc - splenium of the corpus callosum
scp - superior cerebellar peduncle
SFG - superior frontal gyrus
SFG-WM - superior frontal gyrus white matter
SG - straight gyrus
Th - thalamus
VA - ventroanterior thalamic nucleus
VM - ventromedial thalamic nucleus

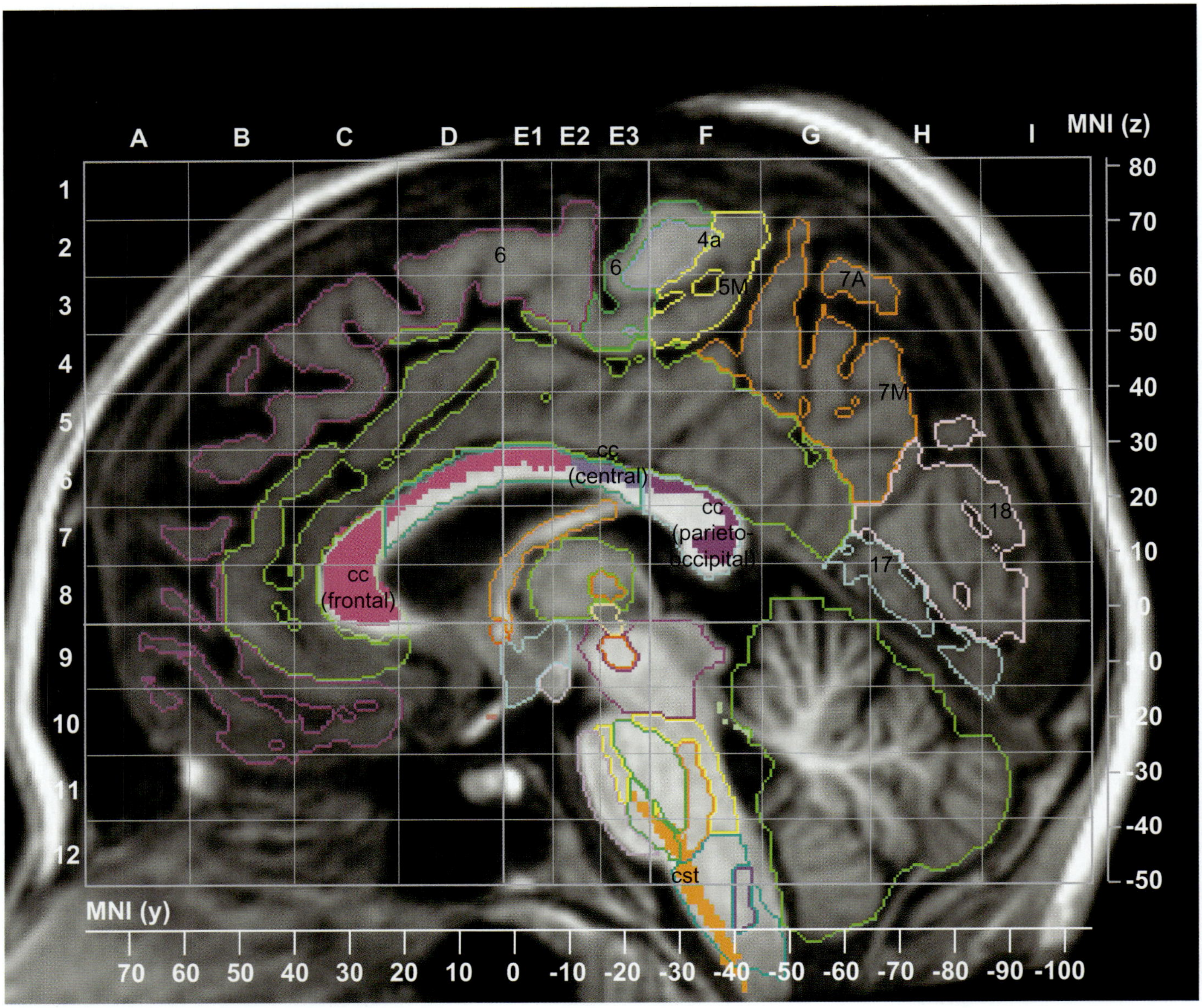

cc (central) - corpus callosum (connecting bilateral central areas)
cc (frontal) - corpus callosum (connecting bilateral frontal areas)
cc (parieto-occipital) - corpus callosum (connecting bilateral parieto-occipital areas)
cst - corticospinal tract

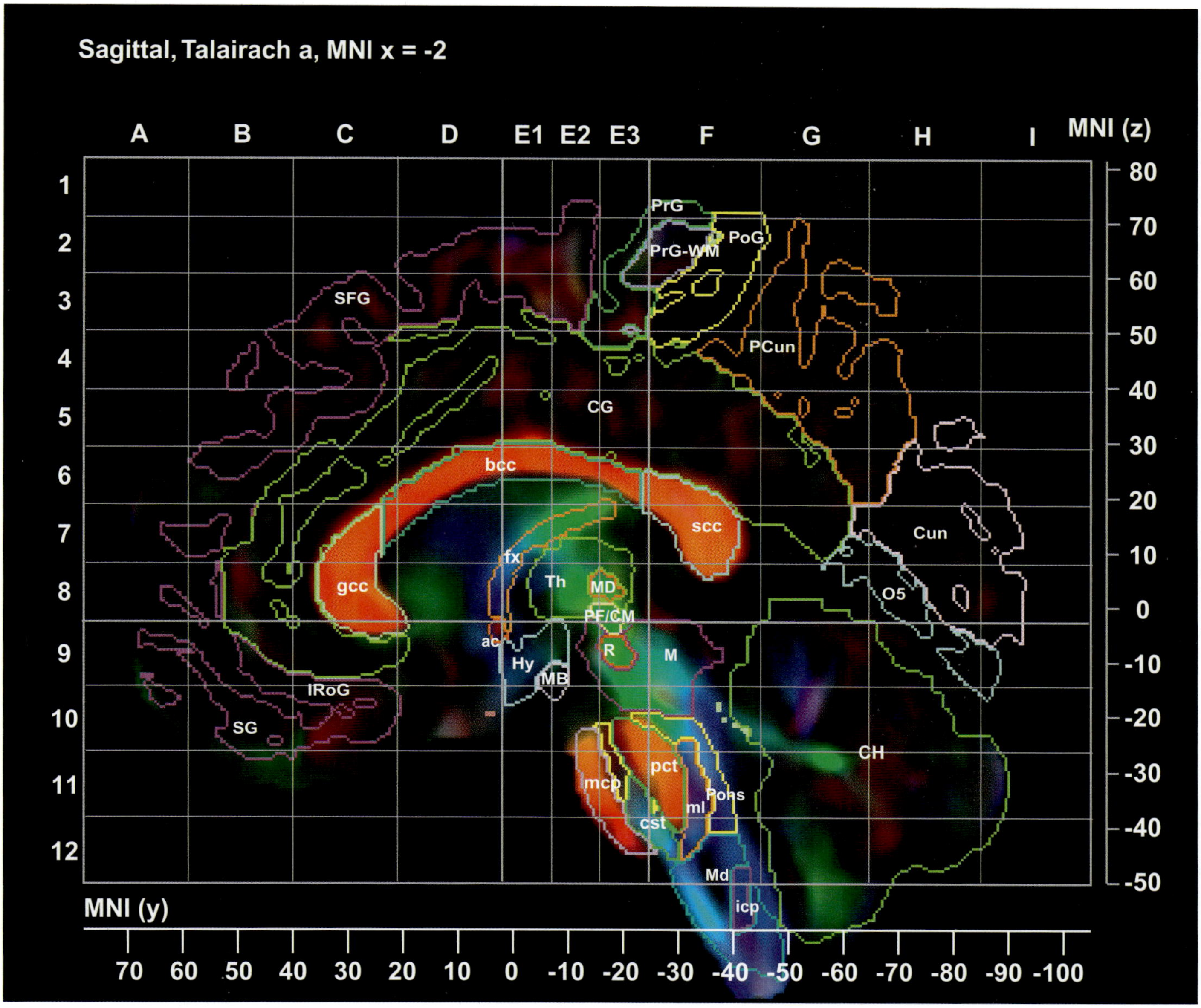

ac - anterior commissure	M - midbrain	PoG - postcentral gyrus
bcc - body of the corpus callosum	MB - mammillary body	Pons - pons
CG - cingulate gyrus	mcp - middle cerebellar peduncle	PrG - precentral gyrus
CH - cerebellar hemisphere	Md - medulla oblongata	PrG-WM - precentral gyrus white matter
cst - corticospinal tract	MD - mediodorsal thalamic nucleus	R - red nucleus
Cun - cuneus	ml - medial lemniscus	scc - splenium of the corpus callosum
fx - fornix (column and body)	O5 - lyngual gyrus	SFG - superior frontal gyrus
gcc - genu of the corpus callosum	pct - pontine crossing tract (a part of mcp)	SG - straight gyrus
Hy - hypothalamus	PCun - precuneus	Th - thalamus
IRoG - inferior rostral gyrus	PF/CM - parafascicular nucleus / centromedian nucleus (thalamus)	
icp - inferior cerebellar peduncle		

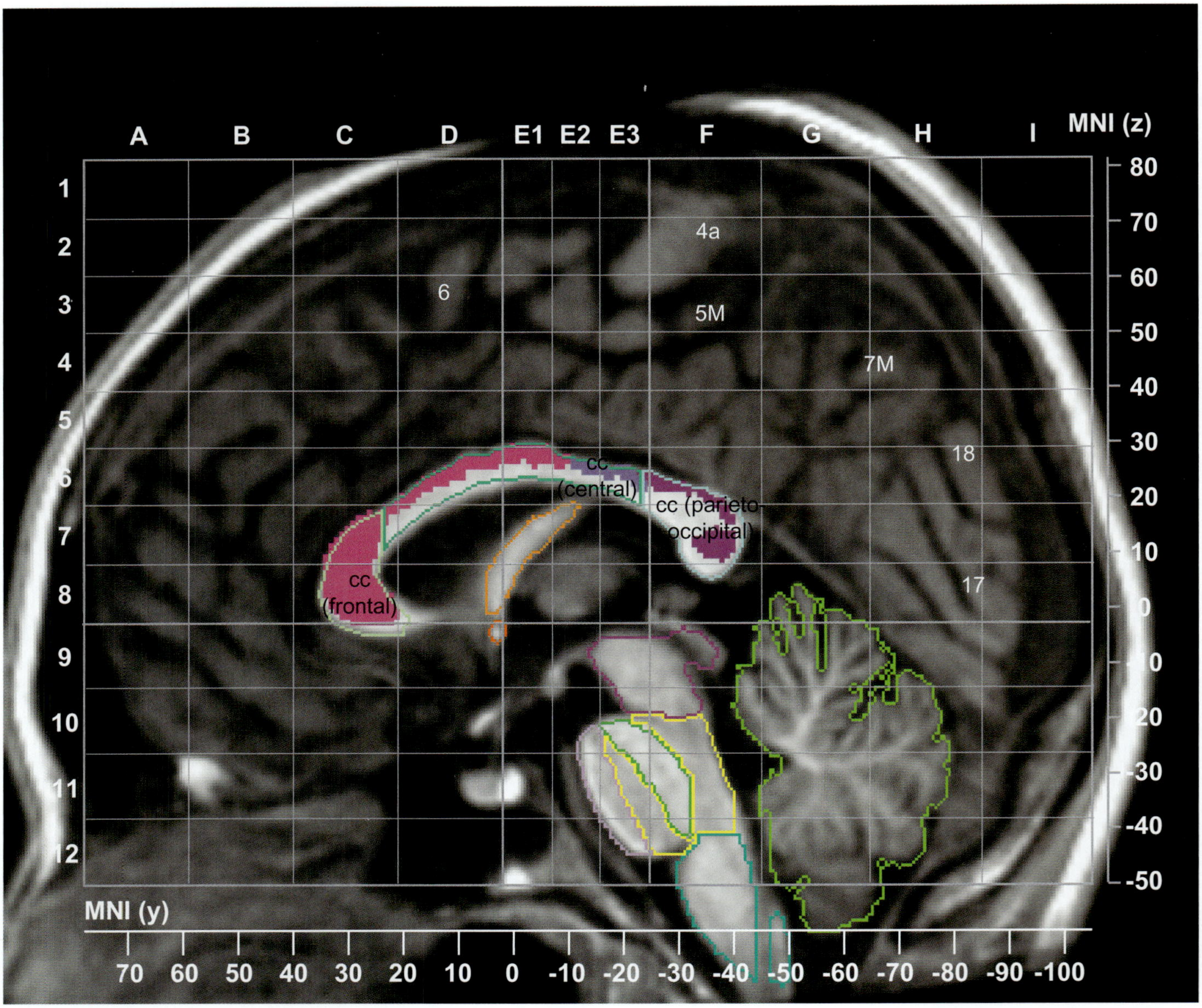

cc (central) - corpus callosum (connecting bilateral central areas)

cc (frontal) - corpus callosum (connecting bilateral frontal areas)

cc (parieto-occipital) - corpus callosum (connecting bilateral parieto-occipital areas)

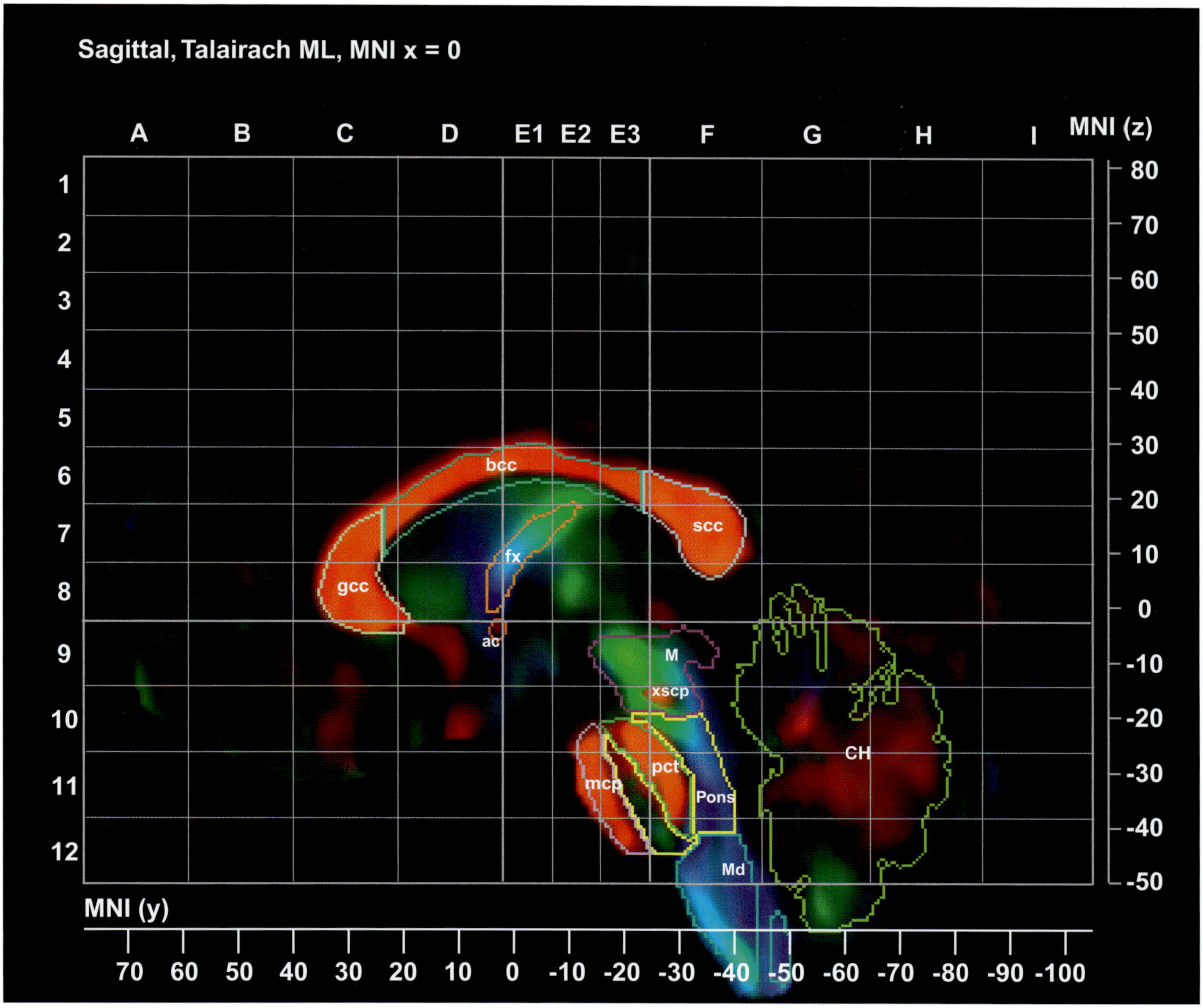

ac - anterior commissure
bcc - body of the corpus callosum
CH - cerebellar hemisphere
fx - fornix (column and body)
gcc - genu of the corpus callosum
M - midbrain
mcp - middle cerebellar peduncle
Md - medulla oblongata
pct - pontine crossing tract (a part of mcp)
Pons - pons
scc - splenium of the corpus callosum
xscp - decussation of the superior cerebellar peduncle

Subject Index